SINGULAR-PERTURBATION THEORY
AN INTRODUCTION WITH APPLICATIONS

T0296389

Singular-perturbation theory

AN INTRODUCTION WITH APPLICATIONS

DONALD R. SMITH

Department of Mathematics
University of California at San Diego

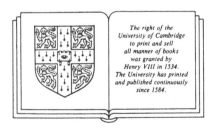

The right of the
University of Cambridge
to print and sell
all manner of books
was granted by
Henry VIII in 1534.
The University has printed
and published continuously
since 1584.

CAMBRIDGE UNIVERSITY PRESS

Cambridge

London   New York   New Rochelle

Melbourne   Sydney

CAMBRIDGE UNIVERSITY PRESS
Cambridge, New York, Melbourne, Madrid, Cape Town, Singapore, São Paulo, Delhi

Cambridge University Press
The Edinburgh Building, Cambridge CB2 8RU, UK

Published in the United States of America by Cambridge University Press, New York

www.cambridge.org
Information on this title: www.cambridge.org/9780521103077

First published 1985
This digitally printed version 2009

A catalogue record for this publication is available from the British Library

Library of Congress Cataloguing in Publication data
Smith, Donald R. (Donald Ray), 1939 Jan. 23–
Singular-perturbation theory.
1. Differential equations–Numerical solutions.
2. Perturbation (Mathematics)   I. Title.
QA372.S63   1985   515.3′5   84-23298

ISBN 978-0-521-30042-1 hardback
ISBN 978-0-521-10307-7 paperback

To
CLOYD VIRGIL SMITH
(1909–1982)
and
THELMA VANANTWERP SMITH
(1910–1983)

IN GRATITUDE

Contents

Preface   xi
Acknowledgments   xv

0   Preliminary results   **1**
0.1 Variation of parameters   1
0.2 The linear Volterra integral equation, and
     Gronwall's inequality   7
0.3 Elementary differential inequalities   12
0.4 The Banach/Picard fixed-point theorem   19
0.5 Divergent series   22

PART I
Initial-value problems of oscillatory type

1   Precession of the planet Mercury   **29**
1.1 The Einstein equation for Mercury   29
1.2 The exact solution   31
1.3 A classical perturbation approach   34

2   Krylov/ Bogoliubov averaging   **40**
2.1 Introduction   40
2.2 The averaged system   42
2.3 Precession of Mercury, revisited   47
2.4 An oscillator with weak cubic damping   50
2.5 Two weakly coupled electric circuits   54

3 The multiscale technique   **60**
3.1 Introduction   60
3.2 A linear oscillator with weak damping   62
3.3 Precession of Mercury, once again   69
3.4 Additional applications of two-timing   74
3.5 Two-timing for systems: transference of vibrations   84

4 Error estimates for perturbed-oscillation problems   **91**
4.1 Estimates on expanding intervals   91
4.2 The validity of two-timing   103
4.3 Estimates via integral inequalities   106
4.4 Global estimates for damped oscillators   116

PART II
Initial-value problems of overdamped type

5 Linear overdamped initial-value problems   **133**
5.1 Introduction   133
5.2 A matching approach   136
5.3 Error estimates   141
5.4 A multivariable approach   145
5.5 The telegraph equation with large resistance   153

6 Nonlinear overdamped initial-value problems   **163**
6.1 Introduction   163
6.2 The O'Malley/Hoppensteadt construction   177
6.3 Existence, uniqueness, and error estimates   197
6.4 Numerical methods   207

7 Conditionally stable problems   **219**
7.1 Introduction   219
7.2 The stable initial manifolds   229
7.3 The multivariable expansion   245
7.4 Error estimates   254
7.5 The fundamental solution   257

PART III
Boundary-value problems

8 Linear scalar problems   **263**
8.1 Introduction   263
8.2 A problem with boundary layers at both endpoints   275

8.3 A problem with a single boundary layer   287
8.4 Higher-order problems   302
8.5 Examples on interior transition points   318

9   Linear first-order systems   **330**
9.1 Introduction   330
9.2 The Riccati transformation   338
9.3 The Green function   353
9.4 The linear state regulator in optimal control   370

10   Nonlinear problems   **382**
10.1 Introduction   382
10.2 A problem with boundary layers at both endpoints   389
10.3 The Green function   411
10.4 Problems with a single boundary layer   426
10.5 A problem in the physical theory of semiconductors   447
10.6 Examples on interior layers   463
10.7 A numerical algorithm of Maier   473

References   **479**

Name index   **495**

Subject index   **498**

Preface

A singular-perturbation problem is a problem that depends on a parameter (or parameters) in such a way that solutions behave nonuniformly as the parameter tends toward some limiting value of interest. The nature of the nonuniformity can vary from problem to problem. In practice, one seeks a uniformly valid, easily interpretable approximation to the nonuniformly behaving solution.

Singular-perturbation theory has assumed vast proportions, and this book is not a comprehensive survey of it. I present an introduction to singular-perturbation theory, mostly for ordinary differential equations, with the selection of material conditioned strongly by my own interests. Within the limitations of a single volume it has seemed necessary to omit some topics altogether and to include only brief coverage of some others. For example, there is little coverage of problems for partial differential equations, and none for the singularly perturbed eigenvalue problem or for the abstract Cauchy problem in a Banach space. However, the book does contain points of contact with these and other omitted topics, along with references to the literature, so that an instructor can easily include such omitted topics in a course based on this book. Numerous applications are included from a variety of areas of science and engineering.

This book has evolved from a course of lectures I have given regularly since 1974 at the University of California at San Diego, the Technical University of Munich, and the Mathematics Research Center at the University of Wisconsin at Madison. Versions of the lectures have been given on different occasions to a variety of audiences, including audiences of professional mathematicians and scientists, audiences of graduate students of mathematics and of the pure and applied sciences and engineering, and audiences of advanced undergraduate students.

The book is organized into three main parts, covering various typical classes of singularly perturbed problems for differential equations. Part I deals with oscillation problems of a type arising in many areas, including classical and relativistic mechanics, mathematical astronomy, electric-circuit theory, and biology. Solutions of such problems are quasi-periodic (if damping is not involved) and exhibit certain nonuniformities, for small values of a parameter, as an independent variable ranges over a lengthy interval. Approximate solutions that are valid on such a lengthy interval are constructed here by the method of averaging and by the method of multiple scales.

Part II deals with initial-value problems of nonoscillatory type whose solutions exhibit rapid variation in a thin initial layer. Some typical problems of this type are electric-circuit problems with large resistance and/or small inductance, mechanical problems with small masses and/or large damping, the propagation of radiation through a highly absorbing medium, and chemical and biochemical processes involving simultaneous multiple reactions with widely different reaction rates. Approximate solutions are constructed for such problems here primarily with the O'Malley/Hoppensteadt multivariable method, although a brief discussion of the method of matched asymptotic expansions is included. Part II also includes brief coverage of conditionally stable problems, singular singularly perturbed problems, and numerical methods of singular-perturbation type.

Part III deals with singularly perturbed boundary-value problems of several types, including problems arising in chemical-reactor theory, fluid dynamics, elasticity theory, optimal-control theory, and the physical theory of semiconductors and transistors, as well as nonlinear problems with multiple stable states. Solutions of the boundary-layer type are considered primarily, although solutions of the interior-layer type are also discussed briefly. Approximate solutions are again obtained with a multivariable method. Numerical solutions of such problems are discussed only briefly.

The general approach followed here for problems of every type is to construct a proposed approximate solution and then linearize the original problem about that intended approximate solution. The resulting linearization, which is itself generally a nonlinear problem, is then studied in order to obtain an existence and local-uniqueness result for the original problem, along with useful error estimates for the difference between the resulting exact solution and the given approximate solution. There are several approaches available for study of the linearization, including, but not limited to, the fixed-point method and the method of differential inequalities. Which approach one employs is often a matter of personal taste. I emphasize primarily the Banach/Picard fixed-point method, although I briefly discuss various other approaches as well. Here, as elsewhere, the book allows for flexibility, so that an instructor can emphasize a different approach (or approaches) as preferred.

There is enough material in the book for a one-year course, although I have generally given shorter courses of a semester or even a quarter. The book allows an instructor considerable flexibility in choosing topics to be included or excluded, so that the book can be used for many different courses. I have often omitted portions of the book altogether and skimmed over other portions lightly in different courses, depending on the interests of the audience and the length of the course. Naturally, a one-quarter course would require the instructor to exercise the greatest selectivity.

I hope the book will be useful to students of science and engineering at the graduate and advanced undergraduate levels. The interest inherent in the many applications of the subject makes the book useful not only as an introduction to singular-perturbation theory for differential equations but also as a vehicle by which the student can obtain a broader appreciation of certain important and useful results and procedures from classical and modern analysis.

Del Mar, California   Donald R. Smith

Acknowledgments

This book is my presentation of the work of many people, and it is a pleasure for me to acknowledge their work here. Many authors were instrumental in the early development of the study of differential equations and in the use of differential equations to construct mathematical models of natural phenomena, including Isaac Newton (1642–1727), Gottfried Leibniz (1646–1716), Jakob Bernoulli (1654–1705), Johann Bernoulli (1667–1748), Leonhard Euler (1707–83), Joseph-Louis Lagrange (1736–1813), Augustin Cauchy (1789–1857), and Henri Poincaré (1854–1912). Perturbation theory for differential equations is a vast and growing subject.

I have had to be very selective in choosing topics to include in this single volume. The list of references at the end of the book includes some 300 listings, only a fraction of the literature on the subject. Throughout the book I have acknowledged the authors of the work discussed here, as far as I know of them. It is likely that I have, out of ignorance, slighted some authors who should have been mentioned or mentioned more prominently. I ask their forgiveness.

My general dependence on other authors, even to the limited extent to which I am aware of it, is too vast for complete citation here. Of the many recent authors responsible for the work discussed in this book, I mention here only Robert E. O'Malley, Jr. His importance to this book is reflected in the fact that some 10 percent of the references bear his name. In particular, O'Malley is responsible for much of the work discussed in Part II and Part III.

I wish to thank R. James Milgram of the Department of Mathematics at Stanford University for creating and making available to me TECPRINT, a technical word processor that contributed greatly in the writing of this book. The staff of Cambridge University Press did an outstanding job in seeing the book through press; I wish especially to thank David Tranah and Michael Gnat for their many contributions to this project. The copy editor, Jim Mobley, did an exceptionally fine job, for which I am grateful.

Parts of this book were written while I was visiting the Technical University of Munich; I thank Roland Bulirsch for making that visit possible and for encouraging me to include the material in Chapter 9 on the linear-state regulator in optimal-control theory. I thank Carolyn Geiman Smith for her encouragement, and thank the Department of Mathematics, University of California at San Diego, for providing an environment in which this work could be pursued and completed.

D. R. S.

0

Preliminary results

This chapter contains several elementary results from the theory of differential and integral equations, along with certain other preliminary results from classical analysis, and can be skimmed lightly during a first reading.

0.1 Variation of parameters

Consider the single (scalar) second-order linear differential equation

$$\mathscr{L}v = f(t) \quad \text{for real } t \text{ in the interval } [t_1, t_2], \qquad (0.1.1)$$

where f is a given piecewise-continuous function on $[t_1, t_2]$, and the differential operator $\mathscr{L}$ is defined as

$$\mathscr{L}v(t) := \frac{d^2v}{dt^2} + a(t)\frac{dv}{dt} + b(t)v \qquad (0.1.2)$$

for any twice-differentiable function v, where a and b are fixed, given piecewise-continuous functions on $[t_1, t_2]$. (When the notation $[t_1, t_2]$ is used to denote an interval, it will always be assumed that $t_1 \leq t_2$.)

If v_1 and v_2 are any two linearly independent solutions of the homogeneous equation $\mathscr{L}v = 0$, then *variation of parameters* can be used to represent the general solution of (0.1.1) in the form (for any fixed t_0 in $[t_1, t_2]$)

$$v(t) = c_1v_1(t) + c_2v_2(t) + \int_{t_0}^{t} K(t,s)f(s)\, ds, \qquad (0.1.3)$$

where c_1 and c_2 are suitable constants, and the kernel K is defined as

$$K(t,s) := \frac{v_1(t)v_2(s) - v_2(t)v_1(s)}{v_1'(s)v_2(s) - v_2'(s)v_1(s)}. \qquad (0.1.4)$$

1

The idea of variation of parameters was introduced by Isaac Newton, Gottfried Leibniz, and Johann Bernoulli independently in certain special situations, and the technique was later developed into a general procedure by Leonhard Euler and Joseph Lagrange. The procedure is much used in perturbation theory, as illustrated repeatedly in every chapter of this book (cf. Exercise 0.1.1).

Example 0.1.1: If $\mathcal{L}$ is the operator $\mathcal{L} = (d^2/dt^2) + 1$ (with $a = 0$, $b = 1$), then the functions v_1 and v_2 can be taken as $v_1(t) = \cos t$ and $v_2(t) = \sin t$, and the kernel K becomes $K(t, s) = \cos s \sin t - \sin s \cos t = \sin(t - s)$. The result (0.1.3) then leads to the identity (with $t_0 = 0$)

$$v(t) = c_1 \cos t + c_2 \sin t + \int_0^t \sin(t - s)[v''(s) + v(s)] \, ds$$

for suitable constants c_1 and c_2 depending on v.

Example 0.1.2: If $\mathcal{L}$ is the operator $\mathcal{L} = (d^2/dt^2) + a(t)(d/dt)$ (with $b = 0$), then the functions v_1 and v_2 can be taken as $v_1(t) = 1$ (for all t) and

$$v_2(t) = \int_{t_0}^t \exp\left(-\int_{t_0}^r a\right) dr,$$

and (0.1.3) leads to the result

$$v(t) = c_1 + c_2 K(t, t_0) + \int_{t_0}^t K(t, s)[v''(s) + a(s)v'(s)] \, ds,$$

with kernel

$$K(t, s) = \int_s^t \exp\left(-\int_s^r a\right) dr, \qquad (0.1.5)$$

where the constants c_1 and c_2 depend on v.

We can use Example 0.1.2 to obtain an integral relation for any solution of the differential equation (0.1.1). Indeed, we can put $v''(s) + a(s)v'(s) = f(s) - b(s)v(s)$ in the result of Example 0.1.2 and find the equation

$$v(t) = c_1 + c_2 K(t, t_0) + \int_{t_0}^t K(t, s) f(s) \, ds - \int_{t_0}^t K(t, s) b(s) v(s) \, ds$$

$$(0.1.6)$$

for any solution of the differential equation

$$v'' + a(t)v' + b(t)v = f(t), \qquad (0.1.7)$$

where K is given by (0.1.5). The constants c_1 and c_2 can be represented as

$$c_1 = v(t_0), \qquad c_2 = v'(t_0). \tag{0.1.8}$$

If $a = 0$, then the kernel K of (0.1.5) reduces to

$$K(t,s) = t - s \quad \text{if } a = 0. \tag{0.1.9}$$

Exercises

Exercise 0.1.1: Let $M_{m,n}(\mathbb{R})$ denote the collection of $m \times n$ matrices over the real numbers $\mathbb{R}$, let $A = A(t)$ be a continuous matrix-valued function on $[t_1, t_2]$ taking values in $M_{n,n}(\mathbb{R})$, let $f = f(t)$ be a continuous vector-valued function on $[t_1, t_2]$ taking values in $M_{n,1}(\mathbb{R})$, and consider the nonhomogeneous vector differential equation*

$$\frac{dx}{dt} = A(t)x + f(t) \quad \text{for } t \in [t_1, t_2] \tag{0.1.10}$$

for a solution vector $x = x(t)$ taking values in $M_{n,1}(\mathbb{R})$. Let $x^1 = x^1(t), x^2 = x^2(t), \ldots, x^n = x^n(t)$ be any fundamental solution set (of solution vectors) for the homogeneous equation

$$\frac{dx}{dt} = A(t)x \quad \text{for } t \in [t_1, t_2] \tag{0.1.11}$$

(that is, the n solution vectors $x^1, x^2, \ldots, x^n$ constitute a linearly independent set), and let t_0 be any fixed number in the given interval $[t_1, t_2]$. Derive the formula

$$x(t) = X(t)c + \int_{t_0}^{t} X(t)X(s)^{-1}f(s)\,ds \tag{0.1.12}$$

for the general solution of (0.1.10), where X is a fundamental matrix for (0.1.11) given as $X := [x^1, x^2, \ldots, x^n]$, and c is a suitable constant vector depending on t_0, x, and X, with $c = X(t_0)^{-1}x(t_0)$. *Hint:* Seek x as a suitable linear combination of $x^1, x^2, \ldots, x^n$ as

$$x = \sum_{j=1}^{n} x^j(t)c_j(t),$$

with scalar coefficients $c_j = c_j(t)$ depending on t, or equivalently $x = X(t)c(t)$ for some suitable vector-valued function $c = c(t)$.

Exercise 0.1.2: Let $A = A(t)$ and $f = f(t)$ be as in Exercise 0.1.1, and consider the *boundary-value problem* consisting of the differential equa-

---

* Note that vector-valued quantities are indicated in italics (i.e., x) rather than boldface (i.e., x) throughout this book.

tion (0.1.10) along with the following specified boundary condition on
$x = x(t)$ at the endpoints t_1 and t_2:

$$Lx(t_1) + Rx(t_2) = \alpha, \tag{0.1.13}$$

where L and R are given (constant) $n \times n$ matrices and α is a given
(constant) vector. Let $X(t)$ be a fundamental solution matrix for the
homogeneous equation (0.1.11) as in Exercise 0.1.1. *Show that this
boundary-value problem has a unique solution* (for every α and f) *if and
only if the matrix*

$$M: = LX(t_1) + RX(t_2) \tag{0.1.14}$$

is nonsingular, and in this case *show that the unique solution $x(t)$ can be
given as*

$$x(t) = X(t)M^{-1}\alpha + \int_{t_1}^{t_2} G(t,s)f(s)\,ds, \tag{0.1.15}$$

where the *Green function* $G = G(t, s)$ is the matrix-valued function
defined as

$$G(t,s): = \begin{cases} X(t)M^{-1}LX(t_1)X(s)^{-1} & \text{for } s < t, \\ -X(t)M^{-1}RX(t_2)X(s)^{-1} & \text{for } s > t. \end{cases} \tag{0.1.16}$$

If the matrix M is singular, *show that the boundary-value problem may
have no solution or infinitely many solutions, depending on the data.* (This
is the **Fredholm alternative**.) *Hint*: Put $t_0 = t_1$ in the general representation (0.1.12), and then insert this result for $x(t)$ into the boundary
condition (0.1.13), so as to obtain for the constant vector c the result

$$Mc = \alpha - RX(t_2)\int_{t_1}^{t_2} X(s)^{-1}f(s)\,ds.$$

This last equation can be solved for c if the matrix M is nonsingular. The
resulting value for c can be inserted back into (0.1.12), and one finds the
stated representation (0.1.15). As a function of t, the Green function
satisfies the homogeneous equation $\partial G(t,s)/\partial t = A(t)G(t,s)$ for $t \neq s$,
along with the homogeneous boundary condition $LG(t_1,s) + RG(t_2,s)$
$= 0$. The Green function has a jump discontinuity at $s = t$ characterized
as $G(t,t-) - G(t,t+) = I =$ identity matrix. Finally, as a function of
s, the Green function satisfies the adjoint equation $\partial G(t,s)/\partial s =
-G(t,s)A(s)$. [See pp. 145–51 of R. Cole (1968) for further details.]

Exercise 0.1.3: The Bessel function J_0 satisfies the differential equation

$$v'' + \frac{1}{t}v' + v = 0 \tag{0.1.17}$$

subject to the initial conditions

$$v(0) = 1, \qquad v'(0) = 0. \tag{0.1.18}$$

Show that the solution v of (0.1.17)–(0.1.18) can be characterized equivalently as the solution of the integral equation

$$v(t) = 1 + \int_0^t s\left(\log\frac{s}{t}\right)v(s)\,ds. \tag{0.1.19}$$

Exercise 0.1.4: Derive the identity

$$v(t) = v(t_0) + v'(t_0)(t - t_0) + v''(t_0)K(t, t_0)$$

$$+ \int_{t_0}^t K(t, s)\left[\frac{d^3v(s)}{ds^3} + a(s)\frac{d^2v(s)}{ds^2}\right]ds \tag{0.1.20}$$

for any function v of class C^3, where the Kernel $K = K(t, s)$ is defined as

$$K(t, s): = \int_s^t (t - r)\exp\left(-\int_s^r a\right)dr. \tag{0.1.21}$$

Hint: The functions

$$v_1(t) = 1, \qquad v_2(t) = t, \qquad v_2(t) = \int_{t_0}^t (t - r)\exp\left(-\int_{t_0}^r a\right)dr,$$

provide three linearly independent solutions of the homogeneous equation

$$\frac{d^3v}{dt^3} + a(t)\frac{d^2v}{dt^2} = 0.$$

Use these solutions along with variation of parameters to obtain the general solution of the inhomogeneous equation

$$\frac{d^3v}{dt^3} + a(t)\frac{d^2v}{dt^2} = f(t).$$

Exercise 0.1.5: Show that the initial-value problem

$$\frac{d^3v}{dt^3} + a(t)\frac{d^2v}{dt^2} + b(t)\frac{dv}{dt} + c(t)v = f(t) \quad \text{for } t \ge t_0,$$

$$v(t_0) = \alpha, \qquad v'(t_0) = \beta, \qquad v''(t_0) = \gamma, \tag{0.1.22}$$

is equivalent to the following system of integral equations:

$$v(t) = \alpha + \beta \cdot (t - t_0) + \gamma \cdot K(t, t_0) + \int_{t_0}^{t} K(t, s) f(s)\, ds$$

$$- \int_{t_0}^{t} K(t, s)[b(s)v'(s) + c(s)v(s)]\, ds \qquad (0.1.23)$$

and

$$v'(t) = \beta + \gamma \cdot K_t(t, t_0) + \int_{t_0}^{t} K_t(t, s) f(s)\, ds$$

$$- \int_{t_0}^{t} K_t(t, s)[b(s)v'(s) + c(s)v(s)]\, ds, \qquad (0.1.24)$$

where $K = K(t, s)$ is given by (0.1.21), and $K_t(t, s) = \partial K(t, s)/\partial t$. *Hint*: The equation (0.1.23) follows from (0.1.20) and (0.1.22), and (0.1.24) follows from (0.1.23) on differentiation.

Exercise 0.1.6: Show that the solution v of the initial-value problem (0.1.22) can be given as

$$v(t) = \alpha + \beta \cdot (t - t_0) + \gamma \cdot K(t, t_0)$$

$$+ \int_{t_0}^{t} K(t, s) f(s)\, ds - \int_{t_0}^{t} K(t, s) u(s)\, ds, \qquad (0.1.25)$$

where the function $u = u(t)$ is the solution of the equation

$$u(t) = P(t) + \int_{t_0}^{t} H(t, s) u(s)\, ds, \qquad (0.1.26)$$

with kernel $H = H(t, s)$ defined as

$$H(t, s) := -[b(t) K_t(t, s) + c(t) K(t, s)], \qquad (0.1.27)$$

and with the function $P = P(t)$ defined as

$$P(t) := b(t) p'(t) + c(t) p(t), \qquad (0.1.28)$$

where $p = p(t)$ is given as

$$p(t) := \alpha + \beta \cdot (t - t_0) + \gamma \cdot K(t, t_0) + \int_{t_0}^{t} K(t, s) f(s)\, ds.$$

$$(0.1.29)$$

Hint: Define $u := bv' + cv$, and then take a suitable combination of equations (0.1.23) and (0.1.24) so as to obtain (0.1.26). Note that this linear Volterra integral equation (0.1.26) can be solved by Picard iteration, as described in Section 0.2. The approach of Exercises 0.1.4–0.1.6 can be extended in the obvious way to the study of the initial-value problem for the scalar nth-order linear differential equation.

0.2 The linear Volterra integral equation, and Gronwall's inequality

Consider the (scalar) integral equation

$$v(t) = p(t) + \int_{t_0}^{t} V(t,s)v(s)\,ds \qquad (0.2.1)$$

for a real function v, where $p = p(t)$ and $V = V(t,s)$ are given piece-wise-continuous functions for $t_0 \le t \le t_1$ and $t_0 \le s \le t \le t_1$, respectively.

The uniqueness of solutions to (0.2.1) follows immediately from the important **Gronwall's inequality**:

Theorem 0.2.1 (Gronwall 1919): *Let a, b, and c be nonnegative-valued continuous functions on the interval $[t_0, t_1)$, and let Z be a continuous function satisfying the integral inequality*

$$|Z(t)| \le a(t)\int_{t_0}^{t} b(s)|Z(s)|\,ds + c(t) \quad \text{on } [t_0, t_1). \qquad (0.2.2)$$

Then Z also satisfies

$$|Z(t)| \le a(t)\int_{t_0}^{t} b(s)c(s)\exp\left(\int_{s}^{t} ab\right)ds + c(t) \qquad (0.2.3)$$

on $[t_0, t_1)$.

Proof: Put

$$S(t) := \int_{t_0}^{t} b(s)|Z(s)|\,ds,$$

with $|Z(t)| \le a(t)S(t) + c(t)$, and find $S(t_0) = 0$ and $S'(t) = b(t)|Z(t)| \le a(t)b(t)S(t) + b(t)c(t)$. Hence, we find

$$\frac{d}{dt}\left\{\left[\exp\left(-\int_{t_0}^{t} ab\right)\right]S(t)\right\} \le \left[\exp\left(-\int_{t_0}^{t} ab\right)\right]b(t)c(t),$$

and integration leads to the inequality

$$\left[\exp\left(-\int_{t_0}^{t} ab\right)\right]S(t) \le \int_{t_0}^{t}\left[\exp\left(-\int_{t_0}^{s} ab\right)\right]b(s)c(s)\,ds,$$

from which the stated result follows easily. ∎

A key step in this proof of Gronwall's inequality involves the integration of a certain *differential inequality*. We shall see in Section 0.3 and in later chapters that related integrations of various differential inequalities play important roles in the study of a wide range of linear and nonlinear singularly perturbed problems.

Gronwall's inequality implies that *(0.2.1) has at most one solution.* Indeed, if v_1 and v_2 are solutions, then $Z := v_1 - v_2$ is seen to satisfy (0.2.2), with $b = 1$, $c = 0$, and $a = M$ for some suitable positive *constant* M. Then (0.2.3) implies the result $|v_1 - v_2| \le 0$, or $v_1 = v_2$, so that any two solutions coincide. [Note that any solution of (0.2.1) must be piecewise-continuous, and the required version of Gronwall's inequality for piecewise-continuous Z is easily seen to be valid.]

The unique solution of the Volterra equation (0.2.1) can be obtained by **Picard's method of successive approximations** as (see Exercises 0.2.1–0.2.2)

$$v(t) = p(t) + \int_{t_0}^t V^*(t,s)p(s)\,ds, \qquad (0.2.4)$$

where V^* is the Volterra *resolvent kernel* for V, defined as

$$V^*(t,s) := \sum_{j=1}^\infty V_j(t,s), \qquad (0.2.5)$$

with

$$V_j(t,s) := \begin{cases} V(t,s) & \text{if } j = 1, \\ \int_s^t V(t,r)V_{j-1}(r,s)\,dr & \text{for } j = 2,3,\dots. \end{cases} \qquad (0.2.6)$$

The series (0.2.5), which defines the resolvent kernel, can be shown to be absolutely and uniformly convergent for $t_0 \le s \le t \le t_1$, and there holds

$$|V^*(t,s)| \le M\exp(M|t - s|), \qquad (0.2.7)$$

where the constant M can be taken to be any upper bound on $|V|$.

The method of successive approximations seems to have been introduced into the study of differential and integral equations by Liouville (1837). Many workers, including E. Picard, C. Neumann, V. Volterra, and S. Banach, have forged the technique into a powerful, general method culminating in the important *Banach/Picard fixed-point theorem* discussed in Section 0.4. [See pp. 719–20, 1054–7, and 1089–90 of Kline (1972).]

The unique solvability of the integral equation (0.2.1) implies also the unique solvability of the Cauchy problem

$$v'' + a(t)v' + b(t)v = f(t) \quad \text{for } t > t_0,$$
$$v(t_0) = c_1, \qquad v'(t_0) = c_2, \qquad (0.2.8)$$

for given constants c_1 and c_2 and for given piecewise-continuous functions a, b, and f. Indeed, the problem (0.2.8) is equivalent to the integral equation (0.1.6), and this latter integral equation is of the form (0.2.1), with

$$p(t) := c_1 + c_2 K(t, t_0) + \int_{t_0}^{t} K(t, s) f(s) \, ds,$$

$$V(t, s) := -K(t, s) b(s),$$

(0.2.9)

with K given by (0.1.5). Hence, there is one and only one solution v to (0.2.8), and this solution can be given by the appropriate representation (0.2.4).

The situation is somewhat more subtle regarding the existence and uniqueness of solutions for *boundary-value problems*. For example, the differential equation

$$v'' + v = 1$$

(0.2.10)

always has a unique solution on the interval $[0, \pi/2]$ subject to the following general Dirichlet boundary conditions:

$$v(0) = v_0, \qquad v(\pi/2) = v_1,$$

(0.2.11)

for any given constants v_0 and v_1. However, the same differential equation (0.2.10) has *no solution* on the interval $[0, \pi]$ subject to the boundary conditions

$$v(0) = 0, \qquad v(\pi) = 1,$$

(0.2.12)

whereas (0.2.10) has *infinitely many solutions* on $[0, \pi]$ subject to the conditions

$$v(0) = 0, \qquad v(\pi) = 2.$$

(0.2.13)

By way of comparison, the Dirichlet problem for the equation

$$v'' - v = f(t)$$

(0.2.14)

always has one and only one solution on any given interval.

Cochran (1968) has shown that the earlier results of this section on the initial-value problem can also be used in the study of boundary-value problems. An example of Cochran's results is given in Section 8.1.

Exercises

Exercise 0.2.1: Prove that the infinite series (0.2.5) converges absolutely and uniformly for $t_0 \leq s \leq t \leq t_1$, and derive the inequality (0.2.7).

Exercise 0.2.2: Prove that the function defined by the right side of (0.2.4) provides a solution to the Volterra integral equation (0.2.1). (Gronwall's inequality implies that the solution is unique.)

Exercise 0.2.3: Use the definition (0.2.5)–(0.2.6) to compute the resolvent kernel V^* for $V(t, s)$: $= t - s$. Use this result along with (0.2.4) to solve the equation

$$v(t) = 1 + t + \int_0^t (t - s) v(s) \, ds.$$

[This integral equation is equivalent to the initial-value problem $v'' - v = 0$, $v(0) = v'(0) = 1$; hence, v can be obtained more simply by any one of several other methods, such as Laplace transformation or variation of parameters.]

Exercise 0.2.4: Consider the function V defined as $V(t, s)$: $= a(t)b(s)$ for two given piecewise-continuous functions a and b. Show that the Volterra resolvent for this kernel can be given as

$$V^*(t, s) = a(t)b(s)\exp\left(\int_s^t ab\right).$$

[This shows that Gronwall's inequality is sharp, in the sense that equality in (0.2.2) implies also equality in (0.2.3).]

Exercise 0.2.5: The Bessel function $J_0(t)$ can be characterized as the solution of the Volterra equation [see (0.1.19)]

$$v(t) = 1 + \int_0^t s(\log s - \log t) v(s) \, ds. \tag{0.2.15}$$

Solve (0.2.15) directly by iteration or successive approximations to find

$$v(t) = J_0(t) = \sum_{k=0}^{\infty} \frac{(-1)^k (t/2)^{2k}}{(k!)^2}.$$

Hint: Seek v as the limit

$$v(t) = \lim_{k \to \infty} v_k(t),$$

with $v_1(t)$: $= 1$ and

$$v_{k+1}(t): = 1 + \int_0^t s(\log s - \log t) v_k(s) \, ds$$

for $k = 1, 2, \ldots$. The indefinite integral

$$\int x^k (\log x) \, dx = x^{k+1}\left[-(k + 1)^{-2} + (k + 1)^{-1}(\log x) \right]$$

will be helpful.

Exercise 0.2.6: Verify directly the validity of the assertions given in the text regarding (0.2.10)–(0.2.13).

Exercise 0.2.7: The second-order equation [see (0.2.10)]

$$\frac{d^2v}{dt^2} + v = f(t) \tag{0.2.16}$$

can be rewritten as the first-order system

$$\frac{dx}{dt} = Ax + \begin{bmatrix} 0 \\ f(t) \end{bmatrix}, \tag{0.2.17}$$

with

$$x = \begin{bmatrix} x_1(t) \\ x_2(t) \end{bmatrix} := \begin{bmatrix} v(t) \\ v'(t) \end{bmatrix}, \qquad A := \begin{bmatrix} 0 & 1 \\ -1 & 0 \end{bmatrix} \tag{0.2.18}$$

The Dirichlet problem for (0.2.16) on $[t_1, t_2]$ with boundary conditions $v(t_1) = \alpha_1$, $v(t_2) = \alpha_2$ (for given constants α_1, α_2) then amounts to the problem of solving the system (0.2.17) on $[t_1, t_2]$ subject to the boundary condition [see (0.1.13)]

$$Lx(t_1) + Rx(t_2) = \alpha \equiv \begin{pmatrix} \alpha_1 \\ \alpha_2 \end{pmatrix}, \tag{0.2.19}$$

with

$$L := \begin{bmatrix} 1 & 0 \\ 0 & 0 \end{bmatrix}, \qquad R := \begin{bmatrix} 0 & 0 \\ 1 & 0 \end{bmatrix}. \tag{0.2.20}$$

Show that the matrix M of (0.1.14) is nonsingular for this problem on the interval $[t_1, t_2] = [0, \pi/2]$, and hence the Dirichlet problem is always uniquely solvable in this case (see Exercise 0.1.2). *Hint:* Use

$$x^1 = \begin{bmatrix} \cos t \\ -\sin t \end{bmatrix} \quad \text{and} \quad x^2 = \begin{bmatrix} \sin t \\ \cos t \end{bmatrix}$$

as the column vectors of the fundamental matrix $X(t)$.

Exercise 0.2.8: Compute the matrix M of (0.1.14) for the boundary-value problem (0.2.17)–(0.2.19) on the interval $[t_1, t_2] = [0, \pi]$, and discuss this boundary-value problem with $f = 1$ in the two cases

$$\alpha = \begin{bmatrix} 0 \\ 1 \end{bmatrix} \quad \text{and} \quad \alpha = \begin{bmatrix} 0 \\ 2 \end{bmatrix}$$

[see (0.2.12) and (0.2.13)].

Exercise 0.2.9: Use the approach of Exercise 0.1.2 to study the Dirichlet problem for (0.2.14) by writing this equation as a suitable first-order system. In particular, show that the corresponding matrix M of (0.1.14) is nonsingular for *every* interval $[t_1, t_2]$.

0.3 Elementary differential inequalities*

In this section, certain differential inequalities are given that will be useful in establishing the validity of various asymptotic expansions discussed in later chapters. The inequalities will be given in terms of the previous differential operator $\mathscr{L}$ defined by (0.1.2), which is repeated here for convenience:

$$\mathscr{L}v(t) := \frac{d^2v}{dt^2} + a(t)\frac{dv}{dt} + b(t)v \qquad (0.3.1)$$

for any twice-differentiable function v on a specified interval.

Throughout this section the operator $\mathscr{L}$ is assumed to be *nonnegatively damped*, with

$$a(t) \geq 0 \quad \text{for all } t, \qquad (0.3.2)$$

and the operator is also assumed to be *nonoscillatory*, with

$$2a'(t) + a(t)^2 - 4b(t) \geq 0 \quad \text{for all } t. \qquad (0.3.3)$$

These conditions (0.3.2) and (0.3.3) are discussed further in Chapters 5 and 6.

Theorem 0.3.1: *Let the operator $\mathscr{L}$ satisfy (0.3.2) and (0.3.3), and let v be any twice-differentiable function on the interval $[t_0, t_1)$ satisfying the conditions*

$$\mathscr{L}v(t) \geq 0 \quad \text{on } [t_0, t_1), \quad \text{with } v(t_0) \geq 0 \quad \text{and} \quad v'(t_0) \geq 0.$$

Then v must also satisfy the first-order differential inequality

$$v'(t) + \tfrac{1}{2}a(t)v(t) \geq 0 \quad \text{on } [t_0, t_1),$$

from which there follows, on integration,

$$v(t) \geq \left[\exp\left(-\tfrac{1}{2}\int_{t_0}^t a\right)\right]v(t_0) \geq 0 \quad \text{on } [t_0, t_1).$$

(The quantity t_1 may be any finite number, with $t_1 \geq t_0$, or there may hold $t_1 = \infty$.)

Proof: For the operator $\mathscr{L}$ it is convenient to introduce the *Sturm transformation* $v \mapsto \bar{v}$, defined as

$$\bar{v}(t) := \left[\exp\left(\tfrac{1}{2}\int_{t_0}^t a\right)\right]v(t) \quad \text{for all } t \qquad (0.3.4)$$

* This section can be skipped or skimmed over lightly in a first reading.

and for any function v. For any smooth function v, the equation $\mathcal{L}v = f$ transforms into $\bar{\mathcal{L}}\bar{v} = \bar{f}$, with the transformed operator $\bar{\mathcal{L}}$ given as

$$\bar{\mathcal{L}}v(t) := \frac{d^2v}{dt^2} + c(t)v \qquad (0.3.5)$$

for any twice-differentiable v, where

$$c(t) := \tfrac{1}{4}\left[-2a'(t) - a(t)^2 + 4b(t)\right]. \qquad (0.3.6)$$

The initial values transform as

$$\bar{v}(t_0) = v(t_0), \qquad \bar{v}'(t_0) = v'(t_0) + \tfrac{1}{2}a(t_0)v(t_0). \qquad (0.3.7)$$

The previous results (0.1.6)–(0.1.9) can be applied to the operator $\bar{\mathcal{L}}$ [with $a = 0$ and $b = c$ in (0.1.6)] to find the integral relation

$$\bar{v}(t) = p(t) + \int_{t_0}^{t} V(t,s)\bar{v}(s)\,ds, \qquad (0.3.8)$$

with

$$p(t) := \bar{v}(t_0) + \bar{v}'(t_0)(t - t_0) + \int_{t_0}^{t}(t - s)\bar{f}(s)\,ds, \qquad (0.3.9)$$

$$V(t,s) := -(t - s)c(s),$$

for any solution of $\bar{\mathcal{L}}\bar{v} = \bar{f}$. As in Section 0.2, equation (0.3.8) can be solved to give

$$\bar{v}(t) = p(t) + \int_{t_0}^{t} V^*(t,s)p(s)\,ds, \qquad (0.3.10)$$

where the resolvent kernel V^* is given by (0.2.5)–(0.2.6).

The function v is assumed to satisfy $f = \mathcal{L}v \geq 0$, $v(t_0) \geq 0$, and $v'(t_0) \geq 0$, and these conditions, along with (0.3.2), (0.3.3), (0.3.4), (0.3.6), (0.3.7), and (0.3.9), imply now

$$p(t) \geq 0 \quad \text{and} \quad p'(t_0) \geq 0 \quad \text{for } t_0 \leq t < t_1, \qquad (0.3.11)$$

and also

$$V(t,s) \geq 0 \quad \text{and} \quad V_t(t,s) \geq 0 \quad \text{for } t_0 \leq s \leq t < t_1. \qquad (0.3.12)$$

This result (0.3.12), along with the definition of V^*, yields also

$$V^*(t,s) \geq 0 \quad \text{for } t_0 \leq s \leq t < t_1,$$

and this last inequality, along with (0.3.10)–(0.3.11), yields directly for $\bar{v}$ the result

$$\bar{v}(t) \geq 0 \quad \text{on } [t_0, t_1]. \qquad (0.3.13)$$

Equation (0.3.8) can be differentiated to give [note that $V(t,t) = 0$]

$$\bar{v}'(t) = p'(t) + \int_{t_0}^{t} V_t(t,s)\bar{v}(s)\,ds,$$

and this result, along with (0.3.11)–(0.3.13), implies $\bar{v}'(t) \geq 0$ everywhere on $[t_0, t_1)$. But now the Sturm transformation (0.3.4) can be used to write this last inequality in terms of v and v', resulting in the first-order differential inequality stated in the theorem, namely, $v' + (1/2)av \geq 0$. The remaining details are omitted. ■

The Sturm transformation (0.3.4) seems to have been first published in Sturm (1836). It has appeared often in the recent literature on perturbation theory for ordinary and partial differential equations [see Langer (1949), Erdélyi (1956), Weinstein and Smith (1975, 1976), and Olver (1978)].

Theorem 0.3.1 remains true without the damping condition (0.3.2) and without the initial inequality $v'(t_0) \geq 0$ if we add instead the initial inequality

$$v'(t_0) + \tfrac{1}{2}a(t_0)v(t_0) \geq 0.$$

This last condition allows the preceding proof to go through unchanged, and in this case the theorem simply states that the given initial inequalities

$$v \geq 0 \quad \text{and} \quad v' + \tfrac{1}{2}av \geq 0 \quad \text{at } t = t_0$$

propagate throughout the entire interval $[t_0, t_1)$. An analogous result is given for functions of two independent variables in Smith and Weinstein (1976).

Theorem 0.3.1 can be used to give various estimates for solutions of the initial-value problem

$$\mathscr{L}v = f \quad \text{on } [t_0, t_1), \quad \text{with } v(t_0) \text{ and } v'(t_0) \text{ specified.} \quad (0.3.14)$$

The following theorem gives a typical result.

Theorem 0.3.2: *Let the operator $\mathscr{L}$ satisfy condition (0.3.3) along with the following strengthened version of (0.3.2),*

$$a(t) \geq a_0 > 0 \quad \text{everywhere on } [t_0, t_1),$$

for a fixed positive constant a_0. Then the solution v of the initial-value problem (0.3.14) satisfies the bounds

$$|v(t)| \leq Q(t), \qquad |v'(t)| \leq Q'(t) + a(t)Q(t) \quad (0.3.15)$$

on $[t_0, t_1)$, *where* $Q = Q(t)$ *is defined as*

$$Q(t) := \frac{1}{\lambda_1 - \lambda_2}\left[\int_{t_0}^t (e^{\lambda_1(t-s)} - e^{\lambda_2(t-s)})|f(s)|\,ds\right.$$

$$+ |v(t_0)|\left(\lambda_1 e^{\lambda_2(t-t_0)} - \lambda_2 e^{\lambda_1(t-t_0)}\right)$$

$$\left. + |v'(t_0)|\left(e^{\lambda_1(t-t_0)} - e^{\lambda_2(t-t_0)}\right)\right],$$

with $\lambda_1 := \frac{1}{2}(-a_0 + [a_0^2 + 4\|b\|]^{1/2})$ *and* $\lambda_2 := \frac{1}{2}(-a_0 - [a_0^2 + 4\|b\|]^{1/2})$, *and where*

$$\|b\| := \sup_{[t_0, t_1)} |b(t)|.$$

The proof follows directly by two applications of Theorem 0.3.1, first with v replaced in Theorem 0.3.1 by $Q(t) - v(t)$, and then with v replaced by $Q(t) + v(t)$. Both of these two functions ($Q - v$ and $Q + v$) can be shown to satisfy the assumptions of Theorem 0.3.1, and the stated results of (0.3.15) can be shown to follow. The details are omitted. *Hint*: Use the facts $Q(t) \geq 0$ and $Q'(t) \geq 0$. Note that Q is simply the solution of the initial-value problem $Q'' + a_0 Q' - \|b\|Q = |f(t)|$ for $t \geq 0$, $Q(t_0) = |v(t_0)|$, $Q'(t_0) = |v'(t_0)|$. The result (0.3.15) is then intuitively expected if we interpret the initial-value problems as models of vibrating mechanical or electrical systems.

In some applications it is known that the given function b appearing in the operator $\mathcal{L}$ is *nonnegative*, in which case we have the simpler estimates of the following theorem.

Theorem 0.3.3: *Let the operator $\mathcal{L}$ satisfy the conditions* (0.3.2) *and* (0.3.3), *and assume that the function $b = b(t)$ is nonnegative, with*

$$b(t) \geq 0 \quad \text{everywhere on } [t_0, t_1).$$

Then the solution v of the initial-value problem (0.3.14) *satisfies the bounds of* (0.3.15), *with the function Q defined as*

$$Q(t) := |v(t_0)| + |v'(t_0)|\int_{t_0}^t \exp\left(-\int_{t_0}^s a\right)ds$$

$$+ \int_{t_0}^t \left\{\int_s^t \left[\exp\left(-\int_s^\tau a\right)\right]d\tau\right\}|f(s)|\,ds.$$

If, moreover, the function f/b is bounded, with finite infinity norm

$$\left\|\frac{f}{b}\right\| := \sup_{[t_0,\, t_1)} \left|\frac{f(t)}{b(t)}\right|,$$

then v satisfies the bounds of (0.3.15), *with*

$$Q(t) := \left\|\frac{f}{b}\right\| + |v(t_0)| + |v'(t_0)| \int_{t_0}^{t} \exp\left(-\int_{t_0}^{s} a\right) ds.$$

The proof follows the same lines as in the previous theorem and is omitted. In both cases the *comparison function* Q is the solution of a companion initial-value problem related naturally to the given problem.

The results of the preceding theorems can be obtained under weaker regularity conditions on the data [see Weinstein and Smith (1975)], but such refinements are often of only limited value in applications to singular-perturbation theory.

The following theorem is an example of certain well-known *maximum/minimum principles* for two-point boundary-value problems. The notation $C^k[t_1, t_2]$, which was used earlier, denotes the class of all functions that have continuous derivatives of all orders through k on the interval $[t_1, t_2]$, and the class $C^k(t_1, t_2)$ is defined similarly for the open interval, for any nonnegative integer k.

Theorem 0.3.4: *Let the operator $\mathscr{L}$ satisfy the conditions* (0.3.2) *and* (0.3.3), *and let v be any function of class $C^0[t_1, t_2] \cap C^2(t_1, t_2)$ satisfying the conditions*

$$\mathscr{L}v(t) \leq 0 \quad \text{on the open interval } (t_1, t_2),$$

$$\text{and} \quad v(t_1) \geq 0 \quad \text{and} \quad v(t_2) \geq 0.$$

Then v is nonnegative everywhere on the closed interval $[t_1, t_2]$, $v(t) \geq 0$.

Proof (by contradiction): If the result is false, then $v < 0$ holds for some points in (t_1, t_2), so that v must have a negative minimum point at some point t_0 between t_1 and t_2, with

$$v(t_0) < 0 \quad \text{and} \quad v'(t_0) = 0.$$

But then Theorem 0.3.1 can be applied to the function $-v$ on the interval $[t_0, t_2]$, from which there follows the result $v < 0$ everywhere on $[t_0, t_2]$. This contradicts the assumption $v(t_2) \geq 0$. ∎

Theorem 0.3.4 can be used to give estimates for solutions of the Dirichlet problem $\mathscr{L}v = f$ on $[t_1, t_2]$, with $v(t_1)$ and $v(t_2)$ specified. The next theorem gives a typical example.

Theorem 0.3.5: *Let b and f be piecewise-continuous functions on the bounded interval $[t_1, t_2]$, with b nonpositive,*

$$b(t) \leq 0 \quad \text{on } [t_1, t_2],$$

and assume that the function f/b has a finite infinity norm. Then any function u of class $C^0[t_1, t_2] \cap C^2(t_1, t_2)$ that satisfies the equation

$$\frac{d^2u}{dt^2} + b(t)u = f(t) \quad \text{on the open interval } (t_1, t_2)$$

must also satisfy the bound

$$\|u\| \leq \max\{|u(t_1)|, |u(t_2)|, \|f/b\|\}, \tag{0.3.16}$$

where

$$\|u\| = \max_{[t_1, t_2]} |u(t)|.$$

Proof: The nonpositiveness of the function b implies that the operator

$$\mathscr{L} := (d^2/dt^2) + b(t)$$

satisfies the previous two conditions (0.3.2) and (0.3.3) (with $a = 0$). The stated inequality (0.3.16) then follows directly by two applications of Theorem 0.3.4, first with

$$v(t) := -u(t) + \text{right side of (0.3.16)},$$

and then with

$$v(t) := +u(t) + \text{right side of (0.3.16)}.$$

The details are omitted. ■

Estimates can also be given in terms of *integrals* of f, analogous to Theorem 0.3.3. A typical result follows.

Theorem 0.3.6: *Let b and f be piecewise-continuous functions on the bounded interval $[t_1, t_2]$, with b nonpositive, $b(t) \leq 0$ for all t. Then any function u of class $C^0[t_1, t_2] \cap C^2(t_1, t_2)$ that satisfies*

$$\frac{d^2u}{dt^2} + b(t)u = f(t) \quad \text{on } (t_1, t_2)$$

must satisfy the bound

$$|u(t)| \leq \int_{t_1}^{t_2} G(t, s)|f(s)| \, ds$$

$$+ G_t(t_1, t)|u(t_1)| + G_t(t_2, t)|u(t_2)| \tag{0.3.17}$$

on $[t_1, t_2]$, where G is the (nonnegative-valued) Green function for the *operator d^2/dt^2, given as*

$$G(t, s) = \begin{cases} (t_2 - t)(s - t_1)/(t_2 - t_1) & \text{for } s \le t, \\ (t_2 - s)(t - t_1)/(t_2 - t_1) & \text{for } s \ge t, \end{cases}$$

with $G_t(t_1, t) = (t_2 - t)/(t_2 - t_1)$ and $G_t(t_2, t) = (t_1 - t)/(t_2 - t_1)$.

The function appearing on the right side of the inequality (0.3.17) is simply the (nonnegative) solution of the (comparison) Dirichlet problem

$$\frac{d^2 w}{dt^2} = -|f(t)| \quad \text{on } (t_1, t_2),$$

$$w(t_1) = |u(t_1)|, \qquad w(t_2) = |u(t_2)|.$$

The proof of the theorem follows by two applications of Theorem 0.3.4, first with $v = w - u$, and then with $v = w + u$. The details are omitted.

A good general discussion of maximum principles, differential inequalities, and comparison techniques is given in Protter and Weinberger (1967). Applications of the maximum principle to singular-perturbation problems are given in Dorr, Parter, and Shampine (1973).

Exercises

Exercise 0.3.1: Give the details of the proof of Theorem 0.3.3.

Exercise 0.3.2: Give the details of the proof of Theorem 0.3.6.

Exercise 0.3.3: Let the operator

$$\mathscr{L} := \frac{d^2}{dt^2} + a(t)\frac{d}{dt} + b(t)$$

satisfy the damping condition

$$a(t) \ge a_0 > 0 \quad \text{everywhere on } [t_0, t_1)$$

for a fixed positive constant a_0. Show that the solution of the initial-value problem $\mathscr{L}v = f(t)$ on $[t_0, t_1)$, with $v(t_0)$ and $v'(t_0)$ specified, must satisfy the a priori bounds

$$|v(t)|, \frac{1}{a_0}|v'(t)|$$

$$\le \left\{ |v(t_0)| + \frac{1}{a_0}|v'(t_0)| + \frac{1}{a_0}\int_{t_0}^{t}|f| \right\} \exp\left[(t - t_0)\|b\|/a_0 \right]$$

$$(0.3.18)$$

for $t \in [t_0, t_1)$, with

$$\|b\| := \sup_{[t_0, t_1)} |b(t)|.$$

Hint: Apply Gronwall's inequality to the integral representation (0.1.5)–(0.1.8). The bound for $v'(t)$ is obtained by differentiating (0.1.6) and using the bound for v. Note that (0.3.18) provides estimates analogous to those of Theorem 0.3.2 *without* the nonoscillatory requirement (0.3.3).

Exercise 0.3.4: Let the third-order operator

$$\mathscr{L} := \frac{d^3}{dt^3} + a(t)\frac{d^2}{dt^2} + b(t)\frac{d}{dt} + c(t)$$

satisfy the condition

$$a(t) \geq a_0 > 0 \quad \text{everywhere on } [t_0, t_1) \qquad (0.3.19)$$

for a fixed positive constant a_0. Show that the solution of the initial-value problem

$$\mathscr{L}v = f(t) \quad \text{on } [t_0, t_1), \quad \text{with } v(t_0), v'(t_0), \quad \text{and} \quad v''(t_0) \text{ specified},$$
$$(0.3.20)$$

satisfies the bound

$$|v(t)| \leq (1 + \kappa e^\kappa)\left\{ |v(t_0)| + T|v'(t_0)| + \frac{T}{a_0}|v''(t_0)| + \frac{T}{a_0}\int_{t_0}^t |f| \right\}$$
$$(0.3.21)$$

for $t \in [t_0, t_1]$, with $T := t_1 - t_0$ and $\kappa := (T/a_0)(\|b\| + T\|c\|)$, and where the norm is here the supremum (infinity) norm,

$$\|h\| := \sup_{[t_0, t_1]} |h(t)|.$$

Similar bounds hold also for derivatives of v. [The bound (0.3.21) is not the best possible, but it often suffices.] *Hint*: Use the result of Exercise 0.1.6. In particular, apply Gronwall's inequality to (0.1.26), and then use the result back in (0.1.25). Note that the kernel K of (0.1.21) satisfies $0 \leq K(t, s) \leq (t - s)/a_0$ and $0 \leq K_t(t, s) \leq 1/a_0$ if (0.3.19) holds.

0.4 The Banach/Picard fixed-point theorem

Let $\mathscr{V}$ be a Banach space (complete normed vector space), let ρ be a positive number, and let B_ρ denote the closed ball in $\mathscr{V}$ of radius ρ

centered at the origin:

$$B_\rho := \left\{ v \in \mathscr{V} \,\middle|\, \|v\| \le \rho \right\}. \tag{0.4.1}$$

The problem of verifying the validity of a given, proposed asymptotic solution for a typical perturbation problem can often be cast in the form of a fixed-point problem for a suitable mapping T defined on B_ρ and taking values in $\mathscr{V}$ (for some suitable Banach space $\mathscr{V}$), so that we seek an element v satisfying

$$Tv = v, \quad v \in B_\rho. \tag{0.4.2}$$

In practice, we can usually arrange matters so that the mapping T is contracting, and then the following Banach/Picard (contraction-mapping) fixed-point theorem can be used in the study of (0.4.2).

Theorem 0.4.1 (Banach/Picard fixed-point theorem): *Let T map the closed ball B_ρ of (0.4.1) into itself ($Tv \in B_\rho$ for all $v \in B_\rho$), and let T be a contraction mapping on B_ρ ($\|Tv_1 - Tv_2\| \le \gamma \|v_1 - v_2\|$ for all $v_1, v_2 \in B_\rho$, for some fixed positive constant $\gamma < 1$). Then (0.4.2) has a unique solution in B_ρ, and this solution can be given by successive approximation as $\lim_{j \to \infty} v_j$, with $v_0 := 0$ and $v_j := Tv_{j-1}$ for $j = 1, 2, \ldots$.*

Proof: The uniqueness of solutions in B_ρ follows directly from the contracting property of T, and the existence of a solution follows from the completeness of the Banach space along with the continuity of T and the inequality $\|v_{j+1} - v_j\| \le \gamma^j \|v_1\|$ (for $j = 0, 1, 2, \ldots$). [See, for example, Cronin (1964) for further details.] ∎

Example 0.4.1: Let v be the solution of the nonlinear initial-value problem

$$\frac{d^2v}{dt^2} + v = v^2 + \rho(t, \varepsilon) \quad \text{for } 0 \le t \le t_1,$$

$$v = \frac{dv}{dt} = 0 \quad \text{at } t = 0, \tag{0.4.3}$$

where $\rho = \rho(t, \varepsilon)$ is a given continuous function depending on t and on a small parameter $\varepsilon > 0$, and where it is known that ρ is uniformly small, of order ε, with

$$|\rho(t, \varepsilon)| \le \kappa \cdot \varepsilon \quad \text{for } 0 \le t \le t_1 \tag{0.4.4}$$

as $\varepsilon \to 0+$, for a given fixed positive constant κ. We wish to prove that the problem (0.4.3) has a unique "small" solution $v = v(t, \varepsilon)$ satisfying some suitable bound, perhaps of the type $|v(t, \varepsilon)| \leq$ const. ε as $\varepsilon \to 0+$, uniformly for $0 \leq t \leq t_1$. Instead of appealing to some known theorem from the theory of differential equations, we shall give a direct study of (0.4.3) using the Banach/Picard fixed-point theorem applied to a suitable integral equation that is equivalent to the given initial-value problem.

In this case it follows from Example 0.1.1 that the present initial-value problem is equivalent to the equation (see Exercise 0.4.1)

$$v(t) = \int_0^t [\sin(t - s)] \rho(s, \varepsilon) \, ds + \int_0^t [\sin(t - s)] v(s)^2 \, ds, \qquad (0.4.5)$$

where the dependence of $v(t) = v(t, \varepsilon)$ on ε is being suppressed in (0.4.5). This latter integral equation can be studied with the Banach/Picard fixed-point theorem.

To this end we introduce an operator T on the vector space $C^0[0, t_1]$, defined as

$$Tv(t) := \text{right side of } (0.4.5), \quad \text{for any } v \in C^0[0, t_1]. \quad (0.4.6)$$

We use the maximum (infinity) norm on C^0, with

$$\|v\| := \max_{[0, t_1]} |v(t)|$$

for any continuous function v on $[0, t_1]$. As is well known, the resulting normed vector space, again denoted as $C^0[0, t_1]$, is complete. The equation (0.4.5) now amounts to the equation

$$v = Tv. \qquad (0.4.7)$$

An easy calculation shows that the operator T maps the ball of radius $2\kappa t_1 \varepsilon$ centered at the origin into itself for all ε satisfying

$$|\varepsilon| < 1 / \left[4\kappa \cdot (t_1)^2 \right]. \qquad (0.4.8)$$

Moreover, it is easily seen that T also satisfies (see Exercise 0.4.2)

$$\|Tv_1 - Tv_2\| \leq 4\kappa t_1^2 \varepsilon \|v_1 - v_2\| \qquad (0.4.9)$$

for all v_1, v_2 in this ball of radius $2\kappa t_1 \varepsilon$, so that T is a contraction map for ε satisfying (0.4.8). It follows from Theorem 0.4.1 that the given problem has a unique solution v satisfying

$$|v(t, \varepsilon)| \leq 2\kappa t_1 \varepsilon \qquad (0.4.10)$$

uniformly for $t \in [0, t_1]$ and for all ε satisfying (0.4.8).

In the following chapters, we shall find many situations that invite use of the Banach/Picard fixed-point theorem.

Exercises

Exercise 0.4.1: Show that the initial-value problem (0.4.3) is equivalent to the integral equation (0.4.5). *Hint*: Use variation of parameters as in Example 0.1.1). This shows that any solution of (0.4.3) must also satisfy the integral equation (0.4.5). A direct calculation shows that any solution of (0.4.5) must also satisfy (0.4.3).

Exercise 0.4.2: Let B denote the ball in $C^0[0, t_1]$ of radius $2\kappa t_1 \varepsilon$, centered at the origin, so that B consists of all continuous functions on $[0, t_1]$ satisfying

$$\|v\| \leq 2\kappa t_1 \varepsilon.$$

Show that the operator T of (0.4.6) maps B into itself for all ε satisfying (0.4.8), and show that T is contracting on B for all such ε. *Hint*: The result $(v_1)^2 - (v_2)^2 = (v_1 + v_2)(v_1 - v_2)$ will be useful in proving contraction.

0.5 Divergent series

As was mentioned in the Preface, a singular-perturbation problem is a problem that depends on a parameter (or parameters) in such a way that solutions of the problem behave nonuniformly as the parameter tends toward some limiting value of interest. The nature of the nonuniformity can vary from problem to problem. In practice, one seeks a uniformly valid, easily interpretable approximation to the nonuniformly behaving solution. The approximation often takes the form of an asymptotic expansion, which may diverge (in the sense of an infinite series) but which nevertheless provides the required information on the exact solution to the problem. The following example from Olver (1974) illustrates the point that a divergent series may provide useful information on the exact solution of a given problem.

Consider the real function

$$F(x) := \int_0^\infty e^{-xt} \cos t \, dt \quad \text{for } x > 0. \tag{0.5.1}$$

If we replace $\cos t$ in the integrand with its Taylor expansion about the origin, we find

$$F(x) = \int_0^\infty e^{-xt} \sum_{k=0}^\infty \frac{(-1)^k t^{2k}}{(2k!)} \, dt = \sum_{k=0}^\infty \frac{(-1)^k}{(2k)!} \int_0^\infty e^{-xt} t^{2k} \, dt$$

$$= \sum_{k=0}^\infty \frac{(-1)^k}{(2k)!} \frac{(2k)!}{x^{2k+1}} = \frac{1}{x} \left[1 - \frac{1}{x^2} + \left(\frac{1}{x^2}\right)^2 - \cdots \right]$$

$$= \frac{1}{x} \frac{1}{1 + x^{-2}} = \frac{x}{1 + x^2} \quad \text{for } x > 1, \tag{0.5.2}$$

where we have used the known sum of the geometric series, and where all series here converge nicely for the indicated values $x > 1$. Of course, $F(x)$ as defined by (0.5.1) is just the Laplace transform of $\cos t$, and the result $F(x) = x/(1 + x^2)$ is actually valid for $x > 0$, as can be derived directly from (0.5.1) by two integrations by parts.

Partial sums of the convergent series obtained in (0.5.2) can, of course, be used to provide approximate values for $F(x)$. For example, if we define the remainder $R_N = R_N(x)$ as

$$R_N(x) := F(x) - \sum_{k=0}^N \frac{(-1)^k}{x^{2k+1}} = \frac{x}{1 + x^2} - \sum_{k=0}^N \frac{(-1)^k}{x^{2k+1}}, \tag{0.5.3}$$

then for, say, $x = 10$ and $N = 2$, we find $R_2(10) = (10/101) - 0.09901$, from which we easily find $|R_2(10)| < 10^{-7}$. Hence, the partial sum

$$\sum_{k=0}^2 \frac{(-1)^k}{x^{2k+1}}$$

provides a useful approximation to $F(x)$ at $x = 10$.

Instead of $F(x)$, consider now the function

$$G(x) := \int_0^\infty e^{-xt} \frac{1}{1 + t} \, dt \quad \text{for } x > 0, \tag{0.5.4}$$

where the cosine function appearing on the right side of (0.5.1) has been replaced now with the function $1/(1 + t)$. If we replace this latter function with its Taylor (geometric) series $\sum_{k=0}^\infty (-t)^k$ and proceed formally as in (0.5.2), we find in this case, from (0.5.4),

$$G(x) \stackrel{?}{=} \sum_{k=0}^\infty (-1)^k \int_0^\infty e^{-xt} t^k \, dt = \sum_{k=0}^\infty \frac{(-1)^k k!}{x^{k+1}}. \tag{0.5.5}$$

But, of course, the geometric series used formally in going from (0.5.4) to (0.5.5) converges only for $|t| < 1$, and the resulting two infinite series appearing on the right side of (0.5.5) *diverge* for all values of x. Even so, partial sums can be used to provide useful approximations to $G(x)$.

Indeed, if we now define the remainder R_N as

$$R_N(x) := G(x) - \sum_{k=0}^{N} \frac{(-1)^k k!}{x^{k+1}} \quad \text{for } x > 0, \qquad (0.5.6)$$

and if we use (0.5.4) along with the result

$$\frac{1}{1+t} = \sum_{k=0}^{N} (-t)^k + \frac{(-t)^{N+1}}{1+t}$$

(for all $t \neq -1$), we find directly the result

$$R_N(x) = (-1)^{N+1} \int_0^{\infty} e^{-xt} \frac{t^{N+1}}{1+t} \, dt, \qquad (0.5.7)$$

from which follows the estimate (for $x > 0$)

$$|R_N(x)| \leq \int_0^{\infty} e^{-xt} t^{N+1} \, dt = \frac{(N+1)!}{x^{N+2}}. \qquad (0.5.8)$$

If we set $x = 10$ and $N = 3$, we find from (0.5.8) the inequality $|R_3(10)| \leq 0.00024$, so that the easily computed partial sum

$$\sum_{k=0}^{3} \frac{(-1)^k k!}{(10)^{k+1}} = 0.0914$$

accurately provides the first few digits for the unknown exact value $G(10)$.

The function G of (0.5.4) is the solution of the terminal-value problem $G'(x) = G(x) - 1/x$ for $x > 0$, $G(\infty) = 0$. This latter problem has been used by Ince (1927) and Turrittin (1973) to illustrate the usefulness of diverging asymptotic series. At $x = 10$, one finds $G(10) = 0.091563\ldots$.

The crux of the matter here and elsewhere is to find a suitable estimate of the difference between an exact (usually unknown) quantity and a suitable, easily interpretable expression intended to approximate the exact quantity.

Notation: Throughout the book we use the customary Landau order symbols O and o as commonly defined (see pp. 1–2 of O'Malley 1984*a*).

Also, an asymptotic relation such as

$$\alpha(\varepsilon) \sim \sum_{k=0}^{\infty} \alpha_k \varepsilon^k$$

is understood to mean

$$\alpha(\varepsilon) = \sum_{k=0}^{n} \alpha_k \varepsilon^k + O(\varepsilon^{n+1}) \quad (\text{as } \varepsilon \to 0).$$

Exercises

Exercise 0.5.1: Give the details of the derivation of the estimate (0.5.8).

Exercise 0.5.2: **(a)** Use repeated integration by parts to obtain the expansion

$$\int_x^{\infty} \frac{\sin t}{t} \, dt = \sum_{k=1}^{N} \frac{(k-1)! u_k(x)}{x^k} + R_N(x) \quad \text{for } x > 0,$$

with

$$u_{2k}(x) = (-1)^{k+1} \sin x \quad \text{and} \quad u_{2k+1}(x) = (-1)^k \cos x$$

for $k = 0, 1, \ldots$ and with a suitable remainder $R_N(x)$.
   (b) Show that the remainder $R_N(x)$ of **(a)** satisfies the estimate $|R_N(x)| \leq (N-1)! x^{-N}$ for $x > 0$.
   (c) Use the expansion of **(a)** to obtain a numerical approximation for the integral

$$\int_8^{\infty} \frac{\sin t}{t} \, dt.$$

How many terms should be used in the (divergent) expansion to obtain the best approximation?

PART I

Initial-value problems of oscillatory type

1

Precession of the planet Mercury

This chapter introduces the Einstein equation for the planet Mercury as an illustration of an important class of initial-value problems of oscillatory type that depend on a parameter in such a way that solutions behave nonuniformly for small values of the parameter as the independent variable tends toward infinity. The typical perturbation problems of celestial mechanics, including Newtonian and Einsteinian gravitational mechanics, are of this type, as are various other oscillation and vibration problems in other fields. This class of problems is considered more generally in Chapters 2, 3, and 4.

1.1 The Einstein equation for Mercury

Consider first a Keplerian model for the motion of the planet Mercury about the sun. It is convenient to introduce plane polar coordinates r, θ in the plane of Mercury's orbit, with the sun placed at the origin. Then the motion of Mercury can be represented by a smooth curve γ given parametrically in terms of time t as

$$\gamma: r = R(t), \qquad \theta = \Theta(t) \quad \text{for } t \geq 0,$$

for suitable functions R and Θ that are to be determined by the laws of physics.

The function Θ is assumed to be a strictly monotonic function of t (in agreement with Newton's second law of motion and the Newtonian inverse-square law of gravity, or, more simply, in agreement with observation), and so θ can be introduced as the independent variable in $r = R[\Theta^{-1}(\theta)]$, with $t = \Theta^{-1}(\theta)$. We let $\bar{r} = 5.83 \times 10^{12}$ cm be a constant *mean value* for the typical distance between Mercury and the sun;

it is then convenient to introduce a new dimensionless function $u = u(\theta)$ by the formula

$$u(\theta): = \frac{\bar{r}}{R[\Theta^{-1}(\theta)]}.$$

If we neglect all bodies except the sun and Mercury, then Newtonian physics leads to the two-body equation

$$\frac{d^2u}{d\theta^2} + u = a \quad \text{for } \theta > 0, \tag{1.1.1}$$

where the dimensionless constant a is given as

$$a = GM\bar{r}/h^2,$$

and where G is the Newtonian gravitational constant, M is the mass of the sun, and h is the angular momentum per unit mass of Mercury. For Mercury, there holds approximately (Danby 1962)

$$a \doteq 0.98.$$

We choose the initial time so that initially Mercury is at perihelion (closest approach to the sun), with $du/d\theta = 0$ at $\theta = 0$. Then the initial conditions are

$$u = b, \quad \frac{du}{d\theta} = 0 \quad \text{at } \theta = 0, \tag{1.1.2}$$

where b is a fixed constant of order unity; $b \doteq 1.01$ for Mercury.

The solution of (1.1.1)–(1.1.2) is given as

$$u = a + (b - a)\cos\theta,$$

which leads to a fixed Keplerian elliptical orbit for Mercury.

When the gravitational effects of the other planets are taken into account, the Newtonian model predicts that the orbit of Mercury should rotate slowly in space, so that the major axis of the ellipse should advance (precess) by about 530 seconds of arc per century, leading to a complete revolution of the major axis in some 250,000 years. However, the major axis is actually observed to precess by about 570 seconds per century, or about 40 seconds more in a century than the Newtonian model predicts. Nineteenth-century astronomers studied this matter very carefully, and it was finally interpreted as a serious, though small, discrepancy between observation and Newtonian theory [see pp. 364–5 of Pannekoek (1961)].

Later, as a consequence of Einstein's theory of gravity, equation (1.1.1) was replaced with the modified equation

$$\frac{d^2u}{d\theta^2} + u = a + \varepsilon u^2, \tag{1.1.3}$$

where the constant ε is given as

$$\varepsilon = 3GM/c^2\bar{r},$$

with G, M, and $\bar{r}$ as in Newton's theory, and c is the speed of light in a vacuum. This equation (1.1.3) is to be solved for $\theta > 0$ subject to the same initial conditions (1.1.2).

The quantities a, b, and u appearing in (1.1.2)–(1.1.3) are all of order unity, and the parameter ε is found to be small, of order 10^{-7} for Mercury. Hence, the initial-value problem for (1.1.3) amounts to a perturbation of the same problem for (1.1.1). However, (1.1.3) is nonlinear and somewhat more difficult to solve than (1.1.1). The exact solution of (1.1.3) is discussed in the next section.

This slow precession of the orbit of Mercury continues steadily in the *same direction*, rather than oscillating back and forth. Such effects that accumulate over long time periods are called in astronomy *secular effects*. (The 1977 edition of *Webster's New Collegiate Dictionary* gives seven definitions of *secular*, the last of which is "relating to a long term of indefinite duration.") Another example of such a secular effect is the gradual diminishing of the moon's period of revolution about the earth by an amount of about 10 seconds per century, first noticed by Edmund Halley in 1693 [see pp. 304 and 366 of Pannekoek (1961)].

1.2 The exact solution

A first integral of the Einstein equation for Mercury can be obtained easily by multiplying (1.1.3) by $du/d\theta$ and integrating the resulting equation to find the (energy) equation

$$\left(\frac{du}{d\theta}\right)^2 + u^2 = 2au + \frac{2}{3}\varepsilon u^3 + \kappa \tag{1.2.1}$$

for some suitable constant of integration κ. This constant κ can be determined by putting $\theta = 0$ in (1.2.1) and using the initial conditions of (1.1.2) to find $\kappa = b^2 - 2ab - \frac{2}{3}\varepsilon b^3$. Hence, the differential equation (1.2.1) can be written as

$$\left(\frac{du}{d\theta}\right)^2 = \frac{2}{3}\varepsilon(u - b)\left[u^2 + \left(b - \frac{3}{2\varepsilon}\right)u + b^2 + \frac{3}{2\varepsilon}(2a - b)\right]$$

or

$$\left(\frac{du}{d\theta}\right)^2 = \frac{2}{3}\varepsilon(u - k_1)(u - k_2)(u - k_3), \tag{1.2.2}$$

where the parameters k_1, k_2, and k_3 are given as $k_2 := b$, with k_1 and

k_3 the two roots of the quadratic equation

$$k^2 + \left(b - \frac{3}{2\varepsilon} \right) k + \left[b^2 + \frac{3}{2\varepsilon} (2a - b) \right] = 0.$$

In particular, up to an arbitrary assignment or labelling of k_1 and k_3,

$$k_1 = 2a - b + O(\varepsilon), \quad k_2 = b, \quad k_3 = \frac{3}{2\varepsilon} + O(1) \quad \text{as } \varepsilon \to 0,$$

$$(1.2.3)$$

and for Mercury one finds the three distinct values

$$k_1 \doteq 0.95, \quad k_2 \doteq 1.01, \quad \text{and} \quad k_3 \doteq \frac{3}{2\varepsilon} = \text{order}(10^7).$$

The differential equation (1.2.2) is to be solved subject to the initial conditions

$$u = k_2 = b, \quad \text{and} \quad \frac{d^2 u}{d\theta^2} = a - b + \varepsilon b^2 \quad \text{for } \theta = 0. \quad (1.2.4)$$

The second condition of (1.2.4) follows from (1.1.2)–(1.1.3); this condition excludes the *trivial* solution $u \equiv k_2$ (for all θ) in (1.2.2), because, for Mercury, $a - b + \varepsilon b^2 < 0$.

The initial-value problem of (1.2.2) and (1.2.4) can be integrated in terms of the *Jacobi elliptic function* sn as

$$u(\theta, \varepsilon) = k_2 - (k_2 - k_1) \left\{ \text{sn} \left[\theta \left(\frac{\varepsilon (k_2 - k_1)}{6} \right)^{1/2}, i \left(\frac{k_2 - k_1}{k_3 - k_2} \right)^{1/2} \right] \right\}^2,$$

$$(1.2.5)$$

where $i = \sqrt{-1}$ is the imaginary unit and where we have written $u = u(\theta, \varepsilon)$ to emphasize the dependence of u on both θ and ε. The Jacobi elliptic function sn $= \text{sn}(x, y)$ is a function of two variables, say x and y, and can be defined by the relation [see pp. 145 and 210 of Davis (1962)]

$$\text{sn}(x, y) = \sin z, \quad \text{with } x = \int_0^z (1 - y^2 \sin^2 t)^{-1/2} dt.$$

The solution (1.2.5) can be fully interpreted for small $\varepsilon > 0$ and for $\theta \geq 0$ only with some effort, based on a careful and nontrivial analysis. Of course, the sharp bonds

$$k_1 \leq u \leq k_2$$

follow easily from (1.2.5). We shall not pursue the study of (1.2.5) here. Rather, we shall be interested in alternative approaches that yield results that are somewhat easier to interpret and that are moreover easy to generalize to other related oscillation problems for which exact representations such as (1.2.5) are not readily available. Chapters 2 and 3 are devoted to two such methods that are strikingly effective for such problems.

One such alternative approach that yields certain limited information without much effort is to replace the original second-order initial-value problem (1.1.2)–(1.1.3) with the integral equation

$$u(\theta) = a + (b - a)\cos\theta + \varepsilon \int_0^\theta \sin(\theta - s) u(s)^2 \, ds, \qquad (1.2.6)$$

which is equivalent to (1.1.2)–(1.1.3); see the argument used earlier in going from (0.4.3) to (0.4.5). Then, from (1.2.6) follows easily the inequality

$$|u(\theta)| \le |a| + |b - a| + \varepsilon \int_0^\theta |u(s)|^2 \, ds, \qquad (1.2.7)$$

and now a Gronwall-type argument similar to the proof of Theorem 0.2.1 can be applied to (1.2.7) as in Exercise 1.2.1 to give the estimate

$$|u(\theta)| \le 2(|a| + |b - a|) \quad \text{for } 0 \le \theta \le \frac{1}{2\varepsilon(|a| + |b - a|)}. \qquad (1.2.8)$$

This last estimate shows that the solution of the given nonlinear problem remains uniformly bounded as the independent variable θ ranges over a large interval. For example, for Mercury the result (1.2.8) leads to the bound

$$|u(\theta)| \le 2.02 \quad \text{for } 0 \le \theta \le \frac{1}{2.02\varepsilon} = \text{order}(10^7). \qquad (1.2.9)$$

Because the (quasi) period of Mercury's motion is about 88 (earth) days, one sees that the bound (1.2.9) has been shown to remain valid over a lengthy time interval of several billion years.

As was mentioned in the paragraph preceding (1.2.6), the "energy method" based on (1.2.1) actually leads to the bounds

$$0.95 \doteq k_1 \le u \le k_2 \doteq 1.01, \quad \text{for all } \theta. \qquad (1.2.10)$$

This last result not only gives tighter, sharper bounds on u as compared with (1.2.9) but also is a global result, because it is known to be valid for all θ, whereas (1.2.9) has been obtained only for all θ on a large interval with length on the order of ε^{-1}.

The method used to obtain (1.2.9) is quite elementary and simple to apply in practice. Indeed, it will be seen in Chapter 4 that this method provides an important approach in obtaining error estimates for singularly perturbed oscillation problems. However, the comments of the previous paragraph show clearly that there is often a price to be paid in using this elementary method: The approach typically leads to useful estimates only on large intervals with lengths typically on the order of the reciprocal of some small parameter. On the other hand, the energy method has been used by Greenlee and Snow (1975) to obtain global error estimates for several important classes of oscillation problems, as will be discussed in Section 4.4.

In the next section we shall consider whether a classical, "regular" perturbation approach can be used to give further detailed information on the exact solution u for large θ and small ε, beyond the information contained in the bounds (1.2.9) or (1.2.10).

Exercise

Exercise 1.2.1: Show that any solution of the integral equation (1.2.6) must satisfy the estimate (1.2.8). *Hint:* Put

$$S(\theta) := |a| + |b - a| + \varepsilon \int_0^\theta |u|^2,$$

and find $S(0) = |a| + |b - a|$ along with the differential inequality

$$S'(\theta) \le \varepsilon S(\theta)^2,$$

which can be integrated by separation of variables to give

$$(|a| + |b - a|)^{-1} - S(\theta)^{-1} \le \varepsilon\theta.$$

This last inequality can be manipulated to yield the stated result.

1.3 A classical perturbation approach

It is known from a basic result in the theory of differential equations that the solution $u = u(\theta, \varepsilon)$ of the initial-value problem

$$\frac{d^2u}{d\theta^2} + u = a + \varepsilon u^2 \qquad \text{for } \theta > 0,$$

$$u = b \quad \text{and} \quad \frac{du}{d\theta} = 0 \quad \text{at } \theta = 0,$$

$$(1.3.1)$$

depends analytically on the parameter ε; hence, u can be represented in the form

$$u(\theta, \varepsilon) = \sum_{k=0}^{\infty} u_k(\theta) \varepsilon^k \qquad (1.3.2)$$

for suitable functions u_k that depend only on θ and not on ε. The resulting infinite series will converge nicely for all suitable values of ε and θ; a direct verification of this fact will be given later.

Such analytic dependence of solutions of differential equations on suitable parameters was assumed to be valid and was used successfully in astronomical calculations in the eighteenth and nineteenth centuries by several workers, including Poisson. The validity of this assumption was later verified by Poincaré [see pp. 212 and 228 of Minorsky (1962)]. Here the analyticity follows directly from the estimate (1.3.6). Of course, Poincaré's result is in general a *local* result. For example, the solution $w(t, \varepsilon) = 1/(1 + \varepsilon t)$ of the initial-value problem $dw/dt = -\varepsilon w^2$, $w(0, \varepsilon) = 1$, can be represented in terms of the power series

$$w(t, \varepsilon) = \sum_{k=0}^{\infty} (-t)^k \varepsilon^k,$$

but only locally for $|\varepsilon t| < 1$.

In the case of (1.3.1)–(1.3.2), the functions u_k are determined by substituting (1.3.2) into (1.3.1) and equating coefficients of like powers of ε in the resulting equations so as to obtain a sequence of (in this case *linear*) problems that can be easily solved recursively for $u_0, u_1, u_2, \ldots$. Here we can work just as well with the equivalent integral equation (1.2.6), in which case we substitute (1.3.2) into (1.2.6) to find

$$\sum_{k=0}^{\infty} u_k(\theta) \varepsilon^k = a + (b - a)\cos\theta + \varepsilon \int_0^{\theta} \sin(\theta - s) \sum_{k=0}^{\infty} \left(\sum_{j=0}^{k} u_j u_{k-j} \right) \varepsilon^k \, ds$$

$$= a + (b - a)\cos\theta + \sum_{k=1}^{\infty} \varepsilon^k \int_0^{\theta} \sin(\theta - s)$$

$$\times \sum_{j=0}^{k-1} u_j(s) u_{k-1-j}(s) \, ds.$$

Equating coefficients of like powers of ε in this last equation, we find

$$u_k(\theta) = \begin{cases} a + (b - a)\cos\theta & \text{for } k = 0, \\ \sum_{j=0}^{k-1} \int_0^{\theta} u_j(s) u_{k-1-j}(s) \sin(\theta - s) \, ds & \text{for } k = 1, 2, \ldots. \end{cases}$$

$$(1.3.3)$$

This result (1.3.3) can be used recursively to determine the required coefficient functions $u_0, u_1, \ldots, u_k, \ldots$. For example, we find

$$u_0(\theta) = a + (b - a)\cos\theta,$$

$$u_1(\theta) = a^2(1 - \cos\theta) + a(b - a)\theta\sin\theta \qquad (1.3.4)$$

$$+ \frac{(b - a)^2}{3}(2 - \cos^2\theta - \cos\theta),$$

$$\vdots$$

From (1.3.4) we see that the function u_1 becomes *unbounded* for large θ, with

$$\begin{array}{c} \text{unbounded} \\ \text{term} \\ \downarrow \end{array}$$

$$u_1(\theta) = a(b - a)\theta\sin\theta + \text{bounded terms.}$$

Similarly, the later coefficient functions u_k are all found to contain unbounded terms (for $k \geq 1$) with

$$u_k(\theta) = O(\theta^k) \quad \text{as } \theta \to \infty; \qquad (1.3.5)$$

compare this with (1.3.6).

From (1.3.3) we find easily by induction the bounds

$$|u_k(\theta)| \leq (|a| + |b - a|)^{k+1}|\theta|^k \quad \text{for } k = 0, 1, 2, \ldots . \qquad (1.3.6)$$

It follows from (1.3.6) and the comparison test that the series (1.3.2)–(1.3.3) converges absolutely (and uniformly on compact subsets) for all θ, ε satisfying

$$|\varepsilon\theta| < (|a| + |b - a|)^{-1}. \qquad (1.3.7)$$

It is easily seen that the resulting series actually solves the given initial-value problem (1.3.1).

This perturbation approach, based on an expansion such as (1.3.2), is said to be a *regular* perturbation approach, because the resulting expansion is regular, or convergent. There are good reasons to weaken this requirement of convergence in order to include related perturbation expansions that may diverge but are nevertheless useful in practice, as illustrated in Section 0.5.

In the present case it is difficult to interpret the infinite series (1.3.2), which is a series representation of the previous solution (1.2.5) involving the Jacobi elliptic function. Moreover, the amount of effort required in practice to compute explicitly the coefficients u_k increases rapidly with increasing k, and in fact one usually succeeds in computing only the first few coefficients. The classical perturbation procedure would be to use a partial sum of the infinite series as an approximation to the exact $u(\theta, \varepsilon)$, as

$$u(\theta, \varepsilon) = \sum_{k=0}^{N} u_k(\theta)\varepsilon^k + R_N(\theta, \varepsilon), \qquad (1.3.8)$$

where the remainder term $R_N(\theta, \varepsilon)$, which is defined by (1.3.8), would in practice be neglected (omitted) to give the approximation

$$u(\theta, \varepsilon) \doteq \sum_{k=0}^{N} u_k(\theta)\varepsilon^k. \qquad (1.3.9)$$

Of course, the usefulness of the approximation (1.3.9) hinges on the size of the remainder R_N on the right side of (1.3.8). Unfortunately, this remainder is in general acceptably small only on a certain limited θ interval; so (1.3.9) cannot be used for large θ.

For example, in the case $N = 0$ we can introduce the residual $\rho_0(\theta, \varepsilon)$ defined as

$$\rho_0(\theta, \varepsilon) := -\left[u_0''(\theta) + u_0(\theta) - a - \varepsilon u_0(\theta)^2 \right], \qquad (1.3.10)$$

so that ρ_0 is a measure of how well u_0 satisfies the differential equation of (1.3.1). From (1.3.4) and (1.3.10) we find

$$\rho_0(\theta, \varepsilon) = \varepsilon u_0(\theta)^2 = \varepsilon \left[a + (b - a)\cos\theta \right]^2, \qquad (1.3.11)$$

so that ρ_0 is of order ε, uniformly for all θ. On the other hand, the remainder R_0 of (1.3.8) (with $N = 0$) is found to satisfy

$$\frac{d^2 R_0}{d\theta^2} + \left[1 - 2\varepsilon u_0(\theta) \right] R_0 = \varepsilon R_0^2 + \rho_0(\theta, \varepsilon) \quad \text{for } \theta > 0,$$

$$\text{with } R_0 = \frac{dR_0}{d\theta} = 0 \quad \text{at } \theta = 0. \qquad (1.3.12)$$

We wish to obtain an estimate for R_0 from (1.3.12). Indeed, this can be done by replacing (1.3.12) with a suitable equivalent integral equation and applying either a direct Gronwall-type argument or an argument related to that used in Example 0.4.1 based on the Banach/Picard fixed-point theorem. In this way we find an estimate of the type

$$|R_0(\theta, \varepsilon)| \leq \text{const. } \varepsilon \quad \text{as } \varepsilon \to 0+, \qquad (1.3.13)$$

uniformly for all θ on any fixed interval of the type

$$0 \leq \theta \leq \theta_1 \qquad (1.3.14)$$

for any fixed positive constant θ_1. [The details of a proof of (1.3.13) can be based on the results of Section 4.1.]

Unfortunately, the result (1.3.13)–(1.3.14) is inadequate for the problem at hand, because we require an estimate of the type (1.3.13) for a lengthy interval, say for all θ on an "expanding" interval (as $\varepsilon \to 0+$) of the type

$$0 \leq \theta \leq \frac{\theta_1}{\varepsilon} \quad \text{as } \varepsilon \to 0+, \qquad (1.3.15)$$

for a fixed positive θ_1. Such a result on an expanding interval would follow if the residual (1.3.11) were of order(ε^2) uniformly on such an interval as (1.3.15); see Exercise 3.2.2, where such a result is proved for a related problem. Here, however, the remainder R_0 fails to satisfy (1.3.13) on an expanding interval of this latter type. Moreover, the situation cannot be helped by taking more terms in the partial sum (1.3.9). Indeed, we find for every partial sum [for every nonnegative integral value of N in (1.3.9)] that the classical approximation (1.3.9) fails to provide useful information for large θ. This is because of the unbounded component, of order(θ^k), in the coefficient function $u_k(\theta)$ for $k \geq 1$, as indicated in (1.3.5).

It was this kind of difficulty in celestial mechanics that provided the early impetus for the development of other, more useful approximation procedures for such problems.

The difficulty here can be viewed as stemming from the use of only a finite number of terms in the Taylor expansion for the approximation of such functions as [see (2.3.8) in the next chapter]

$$\cos(1 - \varepsilon)\theta = \cos\theta + \varepsilon\theta \sin\theta - \tfrac{1}{2}(\varepsilon\theta)^2 \cos\theta + \cdots . \qquad (1.3.16)$$

Any partial sum of such a Taylor expansion fails to exhibit certain of the essential global features of the original function, and therefore the approximation is poor for certain purposes. This points to a serious limitation of the classical perturbation approach.

Two alternative perturbation approaches are discussed in Chapters 2 and 3. These approaches prove to be strikingly more successful than the classical approach for such oscillation problems as are considered here.

The unbounded (as $\theta \to \infty$) components of such coefficient functions as the functions u_k are commonly referred to as (analytical) *secular terms*, because these terms may possibly represent a physical effect that increases steadily in amplitude with increasing θ. Of course, in the present case these terms are actually spurious (false) secular terms, because the true quantity $u(\theta, \varepsilon)$ actually remains bounded. In more complicated situations, the question whether or not such analytical secular terms are spurious is often difficult to answer, because we often have no firm information regarding the true quantity beyond the information contained in the first few coefficients on the right side of some expansion such as (1.3.8).

Exercises

Exercise 1.3.1: Prove the validity of the bounds (1.3.6), and prove that the resulting series (1.3.2)–(1.3.3) provides a solution to the initial-value problem (1.3.1) for suitable values of θ. (That the given problem has a unique solution follows from a well-known basic result in the theory of differential equations, or in this case directly from the results of Section 1.2.)

Exercise 1.3.2: The initial-value problem

$$\frac{d^2u}{dt^2} + \varepsilon\frac{du}{dt} + u = 0 \quad \text{for } t > 0,$$

$$u = 0 \quad \text{and} \quad \frac{du}{dt} = 1 \quad \text{at } t = 0, \tag{1.3.17}$$

provides a mathematical model for an oscillator with linear restoring force and weak linear damping (for small positive ε) subject to an initial impulse.

(a) Find $u_0(t)$ and $u_1(t)$ for the expansion

$$u(t,\varepsilon) = \sum_{k=0}^{\infty} u_k(t)\varepsilon^k, \tag{1.3.18}$$

where $u(t,\varepsilon)$ denotes the solution of (1.3.17).

(b) Derive the result

$$u_k(t) = -\int_0^t \sin(t-s)u'_{k-1}(s)\,ds \quad \text{for } k = 1,2,\ldots \tag{1.3.19}$$

for the coefficient functions in (1.3.18), and use (1.3.19) to prove (by induction) the estimates

$$|u_k(t)|, |u'_k(t)| \le |t|^k/k! \quad \text{for } k = 0,1,2,\ldots.$$

(c) Prove that the expansion (1.3.18) converges for all t and all ε, uniformly on compact sets.

(d) How well do you expect the partial sum $u_0(t) + \varepsilon u_1(t)$ to approximate the exact solution $u(t,\varepsilon)$ for fixed $\varepsilon > 0$ when t is large?

2

Krylov/Bogoliubov averaging

This chapter presents an approximation procedure developed in the 1930s by N. M. Krylov and N. N. Bogoliubov to handle various oscillation problems in mechanics of the type illustrated by the Einstein equation for Mercury. The method has found important applications in many areas involving oscillatory systems, and various of these applications are discussed briefly here.

2.1 Introduction

In this section we introduce a class of differential equations that includes the Einstein equation (1.1.3) as a special case. The independent variable will be labelled here as t rather than the previous θ of (1.1.3), and the class of differential equations considered will consist of those equations that can be written in the form

$$\frac{d^2u}{dt^2} + k^2u = a + \varepsilon f\left(u, \frac{du}{dt}\right) \qquad (2.1.1)$$

for some suitable smooth function $f = f(u, v)$ depending on two variables, where $f(u, v)$ is evaluated in (2.1.1) at $u = u(t)$ and $v = du(t)/dt$. The constant ε is a small positive parameter satisfying

$$0 < \varepsilon \ll |k|.$$

Elementary transformation of the dependent and independent variables can be used, as in Exercise 2.1.1, to reduce (2.1.1) to the special case $k = 1$ and $a = 0$, but we shall retain the parameters k and a as in (2.1.1).

The independent variable t will be referred to as the *time*, and the initial time will be taken to be $t = 0$. The differential equation (2.1.1) is to be solved for $t > 0$ subject to the initial conditions

$$u = b \quad \text{and} \quad \frac{du}{dt} = c \quad \text{at } t = 0, \tag{2.1.2}$$

where b and c are specified constants.

Example 2.1.1: If $k = 1$ and $f(u, v) := u^2$ (with f independent of the second argument v), then (2.1.1) becomes

$$u'' + u = a + \varepsilon u^2,$$

which is just the Einstein equation for Mercury (with $t = \theta$).

Example 2.1.2: If $k = 1$, $a = 0$, and $f(u, v) := v - \frac{1}{3}v^3$, then (2.1.1) becomes

$$u'' + u = \varepsilon\left[u' - \frac{1}{3}(u')^3\right], \tag{2.1.3}$$

which is the differential equation for the *Rayleigh oscillator* first studied by the English mathematical physicist John William Strutt (1842–1919) as a mathematical model for certain problems in acoustics. (Strutt became the third Lord Rayleigh in 1873.) For Strutt's work in acoustics, see Rayleigh (1877–78). The equation (2.1.3) later became important also in electric-circuit theory through the work of B. van der Pol. The *van der Pol oscillator* is described in the next example.

Example 2.1.3: In (2.1.1), if $k = 1$, $a = 0$, and $f(u, v) := (1 - u^2)v$, then the differential equation becomes

$$\frac{d^2u}{dt^2} + u = \varepsilon(1 - u^2)\frac{du}{dt}, \tag{2.1.4}$$

which is the equation of the van der Pol oscillator. This equation provides a useful model for certain mechanical and electrical systems capable of self-excited or self-sustained oscillations called *relaxation oscillations* by van der Pol [see van der Pol (1926) and the references given there]. A good discussion of self-sustained oscillations and relaxation oscillations is given in Stoker (1950). Equation (2.1.4) also seems useful as a model of heartbeats, as was observed by van der Pol.

Rayleigh's equation (2.1.3) can be transformed into van der Pol's equation, as indicated in Exercise 2.1.2.

Example 2.1.4: In (2.1.1), if $a = 0$ and $f(u, v) := -v^3$, then the equation becomes

$$\frac{d^2u}{dt^2} + \varepsilon\left(\frac{du}{dt}\right)^3 + k^2 u = 0, \tag{2.1.5}$$

which is a model for an oscillator with linear restoring force and weak, nonlinear, cubic damping.

Exercises

Exercise 2.1.1: Show that the change of variables $\tau: = kt$, $U(\tau): = u(\tau/k) - a/k^2$ reduces (2.1.1) to the form

$$\frac{d^2U}{d\tau^2} + U = \varepsilon F\left(U, \frac{dU}{d\tau}\right)$$

for some suitable function $F = F(U,V)$. Find F.

Exercise 2.1.2: If u satisfies the Rayleigh equation (2.1.3), show that $w: = du/dt$ satisfies the van der Pol equation (2.1.4). *Hint*: Differentiate (2.1.3).

2.2 The averaged system

In the early twentieth century, a certain averaging technique was used formally by van der Pol (1926, 1927) in the study of the van der Pol equation (2.1.4). A similar technique was developed more fully in Krylov and Bogoliubov (1934, 1947) and was shown to be a useful tool for the approximate solution of the general equation (2.1.1). The technique of averaging for (2.1.1) has as its point of departure the well-known method of variation of parameters, which had been used by astronomers since the late eighteenth century in the study of nonlinear oscillation problems in celestial mechanics.

The method of variation of parameters for (2.1.1) is based initially on the observation that if $\varepsilon = 0$, then the differential equation (2.1.1) reduces to

$$\frac{d^2u}{dt^2} + k^2u = a \quad (\varepsilon = 0), \tag{2.2.1}$$

where the most general solution of (2.2.1) can be represented as

$$u(t) = \frac{a}{k^2} + A\sin(kt + B) \tag{2.2.2}$$

for suitable constants A and B. From (2.2.2) it follows also that

$$\frac{du}{dt} = kA\cos(kt + B). \tag{2.2.3}$$

We now seek the general solution $u(t, \varepsilon)$ of the perturbed equation (2.1.1) (for nonzero ε) in the same form (2.2.2)–(2.2.3), where now

$A = A(t, \varepsilon)$ and $B = B(t, \varepsilon)$ must be taken to be suitable functions determined so that the given equations (2.1.1) and (2.2.2)–(2.2.3) all hold. A direct calculation (omitted here) shows that these equations hold if and only if $A(t, \varepsilon)$ and $B(t, \varepsilon)$ satisfy the system

$$\frac{d}{dt}\begin{bmatrix} A(t, \varepsilon) \\ B(t, \varepsilon) \end{bmatrix} = \frac{\varepsilon}{k} f\left(\frac{a}{k^2} + A\sin\varphi, kA\cos\varphi\right)\begin{bmatrix} \cos\varphi \\ -\frac{1}{A}\sin\varphi \end{bmatrix} \quad (2.2.4)$$

where the quantity φ is here defined as

$$\varphi = \varphi(t, \varepsilon) := kt + B(t, \varepsilon). \quad (2.2.5)$$

These equations (2.2.4)–(2.2.5) are *exact* and are equivalent to the original equation (2.1.1). No approximation has yet been made. However, this latter system (2.2.4)–(2.2.5) is generally a nonlinear, nonautonomous system that is quite difficult to solve in practice, and so it may seem that the difficult equation (2.1.1) has been replaced with an equally difficult system (2.2.4)–(2.2.5). However, van der Pol, and more clearly Krylov and Bogoliubov, showed that this latter system is well suited for an averaging procedure that provides an effective technique for the problems at hand.

Before we look at the averaged equation of Krylov and Bogoliubov, it will be useful to examine (2.2.4)–(2.2.5) in several examples.

Example 2.2.1: In the case of the Einstein equation for Mercury, $f(u, v) = u^2$ (see Example 2.1.1), and so the system (2.2.4) becomes (with $k = 1$)

$$\frac{d}{dt}\begin{bmatrix} A \\ B \end{bmatrix} = \varepsilon(a + A\sin\varphi)^2\begin{bmatrix} \cos\varphi \\ -\frac{1}{A}\sin\varphi \end{bmatrix}, \quad (2.2.6)$$

with $\varphi = t + B$.

Example 2.2.2: For the Rayleigh oscillator of Example 2.1.2, $k = 1$, $a = 0$, and $f(u, v) = v - \frac{1}{3}v^3$. Hence, the system (2.2.4) becomes

$$\frac{d}{dt}\begin{bmatrix} A \\ B \end{bmatrix} = \varepsilon(\cos\varphi)\left(1 - \frac{1}{3}A^2\cos^2\varphi\right)\begin{bmatrix} A\cos\varphi \\ -\sin\varphi \end{bmatrix}, \quad (2.2.7)$$

with φ again given by (2.2.5).

These and similar examples illustrate the point that the system (2.2.4)–(2.2.5) is generally a nonlinear, nonautonomous system whose exact solution can be expected to be quite difficult to obtain. However, for small ε, this system has the merit of exhibiting clearly the fact that the quantities A and B *should vary only slowly with increasing t*, because both dA/dt and dB/dt are proportional to the small quantity ε in (2.2.4). But if A and B are slowly varying quantities, then the major variation on the right side of (2.2.4) is due to the quantity $\varphi = kt + B$, where φ occurs in (2.2.4) always as an argument of the 2π-periodic sine and cosine functions. This led Krylov and Bogoliubov to consider the following averaged system obtained from (2.2.4) by averaging out the φ dependence on the right side:

$$
\frac{d}{dt}\begin{bmatrix} A_0 \\ B_0 \end{bmatrix} = \frac{\varepsilon}{2\pi k} \int_0^{2\pi} f\left(\frac{a}{k^2} + A_0 \sin\theta,\, kA_0 \cos\theta \right) \begin{bmatrix} \cos\theta \\ -\dfrac{1}{A_0}\sin\theta \end{bmatrix} d\theta,
$$

$$\tag{2.2.8}$$

where A_0 and B_0 are held fixed (constant) during the θ integration on the right side of (2.2.8), as illustrated in the following examples. The quantity φ of (2.2.5) no longer appears in the averaged system (2.2.8).

The quantities A_0 and B_0 obtained by solving (2.2.8) subject to suitable initial conditions are intended to provide approximations for the exact quantities A and B in (2.2.2). Of course, $A_0 = A_0(t, \varepsilon)$ and $B_0 = B_0(t, \varepsilon)$ are still functions of t and ε. The resulting functions A_0 and B_0 are used to define a function $u_0 = u_0(t, \varepsilon)$ as [see (2.2.2)]

$$
u_0(t, \varepsilon) := \frac{a}{k^2} + A_0(t, \varepsilon)\sin\left[kt + B_0(t, \varepsilon) \right], \tag{2.2.9}
$$

where this latter function is intended to provide an approximation to the corresponding exact solution u of the original second-order differential equation (2.1.1).

In 1958 it was shown by Bogoliubov and Mitropolsky (1961) that this is indeed the case, at least over a time interval of length order$(1/\varepsilon)$, so that one has an estimate of the form

$$
|u(t, \varepsilon) - u_0(t, \varepsilon)| \le C_1 \varepsilon \tag{2.2.10}
$$

uniformly for all t and ε satisfying

$$
0 \le t \le \frac{C_2}{\varepsilon},
$$

for fixed positive constants C_1 and C_2 that are independent of ε as $\varepsilon \to 0+$. We shall discuss the proofs of such estimates as (2.2.10) in Chapter 4.

The approximation (2.2.9) is only the leading, or first-order, term in an asymptotic expansion that can be obtained by "higher-order averaging," as in Bogoliubov and Mitropolsky (1961); see also Minorsky (1962) and

Perko (1968). The system (2.2.8) is the first-order Krylov/Bogoliubov averaged system, and (2.2.9) is the first-order Krylov/Bogoliubov approximation. Higher-order Krylov/Bogoliubov/Mitropolsky averaging is somewhat more complicated than first-order averaging, and only the latter, based on (2.2.8), is considered here.

Note that the Krylov/Bogoliubov averaged system (2.2.8) is autonomous, and hence somewhat simpler than the nonautonomous exact system (2.2.4)–(2.2.5). Of course, (2.2.8) is still generally nonlinear. Several illustrative examples of first-order averaging are given in the next few sections.

The method of averaging has been extended to various other suitable oscillation systems more general than (2.2.4), leading ordinarily to a first-order system of the following form [see (2.4.9) and (2.5.9)]

$$\frac{dx}{dt} = \varepsilon f(\varepsilon, t, x), \tag{2.2.11}$$

where here the solution is an n-vector $x = x(t, \varepsilon)$, and where f is a given vector function subject to suitable conditions. The analogue of the previous first-order averaged system (2.2.8) becomes

$$\frac{dx}{dt} = \varepsilon f_0(x), \tag{2.2.12}$$

where

$$f_0(x) := \lim_{T \to \infty} \frac{1}{T} \int_0^T f(0, t, x) \, dt. \tag{2.2.13}$$

An illustration is given in Exercise 2.4.7. These generalizations are discussed with references in Nayfeh and Mook (1979); see also Perko (1968). Note also that averaging is sometimes discussed entirely in terms of first-order systems, where the original oscillation problem is then written as such a first-order system; see Exercise 2.2.3 for an example.

The applied literature is replete with applications of the method of averaging. For example, Struble (1961) used averaging techniques to study the motion of a satellite of an oblate spheroidal body, with application in the study of near-earth satellites. Additional applications of averaging techniques are cited in Nayfeh and Mook (1979). Several problems involving nonlinear oscillations in biology are cited in Hoppensteadt (1979). Averaging techniques have also been applied to the partial differential equations that model the macroscopic properties of continuous media with spatial periodic structure at the microscopic level [see Bensoussan, Lions, and Papanicolaou (1978)], and to various perturbed wave equations (see Kurzweil (1963, 1966, 1967)].

Exercises

Exercise 2.2.1: Give a careful derivation of the exact system (2.2.4).

Exercise 2.2.2: Give the exact system (2.2.4) for van der Pol's equation (2.1.4).

Exercise 2.2.3: Consider the two-dimensional system

$$\frac{dx}{dt} = \begin{bmatrix} 0 & 1 \\ -1 & 0 \end{bmatrix} x + \varepsilon f(x) \quad \text{for } t \geq 0, \tag{2.2.14}$$

for a solution vector-valued function

$$x = x(t) = \begin{bmatrix} x_1(t) \\ x_2(t) \end{bmatrix}$$

subject to the initial condition

$$x = c \quad \text{at } t = 0, \tag{2.2.15}$$

where

$$c = \begin{bmatrix} c_1 \\ c_2 \end{bmatrix}$$

is a given initial point in $\mathbb{R}^2$, and where

$$f = f(x) = \begin{bmatrix} f_1(x) \\ f_2(x) \end{bmatrix}$$

is a given vector-valued function of x taking values in $\mathbb{R}^2$. Show that the solution of (2.2.14)–(2.2.15) can be given in the form (variation of parameters)

$$x(t) = \begin{bmatrix} x_1(t) \\ x_2(t) \end{bmatrix} = \begin{bmatrix} \cos t & \sin t \\ -\sin t & \cos t \end{bmatrix} \gamma(t) \tag{2.2.16}$$

if and only if the vector-valued function $\gamma(t)$ satisfies the conditions

$$\frac{d\gamma}{dt} = \varepsilon \begin{bmatrix} \cos t & -\sin t \\ \sin t & \cos t \end{bmatrix} f[\xi(t,\gamma)], \tag{2.2.17}$$

with

$$\xi(t,\gamma) \equiv \begin{bmatrix} \xi_1(t,\gamma) \\ \xi_2(t,\gamma) \end{bmatrix} := \begin{pmatrix} \cos t & \sin t \\ -\sin t & \cos t \end{pmatrix} \gamma, \tag{2.2.18}$$

and where

$$\gamma = c \quad \text{at } t = 0. \tag{2.2.19}$$

The solution $\gamma = \gamma(t)$ of (2.2.17)–(2.2.19) must be a slowly varying function, because its derivative $d\gamma/dt$ is proportional to the small parameter ε on the right side of (2.2.17). The first-order Krylov/Bogoliubov system becomes

$$\frac{d\hat\gamma}{dt} = \varepsilon F(\hat\gamma), \tag{2.2.20}$$

with the function $F = F(\gamma)$ defined by the formula

$$F(\gamma) := \frac{1}{2\pi} \int_0^{2\pi} \begin{bmatrix} \cos\tau & -\sin\tau \\ \sin\tau & \cos\tau \end{bmatrix} f[\xi(\tau,\gamma)]\, d\tau, \tag{2.2.21}$$

where γ is held fixed during the integration on the right side of (2.2.21), so that the right side $\varepsilon F(\hat\gamma)$ of (2.2.20) is obtained as the average or mean value, with respect to the independent variable, of the right side of (2.2.17), where the average is taken over a fixed interval of length 2π. If $\hat\gamma(t)$ is the solution of (2.2.20) subject to the initial condition

$$\hat\gamma = c \quad \text{at } t = 0, \tag{2.2.22}$$

then one obtains the corresponding first-order Krylov/Bogoliubov approximate solution $\hat x(t)$ to (2.2.14)–(2.2.15) as

$$\hat x(t) \equiv \begin{bmatrix} \hat x_1(t) \\ \hat x_2(t) \end{bmatrix} := \begin{bmatrix} \cos t & \sin t \\ -\sin t & \cos t \end{bmatrix} \hat\gamma(t). \tag{2.2.23}$$

2.3 Precession of Mercury, revisited

For the Einstein two-body problem, $k = 1$ and $f(u,v) = u^2$, as in Example 2.2.1. The averaged equation for dA_0/dt is given by the first (component) equation of the system (2.2.8) and becomes [compare with the first equation of the exact system (2.2.6)]

$$\frac{dA_0}{dt} = \frac{\varepsilon}{2\pi} \int_0^{2\pi} (a + A_0\sin\theta)^2 \cos\theta\, d\theta,$$

where A_0 is held constant during the integration on the right side here. The given integral can be easily evaluated, and the resulting value is seen to be zero, so that the averaged equation becomes

$$\frac{dA_0}{dt} = 0 \quad \text{for } t > 0. \tag{2.3.1}$$

Similarly, the averaged equation for dB_0/dt becomes

$$\frac{dB_0}{dt} = -\frac{\varepsilon}{2\pi A_0} \int_0^{2\pi} (a + A_0\sin\theta)^2 \sin\theta\, d\theta,$$

where this integral can be evaluated as

$$\int_0^{2\pi} (a + A_0\sin\theta)^2 \sin\theta \, d\theta = 2\pi a A_0.$$

Hence, the averaged equation for dB_0/dt becomes

$$\frac{dB_0}{dt} = -\varepsilon a \quad \text{for } t > 0. \qquad (2.3.2)$$

The averaged equations (2.3.1)–(2.3.2) are easily solved to give

$$A_0 = \alpha,$$
$$B_0 = -\varepsilon a t + \beta, \qquad (2.3.3)$$

for suitable constants of integration α and β, where then (2.3.3) and (2.2.9) give (with $k = 1$)

$$u_0(t, \varepsilon) = a + \alpha \sin[(1 - \varepsilon a)t + \beta], \qquad (2.3.4)$$

with

$$\frac{du_0}{dt} = \alpha(1 - \varepsilon a)\cos[(1 - \varepsilon a)t + \beta]. \qquad (2.3.5)$$

The integration constants α and β are determined for the problem of Mercury by imposing the initial conditions of (1.1.2), which are listed here again for reference, but with the previous independent variable θ of Chapter 1 replaced here by t:

$$u_0 = b \quad \text{and} \quad \frac{du_0}{dt} = 0 \quad \text{at } t = 0. \qquad (2.3.6)$$

From (2.3.4)–(2.3.6) it follows that

$$\alpha\cos\beta = 0, \quad \alpha\sin\beta = b - a,$$

and these results, along with (2.3.4), imply that u_0 is given uniquely as

$$u_0(t, \varepsilon) = a + (b - a)\cos(1 - \varepsilon a)t. \qquad (2.3.7)$$

(The trigonometric addition formula for the cosine function has been used here.)

For the purpose of interpreting (2.3.7) in terms of the motion of the planet Mercury, it is convenient now to relabel the independent variable t again as the angular variable θ, as in Chapter 1. The first-order Krylov/Bogoliubov approximate solution can be given as

$$u_0(\theta, \varepsilon) = a + (b - a)\cos(1 - \varepsilon a)\theta. \qquad (2.3.8)$$

By way of comparison, the classical perturbation approach of Section 1.3 [see (1.3.4)] gives the leading, or first-order, approximation

$$u_0(\theta) = a + (b - a)\cos\theta. \qquad (2.3.9)$$

Of course, this classical first-order approximation (2.3.9) simply reproduces the fixed Keplerian orbit for Mercury, as discussed in Section 1.1, obtained from Newton's model of gravity.

The Krylov/Bogoliubov approximation (2.3.8) provides a very accurate approximation to the exact solution given in Section 1.2 in terms of the Jacobi elliptic function; moreover, this approximation remains uniformly valid over a lengthy θ interval. For example, it will be shown in Chapter 4 that the exact solution $u(\theta, \varepsilon)$ satisfies the estimate [see (4.3.24)]

$$\left| u(\theta, \varepsilon) - \left[a + (b - a)\cos(1 - \varepsilon a)\theta \right] \right| \leq 17\varepsilon \qquad (2.3.10)$$

for all θ satisfying

$$0 \leq \theta \leq 1/\varepsilon.$$

Hence, for Mercury, we find that the exact solution u can differ from the Krylov/Bogoliubov first-order approximate solution (2.3.8) by at most only about 10^{-6} during several million years! The estimate (2.3.10) is obtained in Chapter 4 without using any explicit representation for u such as that given in Section 1.2 in terms of the Jacobi elliptic function.

Not only does (2.3.8) provide an accurate approximation for u, but the approximation is very easy to interpret. According to (2.3.8), the perihelion precesses or advances by an amount equal to $2\pi a\varepsilon$ radians per orbit; for Mercury, this amounts to about 40 seconds of arc per century, in agreement with observation.

For $\varepsilon = 0$, the Krylov/Bogoliubov approximation (2.3.8) reduces to the classical approximation (2.3.9), which corresponds to a fixed Keplerian elliptical orbit with fixed major axis. However, for any fixed positive ε (no matter how small), the major axis rotates slowly and eventually undergoes large (secular) changes as $\theta \to \infty$. In this sense *the perturbation is nonuniform for small ε as $\theta \to \infty$*, and so the given problem is said to be singularly perturbed.

Exercises

Exercise 2.3.1: Show that the first-order system (2.2.14) subject to the initial condition (2.2.15) corresponds to the Einstein two-body problem for $u: = a + x_1$, with the choices

$$c_1: = b - a, \qquad c_2: = 0,$$

$$f_1(x): = 0, \qquad f_2(x): = (a + x_1)^2. \qquad (2.3.11)$$

Exercise 2.3.2: **(a)** Compute the vector field

$$F(\gamma) = \begin{bmatrix} F_1(\gamma) \\ F_2(\gamma) \end{bmatrix}$$

of (2.2.21) for the system (2.2.14), with f given as in (2.3.11).

(b) Solve the averaged system (2.2.20)–(2.2.21) subject to the initial condition (2.2.22) for $\hat{\gamma}(t)$, with c and f given by (2.3.11). Compare (2.3.7) with the result given here by (2.2.23) for $\hat{u} := a + \hat{x}_1$.

2.4 An oscillator with weak cubic damping

The initial-value problem (see Example 2.1.4)

$$u'' + \varepsilon(u')^3 + u = 0 \quad \text{for } t > 0, \tag{2.4.1}$$

with

$$u = b \quad \text{and} \quad u' = c \quad \text{at } t = 0, \tag{2.4.2}$$

provides a mathematical model for an oscillator with linear restoring force and weak (small $\varepsilon > 0$) cubic damping.

The equation (2.4.1) is of the form (2.1.1), with $k = 1$, $a = 0$, and $f(u,v) = -v^3$, and the averaged system (2.2.8) becomes

$$A_0' = -\frac{\varepsilon}{2\pi} A_0^3 \int_0^{2\pi} \cos^4\theta \, d\theta,$$

$$B_0' = \frac{\varepsilon}{2\pi} A_0^2 \int_0^{2\pi} \cos^3\theta \sin\theta \, d\theta.$$

The two integrals here are easily evaluated, with

$$\int_0^{2\pi} \cos^4\theta \, d\theta = \tfrac{3}{4}\pi, \qquad \int_0^{2\pi} \cos^3\theta \sin\theta \, d\theta = 0,$$

so that the averaged equations become

$$A_0' = -\tfrac{3}{8}\varepsilon A_0^3, \qquad B_0' = 0.$$

Note that the equation for A_0 is nonlinear. These averaged equations can be integrated to give

$$A_0(t, \varepsilon) = \frac{\alpha}{\left(1 + \tfrac{3}{4}\alpha^2\varepsilon t\right)^{1/2}}, \qquad B_0(t, \varepsilon) = \beta$$

for suitable integration constants α and β, and then (2.2.9) yields

$$u_0(t, \varepsilon) = \frac{\alpha \sin(t + \beta)}{\left(1 + \tfrac{3}{4}\alpha^2\varepsilon t\right)^{1/2}}. \tag{2.4.3}$$

The constants α and β can be determined so that u_0 from (2.4.3) satisfies the initial conditions (2.4.2), and we find

$$u_0(t,\varepsilon) = \frac{b\cos t + \left[c + \frac{3}{8}\alpha^2\varepsilon b\right]\sin t}{\left(1 + \frac{3}{4}\alpha^2\varepsilon t\right)^{1/2}}$$

with α^2 given as

$$\alpha^2 = \frac{1 - \frac{3}{4}bc\varepsilon - \left[1 - \frac{3}{2}bc\varepsilon - (3\varepsilon b^2/4)^2\right]^{1/2}}{2\left(\frac{3}{8}b\varepsilon\right)^2}$$

$$= b^2 + c^2 + \text{order}(\varepsilon).$$

However, u_0 is only the first-order Krylov/Bogoliubov approximation to u (with $u - u_0$ of order ε); so one may as well impose the initial conditions on u_0 only up to the first order in ε, so as to obtain the simpler approximation

$$u_0(t,\varepsilon) = \frac{b\cos t + c\sin t}{\left[1 + \frac{3}{4}(b^2 + c^2)\varepsilon t\right]^{1/2}}. \tag{2.4.4}$$

This latter approximation u_0 satisfies the first initial condition of (2.4.2) exactly, but it satisfies the second initial condition only up to the first order, with

$$u_0 = b \quad\text{and}\quad \frac{du_0}{dt} = c - \frac{3}{8}(b^2 + c^2)b\varepsilon = c + \text{order}(\varepsilon) \quad\text{at } t = 0.$$

The result (2.4.4) is consistent with (2.4.3) along with the choices

$$\alpha\sin\beta = b, \qquad \alpha\cos\beta = c.$$

In an important article, Greenlee and Snow (1975) proved that (2.4.4) does in fact provide a suitable approximation to the exact solution u for all $t \geq 0$, with

$$|u(t,\varepsilon) - u_0(t,\varepsilon)| \leq \text{const.}\frac{\varepsilon}{(1 + \varepsilon t)^\gamma} \quad\text{for all } t \geq 0, \tag{2.4.5}$$

where γ can be any positive constant less than $\frac{1}{2}$. [This is a special case of a theorem in Greenlee and Snow (1975).] Such error estimates are discussed in Chapter 4.

From (2.4.4) and (2.4.5) we clearly see the effect of the cubic damping in the given problem. Over any time interval of length 2π, the function u represents approximately a harmonic motion of period 2π, because the denominator on the right side of (2.4.4) is almost constant over any such period. However, as t increases through many such (quasi) periods, the denominator slowly increases, so that the motion slowly decays because of the damping. The damping factor depends on the initial conditions because of the nonlinearity of the problem.

Exercises

Exercise 2.4.1: **(a)** Give the Krylov/Bogoliubov first approximation for the solution of the problem

$$u'' + u + 0.001(u')^3 = 0 \quad \text{for } t > 0,$$

$$u(0) = b, \qquad \frac{du(0)}{dt} = 0.$$

(b) If the time t is measured in seconds in (a), calculate about how long it takes for the damping to reduce the amplitude of oscillation by 10 percent from its initial amplitude if the initial displacement is $b = 1$, and again if $b = 2$.

Exercise 2.4.2: Find the Krylov/Bogoliubov approximation for the linear oscillator

$$u'' + \varepsilon u' + u = 0 \quad \text{for } t > 0,$$
$$u = 0 \quad \text{and} \quad u' = 1 \quad \text{at } t = 0.$$

Compare this approximation with the classical first-order approximation obtained in Exercise 1.3.2.

Exercise 2.4.3: Find the Krylov/Bogoliubov first approximation to the Duffing equation

$$u'' + u + \varepsilon u^3 = 0 \qquad\qquad (2.4.6)$$

subject to the initial conditions $u(0) = 1$ and $u'(0) = 0$. [The Duffing equation (2.4.6) is a mathematical model for a nonlinear spring.]

Exercise 2.4.4: Find the Krylov/Bogoliubov approximation $u_0(t, \varepsilon)$ for the van der Pol oscillator

$$u'' + u + \varepsilon(u^2 - 1)u' = 0$$

subject to the initial conditions

$$u = a \quad \text{and} \quad u' = b \quad \text{at } t = 0.$$

Show that u_0 satisfies

$$\lim_{t \to \infty} \left[u_0^2 + (du_0/dt)^2 \right] = 4,$$

which indicates a limit cycle in the phase plane.

Exercise 2.4.5: Show that the relations

$$u(t, \varepsilon) = A(t, \varepsilon)\cos t + B(t, \varepsilon)\sin t,$$
$$u'(t, \varepsilon) = -A(t, \varepsilon)\sin t + B(t, \varepsilon)\cos t,$$

(2.4.7)

provide a solution to the equation

$$u'' + u = \varepsilon f(t, u, u') \qquad (2.4.8)$$

if and only if A and B satisfy the system

$$\frac{d}{dt}\begin{bmatrix} A \\ B \end{bmatrix} = \varepsilon \begin{bmatrix} -\sin t \\ \cos t \end{bmatrix} f(t, A\cos t + B\sin t, -A\sin t + B\cos t).$$

(2.4.9)

Remark: If $f = f(t, u, v)$ is 2π-periodic in the first variable t, then the first-order Krylov/Bogoliubov averaged system is given as

$$\frac{d}{dt}\begin{bmatrix} A_0 \\ B_0 \end{bmatrix} = \frac{\varepsilon}{2\pi} \int_0^{2\pi} \begin{bmatrix} -\sin\theta \\ \cos\theta \end{bmatrix} F(\theta, A_0, B_0)\, d\theta$$

with $F(\theta, A_0, B_0) := f(\theta, A_0\cos\theta + B_0\sin\theta, -A_0\sin\theta + B_0\cos\theta)$,

(2.4.10)

where $A_0 = A_0(t, \varepsilon)$ and $B_0 = B_0(t, \varepsilon)$ are held fixed during the integrations on the right side of (2.4.10). The system (2.4.10) is solved subject to suitable initial conditions, and then the function

$$u_0(t, \varepsilon) := A_0(t, \varepsilon)\cos t + B_0(t, \varepsilon)\sin t \qquad (2.4.11)$$

gives the first-order Krylov/Bogoliubov approximation for the equation (2.4.8) subject to the specified initial conditions.

Exercise 2.4.6: Use the procedure of the remark in Exercise 2.4.5 to find the first-order Krylov/Bogoliubov approximation for the problem

$$u'' + \varepsilon(\cos t)(u')^2 + u = 0 \qquad \text{for } t > 0,$$
$$u(0) = 0, \quad u'(0) = -1 \quad \text{at } t = 0.$$

(2.4.12)

[This problem is given as an example in Greenlee and Snow (1975), where an approximate solution is obtained by two-timing; see also Exercise 3.4.4.]

Exercise 2.4.7: If $f = f(t, u, v)$ is not 2π-periodic in t in (2.4.8), then the first-order Krylov/Bogoliubov averaged system (2.4.10) is replaced by [compare with (2.2.11)–(2.2.13)]

$$\frac{d}{dt}\begin{bmatrix} A_0 \\ B_0 \end{bmatrix} = \varepsilon \lim_{T \to \infty} \frac{1}{T} \int_0^T \begin{bmatrix} -\sin\theta \\ \cos\theta \end{bmatrix} F(\theta, A_0, B_0)\, d\theta$$

with $F(\theta, A_0, B_0) := f(\theta, A_0\cos\theta + B_0\sin\theta, -A_0\sin\theta + B_0\cos\theta)$,

$$(2.4.13)$$

provided that this limit exists. Use (2.4.13) to find an approximate solution to the problem

$$u'' + \varepsilon(2c + 5de^{-t})u' + \left[1 + \varepsilon(2C + 5De^{-t})\right]u = 0,$$

$$u(0) = 1, \qquad u'(0) = 0 \qquad \text{at } t = 0 \quad (2.4.14)$$

where c, d, C, and D are given constants. (The validity of the resulting approximation is justified in Chapter 4 for the case of predominantly positive damping with $c > 0$; see Example 4.4.3.)

2.5 Two weakly coupled electric circuits

In this section we give an illustration of the use of the Krylov/ Bogoliubov technique in the study of coupled systems of differential equations describing various oscillatory systems. Such systems appear in numerous areas, including mechanics and electric-circuit theory. We shall consider only an elementary example involving the two LC circuits diagrammed in Figure 2.1. The circuits are taken to be identical, with inductance L and capacitance C, and we let $Q_j = Q_j(t)$ denote the electric charge on the capacitor in the jth circuit at time t ($j = 1, 2$). We assume that the circuits are weakly coupled (inductively) with coupling factor ε. Then Kirchhoff's voltage law along with the Faraday/Henry inductance model and the Coulomb capacitance model give the equations

$$\begin{aligned} Q_1'' + \varepsilon Q_2'' + k^2 Q_1 &= 0, \\ Q_2'' + \varepsilon Q_1'' + k^2 Q_2 &= 0, \end{aligned} \qquad (2.5.1)$$

with

$$k^2 := \frac{1}{LC},$$

or, equivalently,

$$\begin{aligned} (1 - \varepsilon^2)Q_1'' + k^2 Q_1 &= \varepsilon k^2 Q_2, \\ (1 - \varepsilon^2)Q_2'' + k^2 Q_2 &= \varepsilon k^2 Q_1. \end{aligned} \qquad (2.5.2)$$

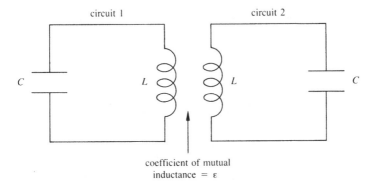

circuit 1   circuit 2

C L L C

coefficient of mutual
inductance $= \varepsilon$

Figure 2.1

We put

$$t_{\text{new}} = \frac{t_{\text{old}}}{(1 - \varepsilon^2)^{1/2}}$$

and find, after relabelling t_{new} again as t,

$$Q_1''(t) + k^2 Q_1(t) = \varepsilon k^2 Q_2(t),$$
$$Q_2''(t) + k^2 Q_2(t) = \varepsilon k^2 Q_1(t) \quad \text{for } t > 0. \tag{2.5.3}$$

For definiteness and simplicity we consider only the special initial conditions

$$Q_1(0) = q, \quad Q_1' = 0,$$
$$Q_2(0) = 0, \quad Q_2' = 0 \quad \text{at } t = 0, \tag{2.5.4}$$

which correspond to a given initial charge q on capacitor 1, with zero initial charge on capacitor 2, and zero initial currents.

The initial-value problem (2.5.3)–(2.5.4) is linear with constant coefficients, and it can be easily solved in closed form to give

$$Q_1(t) = \frac{q}{2}\left[\cos(kt\sqrt{1-\varepsilon}) + \cos(kt\sqrt{1+\varepsilon})\right],$$
$$Q_2(t) = \frac{q}{2}\left[\cos(kt\sqrt{1-\varepsilon}) - \cos(kt\sqrt{1+\varepsilon})\right]. \tag{2.5.5}$$

We shall use the Krylov/Bogoliubov averaging method to obtain an approximation to this solution (for small values of ε) that will be uniformly valid over a long time interval, as well as easy to interpret. The approximation procedure can also be used in more complicated situations involving nonlinear circuits or circuits with variable characteristics for which such exact solutions as (2.5.5) are not readily available.

If $\varepsilon = 0$, then the system (2.5.3) reduces to

$$Q_1'' + k^2 Q_1 = 0,$$
$$Q_2'' + k^2 Q_2 = 0 \quad (\varepsilon = 0),$$

(2.5.6)

and the most general solution of (2.5.6) can be represented as

$$Q_1(t) = A \sin kt + B \cos kt,$$
$$Q_2(t) = C \sin kt + D \cos kt,$$

(2.5.7)

for suitable constants of integration A, B, C, and D. From (2.5.7) we also have

$$Q_1'(t) = k(A \cos kt - B \sin kt),$$
$$Q_2'(t) = k(C \cos kt - D \sin kt).$$

(2.5.8)

Using variation of parameters, we can obtain the general solution of the perturbed system (2.5.3) (for nonzero ε) in the same form as (2.5.7)–(2.5.8), but now the quantities A, B, C, and D must be taken to be functions of t and ε subject to the conditions

$$\frac{dA}{dt} = \varepsilon k(C \sin \varphi + D \cos \varphi) \cos \varphi$$

$$\frac{dB}{dt} = -\varepsilon k(C \sin \varphi + D \cos \varphi) \sin \varphi$$

$$\frac{dC}{dt} = \varepsilon k(A \sin \varphi + B \cos \varphi) \cos \varphi$$

$$\frac{dD}{dt} = -\varepsilon k(A \sin \varphi + B \cos \varphi) \sin \varphi,$$

(2.5.9)

where the quantity φ is defined as $\varphi := kt$. As usual, there is no approximation involved here. The system (2.5.9) is equivalent to (2.5.3) via (2.5.7)–(2.5.8).

The system (2.5.9) shows that the functions A, B, C, and D should vary only slowly with increasing t, because the time derivatives of these functions are proportional to the small parameter ε. Hence, the main source of variation on the right side of (2.5.9) is the quantity φ, which occurs always as an argument of a periodic trigonometric function. The first-order averaged system is obtained by averaging out the φ dependence over a period $0 \leq \varphi \leq 2\pi$, treating A, B, C, and D as constants during the averaging process. Omitting details, we find

$$\frac{d}{dt}A_0 = \frac{\varepsilon}{2}kD_0,$$

$$\frac{d}{dt}B_0 = -\frac{\varepsilon}{2}kC_0,$$

$$\frac{d}{dt}C_0 = \frac{\varepsilon}{2}kB_0,$$ (2.5.10)

$$\frac{d}{dt}D_0 = -\frac{\varepsilon}{2}kA_0,$$

where now A_0, B_0, C_0, and D_0 are expected to provide approximations to the exact quantities A, B, C, and D. The initial conditions (2.5.4) along with (2.5.7)–(2.5.8) lead to the conditions

$$A_0(0) = C_0(0) = D_0(0) = 0 \quad \text{and} \quad B_0(0) = q \quad \text{at } t = 0.$$

(2.5.11)

The solution of (2.5.10)–(2.5.11) is given as

$$B_0(t) = q\cos\frac{\varepsilon kt}{2},$$

$$C_0(t) = q\sin\frac{\varepsilon kt}{2},$$ (2.5.12)

$$A_0(t) = D_0(t) = 0,$$

and this result, along with (2.5.7), leads to the first-order approximations

$$q_1(t, \varepsilon) := A_0(t)\sin kt + B_0(t)\cos kt$$

$$= q\left(\cos\frac{\varepsilon kt}{2}\right)\cos kt,$$

$$q_2(t, \varepsilon) := C_0(t)\sin kt + D_0(t)\cos kt$$ (2.5.13)

$$= q\left(\sin\frac{\varepsilon kt}{2}\right)\sin kt,$$

where the functions q_1 and q_2 are intended to provide approximations to the corresponding exact functions Q_1 and Q_2 for the initial-value problem (2.5.3)–(2.5.4). Indeed, the techniques of Chapter 4 can be used to prove the estimate

$$|Q_j(t, \varepsilon) - q_j(t, \varepsilon)| \le \text{const. } |\varepsilon| \quad (j = 1, 2)$$ (2.5.14)

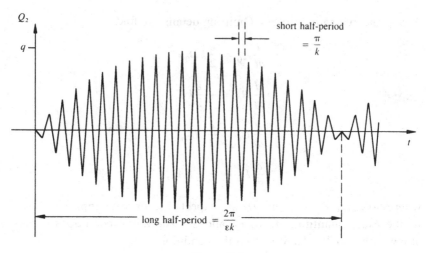

Figure 2.2

uniformly for all t on an expanding interval of the form

$$0 \le t \le \frac{T}{|\varepsilon|}$$

for some fixed positive constant T. Alternatively, the estimate (2.5.14) can be derived directly from the exact solution (2.5.5) along with (2.5.13).

Hence, for small ε, the functions q_1 and q_2 provide useful approximations to Q_1 and Q_2, and these approximate solution functions q_1 and q_2 are quite easy to interpret. We then see that Q_1 and Q_2 are given approximately as harmonic motions $\cos kt$ and $\sin kt$, with periods $2\pi/k$, and with *slowly varying amplitudes* given respectively as

$$q_1 \text{ amplitude} = q\cos\frac{\varepsilon kt}{2},$$

$$q_2 \text{ amplitude} = q\sin\frac{\varepsilon kt}{2}. \tag{2.5.15}$$

These slowly varying amplitudes have (long) period given as

$$\text{long period} = 4\pi/\varepsilon k.$$

The amplitude of Q_1 is a maximum when the amplitude of Q_2 is a minimum, and vice versa. The oscillation is slowly transferred from circuit 1 to circuit 2, and then back again, and so on, because of the weak inductive coupling. We can say that each circuit experiences beats that are driven by the companion circuit. The graph of Q_2 is shown in Figure 2.2.

Exercise

Exercise 2.5.1: Find an easily interpretable approximate solution to the initial-value problem

$$\frac{d^2\Theta}{dt^2} + k^2\Theta = \varepsilon^2 k^2(-2\Theta + \Phi - \Psi),$$

$$\frac{d^2\Phi}{dt^2} + k^2(\Phi - \Theta) = -\varepsilon^2 k^2(-2\Theta + \Phi - \Psi), \qquad (2.5.16)$$

$$\frac{d^2\Psi}{dt^2} + k^2(\Psi + \Theta) = \varepsilon^2 k^2(-2\Theta + \Phi - \Psi) \quad \text{for } t > 0,$$

and

$$\Psi = \alpha, \qquad \Theta = \Phi = \frac{d\Theta}{dt} = \frac{d\Phi}{dt} = \frac{d\Psi}{dt} = 0 \quad \text{at } t = 0.$$
$$(2.5.17)$$

This problem is a mathematical model for a certain compound pendulum discussed in Section 3.5. The quantity k is a given positive constant, α is a fixed constant, and ε is a small positive parameter.

3

The multiscale technique

This chapter presents another approximation procedure for the nonlinear oscillation problem of the previous chapter. The approach here is known variously as *two-timing*, or as the *multivariable* or *multiscale approach*. Several examples are given, and the comparative merits of the multiscale approach and the Krylov/Bogoliubov averaging approach are discussed.

3.1 Introduction

The nonuniformities occurring in typical singular-perturbation problems are related to a multiscale dependence of the solution on more than one scale of the independent variable. For example, the Krylov/Bogoliubov first approximation (2.3.8) for the Einstein two-body problem can be written as

$$u_0(t, \varepsilon) = a + (b - a)\cos(t - a\tau) \quad \text{with } \tau := \varepsilon t, \qquad (3.1.1)$$

where the independent variable is labelled here as t rather than as the previous θ of (2.3.8). The function of (3.1.1), which closely approximates the exact solution $u(t, \varepsilon)$, makes it clear that the solution depends in an essential way on two "times," a *fast time* t and a *slow time* $\tau = \varepsilon t$.

The idea of the multiscale method is to introduce several scaled variables directly into the intended approximation and then choose the dependence of the approximate solution on the variables so as to obtain a uniformly valid approximation to the exact solution.

This idea of explicitly using scaled variables in a perturbation procedure goes back at least one hundred years to the work of Lindstedt (1882, 1883), who introduced and used such scaled variables to eliminate spurious secular (resonant, unbounded) terms in perturbation expansions in celestial mechanics. The work of Lindstedt was carried further by Poincaré (1892–3), and various similar methods were later rediscovered by such workers as Whittaker (1914), Schrödinger (1926), and Lighthill (1949) in the contexts of various different applications. Several variants of the method of Lindstedt and Poincaré have been widely used formally in the applied literature since the early 1950s. The books by Nayfeh (1973), Nayfeh and Mook (1979), and Kevorkian and Cole (1981) contain numerous references to this literature. Particular note can be made of the work of Sturrock (1957), Kuzmak (1959), Kevorkian (1961, 1966), Cole and Kevorkian (1963), J. Cole (1968), and Levey and Mahony (1968), in which multiscale methods were used in the study of various important oscillation problems. Although error estimates are not included in these references, Morrison (1966) and Perko (1968) proved a certain equivalence between the two-variable method of Kevorkian and Cole and the averaging method of Krylov, Bogoliubov, and Mitropolsky, thereby providing such error estimates indirectly, because such estimates had been given for the averaging method in Bogoliubov and Mitropolsky (1961) and Perko (1968). These latter estimates are uniformly valid on *expanding intervals* of the type

$$ 0 \leq t \leq \frac{T}{\varepsilon} \quad \text{as } \varepsilon \to 0+, $$

for a suitable positive constant T.

The development of the multiscale technique for oscillation problems as considered in this chapter has occurred historically in conjunction with the general development of the multiscale approach for a variety of problems, including overdamped nonoscillatory initial-value problems, boundary-value problems, and the presently considered oscillation problems. Within this larger framework, Latta (1951) developed a method of uniform approximation for boundary-value problems based on a direct composite expansion that can be viewed as a form of two-variable expansion. Latta's work can be placed in a long tradition going back to Laplace (1805), Carlini (1817), Liouville (1837), Green (1837), Maxwell (1866), and Kirchhoff (1877), and including also Prandtl (1905), Birkhoff (1908), Rayleigh (1912), Noaillon (1912), Gans (1915), Jeffreys (1924), Wentzel (1926), Kramers (1926), Brillouin (1926), Levinson (1950), and others. Again, similar and related composite expansion methods were later rediscovered in various different contexts by several workers, including Bromberg (1956), Vishik and Lyusternik (1957), and Mahony

(1961–2). Successful use of two-variable techniques in the study of boundary-value problems by Cochran (1962), Erdélyi (1968), and O'Malley (1969) encouraged more general use of such techniques. Multiscale techniques have also been used to study the partial differential equations that model the macroscopic properties of composite media and crystalline or polymer media with microscopic periodic spatial structure [see Bensoussan, Lions, and Papanicolaou (1978)].

For oscillation equations such as [see (2.1.1)]

$$u'' + k^2 u = a + \varepsilon f(u, u'), \tag{3.1.2}$$

The method of Kevorkian and Cole uses a slow time $\tau = \varepsilon t$ as in (3.1.1) and a fast time t^+ of the form [see Kevorkian and Cole (1981)]

$$t^+ = t \cdot \left(1 + c_1 \varepsilon + c_2 \varepsilon^2 + \cdots \right) \tag{3.1.3}$$

for suitable constants $c_1, c_2, \ldots$ determined as part of the approximation procedure. The method is closely related to the approaches of Lindstedt and Poincaré.

Reiss (1971) proposed a simplification of this method in which the fast time t^+ coincides with the original independent variable,

$$t^+ = t,$$

so that all c_j's vanish in (3.1.3), but the previous slow time $\tau = \varepsilon t$ is retained. Kollett (1974) and Smith (1975a) used these variables of Reiss to obtain approximations of the form

$$u(t, \varepsilon) = \sum_{k=0}^{N} u_k(t, \tau) \varepsilon^k + \varepsilon^{N+1} R_N(t, \varepsilon) \quad (\tau = \varepsilon t) \tag{3.1.4}$$

for the initial-value problem for various special cases of the equation [see (3.1.2)]

$$u'' + k^2 u = a + \varepsilon f(\varepsilon, t, u, u'), \tag{3.1.5}$$

where the quantity $R_N(t, \varepsilon)$ in (3.1.4) was shown to be uniformly bounded (as $\varepsilon \to 0+$) for all t on an expanding interval $0 \le t \le T/\varepsilon$. For several important subclasses of equation (3.1.5) with damping, Greenlee and Snow (1975) obtained approximations such as (3.1.4) valid for all $t \ge 0$. These latter results are described in Chapter 4.

Several examples using the multiscale technique will be considered in the next two sections.

3.2 A linear oscillator with weak damping

The initial-value problem

$$\frac{d^2 u}{dt^2} + 2\varepsilon \frac{du}{dt} + u = 0 \quad \text{for } t > 0, \tag{3.2.1}$$

with

$$u = a \quad \text{and} \quad \frac{du}{dt} = b \quad \text{at } t = 0, \qquad (3.2.2)$$

represents an oscillator with linear restoring force and weak linear damping (for small positive ε). It was in the context of this problem that Reiss (1971) proposed a multiscale approach with the slow time $\tau = \varepsilon t$ and fast time $t^+ = t$ as discussed in Section 3.1. Reiss also discussed the simultaneous use of more than one slow time, such as the simultaneous use of both $\tau_1 := \varepsilon t$ and $\tau_2 := \varepsilon^2 t$, in addition to the fast time (see Exercise 3.4.5).

For the problem (3.2.1)–(3.2.2), we can show directly that the classical expansion (see Exercise 1.3.2)

$$u(t, \varepsilon) = \sum_{k=0}^{\infty} u_k(t) \varepsilon^k \qquad (3.2.3)$$

converges for all $t \geq 0$ and for all $\varepsilon \geq 0$, with

$$u_0(t) = a \cos t + b \sin t \qquad (3.2.4)$$

and

$$u_k(t) = -2 \int_0^t [\sin(t - s)] u_{k-1}'(s) \, ds \quad \text{for } k = 1, 2, \ldots . \qquad (3.2.5)$$

In particular,

$$u_1(t) = a(\sin t) - t(a \cos t + b \sin t), \qquad (3.2.6)$$

so that u_1 has an unbounded component (as $t \to \infty$), and, in general, we have

$$u_k(t) = \text{order}(t^k) \quad \text{as } t \to \infty.$$

From (3.2.3)–(3.2.6) we find

$$u(t, \varepsilon) = (1 - \varepsilon t)(a \cos t + b \sin t) + \varepsilon a \sin t + \cdots, \qquad (3.2.7)$$

and this classical expansion suggests that the exact solution depends in an essential way on several scales, including t and $\tau := \varepsilon t$. This suggestion is borne out by the explicit solution

$$u(t, \varepsilon) = e^{-\varepsilon t} \left[\frac{b + \varepsilon a}{\kappa_\varepsilon} \sin(\kappa_\varepsilon \cdot t) + a \cos(\kappa_\varepsilon \cdot t) \right]$$

$$\text{with } \kappa_\varepsilon := [1 - \varepsilon^2]^{1/2}. \qquad (3.2.8)$$

As noted in Section 1.3, partial sums of (3.2.3) fail to provide useful information on $u(t, \varepsilon)$ for large t because of spurious secular terms in $u_k(t)$ for $k \geq 1$. Indeed, u_k satisfies the differential equation

$$u_k'' + u_k = -2u_{k-1}' \quad \text{for } k = 1, 2, \ldots, \qquad (3.2.9)$$

and the forcing term $-2u'_{k-1}$ contains resonance-producing terms that lead to unbounded growth in $u_k(t)$ as $t \to \infty$ [see (3.2.6)].

We seek a more useful approximation by replacing the classical series representation (3.2.3) with a two-variable expansion [see (3.1.4)]:

$$u(t, \varepsilon) = \sum_{k=0}^{N} u_k(t, \tau)\varepsilon^k + \varepsilon^{N+1}R_N(t, \varepsilon) \quad (\tau = \varepsilon t). \quad (3.2.10)$$

Actually, we can either work with the partial sum and remainder of (3.2.10) or introduce a (possibly divergent) formal expansion, as

$$u(t, \varepsilon) \sim \sum_{k=0}^{\infty} u_k(t, \tau)\varepsilon^k \quad (\tau = \varepsilon t). \quad (3.2.11)$$

In the latter case the expansion is used only to organize efficient calculation of the functions u_k. In either case the final result employs only a partial sum, as in (3.2.10), along with a suitable error estimate, to yield an approximation to the exact solution. The functions $u_k = u_k(t, \tau)$ are to be determined as functions of two independent variables t and τ, and then these functions are evaluated in (3.2.10) at $\tau = \varepsilon t$.

The chain rule of differential calculus gives

$$\frac{du_k(t, \tau)}{dt} = \frac{\partial u_k}{\partial t} + \frac{\partial u_k}{\partial \tau}\frac{d\tau}{dt}$$

$$= u_{k,t} + \varepsilon u_{k,\tau} \quad (\tau = \varepsilon t) \quad (3.2.12)$$

where $u_{k,t}$ denotes the partial derivative of $u_k(t, \tau)$ with respect to the first argument t,

$$u_{k,t} = \frac{\partial u_k}{\partial t},$$

and, similarly,

$$u_{k,\tau} = \frac{\partial u_k}{\partial \tau}.$$

Similarly,

$$\frac{d^2 u_k}{dt^2} = u_{k,tt} + 2\varepsilon u_{k,t\tau} + \varepsilon^2 u_{k,\tau\tau}, \quad (3.2.13)$$

where $u_{k,tt} = \partial^2 u_k/\partial t^2$, and so forth. Taking now the derivative of (3.2.11) with respect to t, we find the formal expansion

$$\frac{du}{dt} \sim \sum_{k=0}^{\infty}(u_{k,t} + \varepsilon u_{k,\tau})\varepsilon^k \sim \sum_{k=0}^{\infty}(u_{k,t} + u_{k-1,\tau})\varepsilon^k,$$

$$(3.2.14)$$

and, similarly,

$$\frac{d^2 u}{dt^2} \sim \sum_{k=0}^{\infty} (u_{k,tt} + 2u_{k-1,t\tau} + u_{k-2,\tau\tau}) \varepsilon^k, \qquad (3.2.15)$$

where the convention is adopted here that

$$u_k := 0 \quad \text{for negative } k. \qquad (3.2.16)$$

These formal expansions are now inserted into the differential equation (3.2.1) to give

$$\sum_{k=0}^{\infty} [u_{k,tt} + u_k + 2(u_{k-1,t} + u_{k-1,t\tau}) + (2u_{k-2,\tau} + u_{k-2,\tau\tau})] \varepsilon^k \sim 0,$$

$$(3.2.17)$$

and this asymptotic identity will hold if we impose the requirements

$$u_{k,tt} + u_k = -2(u_{k-1,t} + u_{k-1,t\tau}) - (2u_{k-2,\tau} + u_{k-2,\tau\tau})$$

$$\text{for } k = 0, 1, 2, \dots \qquad (3.2.18)_k$$

Similarly, the initial conditions (3.2.2) along with (3.2.11) and (3.2.14) lead to the conditions (note that $\tau = 0$ when $t = 0$)

$$u_k(0,0) = \begin{cases} a & \text{for } k = 0, \\ 0 & \text{for } k = 1, 2, \dots, \end{cases} \qquad (3.2.19)_k$$

and

$$u_{k,t}(0,0) + u_{k-1,\tau}(0,0) = \begin{cases} b & \text{for } k = 0, \\ 0 & \text{for } k = 1, 2, \dots. \end{cases} \qquad (3.2.20)_k$$

Equation $(3.2.18)_0$ along with (3.2.16) becomes

$$u_{0,tt} + u_0 = 0.$$

This equation can be integrated with respect to t for each fixed τ, and we find the most general solution in the form

$$u_0(t,\tau) = A_0(\tau)\cos t + B_0(\tau)\sin t, \qquad (3.2.21)$$

where the "constants" of integration A_0 and B_0 may still depend on the variable τ. The initial conditions $(3.2.19)_0$ and $(3.2.20)_0$ imply, respectively, with (3.2.21),

$$A_0(0) = a \quad \text{and} \quad B_0(0) = b. \qquad (3.2.22)$$

In order to complete the determination of the expressions $A_0(\tau)$ and $B_0(\tau)$ as functions of τ, we use the general approach of Lindstedt and Poincaré, which requires here that the next term $u_1(t,\tau)$ in the formal expansion (3.2.11) should not exhibit any spurious secular (unbounded) behavior as a consequence of u_0. Alternatively we can set

$$u(t,\varepsilon) = u_0(t,\varepsilon t) + \text{remainder}$$

and require that the remainder be suitably small over a suitably large t interval. This latter approach can also be viewed in terms of the residual corresponding to $u_0(t, \varepsilon t)$ (see Exercise 3.2.2). In any case, we shall obtain (3.2.27).

The general form of u_1 can be obtained by integrating (3.2.18)$_1$ with respect to t for each fixed τ, from which we find (see the result of Example 0.1.1)

$$u_1(t, \tau) = A_1(\tau)\cos t + B_1(\tau)\sin t$$
$$-2\int_0^t \left[u_{0,t}(s, \tau) + u_{0,t\tau}(s, \tau) \right] \sin(t - s)\, ds, \quad (3.2.23)$$

where the quantity in square brackets in this last integral is evaluated at $t = s$ and fixed τ, so that τ is treated as a constant in the integration.

From (3.2.21) it follows that

$$u_{0,t}(t, \tau) + u_{0,t\tau}(t, \tau)$$
$$= \left[A_0'(\tau) + A_0(\tau) \right] \sin t - \left[B_0'(\tau) + B_0(\tau) \right] \cos t, \quad (3.2.24)$$

from which we find that *this expression is resonance-producing* in the integral on the right side of (3.2.23) because of the unboundedness of the integrals

$$\int_0^t \cos s \sin(t - s)\, ds = \tfrac{1}{2} t \sin t,$$
$$\int_0^t \sin s \sin(t - s)\, ds = \tfrac{1}{2}(-t \cos t + \sin t). \quad (3.2.25)$$

It is just such terms involving these integrals (3.2.25) that produce the spurious secular terms in the classical expansion (3.2.3); see (3.2.6). Hence, the integrals (3.2.25) must be removed from the last term on the right side of (3.2.23). In view of (3.2.24), and following the lead of Lindstedt and Poincaré, we are thus led to impose the conditions

$$A_0'(\tau) + A_0(\tau) = 0, \qquad B_0'(\tau) + B_0(\tau) = 0. \quad (3.2.26)$$

The unique solution of this last system of differential equations subject to the initial conditions of (3.2.22) is given as

$$A_0(\tau) = ae^{-\tau}, \qquad B_0(\tau) = be^{-\tau},$$

which, with (3.2.21), gives for u_0 the result

$$u_0(t, \tau) = e^{-\tau}(a\cos t + b\sin t). \quad (3.2.27)$$

This completes the determination of the leading term u_0 in the two-variable expansion (3.2.11).

The multiscale technique can in principle be used systematically to obtain further terms in the expansion (3.2.11) if desired. For example, to obtain u_1, we simply insert (3.2.27) back into (3.2.23) to find

$$u_1(t, \tau) = A_1(\tau)\cos t + B_1(\tau)\sin t. \tag{3.2.28}$$

The functions A_1 and B_1 must satisfy initial conditions obtained from $(3.2.19)_1$ and $(3.2.20)_1$ as

$$A_1(0) = 0, \qquad B_1(0) = a. \tag{3.2.29}$$

Determination of A_1 and B_1 is now completed by requiring the next term in the expansion, namely u_2, to have no unbounded terms as a result of u_1. One integrates $(3.2.18)_2$ with respect to t, always holding τ fixed, so as to obtain, with (3.2.27) and (3.2.28), the result

$$u_2(t, \tau) = A_2(\tau)\cos t + B_2(\tau)\cos t$$
$$+ 2\left[A_1'(\tau) + A_1(\tau) + \frac{b}{2}e^{-\tau} \right] \int_0^t \sin s \sin(t - s) \, ds$$
$$- 2\left[B_1'(\tau) + B_1(\tau) - \frac{a}{2}e^{-\tau} \right] \int_0^t \cos s \sin(t - s) \, ds. \tag{3.2.30}$$

The offending secular terms are eliminated in (3.2.30) with the conditions

$$A_1'(\tau) + A_1(\tau) = -\frac{b}{2}e^{-\tau},$$
$$B_1'(\tau) + B_1(\tau) = \frac{a}{2}e^{-\tau}. \tag{3.2.31}$$

The solution of this last system subject to the initial conditions of (3.2.29) is given as

$$A_1(\tau) = -\frac{b}{2}\tau e^{-\tau}, \qquad B_1(\tau) = a \cdot \left(1 + \frac{\tau}{2}\right)e^{-\tau},$$

which, with (3.2.28), gives

$$u_1(t, \tau) = e^{-\tau}\left[a \cdot \left(1 + \frac{\tau}{2}\right)\sin t - \frac{b\tau}{2}\cos t \right]. \tag{3.2.32}$$

This procedure can be continued recursively. After the Nth stage, we have well-determined expressions for the functions $u_k = u_k(t, \tau)$ for $k = 0, 1, 2, \ldots, N - 1$, and u_N will be given as

$$u_N(t, \tau) = A_N(\tau)\cos t + B_N(\tau)\sin t. \tag{3.2.33}$$

For u_{N+1} the equation $(3.2.18)_{N+1}$ gives

$$u_{N+1}(t, \tau) = A_{N+1}(\tau)\cos t + B_{N+1}(\tau)\sin t$$
$$- \int_0^t [2(u_{N,t} + u_{N,t\tau}) + (2u_{N-1,\tau} + u_{N-1,\tau\tau})]\sin(t - s) \, ds, \tag{3.2.34}$$

and then elimination of unbounded terms from the integral on the right side of (3.2.34) leads to a system of differential equations for A_N and B_N. This system can be solved subject to the appropriate initial conditions provided by (3.2.19)$_N$ and (3.2.20)$_N$, thereby completing the determination of A_N and B_N, and also u_N with (3.2.33).

Error estimates for the resulting two-variable approximation can be obtained as indicated in Exercises 3.2.1–3.2.2. Error estimates for related nonlinear problems are studied in Chapter 4.

Exercises

Exercise 3.2.1: Derive the integral relation

$$v(t) = v(0)\cos t + v'(0)\sin t$$
$$+ \int_0^t \sin(t - s)f(s)\,ds - 2\varepsilon\int_0^t \cos(t - s)v(s)\,ds \quad (3.2.35)$$

for any solution of the differential equation

$$\frac{d^2v}{dt^2} + 2\varepsilon\frac{dv}{dt} + v = f(t). \qquad (3.2.36)$$

Hint: Use the result of Example 0.1.1, and integrate the resulting integral involving v' by parts.

Exercise 3.2.2: **(a)** Let $U = U(t, \varepsilon)$ be defined as

$$U(t, \varepsilon) := u_0(t, \varepsilon t), \qquad (3.2.37)$$

where $u_0 = u_0(t, \tau)$ is given by (3.2.21) as

$$u_0(t, \tau) = A_0(\tau)\cos t + B_0(\tau)\sin t,$$

and let the remainder $R = R(t, \varepsilon)$ be defined as

$$R(t, \varepsilon) := u(t, \varepsilon) - U(t, \varepsilon), \qquad (3.2.38)$$

where u is the exact solution of the initial-value problem (3.2.1)–(3.2.2) and U is the intended approximate solution (3.2.37). Show that R satisfies

$$R'' + 2\varepsilon R' + R = \rho(t, \varepsilon) \quad \text{for } t > 0,$$

$$R(0) = a - A_0(0), \quad R'(0) = b - B_0(0) - \varepsilon A_0'(0) \quad \text{at } t = 0,$$

$$(3.2.39)$$

where the residual $\rho = \rho(t, \varepsilon)$ is defined as

$$\rho(t, \varepsilon) := -\left(\frac{d^2U}{dt^2} + 2\varepsilon\frac{dU}{dt} + U\right). \tag{3.2.40}$$

(b) Show that R satisfies an inequality of the form

$$|R(t, \varepsilon)| \leq \text{const.} \ \varepsilon\left(1 + \int_0^t |R(s, \varepsilon)|\, ds\right) \tag{3.2.41}$$

as $\varepsilon \to 0+$, uniformly for $0 \leq t \leq T/\varepsilon$, for any fixed positive T, *if A_0 and B_0 are taken to satisfy the system* (3.2.26) *along with the initial conditions* (3.2.22). *Hint:* Take $v = R$ in (3.2.35) with $f = \rho$. The conditions (3.2.22) and (3.2.26) imply $R(0, \varepsilon) = 0$, $R'(0, \varepsilon) = \text{order}(\varepsilon)$, and $\rho(t, \varepsilon) = \text{order}(\varepsilon^2)$, from which (3.2.41) can be shown to follow.

(c) Use (3.2.41) along with a Gronwall-type argument to obtain an estimate of the type

$$|R(t, \varepsilon)| \leq \text{const.} \ \varepsilon \quad \text{as } \varepsilon \to 0+, \tag{3.2.42}$$

uniformly for $0 \leq t \leq T/\varepsilon$. (Such error estimates for related nonlinear problems are given in Chapter 4.)

3.3 Precession of Mercury, once again

In this section the two-variable approach is used to study the Einstein two-body equation (1.1.3), which is rewritten here with t as independent variable:

$$\frac{d^2u}{dt^2} + u = a + \varepsilon u^2. \tag{3.3.1}$$

This equation is to be studied subject to the initial conditions

$$u = b, \quad \frac{du}{dt} = 0 \quad \text{at } t = 0. \tag{3.3.2}$$

We seek an approximation to the exact solution u in the form of the two-variable expansion (3.2.10), where we again use the formal series (3.2.11) as an aid in organizing the efficient calculation of the approximation. Along with the expansions for u, du/dt, and d^2u/dt^2 given by (3.2.11), (3.2.14), and (3.2.15), we also have the result

$$\varepsilon u^2 \sim \sum_{k=1}^{\infty} \varepsilon^k \sum_{n=0}^{k-1} u_n u_{k-1-n}, \tag{3.3.3}$$

which follows directly from (3.2.11).

These expansions can be inserted into the differential equation (3.3.1) to give

$$\left(u_{0,tt} + u_0 - a \right) + \left[u_{1,tt} + u_1 + 2u_{0,t\tau} - (u_0)^2 \right] \varepsilon$$
$$+ \sum_{k=2}^{\infty} \left(u_{k,tt} + u_k + 2u_{k-1,t\tau} + u_{k-2,\tau\tau} - \sum_{n=0}^{k-1} u_n u_{k-1-n} \right) \varepsilon^k \sim 0,$$

and this equation will hold if we impose the requirements

$$u_{0,tt} + u_0 = a, \qquad\qquad (3.3.4)_0$$

$$u_{1,tt} + u_1 = -2u_{0,t\tau} + (u_0)^2, \qquad\qquad (3.3.4)_1$$

and

$$u_{k,tt} + u_k = -2u_{k-1,t\tau} - u_{k-2,\tau\tau} + \sum_{n=0}^{k-1} u_n u_{k-1-n} \quad \text{for } k = 2, 3, \ldots .$$
$$(3.3.4)_k$$

Similarly, from (3.3.2), (3.2.11), and (3.2.14) follow the initial relations

$$u_0(0,0) = b, \quad u_{0,t}(0,0) = 0, \qquad\qquad (3.3.5)_0$$

and

$$u_k(0,0) = 0, \quad u_{k,t}(0,0) = -u_{k-1,\tau}(0,0) \quad \text{for } k = 1, 2, \ldots .$$
$$(3.3.5)_k$$

Equation $(3.3.4)_0$ can be integrated with respect to t for each fixed τ, and we find the most general solution in the form

$$u_0(t,\tau) = a + A_0(\tau)\cos t + B_0(\tau)\sin t, \qquad\qquad (3.3.6)$$

where, as usual, the "constants" of integration A_0 and B_0 may still depend on the variable τ. The initial conditions of $(3.3.5)_0$ imply, with (3.3.6),

$$A_0(0) = b - a, \qquad B_0(0) = 0. \qquad\qquad (3.3.7)$$

We complete the determination of A_0 and B_0 by requiring that the next term u_1 in the expansion (3.2.11) not exhibit irregular behavior as a consequence of u_0. The general form of u_1 can be obtained by integrating $(3.3.4)_1$ with respect to t for each fixed τ, which gives

$$u_1(t,\tau) = A_1(\tau)\cos t + B_1(\tau)\sin t$$
$$+ \int_0^t \left[-2u_{0,t\tau}(s,\tau) + u_0(s,\tau)^2 \right] \sin(t - s) \, ds, \qquad (3.3.8)$$

where the quantity in square brackets in this last integral is evaluated at $t = s$ and fixed τ. This quantity in square brackets can be evaluated with

(3.3.6) and then inserted back into (3.3.8) to find

$$u_1(t, \tau) = A_1(\tau)\cos t + B_1(\tau)\sin t + (1 - \cos t)a^2$$
$$+ A_0(\tau)^2 \int_0^t \sin^2 s \sin(t - s) \, ds$$
$$+ A_0(\tau)B_0(\tau) \int_0^t \sin 2s \sin(t - s) \, ds$$
$$+ B_0(\tau)^2 \int_0^t \cos^2 s \sin(t - s) \, ds$$
$$+ 2(A_0' + aB_0) \int_0^t \sin s \sin(t - s) \, ds$$
$$+ 2(-B_0' + aA_0) \int_0^t \cos s \sin(t - s) \, ds. \qquad (3.3.9)$$

The integrals here are easily evaluated, and we see that the following integrals are bounded:

$$\int_0^t \sin^2 s \sin(t - s) \, ds = \tfrac{1}{3}(1 + \cos^2 t - 2\cos t)$$
$$\int_0^t \cos^2 s \sin(t - s) \, ds = \tfrac{1}{3}(2 - \cos^2 t - \cos t)$$
$$\int_0^t \sin 2s \sin(t - s) \, ds = \tfrac{2}{3}(1 - \cos t)\sin t,$$

whereas, according to (3.2.25), the last two integrals appearing on the right side of (3.3.9) are unbounded, of order t as $t \to \infty$. These last two offending integrals must therefore be removed by imposing the Lindstedt/Poincaré conditions

$$\frac{dA_0}{d\tau} + aB_0 = 0, \qquad -\frac{dB_0}{d\tau} + aA_0 = 0. \qquad (3.3.10)$$

The unique solution of this last system of differential equations subject to the initial conditions of (3.3.7) is given as

$$A_0(\tau) = (b - a)\cos a\tau, \qquad B_0(\tau) = (b - a)\sin a\tau, \qquad (3.3.11)$$

which with (3.3.6) gives for u_0 the result

$$u_0(t, \tau) = a + (b - a)(\sin t \sin a\tau + \cos t \cos a\tau)$$
$$= a + (b - a)\cos(t - a\tau). \qquad (3.3.12)$$

If we put $\tau = \varepsilon t$ in (3.3.12), we see that this result agrees with the Krylov/Bogoliubov first approximation obtained in Section 2.3 [see (2.3.7) or (3.1.1)].

The approximate solution $u_0(t, \varepsilon t)$ given by (3.3.12) is already an adequate approximation to the exact solution in this case, as noted in

Section 2.3. However, the multivariable procedure can be used systematically to obtain further terms in the expansion (3.2.11) if desired. To obtain u_1, we simply insert (3.3.11) into (3.3.9) to find

$$u_1(t, \tau) = A_1(\tau)\cos t + B_1(\tau)\sin t + h_1(t, \tau) \qquad (3.3.13)$$

for a suitable known function h_1, which need not be given explicitly here. We see directly that h_1 is bounded. The functions A_1 and B_1 in (3.3.13) satisfy initial conditions obtained from $(3.3.5)_1$, given as

$$A_1(0) = B_1(0) = 0. \qquad (3.3.14)$$

The determination of A_1 and B_1 is completed by eliminating the spurious secular terms in u_2, where u_2 is found from $(3.3.4)_2$ on integration with respect to t. Equivalently, without integration, we need only eliminate the resonance-producing terms on the right side of $(3.3.4)_2$. This elimination of spurious secular terms from u_2 leads to a first-order system for A_1 and B_1 analogous to the system (3.3.10) for A_0 and B_0. This system can be solved for A_1 and B_1 using the initial conditions of (3.3.14). The procedure can be continued through as many terms as desired, though the actual calculations rapidly become prohibitively messy and difficult in practice.

It is convenient and useful to discuss the foregoing procedure of casting our spurious secular terms using Fourier expansion of the right side of $(3.3.4)_k$. Thus, if $(3.3.4)_k$ is written as

$$u_{k,tt} + u_k = G_k(t, \tau), \qquad (3.3.15)_k$$

with

$$G_k(t, \tau) := \begin{cases} a & \text{for } k = 0, \\ -2u_{0,t\tau} + (u_0)^2 & \text{for } k = 1, \\ -2u_{k-1,t\tau} - u_{k-2,\tau\tau} + \displaystyle\sum_{n=0}^{k-1} u_n u_{k-1-n} & \text{for } k = 2, 3, \dots, \end{cases}$$

$$(3.3.16)_k$$

then the Fourier expansion of $G_k(t, \tau)$ as a function of t, for each fixed τ, is

$$G_k(t, \tau) = a_{k;0}(\tau) + \sum_{m=1}^{\infty} \left[a_{k;m}(\tau)\cos mt + b_{k;m}(\tau)\sin mt \right],$$

$$(3.3.17)$$

with coefficients $a_{k;m}$ and $b_{k;m}$ given as

$$\begin{bmatrix} a_{k;m}(\tau) \\ b_{k;m}(\tau) \end{bmatrix} := \frac{1}{\pi} \int_0^{2\pi} G_k(t,\tau) \begin{bmatrix} \cos mt \\ \sin mt \end{bmatrix} dt \quad \text{for } m = 1,2,\ldots,$$

$$(3.3.18)$$

and with

$$a_{k;0}(\tau) := \frac{1}{2\pi} \int_0^{2\pi} G_k(t,\tau)\, dt.$$

The kth-order Lindstedt/Poincaré conditions for elimination of spurious secular terms can then be given in the form

$$a_{k;1}(\tau) = 0,$$
$$b_{k;1}(\tau) = 0 \quad \text{for } k = 1,2,\ldots. \qquad (3.3.19)_k$$

For example, in the case $k = 1$, we find with (3.3.6) and $(3.3.16)_1$ the result

$$G_1(t,\tau) = -2u_{0,t\tau} + (u_0)^2$$
$$= a^2 + [A_0(\tau)\cos t + B_0(\tau)\sin t]^2$$
$$+ 2[A_0' + aB_0]\sin t + 2[-B_0' + aA_0]\cos t, \quad (3.3.20)$$

and then (3.3.18) gives (with $k = m = 1$)

$$a_{1;1}(\tau) = 2[-B_0' + aA_0(\tau)],$$
$$b_{1;1}(\tau) = 2[A_0' + aB_0(\tau)]. \qquad (3.3.21)$$

Hence, the condition $(3.3.19)_1$ simply reproduces the previous first-order result (3.3.10) for $k = 1$.

In the case $k = 2$, the condition $(3.3.19)_2$ provides a system of differential equations for determination of A_1 and B_1. We can continue recursively, and then $(3.3.19)_k$ provides a system of differential equations for determination of A_{k-1} and B_{k-1}.

Use of the multiscale approach, when applied to other nonlinear problems, as discussed in the next section, will generally lead to a nonlinear system of differential equations for A_0 and B_0 as functions of τ. This system happens to be linear in the examples considered in Sections 3.2 and 3.3, but this is generally not the case. However, the differential equations governing A_k and B_k are always linear for $k \geq 1$, so that at most only one nonlinear system need be solved, for A_0 and B_0.

Example 3.4.2 in the next section illustrates the fact that the functions A_0 and B_0 need exist only locally on some interval $0 \leq \tau < T$ for some constant T.

Error estimates for the multiscale approximation are discussed in Chapter 4.

3.4 Additional applications of two-timing

Kollett (1974) used two-timing to study the equation

$$\frac{d^2u}{dt^2} + u = a + \varepsilon f\left(u, \frac{du}{dt}\right), \tag{3.4.1}$$

where the function $f = f(u, v)$ is a polynomial in two variables. (Kollett considered the case $a = 0$, but this is no loss in generality.) The Einstein two-body equation is of this form, with

$$f(u, v) := u^2, \tag{3.4.2}$$

and the Duffing equation is given with

$$f(u, v) := -u^3, \qquad a = 0. \tag{3.4.3}$$

Equation (3.4.1) in various other special cases gives the Rayleigh oscillator $[f(u, v) := v - \frac{1}{3}v^3]$, the van der Pol oscillator $[f(u, v) := v(1 - u^2)]$, the simple oscillator with cubic damping $[f(u, v) := -v^3]$, and so forth.

Kollett (1974) also studied the equation

$$u'' + u + \varepsilon g(t)u' = 0, \tag{3.4.4}$$

where the given function g is periodic with period 2π. Kollett gave error estimates in both cases, valid on expanding intervals of the from $0 \le t < T/\varepsilon$ for a fixed constant T as $\varepsilon \to 0+$, as will be discussed in Chapter 4.

Smith (1975a) independently studied certain equations of the form (3.4.1), including the case in which f is a polynomial in u alone, and certain cases in which f can be a polynomial in u and v, with coefficients depending on t and ε, as in the equation

$$u'' + u + 2\varepsilon p(\varepsilon t)(u')^3 = 0, \tag{3.4.5}$$

which represents a model for a damped oscillator with small, slowly varying cubic damping. The function $p = p(\tau)$ is a given, smooth, positive-valued function of a single variable τ and is evaluated at $\tau = \varepsilon t$ in (3.4.5). The special case $p = $ constant was considered in Section 2.4 using the Krylov/Bogoliubov averaging technique. Because ε is small, the expression $p(\varepsilon t)$ will vary only slowly as t increases. Again, error estimates on expanding intervals are given in Smith (1975a).

Equation (3.4.5) is a special case of the more general equation

$$u'' + u = \varepsilon f(\varepsilon, t, u, u'). \tag{3.4.6}$$

Indeed, (3.4.5) corresponds to this latter equation in the case

$$f(\varepsilon, t, u, v) := -2p(\varepsilon t)v^3. \qquad (3.4.7)$$

Greenlee and Snow (1975) used two-timing to study (3.4.6) in the case in which f is a polynomial in the last two variables whose coefficients are 2π-periodic functions of t having asymptotic power-series representations with respect to ε. Greenlee and Snow (1975) used the method of Kollett (1974) to obtain error estimates on expanding intervals. In addition, Greenlee and Snow also obtained error estimates valid for all $t \geq 0$ for several important subclasses of (3.4.6) subject to damping. Such a result had been given earlier by Reiss (1971) in the special case $f(\varepsilon, t, u, v) := -2v$, that is, for the linear oscillator considered in Section 3.2. These results will be discussed in Section 4.4.

Snow (1976) considered the nonhomogeneous equation

$$u'' + u = \varepsilon f(t, u, u') + F(t), \qquad (3.4.8)$$

where f is a polynomial in the last two variables with coefficients that are 2π-periodic functions of t, and the forcing function F is a 2π-periodic function. In a special linear case including damping, Snow obtained estimates valid for all $t \geq 0$. For more general situations, Snow obtained estimates on expanding intervals of the form $0 \leq t < T/\varepsilon^{1/d}$, where d is the degree of f, and the expansion takes the form

$$u(t, \varepsilon) \sim \sum_{k=-1}^{\infty} u_k(t, \tau)(\varepsilon^{1/d})^k, \qquad (3.4.9)$$

where the slow time is

$$\tau := \varepsilon^{1/d}t,$$

and the summation in (3.4.9) begins with $k = -1$. A related example involving "beats" or "pulsing" is given in Exercise 3.4.3.

Several authors have used the multiscale approach to study generalizations to partial differential equations of the problems we have discussed. For example, Keller and Kogelman (1970) studied an initial-boundary-value problem for a nonlinear wave equation of the form

$$\frac{\partial^2 u}{\partial t^2} - \frac{\partial^2 u}{\partial x^2} + u = \varepsilon f\left(\varepsilon, u, \frac{\partial u}{\partial t}\right), \qquad (3.4.10)$$

where $f = f(\varepsilon, u, v)$ is a given function of three real variables. In particular, Keller and Kogelman illustrated their results in detail for the special case in which

$$f = f(v) := v - \tfrac{1}{3}v^2. \qquad (3.4.11)$$

In this case the equation becomes

$$\frac{\partial^2 u}{\partial t^2} - \frac{\partial^2 u}{\partial x^2} + u = \varepsilon \left[\frac{\partial u}{\partial t} - \frac{1}{3} \left(\frac{\partial u}{\partial t} \right)^3 \right], \qquad (3.4.12)$$

which is a generalization of the Rayleigh equation. These results were extended by Chow (1972), who also justified the results using a theorem of Kurzweil (1963). Reiss (1971) studied a linear problem of this type for a special class of initial data. Additional references to other relevant work involving perturbed wave equations are given in Nayfeh (1973), Nayfeh and Mook (1979), and Kevorkian and Cole (1981).

Here we are content to illustrate these results for equation (3.4.1) subject to the initial conditions

$$u = b \quad \text{and} \quad u' = c \quad \text{at } t = 0. \qquad (3.4.13)$$

We follow the approach of Kollett (1974), Greenlee and Snow (1975), and Smith (1975a).

The coefficients u_k in the two-variable expansion [see (3.2.11)]

$$u(t, \varepsilon) \sim \sum_{k=0}^{\infty} u_k(t, \tau) \varepsilon^k \qquad (\tau := \varepsilon t) \qquad (3.4.14)$$

are found to satisfy the initial conditions [see $(3.2.19)_k$]

$$u_k(0,0) = \begin{cases} b & \text{for } k = 0, \\ 0 & \text{for } k = 1, 2, \ldots, \end{cases} \qquad (3.4.15)_k$$

and

$$u_{k,t}(0,0) + u_{k-1,\tau}(0,0) = \begin{cases} c & \text{for } k = 0, \\ 0 & \text{for } k = 1, 2, \ldots, \end{cases} \qquad (3.4.16)_k$$

along with the differential equations [see (3.4.1) and $(3.3.15)_k$–$(3.3.16)_k$]

$$u_{k,tt} + u_k = G_k(t, \tau) \quad \text{for } k = 0, 1, 2, \ldots, \qquad (3.4.17)_k$$

with

$$G_0(t, \tau) = a, \qquad (3.4.18)_0$$

$$G_1(t, \tau) = -2u_{0,t\tau} + f(u_0, u_{0,t}), \qquad (3.4.18)_1$$

$$G_2(t, \tau) = -2u_{1,t\tau} - u_{0,\tau\tau} + f_u(u_0, u_{0,t})u_1$$
$$+ f_v(u_0, u_{0,t})(u_{1,t} + u_{0,\tau}), \qquad (3.4.18)_2$$

and analogous expressions for $G_3(t, \tau), G_4(t, \tau), \ldots$, where each G_k is determined recursively in terms of *previous* coefficients $u_0, u_1, \ldots, u_{k-1}$, with indices less than k. The terms involving f on the right sides of $(3.4.18)_k$ are obtained by inserting (3.4.14) into $f(u, u')$ on the right side of the differential equation (3.4.1) and using the Taylor expansion of the resulting expression with respect to ε. Note in (3.4.18) that such terms as f_u and f_v denote suitable partial derivatives as

$$f_u(u_0, u_{0,t}) \equiv \frac{\partial f(u, v)}{\partial u}\bigg|_{\substack{u = u_0 \\ v = u_{0,t}}}$$

and

$$f_v(u_0, u_{0,t}) \equiv \frac{\partial f(u, v)}{\partial v}\bigg|_{\substack{u = u_0 \\ v = u_{0,t}}}.$$

From (3.4.17)$_0$–(3.4.18)$_0$ we find the form of u_0 as

$$u_0(t, \tau) = a + A_0(\tau)\cos t + B_0(\tau)\sin t, \tag{3.4.19}$$

and then (3.4.15)$_0$, (3.4.16)$_0$, and (3.4.19) imply the initial conditions

$$A_0(0) = b - a, \qquad B_0(0) = c. \tag{3.4.20}$$

The expression (3.4.19) can be used in (3.4.18)$_1$ to find G_1 as

$$G_1(t, \tau) = -2\left[-A_0'(\tau)\sin t + B_0'(\tau)\cos t\right]$$
$$+ f(a + A_0\cos t + B_0\sin t, -A_0\sin t + B_0\cos t). \tag{3.4.21}$$

The Lindstedt/Poincaré condition for the elimination of spurious secular terms in u_1 can be given as [see (3.3.18) and (3.3.19)$_1$]

$$\int_0^{2\pi} G_1(t, \tau)\begin{bmatrix} \cos t \\ \sin t \end{bmatrix} dt = 0, \tag{3.4.22}$$

and then (3.4.21)–(3.4.22) give the following system of differential equations for the determination of A_0 and B_0:

$$\frac{d}{d\tau}\begin{bmatrix} A_0(\tau) \\ B_0(\tau) \end{bmatrix} = \frac{1}{2\pi}\int_0^{2\pi}\begin{bmatrix} -\sin t \\ \cos t \end{bmatrix}F[t, A_0(\tau), B_0(\tau)]\,dt$$

with $F(t, A_0, B_0) := f(a + A_0\cos t + B_0\sin t, -A_0\sin t + B_0\cos t)$.
$$\tag{3.4.23}$$

The (generally nonlinear) system (3.4.23) is to be solved subject to the initial conditions (3.4.20), where the basic result in the theory of ordinary differential equations guarantees that this initial-value problem always has a unique solution, at least on some interval $0 \le \tau < T$ for some fixed constant T, perhaps with $T = \infty$. In Chapter 4 it is shown that the resulting approximation (3.4.19) satisfies

$$|u(t, \varepsilon) - u_0(t, \varepsilon t)| \le \text{const. } \varepsilon,$$
$$\left|\frac{d}{dt}u(t, \varepsilon) - \frac{d}{dt}u_0(t, \varepsilon t)\right| \le \text{const. } \varepsilon, \tag{3.4.24}$$

uniformly for all t and ε satisfying

$$|\varepsilon t| \leq T_1$$

for any $T_1 < T$; indeed, (3.4.24) holds for all $t \geq 0$ in certain cases involving positive damping (Greenlee and Snow, 1975). Hence, the function $u_0 = u_0(t, \varepsilon t)$ provides a useful approximation to the exact solution u.

The procedure can be continued recursively using the Lindstedt/ Poincaré conditions $(3.3.19)_k$ to obtain further terms in the expansion (3.4.14) if desired, as illustrated in Sections 3.2 and 3.3. The resulting system of differential equations for A_k and B_k analogous to (3.4.23) is always *linear* for $k \geq 1$, with coefficients that involve previous A_j and B_j for $j = 0, 1, \ldots, k - 1$.

As is pointed out in Greenlee and Snow (1975), some generalizations are possible in the foregoing procedure. The initial conditions may be given as asymptotic series in ε, and the dependence in (3.4.6) of f on u and u' need not be of polynomial nature. Also, the left sides of (3.4.1) and (3.4.6) can be replaced with

$$u'' + [k(\varepsilon)]^2 u,$$

where $k = k(\varepsilon)$ is a smooth function of ε near and up to the origin, with $k(0) \neq 0$.

It is of interest to note that the first-order Lindstedt/Poincaré system (3.4.23) coincides with the first-order Krylov/Bogoliubov system (2.4.10), which proves directly the *first-order equivalence of averaging and two-timing* for the given differential equation (3.4.1), in agreement with a more general result of Morrison (1966).

Example 3.4.1: The van der Pol oscillator is given by (3.4.1), with

$$f(u, v) := v(1 - u^2) \qquad (a = 0), \qquad (3.4.25)$$

so that the function F appearing in the integrand on the right side of (3.4.23) becomes

$$F(t, A_0, B_0) = (-A_0 \sin t + B_0 \cos t)\left[1 - (A_0 \cos t + B_0 \sin t)^2\right].$$

$$(3.4.26)$$

The resulting integrals occurring on the right side of (3.4.23) can then be evaluated with the aid of the Fourier expansions

$$\sin^3 t = \tfrac{3}{4}\sin t - \tfrac{1}{4}\sin 3t,$$

$$\sin^2 t \cos t = \tfrac{1}{4}\cos t - \tfrac{1}{4}\cos 3t,$$

$$\sin t \cos^2 t = \tfrac{1}{4}\sin t + \tfrac{1}{4}\sin 3t,$$

$$\cos^3 t = \tfrac{3}{4}\cos t + \tfrac{1}{4}\cos 3t,$$

(3.4.27)

along with such known and related integrals as

$$\int_0^{2\pi} \sin t \sin mt \, dt = \begin{cases} \pi & \text{if } m = 1, \\ 0 & \text{if } m = 0, 2, 3, \ldots, \end{cases}$$

and we find from (3.4.23) and (3.4.26) the system

$$\frac{d}{d\tau} A_0 = \tfrac{1}{2} A_0 \Big[1 - \tfrac{1}{4}\big((A_0)^2 + (B_0)^2 \big) \Big],$$

$$\frac{d}{d\tau} B_0 = \tfrac{1}{2} B_0 \Big[1 - \tfrac{1}{4}\big((A_0)^2 + (B_0)^2 \big) \Big].$$

(3.4.28)

This system implies

$$\frac{d}{d\tau}\Big[(A_0)^2 + (B_0)^2 \Big] = \Big[(A_0)^2 + (B_0)^2 \Big]\Big\{ 1 - \tfrac{1}{4}\Big[(A_0)^2 + (B_0)^2 \Big] \Big\},$$

(3.4.29)

which can be integrated with (3.4.20) to give

$$(A_0)^2 + (B_0)^2 = \begin{cases} 4 & \text{if } b^2 + c^2 = 4, \\ \dfrac{4(b^2 + c^2)}{(b^2 + c^2) - (b^2 + c^2 - 4)e^{-\tau}} & \text{otherwise.} \end{cases}$$

(3.4.30)

Note that, in accordance with (3.4.25), we have taken $a = 0$ in (3.4.20). Also, here and later we are assuming the condition

$$b^2 + c^2 > 0$$

in order to eliminate the trivial case $b = c = 0$. The result (3.4.30) can be inserted into the right side of (3.4.28), and the resulting system is easily solved, subject to the given initial conditions, to give

$$A_0(\tau) = \frac{2b}{\big[(b^2 + c^2) - (b^2 + c^2 - 4)e^{-\tau} \big]^{1/2}},$$

$$B_0(\tau) = \frac{2c}{\big[(b^2 + c^2) - (b^2 + c^2 - 4)e^{-\tau} \big]^{1/2}} \quad \text{if } b^2 + c^2 \neq 4,$$

(3.4.31)

whereas

$$A_0(\tau) = b, \qquad B_0(\tau) = c \quad \text{if } b^2 + c^2 = 4. \qquad (3.4.32)$$

These results (3.4.31) and (3.4.32), along with (3.4.13) and (3.4.19), point to the existence of a *limit cycle* (as $t \to \infty$) for the van der Pol oscillator. This limit cycle is described in the phase plane as

$$(u)^2 + (u')^2 = 4. \qquad (3.4.33)$$

Example 3.4.2: The nonlinearly damped oscillator (3.4.5) is an example of equation (3.4.6), with f given as [see (3.4.7)]

$$f(\varepsilon, t, u, v) = -2p(\tau)v^3 \quad (\tau = \varepsilon t). \qquad (3.4.34)$$

This case is not included in (3.4.1), because of the dependence here of f on τ, but the same two-variable approach used for (3.4.1) suffices for more general equations, including (3.4.5), as we shall now see. We again obtain an equation of the form $(3.4.17)_k$; here we find

$$G_0(t, \tau) = 0, \qquad (3.4.35)_0$$

$$G_1(t, \tau) = -2u_{0,t\tau} - 2p(\tau)(u_{0,t})^3, \qquad (3.4.35)_1$$

and

$$G_k(t, \tau) = -2u_{k-1,t\tau} - u_{k-2,\tau\tau}$$

$$-2p(\tau) \sum_{n=0}^{k-1} (u_{k-1-n} + u_{k-2-n,\tau})$$

$$\times \sum_{m=0}^{n} (u_{n-m,t} + u_{n-1-m,\tau})(u_{m,t} + u_{m-1,\tau})$$

$$\text{for } k = 2, 3, \dots . \qquad (3.4.35)_k$$

We see from $(3.4.17)_0$ and $(3.4.35)_0$ that the form of u_0 is again given by (3.4.19) (with $a = 0$), and then $(3.4.35)_1$ yields

$$G_1(t, \tau) = 2A_0'(\tau)\sin t - 2B_0'(\tau)\cos t$$

$$-2p(\tau)[-A_0(\tau)\sin t + B_0(\tau)\cos t]^3.$$

Hence, the Lindstedt/Poincaré condition (3.4.22) yields for the functions $A_0 = A_0(\tau)$ and $B_0 = B_0(\tau)$ the system [see (3.4.23)]

$$\frac{d}{d\tau}\begin{bmatrix} A_0 \\ B_0 \end{bmatrix} = \frac{1}{2\pi} \int_0^{2\pi} \begin{bmatrix} -\sin t \\ \cos t \end{bmatrix} F(t, A_0, B_0, \tau)\, dt$$

with $F(t, A_0, B_0, \tau) := -2p(\tau)(-A_0\sin t + B_0\cos t)^3. \qquad (3.4.36)$

The integrals on the right side of (3.4.36) are easily evaluated with (3.4.27), and we find the system

$$\frac{d}{d\tau}\begin{bmatrix} A_0 \\ B_0 \end{bmatrix} = -\frac{3}{4}p(\tau)\left[(A_0)^2 + (B_0)^2\right]\begin{bmatrix} A_0 \\ B_0 \end{bmatrix}. \qquad (3.4.37)$$

This system implies

$$\frac{d}{d\tau}\left[(A_0)^2 + (B_0)^2\right] = -\frac{3}{2}p(\tau)\left[(A_0)^2 + (B_0)^2\right]^2,$$

and this latter equation can be integrated subject to the appropriate initial conditions to give

$$(A_0)^2 + (B_0)^2 = \frac{b^2 + c^2}{1 + \frac{3}{2}(b^2 + c^2)\int_0^\tau p(s)\,ds} \quad \text{for } 0 \leq \tau < T, \qquad (3.4.38)$$

where T can be taken to be the smallest positive zero of the denominator on the right side here, with

$$1 + \frac{3}{2}(b^2 + c^2)\int_0^T p = 0. \qquad (3.4.39)$$

Equation (3.4.39) may have one or many positive solutions $T > 0$. On the other hand, there is no positive solution for T if the given function p is nonnegatively valued. In this latter case we can take $T = \infty$ in (3.4.38).

The result (3.4.38) can be inserted back into the right side of (3.4.37), and the resulting system can be integrated subject to the given initial conditions to give

$$A_0(\tau) = b\left[1 + \frac{3}{2}(b^2 + c^2)\int_0^\tau p(s)\,ds\right]^{-1/2},$$
$$B_0(\tau) = c\left[1 + \frac{3}{2}(b^2 + c^2)\int_0^\tau p(s)\,ds\right]^{-1/2}. \qquad (3.4.40)$$

Hence, (3.4.19) and (3.4.40) yield (with $a = 0$)

$$u_0(t, \varepsilon t) = \frac{b\cos t + c\sin t}{\left[1 + \frac{3}{2}(b^2 + c^2)\int_0^{\varepsilon t} p(s)\,ds\right]^{1/2}}. \qquad (3.4.41)$$

Estimates of the form (3.4.24) are obtained for this problem in Chapter 4, so that this function (3.4.41) provides a useful approximation to the exact solution u, provided that the given function p satisfies the conditions

$$p \geq 0, \qquad \int_0^\infty p = \infty.$$

From (3.4.41) it follows that the solution u represents approximately a harmonic motion of period 2π, because the denominator is almost constant on the right side of (3.4.41) as t increases through any interval having length on the order of unity. However, as t increases through many such quasi-periods of length 2π, the denominator slowly increases, so that the motion slowly decays because of the damping. The damping depends on the initial conditions because of the nonlinearity of the problem.

See Hoppensteadt (1979) for references on various nonlinear oscillation problems in biology.

Exercises

Exercise 3.4.1: Use the two-variable approach to find an approximation to the solution of the Duffing equation $u'' + u + \varepsilon u^3 = 0$ subject to the initial conditions $u(0) = 1$, $u'(0) = 0$.

Exercise 3.4.2 (Greenlee and Snow 1975): For the autonomous oscillator (3.4.1), show that the Lindstedt/Poincaré system (3.4.23) can be reduced to the uncoupled system

$$\frac{d}{d\tau}\begin{bmatrix} R \\ \Phi \end{bmatrix} = \frac{1}{2\pi}\int_0^{2\pi} f(a + R\sin t, R\cos t)\begin{bmatrix} \cos t \\ -\frac{1}{R}\sin t \end{bmatrix}dt, \quad (3.4.42)$$

where $R = R(\tau)$ and $\Phi = \Phi(\tau)$ are related to A_0 and B_0 as

$$A_0 = R\sin\Phi, \qquad B_0 = R\cos\Phi. \qquad (3.4.43)$$

Hint: Differentiate (3.4.43) with respect to τ, solve the resulting equations for $R'(\tau)$ and $\Phi'(\tau)$, and simplify the results using (3.4.23) along with the trigonometric addition formulas and the periodicity of the sine and cosine functions. Note that the system (3.4.42) is uncoupled, because the first equation can be solved as a scalar, separable, first-order equation for $R = [(A_0)^2 + (B_0)^2]^{1/2}$. Then the second equation gives Φ up to a simple quadrature. This technique has been used in particular examples by Morrison (1966), J. Cole (1968), Chow (1972), Baum (1972), and others (as in our Examples 3.4.1 and 3.4.2) and was shown to be a general technique for (3.4.1) by Greenlee and Snow (1975). Note also that (3.4.42) is identical with the Krylov/Bogoliubov averaged system (2.2.8).

Exercise 3.4.3: Consider the classical case of beats for a linear undamped

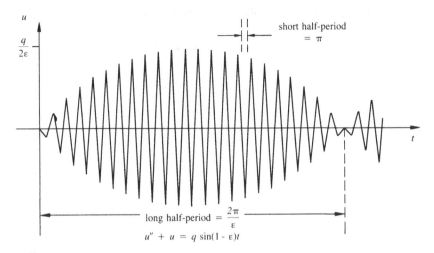

Figure 3.1
Beats

oscillator in which the driving frequency is close to the natural frequency,

$$\frac{d^2u}{dt^2} + u = q\sin(1-\varepsilon)t, \tag{3.4.44}$$

where the constant q is the amplitude of the forcing term, and ε is a small positive parameter. Find the exact solution $u = u(t, \varepsilon)$ of this equation subject to the initial conditions

$$u(0, \varepsilon) = 1, \qquad u'(0, \varepsilon) = 0,$$

and compare the exact solution with the first two terms in the two-variable expansion

$$u(t, \varepsilon) \sim \sum_{k=-1}^{\infty} u_k(t, \tau)\varepsilon^k \qquad (\tau := \varepsilon t).$$

(The solution exhibits "fast" oscillations of period 2π, with slowly varying amplitude of "long" period $4\pi/\varepsilon$, as illustrated in Figure 3.1.)

Exercise 3.4.4 (Greenlee and Snow 1975): Use two-timing to derive the approximate solution

$$u_0 = -\frac{\sin t}{1 - \frac{3}{8}\varepsilon t}$$

for the initial-value problem

$$u'' + \varepsilon(\cos t)(u')^2 + u = 0 \qquad \text{for } t > 0,$$
$$u = 0 \quad \text{and} \quad u' = -1 \quad \text{at } t = 0.$$

[The stated approximate solution u_0 is defined only for $0 \le t < 8/(3\varepsilon)$.]

Exercise 3.4.5 (Kevorkian 1966; Morrison 1966; Rubenfeld 1978):
  (a) Derive the result

$$u_0(t, \tau) = \frac{a\cos t + b\sin t}{\left[1 + \frac{3}{4}(a^2 + b^2)\tau\right]^{1/2}} \qquad (3.4.45)$$

for the leading term in the two-variable expansion

$$u(t, \varepsilon) \sim \sum_{k=0}^{\infty} u_k(t, \tau)\varepsilon^k \qquad (\tau := \varepsilon t)$$

for the problem

$$u'' + u + \varepsilon\left[2\varepsilon + (u')^2\right]u' = 0 \quad \text{for } t > 0,$$
$$u = a \quad \text{and} \quad u' = b \quad \text{at } t = 0. \qquad (3.4.46)$$

(b) Derive the result

$$u_0(t, \tau, \theta) = \frac{(a\cos t + b\sin t)e^{-\theta/3}}{\left[1 + \frac{3}{4}(a^2 + b^2)\tau\left(\dfrac{1 - e^{-2\theta/3}}{2\theta/3}\right)\right]^{1/2}} \qquad (3.4.47)$$

for the leading term in the three-variable expansion

$$u(t, \varepsilon) \sim \sum_{k=0}^{\infty} u_k(t, \tau, \theta)\varepsilon^k \qquad (\tau := \varepsilon t, \ \theta := \varepsilon^2 t)$$

for the problem (3.4.46). See Exercises 4.4.1 and 4.4.2 for further considerations on this problem (3.4.46).

3.5 Two-timing for systems: transference of vibrations

In this section we consider a vibration problem involving a system of three second-order differential equations that exhibits some of the features of the more complicated perturbation problems of celestial mechanics. We consider a heavy bar of mass M that hangs by two cords of lengths L as indicated in Figure 3.2. Two light particles or beads, each of mass m, hang from the ends of the bar by strings of length L, and the system oscillates as a compound pendulum when disturbed from equilibrium. We let Θ, Φ, and Ψ be the angles shown in Figure 3.2, where we consider for simplicity only vibrations in a fixed vertical plane. If we

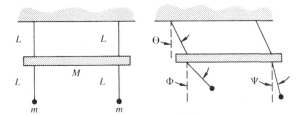

Figure 3.2

neglect the masses of the strings and consider only small vibrations, then
Newton's laws of mechanics lead to the following equations of motion:

$$\frac{d^2\Theta}{dt^2} + \kappa^2\Theta = \varepsilon^2\kappa^2(-2\Theta + \Phi - \Psi),$$

$$\frac{d^2\Phi}{dt^2} + \kappa^2(\Phi - \Theta) = -\varepsilon^2\kappa^2(-2\Theta + \Phi - \Psi), \qquad (3.5.1)$$

$$\frac{d^2\Psi}{dt^2} + \kappa^2(\Psi + \Theta) = \varepsilon^2\kappa^2(-2\Theta + \Phi - \Psi),$$

where

$$\kappa^2 := \frac{g}{L}, \qquad \varepsilon^2 := \frac{m}{M},$$

and where g is a given constant acceleration due to gravity. The
functions Θ, Φ, and Ψ are to be determined as functions of time t
subject to these equations (3.5.1) and also subject to given initial conditions. For definiteness we shall consider only the case in which the system
starts from rest after one of the small particles has been given a fixed
initial angular displacement. Hence, the initial conditions are

$$\Theta = \Phi = 0, \qquad \Psi = \alpha,$$

$$\frac{d\Theta}{dt} = \frac{d\Phi}{dt} = \frac{d\Psi}{dt} = 0 \quad \text{at } t = 0, \qquad (3.5.2)$$

where α is a given constant.

The initial-value problem (3.5.1)–(3.5.2) is linear, with constant coefficients, and it can be solved in closed form. However, the form of the
exact solution is somewhat complicated, and we can use the multiscale
procedure to obtain an approximation to the solution (for small values of
the parameter ε) that will be uniformly valid (over a lengthy expanding
interval) and easy to interpret. The approximation procedure can also be
used in more complicated situations involving nonlinear systems or
systems with variable coefficients for which exact solutions in closed form
are not readily available.

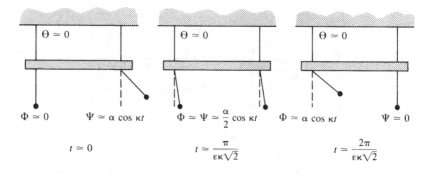

Figure 3.3

The reduced problem obtained by putting $\varepsilon = 0$ in (3.5.1) has the unique solution

$$\Theta(t) \equiv \Phi(t) \equiv 0,$$

$$\Psi(t) = \alpha \cos \kappa t, \tag{3.5.3}$$

so that in this case the heavy bar and initially undisturbed particle remain at rest, and the disturbed particle undergoes a simple harmonic motion. We can say in this case that the heavy bar acts as a rigid motionless bar with infinite mass, and the motions of the two particles remain independent of each other. Of course, in fact the heavy bar is not rigid for any fixed positive value of ε, and we shall see that energy is transmitted back and forth through the bar from one particle to the other, so that initially the disturbed particle will undergo (almost) harmonic motion of period $2\pi/\kappa$ while the other particle remains (almost) at rest. Later, after many quasi-periods, the initially disturbed particle will be (almost) at rest, and the other particle will undergo (almost) harmonic motion. The situation is illustrated in Figure 3.3.

In order to approximate the solution of (3.5.1)–(3.5.2), we introduce the slow time $\tau := \varepsilon t$, and we seek expansions in the form

$$\begin{bmatrix} \Theta(t, \varepsilon) \\ \Phi(t, \varepsilon) \\ \Psi(t, \varepsilon) \end{bmatrix} \sim \sum_{k=0}^{\infty} \begin{bmatrix} \Theta_k(t, \tau) \\ \Phi_k(t, \tau) \\ \Psi_k(t, \tau) \end{bmatrix} \varepsilon^k \qquad (\tau := \varepsilon t). \tag{3.5.4}$$

If we insert these expansions into (3.5.1)–(3.5.2), we obtain differential equations and initial conditions that can be used recursively to determine the functions Θ_k, Φ_k, and Ψ_k.

We omit details and indicate only the final results. For Θ_0, Φ_0, and Ψ_0 we find the equations

$$\Theta_{0,tt} + \kappa^2\Theta_0 = 0,$$
$$\Phi_{0,tt} + \kappa^2\Phi_0 = \kappa^2\Theta_0, \qquad (3.5.5)$$
$$\Psi_{0,tt} + \kappa^2\Psi_0 = -\kappa^2\Theta_0,$$

so that in particular

$$\Theta_0(t,\tau) = A_0(\tau)\cos\kappa t + B_0(\tau)\sin\kappa t. \qquad (3.5.6)$$

It then follows with (3.5.5) that we must choose $A_0 \equiv B_0 \equiv 0$ in order to eliminate spurious secular (resonant, unbounded) terms in Φ_0 and Ψ_0. In this way we find

$$\Theta_0(t,\tau) = 0,$$
$$\Phi_0(t,\tau) = C_0(\tau)\cos\kappa t + D_0(\tau)\sin\kappa t, \qquad (3.5.7)$$
$$\Psi_0(t,\tau) = E_0(\tau)\cos\kappa t + F_0(\tau)\sin\kappa t,$$

for suitable functions C_0, D_0, E_0, and F_0, which are to be determined so as to eliminate further spurious secular terms. Along with (3.5.7), we find the initial conditions

$$C_0(0) = D_0(0) = F_0(0) = 0,$$
$$E_0(0) = \alpha. \qquad (3.5.8)$$

For Θ_1, Φ_1, and Ψ_1 we find the equations

$$\Theta_{1,tt} + \kappa^2\Theta_1 = -2\Theta_{0,t\tau} \overset{(3.5.7)}{=} 0,$$
$$\Phi_{1,tt} + \kappa^2\Phi_1 = \kappa^2\Theta_1 - 2\Phi_{0,t\tau}, \qquad (3.5.9)$$
$$\Psi_{1,tt} + \kappa^2\Psi_1 = -\kappa^2\Theta_1 - 2\Psi_{0,t\tau},$$

so that, in particular, Θ_1 must have the form

$$\Theta_1(t,\tau) = A_1(\tau)\cos\kappa t + B_1(\tau)\sin\kappa t \qquad (3.5.10)$$

for suitable functions A_1 and B_1, which are easily seen to be subject to the initial conditions

$$A_1(0) = B_1(0) = 0. \qquad (3.5.11)$$

Now (3.5.7) and (3.5.10) can be used in the last two equations of (3.5.9) to find

$$\Phi_{1,tt} + \kappa^2\Phi_1 = 2\kappa\left(\frac{\kappa}{2}B_1 + C_0'\right)\sin\kappa t + 2\kappa\left(\frac{\kappa}{2}A_1 - D_0'\right)\cos\kappa t,$$

$$\Psi_{1,tt} + \kappa^2\Psi_1 = -2\kappa\left(\frac{\kappa}{2}B_1 - E_0'\right)\sin\kappa t - 2\kappa\left(\frac{\kappa}{2}A_1 + F_0'\right)\cos\kappa t.$$

$$(3.5.12)$$

The Lindstedt/Poincaré requirement for elimination of spurious secular terms can be applied to each of the differential equations of (3.5.12), and we find the conditions

$$C_0'(\tau) = -\frac{\kappa}{2} B_1(\tau), \qquad D_0'(\tau) = +\frac{\kappa}{2} A_1(\tau),$$

$$E_0'(\tau) = +\frac{\kappa}{2} B_1(\tau), \qquad F_0'(\tau) = -\frac{\kappa}{2} A_1(\tau).$$

These equations can be integrated with (3.5.8) to give

$$C_0(\tau) = -\frac{\kappa}{2} \int_0^\tau B_1, \qquad D_0(\tau) = \frac{k}{2} \int_0^\tau A_1,$$

with $E_0(\tau) = \alpha - C_0(\tau)$ and $F_0(\tau) = -D_0(\tau)$,   (3.5.13)

and then the resulting (homogeneous) equations of (3.5.12) can be integrated to give

$$\Phi_1(t,\tau) = C_1(\tau)\cos\kappa t + D_1(\tau)\sin\kappa t,$$
$$\Psi_1(t,\tau) = E_1(\tau)\cos\kappa t + F_1(\tau)\sin\kappa t,$$

(3.5.14)

for suitable functions C_1, D_1, E_1, and F_1.

For Θ_2 we find the equation

$$\Theta_{2,tt} + \kappa^2\Theta_2 = -2\Theta_{1,t\tau} + \kappa^2(\Phi_0 - \Psi_0)$$

$$= 2\kappa\left[\frac{\kappa}{2}(C_0 - E_0) - B_1'\right]\cos\kappa t$$

$$+ 2\kappa\left[\frac{\kappa}{2}(D_0 - F_0) + A_1'\right]\sin\kappa t,   \quad (3.5.15)$$

and so Θ_2 will contain resonant terms unless the following conditions hold:

$$A_1'(\tau) = -\frac{\kappa}{2}\left[D_0(\tau) - F_0(\tau)\right],$$

$$B_1'(\tau) = +\frac{\kappa}{2}\left[C_0(\tau) - E_0(\tau)\right].$$

(3.5.16)

Hence, we impose these conditions.

The unique solution of (3.5.11), (3.5.13), and (3.5.16) yields

$$A_1(\tau) \equiv 0, \qquad B_1(\tau) = -\frac{\alpha}{\sqrt{2}}\sin\frac{\kappa\tau}{\sqrt{2}},   \quad (3.5.17)$$

with

$$D_0(\tau) \equiv F_0(\tau) \equiv 0,$$

$$C_0(\tau) = \frac{\alpha}{2}\left(1 - \cos\frac{\kappa\tau}{\sqrt{2}}\right),   \quad (3.5.18)$$

$$E_0(\tau) = \frac{\alpha}{2}\left(1 + \cos\frac{\kappa\tau}{\sqrt{2}}\right).$$

Putting these results all together with (3.5.7) and (3.5.10), we have

$$\Theta_0(t,\tau) \equiv 0, \qquad \Theta_1(t,\tau) = -\frac{\alpha}{\sqrt{2}} \sin\frac{\kappa\tau}{\sqrt{2}} \sin\kappa t,$$

$$\Phi_0(t,\tau) = \frac{\alpha}{2}\left(1 - \cos\frac{\kappa\tau}{\sqrt{2}}\right)\cos\kappa t, \qquad (3.5.19)$$

$$\Psi_0(t,\tau) = \frac{\alpha}{2}\left(1 + \cos\frac{\kappa\tau}{\sqrt{2}}\right)\cos\kappa t,$$

and we use these functions to construct the following approximate solutions Θ^*, Φ^*, and Ψ^* obtained by truncating the expansion (3.5.4):

$$\Theta^*(t,\varepsilon) := \Theta_0(t,\varepsilon t) + \varepsilon\Theta_1(t,\varepsilon t) = -\frac{\varepsilon\alpha}{\sqrt{2}} \sin\frac{\kappa\varepsilon t}{\sqrt{2}} \sin\kappa t,$$

$$\Phi^*(t,\varepsilon) := \Phi_0(t,\varepsilon t) = \frac{\alpha}{2}\left(1 - \cos\frac{\kappa\varepsilon t}{\sqrt{2}}\right)\cos\kappa t, \qquad (3.5.20)$$

$$\Psi^*(t,\varepsilon) := \Psi_0(t,\varepsilon t) = \frac{\alpha}{2}\left(1 + \cos\frac{\kappa\varepsilon t}{\sqrt{2}}\right)\cos\kappa t.$$

These latter functions do indeed provide useful approximations to the exact solution functions. The techniques of Chapter 4 permit us to obtain the estimates [see (4.3.53)]

$$|\Theta(t,\varepsilon) - \Theta^*(t,\varepsilon)| \le \text{const. } \varepsilon^2,$$

$$|\Phi(t,\varepsilon) - \Phi^*(t,\varepsilon)| \le \text{const. } \varepsilon, \qquad (3.5.21)$$

$$|\Psi(t,\varepsilon) - \Psi^*(t,\varepsilon)| \le \text{const. } \varepsilon,$$

and these estimates hold uniformly for all small $\varepsilon > 0$ and for all $0 \le t \le T/\varepsilon$, for some fixed constant T.

The estimates (3.5.21) along with (3.5.20) show that $\Theta(t,\varepsilon)$ remains small, of order ε, but $\Phi(t,\varepsilon)$ and $\Psi(t,\varepsilon)$ are given approximately as harmonic motions of the form $\cos\kappa t$, of period $2\pi/\kappa$ and with *slowly varying amplitudes* given, respectively, as

$$\frac{\alpha}{2}\left(1 - \cos\frac{\kappa\varepsilon t}{\sqrt{2}}\right) \quad \text{and} \quad \frac{\alpha}{2}\left(1 + \cos\frac{\kappa\varepsilon t}{\sqrt{2}}\right).$$

These slowly varying amplitudes have a long period $2\pi\sqrt{2}/\kappa\varepsilon$. The amplitude of Φ is a maximum when the amplitude of Ψ is a minimum and vice versa. The oscillation is slowly transferred from the first pendulum to the second, and back again, and so on, as illustrated in Figure 3.3.

Exercise

Exercise 3.5.1: Find the leading terms $Q_0(t, \tau)$ and $P_0(t, \tau)$ in a two-variable expansion of the solution of the initial-value problem

$$Q'' + \kappa^2 Q = \varepsilon \kappa^2 P, \qquad P'' + \kappa^2 P = \varepsilon \kappa^2 Q \quad \text{for } t > 0,$$

and

$$Q(0) = q, \qquad P(0) = Q'(0) = P'(0) = 0 \quad \text{at } t = 0.$$

This problem is a mathematical model for two weakly coupled *LC* circuits as discussed in Section 2.5. The quantities κ and q are fixed constants, and ε is a small parameter.

4

Error estimates for perturbed-oscillation problems

In this chapter we shall discuss certain error estimates that provide a theoretical justification for the approximation procedures of the previous two chapters.

4.1 Estimates on expanding intervals

In this section, error estimates are given for the equation

$$\frac{d^2u}{dt^2} + u = \varepsilon f\left(\varepsilon, t, u, \frac{du}{dt}\right) \quad \text{for } t > 0 \tag{4.1.1}$$

subject to the initial conditions

$$u = a \quad \text{and} \quad \frac{du}{dt} = b \quad \text{at } t = 0, \tag{4.1.2}$$

where $f = f(\varepsilon, t, u, v)$ is a given smooth function of its four arguments, for all small nonnegative values of ε, all $t \geq 0$, and all suitable real values of u and v, and where f is evaluated in (4.1.1) at $u = u(t, \varepsilon)$ and $v = du(t, \varepsilon)/dt = u'(t, \varepsilon)$, where $u(t, \varepsilon)$ denotes the solution of (4.1.1)–(4.1.2).

The basic existence and uniqueness results in the theory of differential equations guarantee that the problem (4.1.1)–(4.1.2) has a unique solution, at least on some fixed interval $0 \leq t < t_1$ for some positive t_1. As a consequence of the error estimates given later, along with the procedures of the previous chapters, it will be seen that this solution exists at least on some expanding interval of the type

$$0 \leq t \leq \frac{T}{\varepsilon} \quad \text{as } \varepsilon \to 0+$$

for some fixed positive constant T, so that the previous t_1 can be taken as

$t_1 = T/\varepsilon$. Moreover, it will be seen that the procedures of the previous two chapters do, in fact, provide useful approximate solutions to (4.1.1)–(4.1.2) on the resulting expanding interval. The possibility of taking $t_1 = \infty$ is considered in Section 4.4.

It is useful first to consider the linear case, in which f has the form

$$f(\varepsilon, t, u, v) := g(t, \varepsilon)u + h(t, \varepsilon)v \qquad (4.1.3)$$

for given bounded continuous functions $g = g(t, \varepsilon)$ and $h = h(t, \varepsilon)$ on the region

$$0 \le \varepsilon t \le T, \qquad 0 \le \varepsilon \le \varepsilon_0 \qquad (4.1.4)$$

for fixed positive constants T and ε_0. In this case, (4.1.1) becomes

$$u'' - \varepsilon h(t, \varepsilon)u' + [1 - \varepsilon g(t, \varepsilon)]u = 0,$$

and in fact we shall consider the corresponding nonhomogeneous equation

$$\mathscr{L}u = F(t, \varepsilon) \qquad (4.1.5)$$

for a given function F, where the differential operator $\mathscr{L}$ is defined as

$$\mathscr{L} := \frac{d^2}{dt^2} - \varepsilon h(t, \varepsilon)\frac{d}{dt} + [1 - \varepsilon g(t, \varepsilon)]. \qquad (4.1.6)$$

Theorem 4.1.1: *The solution of the initial-value problem (4.1.2) and (4.1.5) can be represented as*

$$u(t, \varepsilon) = au_1(t, \varepsilon) + bu_2(t, \varepsilon) + \int_0^t K(t, s, \varepsilon)F(s, \varepsilon)\, ds, \qquad (4.1.7)$$

where u_1 and u_2 are the solutions of the homogeneous equation $\mathscr{L}u = 0$ [$\mathscr{L}$ given by (4.1.6)] subject to the initial conditions

$$u_1(0, \varepsilon) = u_2'(0, \varepsilon) = 1,$$
$$u_1'(0, \varepsilon) = u_2(0, \varepsilon) = 0, \qquad (4.1.8)$$

and where the kernel K is given as

$$K(t, s, \varepsilon) := [u_1(s, \varepsilon)u_2(t, \varepsilon) - u_2(s, \varepsilon)u_1(t, \varepsilon)]\exp\left(-\varepsilon\int_0^s h\right).$$
$$(4.1.9)$$

Moreover, we have the estimates (Kollett 1974)

$$|u_j(t, \varepsilon)|, |u_j'(t, \varepsilon)| \le e^{T(\|g\|+\|h\|)} \quad (j = 1, 2) \qquad (4.1.10)$$

and

$$|K(t, s, \varepsilon)|, |K_t(t, s, \varepsilon)| \le 2e^{T\|h\|}e^{2T(\|g\|+\|h\|)} \qquad (4.1.11)$$

uniformly for all points (t, ε) *and* (s, ε) *in the region* (4.1.4), *where* $K_t(t, s, \varepsilon) = \partial K(t, s, \varepsilon)/\partial t$, *and where the norm* $\| \cdot \| = \| \cdot \|_\varepsilon$ *is here the maximum norm taken with respect to* t *over the interval* $0 \leq t \leq T/\varepsilon$.

Proof: The well-known representation (4.1.7) is given directly by (0.1.3), where (4.1.9) follows from (0.1.4) along with Abel's equation $W' = \varepsilon h W$ for the wronskian $W := u_1 u_2' - u_2 u_1'$. The estimates of (4.1.11) follow directly from (4.1.9) along with the estimates of (4.1.10). Hence, we need only prove the estimates (4.1.10) for the solutions u_1 and u_2 of the homogeneous equation $\mathscr{L}u = 0$.

The Prüfer substitution

$$u = r \sin \theta, \qquad u' = r \cos \theta, \tag{4.1.12}$$

sets up a one-to-one correspondence between solutions of $\mathscr{L}u = 0$ and solutions of the first-order system

$$r' = \varepsilon r (\cos \theta)(g \sin \theta + h \cos \theta),$$
$$\theta' = 1 - \varepsilon (\sin \theta)(g \sin \theta + h \cos \theta). \tag{4.1.13}$$

From the first equation of (4.1.13) there follows, on integration,

$$r(t, \varepsilon) = r(0, \varepsilon) \exp \left[\varepsilon \int_0^t \cos \theta (g \sin \theta + h \cos \theta) \right], \tag{4.1.14}$$

and the stated bounds (4.1.10) follow directly from (4.1.14), with $r = [u^2 + (u')^2]^{1/2}$. ∎

The solution of the inhomogeneous equation (4.1.5) subject to the initial conditions of (4.1.2) can be estimated from (4.1.7), (4.1.10), and (4.1.11), and we find directly (Kollett 1974)

$$|u(t, \varepsilon)|, |u'(t, \varepsilon)| \leq e^{2T(\|g\| + \|h\|)} \left(|a| + |b| + \frac{2Te^{T\|h\|} \|F\|}{\varepsilon} \right) \tag{4.1.15}$$

uniformly for all t and ε in the region of (4.1.4). Mahony (1972) obtained bounds of the type (4.1.15) using energy estimates closely related to the preceding use of the Prüfer substitution (4.1.12).

If $\varepsilon = 0$, the preceding solutions u_1 and u_2 of the homogeneous equation $\mathscr{L}u = 0$ reduce to

$$u_1(t, 0) = \cos t,$$
$$u_2(t, 0) = \sin t \quad (\varepsilon = 0), \tag{4.1.16}$$

and the bounds of (4.1.10) are satisfied for all t, with $g \equiv h \equiv 0$.

However, we cannot generally expect that the solutions u_1 and u_2 should remain globally bounded for all t, because it is easy to give examples with negative damping $[h(t, \varepsilon) > 0]$ such that these solutions become unbounded as $t \to \infty$ for any fixed $\varepsilon > 0$. What is perhaps surprising is the fact that such unbounded behavior can occur even with *zero damping* ($h = 0$), as illustrated by the following example of Perron.

Example 4.1.1 (Perron 1930; cf. p. 113 of Bellman 1953): If we take $h = 0$ and

$$g(t, \varepsilon) = -\frac{2\sin 2(t + 1)}{t + 1} - \frac{\cos^2(t + 1)}{(t + 1)^2}\left[1 - \varepsilon\cos^2(t + 1)\right],$$

$$(4.1.17)$$

then the function $u_3(t, \varepsilon)$, defined as

$$u_3(t, \varepsilon) := \cos(t + 1)\exp\left(\varepsilon\int_0^t \frac{\cos^2(s + 1)}{s + 1}\, ds\right), \quad (4.1.18)$$

is found to satisfy the initial-value problem

$$u'' + (1 - \varepsilon g)u = 0 \quad \text{for } t > 0,$$

$$u = \cos 1 \quad \text{and} \quad u' = -(\sin 1) + \varepsilon(\cos^3 1) \quad \text{at } t = 0,$$

$$(4.1.19)$$

so that u_3 is a linear combination of u_1 and u_2. The function g is uniformly bounded for $t \geq 0$, and $g(t, \varepsilon)$ decays uniformly toward zero as $t \to \infty$, with

$$|g(t, \varepsilon)| \leq 3 + \varepsilon \quad \text{for all } t, \varepsilon \geq 0,$$

with

$$|g(t, \varepsilon)| \leq \frac{3 + \varepsilon}{t + 1} = \text{order}\left(\frac{1}{t}\right) \quad \text{as } t \to \infty.$$

Hence, for small $\varepsilon > 0$, we might expect that all solutions of the given differential equation should remain bounded as $t \to \infty$. This is, however, not the case, as follows from the result

$$\int_0^t \frac{\cos^2(s + 1)}{s + 1}\, ds = \frac{1}{2}\left[\log(t + 1) + \int_0^t \frac{\cos 2(s + 1)}{s + 1}\, ds\right],$$

$$(4.1.20)$$

along with the boundedness of the last integral on the right side of

(4.1.20). Hence, we find, from (4.1.18) and (4.1.20),

$$u_3(t, \varepsilon) = (t + 1)^{\varepsilon/2} \cos(t + 1) \exp\left[\frac{\varepsilon}{2} \int_0^t \frac{\cos 2(s + 1)}{s + 1} \, ds\right],$$

(4.1.21)

which shows that $u_3(t, \varepsilon)$ is *unbounded* as $t \to \infty$ for any fixed $\varepsilon > 0$. Of course, u_3 is bounded on any fixed expanding interval $0 \le t \le T/\varepsilon$, and even on any such interval $0 \le t \le T/\varepsilon^\nu$ for any fixed ν.

We return now to the problem of obtaining error estimates for the original problem (4.1.1)–(4.1.2). More precisely, we wish to obtain bounds for a given, conjectured approximate solution $U = U(t, \varepsilon)$, where U is known to satisfy the given initial-value problem approximately in the sense that

$$U'' + U = \varepsilon f(\varepsilon, t, U, U') - \rho(t, \varepsilon) \quad \text{for } t > 0,$$

(4.1.22)

and

$$U(0, \varepsilon) = a - \alpha(\varepsilon), \qquad U'(0, \varepsilon) = b - \beta(\varepsilon) \quad \text{at } t = 0,$$

(4.1.23)

for certain known residuals $\alpha = \alpha(\varepsilon)$, $\beta = \beta(\varepsilon)$, and $\rho = \rho(t, \varepsilon)$, where these residuals are known to be small for all t and ε satisfying (4.1.4).

Example 4.1.2: The Einstein two-body problem (1.1.2)–(1.1.3) can be written in the form (see Exercise 2.1.1)

$$u'' + u = \varepsilon(\tilde{a} + u)^2 \quad \text{for } t > 0,$$

(4.1.24)

and

$$u = \tilde{b} - \tilde{a}, \qquad u' = 0 \quad \text{at } t = 0,$$

(4.1.25)

where $\tilde{a}$ and $\tilde{b}$ are here the previous constants a and b of (1.1.2)–(1.1.3), and the present solution u corresponds to the previous solution less the constant $\tilde{a}$. The problem (4.1.24)–(4.1.25) is of the form (4.1.1)–(4.1.2), with

$$f(\varepsilon, t, u, v) := (\tilde{a} + u)^2, \qquad a := \tilde{b} - \tilde{a}, \qquad b := 0. \quad (4.1.26)$$

The two-variable approach of Section 3.3 can be carried out through $N + 1$ terms $u_0, u_1, \ldots, u_N$, and we can then take $U = U(t, \varepsilon)$ to be defined as

$$U(t, \varepsilon) := \sum_{k=0}^{N} u_k(t, \varepsilon t) \varepsilon^k.$$

(4.1.27)

For $N \geq 1$, a direct calculation gives

$$
U'' + U - \varepsilon(\tilde{a} + U)^2
$$
$$
= (u_{0,tt} + u_0) + \left[u_{1,tt} + u_1 + 2u_{0,t\tau} - (\tilde{a} + u_0)^2 \right] \varepsilon
$$
$$
+ \sum_{k=2}^{N} \left(u_{k,tt} + u_k + 2u_{k-1,t\tau} + u_{k-2,\tau\tau} \right.
$$
$$
\left. - 2\tilde{a} u_{k-1} - \sum_{m=0}^{k-1} u_m u_{k-1-m} \right) \varepsilon^k
$$
$$
+ \left[2u_{N,t\tau} + u_{N-1,\tau\tau} - 2\tilde{a} u_N + \varepsilon u_{N,\tau\tau} \right.
$$
$$
\left. - \varepsilon F_N(t,\varepsilon) - \sum_{k=0}^{N} u_k u_{N-k} \right] \varepsilon^{N+1} \qquad (4.1.28)
$$

for certain well-determined functions F_N that are uniformly bounded for all t and ε in the region (4.1.4), and where each F_N is specified in terms of $u_0, u_1, \ldots, u_N$. For example, there holds

$$
F_N(t,\varepsilon) = \begin{cases} (u_1)^2 & \text{for } N = 1, \\ 2u_1 u_2 + \varepsilon(u_2)^2 & \text{for } N = 2, \\ \left[2u_1 u_3 + (u_2)^2 \right] + 2\varepsilon u_2 u_3 + \varepsilon^2 (u_3)^2 & \text{for } N = 3, \\ \vdots & (\tau = \varepsilon t). \end{cases}
$$
$$
(4.1.29)
$$

The two-variable construction yields the existence of constants C_N such that

$$
|F_N(t,\varepsilon)| \leq C_N \qquad (4.1.30)
$$

for all t and ε satisfying (4.1.4). Moreover, the two-variable construction (3.3.4) implies that the lower-order terms vanish on the right side of (4.1.28), so that we obtain

$$
U'' + U = \varepsilon(\tilde{a} + U)^2 - \rho(t,\varepsilon), \qquad (4.1.31)
$$

with

$$
\rho(t,\varepsilon) \equiv \rho_N(t,\varepsilon):
$$
$$
= -\left[2u_{N,t\tau} + u_{N-1,\tau\tau} - 2\tilde{a} u_N \right.
$$
$$
\left. + \varepsilon u_{N,\tau\tau} - \varepsilon F_N(t,\varepsilon) - \sum_{k=0}^{N} u_k u_{N-k} \right] \varepsilon^{N+1}. \qquad (4.1.32)
$$

Hence, in this case the residual ρ is found to satisfy

$$|\rho(t, \varepsilon)| \leq \text{const. } \varepsilon^{N+1} \qquad (4.1.33)$$

for a fixed constant depending only on N.

Similarly, we find from the two-variable construction the initial results [compare with (4.1.25)]

$$U(0, \varepsilon) = \tilde{b} - \tilde{a} - \alpha(\varepsilon), \qquad U'(0, \varepsilon) = -\beta(\varepsilon) \quad \text{at } t = 0,$$

$$(4.1.34)$$

with residuals α and β given as

$$\alpha(\varepsilon) = 0, \qquad \beta(\varepsilon) = -u_{N,\tau}(0,0)\varepsilon^{N+1}, \qquad (4.1.35)$$

so that α and β are also of order (at least) ε^{N+1}.

Returning to the general case (4.1.22)–(4.1.23), we shall now seek to estimate the difference between the exact solution u and the proposed approximate solution U. We denote this difference, or remainder, as $R = R(t, \varepsilon)$, so that R is defined as

$$R(t, \varepsilon) := u(t, \varepsilon) - U(t, \varepsilon). \qquad (4.1.36)$$

From (4.1.2), (4.1.23), and (4.1.36) it follows that

$$R(0, \varepsilon) = \alpha(\varepsilon), \qquad R'(0, \varepsilon) = \beta(\varepsilon), \qquad (4.1.37)$$

and similarly from (4.1.1), (4.1.22), and (4.1.36) it follows that

$$R'' + R = \rho(t, \varepsilon) + \varepsilon[f(\varepsilon, t, U + R, U' + R') - f(\varepsilon, t, U, U')].$$

$$(4.1.38)$$

In the following we shall assume that the given function $f = f(\varepsilon, t, u, v)$ is jointly continuous with respect to its four arguments for all u and v near $U(t, \varepsilon)$ and $U'(t, \varepsilon)$, respectively, and for all t and ε in the region (4.1.4), and we also assume that f is of class C^2 with respect to the last two variables u and v, with second derivatives that are Lipschitz-continuous, uniformly for t and ε in the region (4.1.4). In many applications, f is a polynomial in the last two variables, with coefficients depending smoothly on t and ε, so that the required regularity is then assured. In any case, certain regularity is required in the construction of U. For example, if U is given as a two-variable $(N + 1)$st-order approximation as in (4.1.27), then we would ordinarily require that f be of class C^{N+2}. Also, the function U is assumed to be smooth. In practice, the regularity of U follows from the construction of U via either the two-variable construction or the Krylov/Bogoliubov construction.

For a smooth function f as considered here, the Taylor/Cauchy formula can be used to write

$$f\left[\varepsilon, t, U(t, \varepsilon) + u, U'(t, \varepsilon) + v\right] - f\left[\varepsilon, t, U(t, \varepsilon), U'(t, \varepsilon)\right]$$
$$= g(t, \varepsilon)u + h(t, \varepsilon)v + k(\varepsilon, t, u, v) \qquad (4.1.39)$$

for all suitable real values of u and v, with

$$g(t, \varepsilon): = \left.\frac{\partial}{\partial u}f(\varepsilon, t, u, v)\right|_{\substack{u = U(t, \varepsilon) \\ v = U'(t, \varepsilon)}}$$

$$\equiv f_u\left[\varepsilon, t, U(t, \varepsilon), U'(t, \varepsilon)\right],$$
$$\qquad (4.1.40)$$
$$h(t, \varepsilon): = \left.\frac{\partial}{\partial v}f(\varepsilon, t, u, v)\right|_{\substack{u = U(t, \varepsilon) \\ v = U'(t, \varepsilon)}}$$

$$\equiv f_v\left[\varepsilon, t, U(t, \varepsilon), U'(t, \varepsilon)\right],$$

and

$$k(\varepsilon, t, u, v): = \int_0^1 (1 - s)\left[f_{uu} \cdot u^2 + 2f_{uv} \cdot uv + f_{vv} \cdot v^2\right] ds,$$
$$\qquad (4.1.41)$$

where f_{uu}, f_{uv}, and f_{vv} are the second-order partial derivatives of $f(\varepsilon, t, u, v)$ with respect to u and v, evaluated on the right side of (4.1.41) at $[\varepsilon, t, U(t, \varepsilon) + su, U'(t, \varepsilon) + sv]$; for example,

$$f_{uu} = \left.\frac{\partial^2}{\partial u_0^2}f(\varepsilon, t, u_0, v_0)\right|_{\substack{u_0 = U(t, \varepsilon) + su \\ v_0 = U'(t, \varepsilon) + sv}} .$$

The derivation of (4.1.39)–(4.1.41) follows from the identity

$$\Phi(1) = \Phi(0) + \Phi'(0) + \int_0^1 (1 - s)\Phi''(s) \, ds,$$

with

$$\Phi(s): = f\left[\varepsilon, t, U(t, \varepsilon) + su, U'(t, \varepsilon) + sv\right],$$

where ε and t are held fixed here.

We put $u = R(t, \varepsilon)$ and $v = R'(t, \varepsilon)$ in (4.1.39), and then the differential equation (4.1.38) can be written as

$$\mathscr{L}R = \rho(t, \varepsilon) + \varepsilon k(\varepsilon, t, R, R'), \qquad (4.1.42)$$

where the operator $\mathscr{L}$ is defined by (4.1.6), with g and h given by (4.1.40). This operator $\mathscr{L}$ is the linearization about $R = 0$ of the corresponding nonlinear operator of (4.1.38).

The representation (4.1.7) can be used to rewrite the initial-value problem of (4.1.37) and (4.1.42) as the equivalent integral equation

$$R(t) = \xi(t, \varepsilon) + \varepsilon \int_0^t K(t, s, \varepsilon)k\left[\varepsilon, s, R(s), R'(s)\right] ds, \quad (4.1.43a)$$

where

$$\xi(t,\varepsilon) := \alpha(\varepsilon)u_1(t,\varepsilon) + \beta(\varepsilon)u_2(t,\varepsilon) + \int_0^t K(t,s,\varepsilon)\rho(s,\varepsilon)\,ds,$$

(4.1.43b)

where we are here suppressing the dependence of $R = R(t,\varepsilon)$ on ε to lighten the notation. Note that the functions u_1, u_2, and K are defined for the operator $\mathscr{L}$ as in Theorem 4.1.1. We shall now see that the Banach/Picard fixed-point theorem can be used to prove that (4.1.43a) has a unique solution R near the origin in function space. The fixed-point theorem will also give a bound on the resulting solution R, and in this way we shall obtain the existence of a unique solution u to the original initial-value problem near the given approximate solution U, and we shall have an estimate on the difference $u - U$ in terms of the given residuals α, β, and ρ.

The equation (4.1.43a) for R is to be studied in the Banach space $\mathscr{V}_\varepsilon$ consisting of all functions R that are of class C^1 with respect to t on the interval

$$0 \le t \le \frac{T}{\varepsilon},$$

equipped with the norm $\|\cdot\|_1$ defined as

$$\|R\|_1 := \max\{\|R\|, \|R'\|\},$$

(4.1.44)

where $\|\cdot\|$ is the maximum norm taken with respect to t over the interval $0 \le t \le T/\varepsilon$,

$$\|\Phi\| := \max_{0 \le t \le \frac{T}{\varepsilon}} |\Phi(t)|, \quad \text{for any continuous function } \Phi.$$

(4.1.45)

Again the notation has been lightened by suppressing the dependence of $\|\cdot\|$ on ε.

We define the operator M on $\mathscr{V}_\varepsilon$ by the formula

$$MR(t) := \text{right side of the equation (4.1.43a)}$$

$$= \xi(t,\varepsilon) + \varepsilon \int_0^t K(t,s,\varepsilon)k[\varepsilon,s,R(s),R'(s)]\,ds$$

(4.1.46)

for any $R \in \mathscr{V}_\varepsilon$, where we again suppress the dependence of $MR(t)$ on ε. Then the equation (4.1.43a) can be written as

$$R = MR,$$

(4.1.47)

so that we seek a fixed point of the mapping M. In the following we shall first restrict R to the unit ball in $\mathscr{V}_\varepsilon$, so that we consider only R satisfying

$$\|R\|_1 \le 1.$$

(4.1.48)

Because of the regularity assumptions on f and U, it follows that the functions f_{uu}, f_{uv}, and f_{vv} appearing on the right side of (4.1.41) are uniformly bounded for $u = R(t)$ and $v = R'(t)$, uniformly for all functions R satisfying (4.1.48). Hence, there is a constant γ_1 such that

$$|k[\varepsilon, t, R(t), R'(t)]| \le \gamma_1 \|R\|_1^2 \qquad (4.1.49)$$

for all t and ε in the region (4.1.4) and for all R satisfying (4.1.48). Similarly, there is a constant γ_2 such that

$$|g(t, \varepsilon)|, |h(t, \varepsilon)| \le \gamma_2 \qquad (4.1.50)$$

uniformly for all t and ε satisfying (4.1.4), where g and h are given by (4.1.40). From (4.1.10), (4.1.11), and (4.1.50) we have

$$\|u_j\|_1 \le e^{2T\gamma_2} \quad \text{for } j = 1, 2, \qquad (4.1.51)$$

and also

$$\|K(\cdot, s, \varepsilon)\|_1 \le 2e^{5T\gamma_2} \qquad (4.1.52)$$

where the last estimate is uniformly valid for all $0 \le s \le T/\varepsilon$ and for all $0 < \varepsilon \le \varepsilon_0$. Finally, it follows from the definition of $\xi(t, \varepsilon)$ given in (4.1.43b), along with (4.1.51)–(4.1.52), that

$$\|\xi(\cdot, \varepsilon)\|_1 \le \delta(\varepsilon), \qquad (4.1.53)$$

where the quantity $\delta(\varepsilon)$ can be taken to be

$$\delta(\varepsilon) := e^{2T\gamma_2}\big(|\alpha(\varepsilon)| + |\beta(\varepsilon)| + \varepsilon^{-1}2T\|\rho(\cdot, \varepsilon)\|e^{3T\gamma_2}\big).$$

$$(4.1.54)$$

These estimates imply that *the operator M maps the ball*

$$B_r := \{ R \in \mathcal{V}_\varepsilon \mid \|R\|_1 \le r \}$$

into itself if the radius r is suitably chosen. Indeed, if we take

$$r := 2\delta(\varepsilon), \qquad (4.1.55)$$

where $\delta(\varepsilon)$ is given by (4.1.54), and if we restrict R to the resulting ball B_r, then (4.1.46), (4.1.49), (4.1.51), and (4.1.52) give directly the result

$$\|MR\|_1 \le \tfrac{1}{2}r + 2T\gamma_1 e^{5T\gamma_2} r^2 \quad \text{if } \|R\|_1 \le r. \qquad (4.1.56)$$

We assume here that the residuals are such that

$$\delta(\varepsilon) \to 0 \quad \text{as } \varepsilon \to 0+, \qquad (4.1.57)$$

so that, in particular, the choice $r = 2\delta(\varepsilon)$ will yield a ball B_r that is a subset of the unit ball (4.1.48) provided that ε is small. Using (4.1.57), we now assume that ε is small enough to yield the condition

$$\delta(\varepsilon) \le \frac{e^{-5T\gamma_2}}{8T\gamma_1}. \qquad (4.1.58)$$

From (4.1.55), (4.1.56), and (4.1.58) follows directly the desired result

$$\|MR\|_1 \le r \quad \text{if} \quad \|R\|_1 \le r, \tag{4.1.59}$$

so that M does indeed map such a ball into itself.

We now prove that M is a contracting map. To this end, for any two elements R_1 and R_2 in B_r, we use (4.1.46) to compute the difference $MR_1 - MR_2$ as

$$MR_1(t) - MR_2(t)$$

$$= \varepsilon \int_0^t K(t, s, \varepsilon) \{ k [\varepsilon, s, R_1(s), R_1'(s)] - k [\varepsilon, s, R_2(s), R_2'(s)] \} \, ds. \tag{4.1.60}$$

On the other hand, from (4.1.41) we see that $k(\varepsilon, t, R, R')$ exhibits a quasi-quadratic dependence on R and R', and in fact (4.1.41), along with the assumed regularity of f, yields the existence of a positive constant γ_3 such that (see Exercises 4.1.1 and 4.1.2)

$$\left| k [\varepsilon, s, R_1(t), R_1'(t)] - k [\varepsilon, s, R_2(t), R_2'(t)] \right|$$

$$\le \gamma_3 r \|R_1 - R_2\|_1 \quad \text{for} \quad R_1, R_2 \in B_r, \tag{4.1.61}$$

uniformly for all t and ε satisfying (4.1.4). These two results (4.1.60) and (4.1.61) yield now

$$\|MR_1 - MR_2\|_1 \le 2T\gamma_3 e^{5T\gamma_2} r \|R_1 - R_2\|_1 \tag{4.1.62}$$

for $R_1, R_2 \in B_r$, so that

$$\|MR_1 - MR_2\|_1 \le \tfrac{1}{2}\|R_1 - R_2\|_1 \quad \text{for} \quad R_1, R_2 \in B_r \tag{4.1.63}$$

provided that we impose the further condition

$$\delta(\varepsilon) \le e^{-5T\gamma_2}/8T\gamma_3. \tag{4.1.64}$$

We assume that (4.1.64) holds, and then M is a contraction map.

Putting these results together, we have the following theorem.

Theorem 4.1.2 (Mahony 1972; Kollett 1974; Greenlee and Snow 1975):
Let the problem (4.1.1)–(4.1.2) be regular (with, for example, f of class C^3), and let U be a smooth solution of (4.1.22)–(4.1.23) for all t and ε satisfying (4.1.4), with residuals α, β, and ρ satisfying

$$|\alpha(\varepsilon)| + |\beta(\varepsilon)| + \frac{\|\rho(\cdot, \varepsilon)\|}{\varepsilon} \to 0 \quad \text{as } \varepsilon \to 0 + . \tag{4.1.65}$$

Then there are positive constants C_1 and ε_1 such that (4.1.1)–(4.1.2) has a *unique solution* $u = u(t, \varepsilon)$ *for all*

$$0 \le t \le T/\varepsilon, \qquad 0 < \varepsilon \le \varepsilon_1,$$

and u can be approximated by U as

$$\|u - U\|_1 \le C_1\left(|\alpha(\varepsilon)| + |\beta(\varepsilon)| + \frac{\|\rho(\cdot, \varepsilon)\|}{\varepsilon}\right) \qquad (4.1.66)$$

for $0 < \varepsilon \le \varepsilon_1$, *where the norms are given by* (4.1.44) *and* (4.1.45).

Proof: We need only take ε_1 small enough with (4.1.54) and (4.1.65) so that the inequalities (4.1.58) and (4.1.64) both hold for $0 < \varepsilon \le \varepsilon_1$, and then the stated result follows directly from the foregoing discussion, along with Theorem 0.4.1, and the basic uniqueness result from the theory of ordinary differential equations. ∎

A result similar to Theorem 4.1.2 was given by Mahony (1972), and independently by Kollett (1974), for the case in which $f(\varepsilon, t, u, v)$ depends only on u and v. Greenlee and Snow (1975) observed that Kollett's result is valid without this latter restriction. Smith (1975a) gave results equivalent to Theorem 4.1.2 for certain special classes of functions f. Results related to Theorem 4.1.2 had appeared earlier in the literature on Krylov/Bogoliubov/Mitropolsky averaging; see Perko (1968).

Theorem 4.1.2 can be used to justify the validity of the multiscale technique and also the Krylov/Bogoliubov/Mitropolsky technique. This justification is given in the next section for the multiscale technique.

Exercises

Exercise 4.1.1: Let $\Psi = \Psi(u)$ be a given real-valued function, defined and Lipschitz-continuous on the interval $|u| \le r$, with Lipschitz constant L,

$$|\Psi(u_1) - \Psi(u_2)| \le L|u_1 - u_2| \quad \text{for } |u_1|, |u_2| \le r. \qquad (4.1.67)$$

Let a function Φ be defined as

$$\Phi(u) := \Psi(u)u^2 \quad \text{for } |u| \le r. \qquad (4.1.68)$$

Prove that Φ satisfies an inequality of the form

$$|\Phi(u_1) - \Phi(u_2)| \le \text{const. } r|u_1 - u_2| \quad \text{for } |u_1|, |u_2| \le r. \qquad (4.1.69)$$

Hint: $\Phi(u_1) - \Phi(u_2) = \Psi(u_1)(u_1 + u_2)(u_1 - u_2) + [\Psi(u_1) - \Psi(u_2)]$ $(u_2)^2$, from which (4.1.69) follows with const. $= Lr + 2C$, where C is a bound on Ψ, $|\Psi(u)| \le C$ for $|u| \le r$.

Exercise 4.1.2: Derive the inequality (4.1.61). *Hint*: A generalization of Exercise 4.1.1 will be helpful.

4.2 The validity of two-timing

The estimate of Theorem 4.1.2 is used in this section to examine the validity of two-timing for the initial-value problem

$$u'' + u = \varepsilon f(\varepsilon, t, u, u') \quad \text{for } t > 0, \qquad (4.2.1)$$

with

$$u = a \quad \text{and} \quad u' = b \quad \text{at } t = 0. \qquad (4.2.2)$$

We follow the presentation of Greenlee and Snow (1975), and so we assume that the function f has the form

$$f(\varepsilon, t, u, v) := \sum_{m+n=1}^{l} f_{m,n}(\varepsilon, t) u^m v^n \qquad (4.2.3)$$

for some positive integer l and some suitable real-valued functions $f_{m,n}(\varepsilon, t)$ that are assumed to be continuous for $0 \leq \varepsilon \leq \varepsilon_0$, $t \geq 0$ (some fixed positive ε_0), and 2π-periodic in t. Moreover, each $f_{m,n}$ is assumed to have an asymptotic expansion in powers of ε with coefficients that are continuous 2π-periodic functions of t. The two-variable approach of Chapter 3 then permits construction of an asymptotic expansion for the solution u as

$$u(t, \varepsilon) \sim \sum_{k=0}^{\infty} u_k(t, \tau) \varepsilon^k, \qquad \tau := \varepsilon t. \qquad (4.2.4)$$

See Greenlee and Snow (1975) for the details of the construction; the general approach is similar to that given in Section 3.4 for the special case in which each $f_{m,n}$ is a constant. The 2π-periodicity in t of $f_{m,n}(\varepsilon, t)$ is used in writing the Lindstedt/Poincaré conditions involving suitable integral expressions over $0 \leq t \leq 2\pi$ analogous to (3.4.22); see Example 4.4.1.

The leading term $u_0(t, \tau)$ is found to exist as a smooth function for $t \geq 0$, $0 \leq \tau \leq T$, for some positive T (and possibly for $T = \infty$, though we do not use this latter possibility here), and then the later terms $u_1, u_2, \ldots$ are constructed recursively (as solutions of linear problems) and are also found to be smooth functions for $t \geq 0$, $0 \leq \tau \leq T$. We then define $U_N = U_N(t, \varepsilon)$ as

$$U_N(t, \varepsilon) := \sum_{k=0}^{N} u_k(t, \varepsilon t) \varepsilon^k \quad \text{for } 0 \leq t \leq \frac{T}{\varepsilon}. \qquad (4.2.5)$$

The two-variable construction guarantees that U_N satisfies [compare with (4.1.31)–(4.1.35)]

$$U_N'' + U_N = \varepsilon f\left(\varepsilon, t, U_N, U_N'\right) - \rho_N(t, \varepsilon) \quad \text{for } 0 \le t \le \frac{T}{\varepsilon}, \quad (4.2.6)$$

and

$$U_N(0, \varepsilon) = a - \alpha_N(\varepsilon), \qquad U_N'(0, \varepsilon) = b - \beta_N(\varepsilon) \quad \text{at } t = 0,$$

$$(4.2.7)$$

for suitable residuals α_N, β_N, and ρ_N that satisfy

$$|\alpha_N(\varepsilon)| + |\beta_N(\varepsilon)| + \frac{\|\rho_N(\cdot, \varepsilon)\|}{\varepsilon} \le C_N \varepsilon^N \qquad (4.2.8)$$

for a fixed constant C_N as $\varepsilon \to 0+$, where the norm here is the maximum norm with respect to t over the interval $0 \le t \le T/\varepsilon$. [In fact, we have even more for $\alpha_N(\varepsilon)$ and $\beta_N(\varepsilon)$, namely, $\alpha_N(\varepsilon) = 0$ and $\beta_N(\varepsilon) = -u_{N,\tau}(0,0)\varepsilon^{N+1}$.]

It follows directly from (4.2.1)–(4.2.8) and Theorem 4.1.2 that the given problem (4.2.1)–(4.2.3) has a unique solution $u = u(t, \varepsilon)$ for $0 \le t \le T/\varepsilon$. Moreover, it follows that there are constants $\kappa_1, \kappa_2, \ldots$ such that

$$\|u - U_N\|_1 \le \kappa_N \varepsilon^N \quad \text{as } \varepsilon \to 0+, \quad \text{for } N \ge 1, \qquad (4.2.9)$$

where the norm $\|\cdot\|_1$ is defined by (4.1.44)–(4.1.45). Hence, we have the results of the following theorem.

Theorem 4.2.1 (Kollett 1974; Greenlee and Snow 1975): *There are positive constants ε_1 and T such that the initial-value problem (4.2.1)–(4.2.3) has a unique solution $u(t, \varepsilon)$ for $0 \le t \le T/\varepsilon$, $0 \le \varepsilon \le \varepsilon_1$. Moreover, for $N = 0, 1, 2, \ldots$, there are positive constants D_N such that*

$$|u(t, \varepsilon) - U_N(t, \varepsilon)| + |u'(t, \varepsilon) - U_N'(t, \varepsilon)| \le D_N \varepsilon^{N+1} \qquad (4.2.10)$$

uniformly for $0 \le t \le T/\varepsilon$, $0 \le \varepsilon \le \varepsilon_1$, where the two-variable approximation U_N of (4.2.5) exists for all such t and ε.

Proof: The preceding discussion, along with Theorem 4.1.2, gives directly the stated results of Theorem 4.2.1 for $N \ge 1$, *but with ε^{N+1} replaced by ε^N on the right side of (4.2.10).* That is, the discussion yields

$$\|u - U_N\|_1 \le \text{const. } \varepsilon^N \quad \text{for } 0 \le \varepsilon \le \varepsilon_1, N \ge 1, \qquad (4.2.11)$$

where the constant depends on N. (A direct argument shows that the case $\varepsilon = 0$ can be included here.) But from (4.2.5) we have

$$U_{N+1} = U_N + u_{N+1}\varepsilon^{N+1},$$

from which there follows

$$\|u - U_N\|_1 \leq \|u - U_{N+1}\|_1 + \|u_{N+1}\|_1\varepsilon^{N+1}. \qquad (4.2.12)$$

The two-variable construction gives $\|u_{N+1}\|_1 \leq$ constant, so that the last term on the right side of (4.2.12) is of order ε^{N+1}. But the first term on the right side of (4.2.12) is also of order ε^{N+1}, as follows from (4.2.11) with N replaced by $N + 1$. Hence, (4.2.12) leads to the result

$$\|u - U_N\|_1 \leq \text{const. } \varepsilon^{N+1},$$

where this last result is now valid for all nonnegative integers $N \geq 0$. This completes the proof. ∎

Theorem 4.2.1 was proved by Kollett (1974) for the case in which $f(\varepsilon, t, u, v)$ is independent of ε and t, with each $f_{m,n}(\varepsilon, t)$ a constant in (4.2.3), and also for the case in which f has the special form $f(\varepsilon, t, u, v) = g(t)v$ for a give 2π-periodic function g. Greenlee and Snow (1975) extended the results of Kollett to the case (4.2.3) and pointed out several further generalizations as well; see the remarks found a few paragraphs following (3.4.24). Smith (1975a) independently gave related results that partly overlap Theorem 4.2.1 (see Section 4.3). Mahony (1972) had earlier given the corresponding version of Theorem 4.2.1 using the approximations U_N provided by Krylov/Bogoliubov/Mitropolsky averaging.

The steps in the proof of Theorem 4.2.1 can be followed carefully and examined so as to yield information on the actual *sizes* of the various constants ε_1, T, and D_N. However, in practice it is sometimes more efficient to give an alternative, direct derivation of such estimates as (4.2.10) using a Gronwall-type argument, as illustrated in Section 4.3.

Theorem 4.2.1, based on two-timing, can be extended to functions $f(\varepsilon, t, u, v)$ in (4.2.1), which *need not* be 2π-periodic in t, as illustrated in Examples 3.4.2 and 4.4.3. Also, the preceding development of Theorem 4.2.1 for the scalar equation (4.2.1) carries over directly to oscillatory systems of equations, including nonlinear versions of such systems as illustrated by (3.5.1).

The validity of the estimate (4.2.10) cannot *generally* be extended significantly beyond an expanding interval of the type $0 \leq t \leq T/\varepsilon$. For example, the problem [$f = -2u$ in (4.2.1)]

$$
\begin{aligned}
u'' + (1 + 2\varepsilon)u &= 0 \quad \text{for } t > 0, \\
u(0) &= 1, \quad u'(0) = 0 \quad \text{at } t = 0,
\end{aligned}
\qquad (4.2.13)
$$

has the exact solution

$$u(t, \varepsilon) = \cos\sqrt{1 + 2\varepsilon}\, t, \qquad (4.2.14)$$

whereas the leading two-variable approximation U_0 of (4.2.5) becomes

$$U_0(t, \varepsilon) = \cos(1 + \varepsilon)t. \qquad (4.2.15)$$

The difference

$$R(t, \varepsilon) := u(t, \varepsilon) - U_0(t, \varepsilon)$$

satisfies (4.2.10) for $0 \leq t \leq T/\varepsilon$, with $N = 0$. However, the *period* of U_0 suffers an error of amount order(ε^2) as compared with the period of u, so that (4.2.10) is not valid over a larger interval such as

$$0 \leq t \leq \frac{T}{\varepsilon^2}.$$

Rather, as $t \to \infty$, $R(t, \varepsilon)$ fluctuates between extreme values that are of order unity. Of course, $R(t, \varepsilon)$ remains small, of order ε, for t on the smaller interval $0 \leq t \leq T/\varepsilon$. The length of the interval of validity of (4.2.10) can be significantly extended by introducing further slow time scales into the multivariable expansion, such as $\varepsilon^2 t$ in addition to the earlier $\tau = \varepsilon t$ [compare with (3.1.3)]. Moreover, *global* results are available for all $t \geq 0$ for certain problems with damping, as discussed in Section 4.4.

4.3 Estimates via integral inequalities

In this section we illustrate an approach via integral inequalities coupled with a Gronwall-type argument for justification of the approximation procedures of Chapters 2 and 3 in the study of such initial-value problems as

$$u'' + u = \varepsilon f(\varepsilon, t, u, u') \quad \text{for } t > 0, \qquad (4.3.1)$$

with

$$u = a \quad \text{and} \quad u' = b \quad \text{for } t = 0. \qquad (4.3.2)$$

We again assume that we have a given, proposed approximate solution $U = U(t, \varepsilon)$ that satisfies

$$U'' + U = \varepsilon f(\varepsilon, t, U, U') - \rho(t, \varepsilon) \quad \text{for } t > 0, \qquad (4.3.3)$$

and

$$U(0, \varepsilon) = a - \alpha(\varepsilon), \qquad U'(0, \varepsilon) = b - \beta(\varepsilon), \quad \text{at } t = 0, \qquad (4.3.4)$$

for certain given residuals $\alpha = \alpha(\varepsilon)$, $\beta = \beta(\varepsilon)$, and $\rho = \rho(t, \varepsilon)$ known to be small for all t and ε satisfying

$$0 \le t \le \frac{T}{\varepsilon}, \qquad 0 < \varepsilon \le \varepsilon_0, \tag{4.3.5}$$

for given positive constants T and ε_0.

We again introduce the difference or remainder R defined as

$$R(t, \varepsilon) := u(t, \varepsilon) - U(t, \varepsilon), \tag{4.3.6}$$

and then the problem (4.3.1)–(4.3.2) is equivalent to the problem [see (4.1.37)–(4.1.38)]

$$R'' + R = \varepsilon[f(\varepsilon, t, U + R, U' + R') - f(\varepsilon, t, U, U')] + \rho(t, \varepsilon) \tag{4.3.7}$$

for $t > 0$, and

$$R(0, \varepsilon) = \alpha(\varepsilon), \qquad R'(0, \varepsilon) = \beta(\varepsilon), \quad \text{at } t = 0. \tag{4.3.8}$$

As in Section 4.2, we assume that f, U, and ρ are sufficiently smooth, and then the basic result in the theory of ordinary differential equations guarantees the existence of a unique solution R of (4.3.7)–(4.3.8), at least on some interval $0 \le t < t_1$. We shall now see that a Gronwall-type argument can be used to obtain an a priori estimate for R on an expanding interval $0 \le t \le T/\varepsilon$, and it follows from a known continuation result that the previous solution R, and hence also u, exists and is unique on this expanding interval, so that the previous t_1 can be taken as $t_1 = T/\varepsilon$.

In order to pursue this program, it is possible to rewrite (4.3.7) as in (4.1.42) as

$$R'' - \varepsilon h(t, \varepsilon)R' + [1 - \varepsilon g(t, \varepsilon)]'R = \varepsilon k(\varepsilon, t, R, R') + \rho(t, \varepsilon), \tag{4.3.9}$$

where g, h, and k are given by (4.1.40)–(4.1.41), and then we can consider the resulting integral equation (4.1.43). However, in the present case we can just as well work directly with (4.3.7), and so we replace (4.3.7)–(4.3.8) with the integral equation (see Example 0.1.1)

$$R(t) = \alpha(\varepsilon)\cos t + \beta(\varepsilon)\sin t + \int_0^t \sin(t - s)\rho(s, \varepsilon) \, ds$$

$$+ \varepsilon \int_0^t \sin(t - s)\{ f[\varepsilon, s, U(s, \varepsilon) + R(s), U'(s, \varepsilon) + R'(s)]$$

$$- f[\varepsilon, s, U(s, \varepsilon), U'(s, \varepsilon)]\} \, ds, \tag{4.3.10}$$

where we are here suppressing the dependence of $R = R(t, \varepsilon)$ on ε.

In the following we shall be content to illustrate the present approach in several examples taken from Smith (1975a). [For more general problems, it is sometimes more convenient to work with (4.1.43) rather than (4.3.10).]

Example 4.3.1: As the first example, we return to our old friend the Einstein two-body problem (3.3.1)–(3.3.2) or (4.1.24)–(4.1.25), which we take here as

$$u'' + u = a + \varepsilon u^2 \quad \text{for } t > 0, \qquad (4.3.11)$$

and

$$u(0, \varepsilon) = b, \qquad u'(0, \varepsilon) = 0, \quad \text{at } t = 0. \qquad (4.3.12)$$

For definiteness and simplicity we take the approximate solution U to be given as the Krylov/Bogoliubov first-order approximation (2.3.7), which also agrees with the lowest-order two-variable approximation (3.3.12), so that

$$U(t, \varepsilon) = a + (b - a)\cos(1 - \varepsilon a)t. \qquad (4.3.13)$$

The residuals in (4.3.3)–(4.3.4) are found to be given as

$$\alpha(\varepsilon) = \beta(\varepsilon) = 0,$$

$$\rho(t, \varepsilon) = \varepsilon\left[a^2 + (b - a)^2\cos^2(1 - \varepsilon a)t + \varepsilon a^2(b - a)\cos(1 - \varepsilon a)t\right],$$
$$(4.3.14)$$

and the corresponding integral equation (4.3.10) is

$$R(t) = \int_0^t \sin(t - s)\rho(s, \varepsilon)\, ds$$
$$+ \varepsilon\int_0^t \sin(t - s)\left[2U(s, \varepsilon)R(s) + R(s)^2\right] ds, \qquad (4.3.15)$$

with U and ρ given as in (4.3.13) and (4.3.14), respectively.

The integral involving ρ on the right side of (4.3.15) can be evaluated with (4.3.14), and we find

$$\int_0^t \sin(t - s)\rho(s, \varepsilon)\, ds$$

$$= \varepsilon a^2(1 - \cos t) + \varepsilon a(b - a)\frac{-\cos t + \cos(1 - \varepsilon a)t}{2 - \varepsilon a}$$

$$+ \varepsilon\frac{(b - a)^2}{2}\left[1 - \cos t + \frac{\cos t - \cos[2(1 - \varepsilon a)]t}{(1 - 2\varepsilon a)(3 - 2\varepsilon a)}\right],$$
$$(4.3.16)$$

from which follows the uniform bound

$$\left| \int_0^t \sin(t-s)\rho(s,\varepsilon)\,ds \right| \le C \cdot \varepsilon \quad \text{for } 0 \le \varepsilon \le \varepsilon_1, \text{ all } t,$$

(4.3.17)

for fixed positive constants C and ε_1. For example, we can take

$$C := 2(|a| + |b - a|)^2, \qquad \varepsilon_1 := \frac{1}{3|a|}. \qquad (4.3.18)$$

For the function U, from (4.3.13) we have the bound

$$|U(t,\varepsilon)| \le D := |a| + |b - a| \quad \text{for all } t, \text{ all } \varepsilon. \qquad (4.3.19)$$

From (4.3.15), (4.3.17), and (4.3.19) we find the integral inequality

$$|R(t,\varepsilon)| \le \varepsilon C + 2\varepsilon D \int_0^t |R| + \varepsilon \int_0^t |R|^2 \qquad (4.3.20)$$

for all t and all $0 \le \varepsilon \le \varepsilon_1$. This last integral inequality can be easily resolved by a Gronwall-type argument, and we find from (4.3.20) the result (see Exercise 4.3.1)

$$|R(t,\varepsilon)| \le \frac{2\varepsilon CDe^{2D}}{2D - \varepsilon C(e^{2D} - 1)} \quad \text{for } 0 \le t \le \frac{1}{\varepsilon} \qquad (4.3.21)$$

for all small $\varepsilon > 0$. From (4.3.21) we find also the result

$$|R(t,\varepsilon)| \le 2\varepsilon Ce^{2D} \quad \text{for } 0 \le t \le \frac{1}{\varepsilon}, \qquad (4.3.22)$$

provided that there holds, say,

$$0 < \varepsilon \le \frac{D}{C(e^{2D} - 1)}.$$

In the case of the planet Mercury, the actual numerical values of a, b, and ε imply the results (see Chapter 1)

$$|a| + |b - a| \le 1.02, \qquad 0 < \varepsilon \le 10^{-7} \qquad (4.3.23)$$

and so (4.3.21), along with (4.3.6), (4.3.13), (4.3.18), and (4.3.19), yields for the exact solution u the result

$$|u(t,\varepsilon) - [a + (b - a)\cos(1 - \varepsilon a)t]|$$

$$\le \frac{2\varepsilon(|a| + |b - a|)^2 \exp[2(|a| + |b - a|)]}{1 - \varepsilon(|a| + |b - a|)\{\exp[2(|a| + |b - a|)] - 1\}} \le 17\varepsilon$$

(4.3.24)

for $0 \le t \le 1/\varepsilon$, which corroborates the previously stated estimate (2.3.10).

The integral equation (4.3.15) can be differentiated, and the resulting equation, along with the estimate (4.3.22), yields directly the bound

$$|u'(t, \varepsilon) - U'(t, \varepsilon)| = |u'(t, \varepsilon) + (1 - \varepsilon a)(b - a)\sin(1 - \varepsilon a)t|$$

$$\leq \text{const. } \varepsilon \qquad (4.3.25)$$

uniformly on the stated expanding interval.

As mentioned in the paragraph following (4.3.8), these estimates imply the existence of the exact solution u on the given expanding interval, and the estimates prove that the given function U of (4.3.13) provides a useful approximation to u.

This same approach can be used to prove that the two-variable procedure is asymptotically correct to any order, with [compare with (4.2.10)]

$$|u(t, \varepsilon) - U_N(t, \varepsilon)| + |u'(t, \varepsilon) - U_N'(t, \varepsilon)| \leq \text{const. } \varepsilon^{N+1}$$

$$(4.3.26)$$

uniformly on the expanding interval, for a fixed constant depending on N, where

$$U_N(t, \varepsilon) := \sum_{k=0}^{N} u_k(t, \varepsilon t)\varepsilon^k \qquad (4.3.27)$$

with $u_0, u_1, \ldots$ given by the two-variable construction.

Example 4.3.2: Consider the nonlinearly damped oscillator

$$u'' + u + 2\varepsilon p(\varepsilon t)(u')^3 = 0 \quad \text{for } t \geq 0, \qquad (4.3.28)$$

where p is a given smooth function of a single real variable, say τ, evaluated in (4.3.28) at $\tau = \varepsilon t$. This equation (4.3.28) is of the form (4.3.1), with $f(\varepsilon, t, u, v) := -2p(\varepsilon t)v^3$.

The initial-value problem for (4.3.28) was considered in Section 3.4, where two-timing was used to construct the leading term u_0 in an asymptotic expansion for u. Hence, we take U to be this lowest-order two-variable approximation, given as [see (3.4.41)]

$$U(t, \varepsilon) := \frac{a\cos t + b\sin t}{\left[1 + \frac{3}{2}(a^2 + b^2)\int_0^{\varepsilon t} p(s)\, ds\right]^{1/2}}, \qquad (4.3.29)$$

where the exact solution is to be studied here subject to the initial conditions of (4.3.2). The corresponding residuals in (4.3.3)–(4.3.4) are found to be given as

$$\alpha(\varepsilon) = 0, \qquad \beta(\varepsilon) = \tfrac{3}{4}\varepsilon p(0)a(a^2 + b^2), \qquad (4.3.30)$$

and

$$\rho(t, \varepsilon) = \varepsilon^2 \big[C_1(\varepsilon t, \varepsilon)\cos t + S_1(\varepsilon t, \varepsilon)\sin t \big]$$
$$+ \varepsilon \big[C_3(\varepsilon t, \varepsilon)\cos 3t + S_3(\varepsilon t, \varepsilon)\sin 3t \big] \qquad (4.3.31)$$

for certain known, smooth functions $C_j(\tau, \varepsilon)$ and $S_j(\tau, \varepsilon)$ ($j = 1, 3$) evaluated at $\tau = \varepsilon t$ in (4.3.31), where these functions can be given explicitly in terms of the data a, b, and p. The explicit formulas for these functions are somewhat lengthy and need not be given here. However, from the expressions for these functions we find directly the estimates

$$|C_j^{(k)}(\tau, \varepsilon)|, |S_j^{(k)}(\tau, \varepsilon)| \le \kappa_1, \quad \text{for } k = 0, 1 \quad \text{and} \quad j = 1, 3,$$

$$(4.3.32)$$

and uniformly for

$$0 \le \tau \le T, \qquad 0 \le \varepsilon \le \varepsilon_1,$$

for certain fixed positive constants κ_1, T, and ε_1 that depend only on the data, and where

$$C^{(k)}(\tau, \varepsilon) \equiv \frac{d^k}{d\tau^k} C(\tau, \varepsilon), \text{ etc.}$$

The integral equation (4.3.10) for the problem (4.3.1)–(4.3.2) gives here

$$R(t) = \beta(\varepsilon)\sin t + \int_0^t \sin(t - s)\rho(s, \varepsilon)\, ds - 2\varepsilon \int_0^t p(\varepsilon s)\sin(t - s)$$
$$\times \big\{ 3[U'(s, \varepsilon)]^2 R'(s) + 3U'(s, \varepsilon)[R'(s)]^2 + [R'(s)]^3 \big\}\, ds$$

$$(4.3.33)$$

and

$$R'(t) = \beta(\varepsilon)\cos t + \int_0^t \cos(t - s)\rho(s, \varepsilon)\, ds - 2\varepsilon \int_0^t p(\varepsilon s)\cos(t - s)$$
$$\times \big\{ 3[U'(s, \varepsilon)]^2 R'(s) + 3U'(s, \varepsilon)[R'(s)]^2 + [R'(s)]^3 \big\}\, ds,$$

$$(4.3.34)$$

where this latter result for R' is obtained by differentiating (4.3.33). We can now use (4.3.33)–(4.3.34) to obtain useful estimates for R and R'.

From (4.3.31)–(4.3.32) we find [compare with (4.3.17)]

$$\left| \int_0^t \sin(t - s)\rho(s, \varepsilon)\, ds \right| \le \kappa_2 \cdot \varepsilon,$$

$$(4.3.35)$$

$$\left| \int_0^t \cos(t - s)\rho(s, \varepsilon)\, ds \right| \le \kappa_2 \cdot \varepsilon,$$

for a fixed constant κ_2 and uniformly for $0 \le t \le T/\varepsilon$, $0 \le \varepsilon \le \varepsilon_1$. In the derivation of (4.3.35), we use such results as

$$\int_0^t [\sin(t - s)\sin 3s]\,h(\varepsilon s)\,ds$$

$$= \tfrac{1}{8}[3h(0)\sin t - h(\varepsilon t)\sin 3t]$$

$$+ \tfrac{1}{8}\varepsilon \int_0^t [\cos(t - s)\sin 3s + 3\sin(t - s)\cos 3s]\,h'(\varepsilon s)\,ds$$

$$(4.3.36)$$

for any smooth function h, where such results follow by several integrations by parts. Then (4.3.29), (4.3.30), (4.3.33), and (4.3.35) give

$$|R(t, \varepsilon)|, |R'(t, \varepsilon)| \le \varepsilon\kappa_3\left(1 + \int_0^t |R'| + \int_0^t |R'|^2 + \int_0^t |R'|^3\right)$$

$$(4.3.37)$$

for a fixed constant κ_3 and uniformly for $0 \le t \le T/\varepsilon$, $0 \le \varepsilon \le \varepsilon_1$.

The inequality for R' in (4.3.37) can be resolved by a Gronwall-type argument (Exercise 4.3.2), and we find

$$|R'(t, \varepsilon)| \le \text{const. } \varepsilon \qquad (4.3.38)$$

for all small $\varepsilon > 0$ and all t on the given expanding interval. This estimate for R' can then be used in the right side in the inequality of (4.3.37) for R, and we find directly

$$|R(t, \varepsilon)| \le \text{const. } \varepsilon, \qquad (4.3.39)$$

again uniformly on the given expanding interval.

As mentioned earlier, these estimates prove the existence of the exact solution u on the expanding interval, and we have

$$|u(t, \varepsilon) - U(t, \varepsilon)| + |u'(t, \varepsilon) - U'(t, \varepsilon)| \le \text{const. } \varepsilon \quad (4.3.40)$$

for all small $\varepsilon > 0$ and all t on the expanding interval. Hence, the function U of (4.3.29) provides a useful approximation to the (unknown) exact solution. Again, this same approach can be used to prove that the two-variable expansion is correct to any order, as in (4.3.26).

Example 4.3.3: As our final illustration of the present approach, we consider the coupled system of differential equations for the compound pendulum considered in Section 3.5, namely, the system

$$\frac{d^2\Theta}{dt^2} + \kappa^2\Theta = \varepsilon^2\kappa^2(-2\Theta + \Phi - \Psi),$$

$$\frac{d^2\Phi}{dt^2} + \kappa^2(\Phi - \Theta) = -\varepsilon^2\kappa^2(-2\Theta + \Phi - \Psi), \qquad (4.3.41)$$

$$\frac{d^2\Psi}{dt^2} + \kappa^2(\Psi + \Theta) = \varepsilon^2\kappa^2(-2\Theta + \Phi - \Psi),$$

where κ is a fixed positive constant. We impose the earlier initial conditions of (3.5.2), and then we consider the proposed approximate solution functions $\Theta^* = \Theta^*(t, \varepsilon)$, $\Phi^* = \Phi^*(t, \varepsilon)$, and $\Psi^* = \Psi^*(t, \varepsilon)$ provided by the multiscale technique, given by (3.5.20) as

$$\Theta^*(t, \varepsilon) = -\frac{\varepsilon\alpha}{\sqrt{2}}\sin\frac{\kappa\varepsilon t}{\sqrt{2}}\sin\kappa t,$$

$$\Phi^*(t, \varepsilon) = \frac{\alpha}{2}\left(1 - \cos\frac{\kappa\varepsilon t}{\sqrt{2}}\right)\cos\kappa t, \qquad (4.3.42)$$

$$\Psi^*(t, \varepsilon) = \frac{\alpha}{2}\left(1 + \cos\frac{\kappa\varepsilon t}{\sqrt{2}}\right)\cos\kappa t.$$

In this case we use the fact (obtained, for example, by variation of parameters) that the unique solution functions R, S, T of the system

$$\frac{d^2R}{dt^2} + \kappa^2 R = f(t),$$

$$\frac{d^2S}{dt^2} + \kappa^2(S - R) = g(t), \qquad (4.3.43)$$

$$\frac{d^2T}{dt^2} + \kappa^2(T + R) = h(t),$$

satisfying homogeneous initial conditions can be represented as

$$R(t) = \frac{1}{\kappa}\int_0^t \sin\kappa(t - s)f(s)\,ds, \qquad (4.3.44)_R$$

$$S(t) = \frac{1}{\kappa}\int_0^t \sin\kappa(t - s)g(s)\,ds$$

$$+ \frac{1}{2\kappa}\int_0^t \{\sin^3\kappa(t - s) - \cos\kappa(t - s)$$

$$\times [\kappa(t - s) - \tfrac{1}{2}\sin 2\kappa(t - s)]\}f(s)\,ds, \qquad (4.3.44)_S$$

and

$$T(t) = \frac{1}{\kappa}\int_0^t \sin\kappa(t - s)h(s)\,ds$$

$$- \frac{1}{2\kappa}\int_0^t \{\sin^3\kappa(t - s) - \cos\kappa(t - s)$$

$$\times [\kappa(t - s) - \tfrac{1}{2}\sin 2\kappa(t - s)]\}f(s)\,ds. \qquad (4.3.44)_T$$

If we define remainder functions R, S, and T as

$$R(t, \varepsilon) := \frac{1}{\varepsilon} [\Theta(t, \varepsilon) - \Theta^*(t, \varepsilon)],$$

$$S(t, \varepsilon) := \frac{1}{\varepsilon} [\Phi(t, \varepsilon) - \Phi^*(t, \varepsilon)], \qquad (4.3.45)$$

$$T(t, \varepsilon) := \frac{1}{\varepsilon} [\Psi(t, \varepsilon) - \Psi^*(t, \varepsilon)],$$

then a direct calculation shows that these functions R, S, T satisfy homogeneous initial conditions along with the system (4.3.43), with

$$f(t) = \frac{3\kappa^2 \alpha \varepsilon^2}{2\sqrt{2}} \sin \frac{\kappa \varepsilon t}{\sqrt{2}} \sin \kappa t - \kappa^2 \varepsilon^2 (2R - S + T),$$

$$g(t) = \frac{3\kappa^2 \alpha \varepsilon}{4} \cos \frac{\kappa \varepsilon t}{\sqrt{2}} \cos \kappa t - \frac{2\kappa^2 \alpha \varepsilon^2}{\sqrt{2}} \sin \frac{\kappa \varepsilon t}{\sqrt{2}} \sin \kappa t$$

$$+ \kappa^2 \varepsilon^2 (2R - S + T), \qquad (4.3.46)$$

$$h(t) = -\frac{3\kappa^2 \alpha \varepsilon}{4} \cos \frac{\kappa \varepsilon t}{\sqrt{2}} \cos \kappa t + \frac{2\kappa^2 \alpha \varepsilon^2}{\sqrt{2}} \sin \frac{\kappa \varepsilon t}{\sqrt{2}} \sin \kappa t$$

$$- \kappa^2 \varepsilon^2 (2R - S + T).$$

Hence, we have the representation (4.3.44) in terms of these particular functions f, g, and h, and this representation leads directly to the inequalities

$$|R(t, \varepsilon)| \leq \frac{3\kappa |\alpha| \tau_1}{2\sqrt{2}} \varepsilon + \kappa \varepsilon^2 \int_0^t (2|R| + |S| + |T|) \, ds$$

$$(4.3.47)$$

and

$$|S(t, \varepsilon)|, |T(t, \varepsilon)| \leq \left(\frac{3}{4} + \frac{3\kappa \tau_1}{8\sqrt{2}} + \frac{5\varepsilon}{\sqrt{2}} \right) \kappa |\alpha| \tau_1$$

$$+ \kappa \varepsilon \left(\frac{\kappa \tau_1}{2} + 2\varepsilon \right) \int_0^t (2|R| + |S| + |T|) \, ds$$

$$(4.3.48)$$

for any fixed positive τ_1 and for all $0 \leq \varepsilon t \leq \tau_1$. The inequalities of (4.3.47)–(4.3.48) can be added to give

$$2|R(t, \varepsilon)| + |S(t, \varepsilon)| + |T(t, \varepsilon)| \leq \mu + \nu \int_0^t (2|R| + |S| + |T|) \, ds$$

$$(4.3.49)$$

for suitable fixed constants μ and ν, and for all small $\varepsilon > 0$. For example, if we take

$$\mu = \frac{\kappa |\alpha| \tau_1}{5} (10 + 3\kappa \tau_1) \quad \text{and} \quad \nu = \kappa(1 + \kappa \tau_1),$$

then we find that (4.3.49) holds for, say, $0 < \varepsilon \leq 0.05$.

From (4.3.49) and Gronwall's inequality, we find directly the estimate

$$2|R(t,\varepsilon)| + |S(t,\varepsilon)| + |T(t,\varepsilon)| \leq \mu \cdot e^{\nu \tau_1} \qquad (4.3.50)$$

for $0 \leq \varepsilon t \leq \tau_1$. This last estimate can be used in the right sides of (4.3.47) and (4.3.48) to yield

$$|R(t,\varepsilon)| \leq \text{const. } \varepsilon \qquad (4.3.51)$$

and

$$|S(t,\varepsilon)|, |T(t,\varepsilon)| \leq \text{constant}, \qquad (4.3.52)$$

where these estimates are uniformly valid for all small $\varepsilon > 0$ and for all $0 \leq t \leq \tau_1/\varepsilon$.

From (4.3.45), (4.3.51), and (4.3.52) we have the error estimates

$$|\Theta(t,\varepsilon) - \Theta^*(t,\varepsilon)| \leq \text{const. } \varepsilon^2,$$
$$|\Phi(t,\varepsilon) - \Phi^*(t,\varepsilon)| \leq \text{const. } \varepsilon, \qquad (4.3.53)$$
$$|\Psi(t,\varepsilon) - \Psi^*(t,\varepsilon)| \leq \text{const. } \varepsilon,$$

which corroborate (3.5.21).

This same approach can be used to prove the asymptotic correctness of the two-variable or Krylov/Bogoliubov/Mitropolsky expansions for $\Theta(t, \varepsilon)$, $\Phi(t, \varepsilon)$, and $\Psi(t, \varepsilon)$ to any order.

Exercises

Exercise 4.3.1: Show that the integral inequality

$$|R(t,\varepsilon)| \leq \varepsilon C + 2\varepsilon D \int_0^t |R| + \varepsilon \int_0^t |R|^2 \quad \text{for } t \geq 0, \varepsilon \geq 0$$

implies the further inequality

$$|R(t,\varepsilon)| \leq \frac{2\varepsilon C D e^{2D}}{2D - \varepsilon C(e^{2D} - 1)} \quad \text{for } 0 \leq t \leq \frac{1}{\varepsilon},$$

where C and D are fixed positive constants. *Hint:* Put

$$S(t) := C + 2D \int_0^t |R| + \int_0^t |R|^2,$$

and find $S'(t) \leq \varepsilon S \cdot (2D + \varepsilon S)$. This differential inequality can be integrated (by partial fractions) along with the initial condition $S(0) = C$, and the stated result follows directly.

Exercise 4.3.2: Use a Gronwall-type argument as in Exercise 4.3.1 to show that the inequalities

$$0 \le w(t) \le \varepsilon E\left[1 + \int_0^t (w + w^2 + w^3)\right] \quad \text{for } t, \varepsilon \ge 0$$

imply a bound of the type

$$|w(t)| \le \text{const. } \varepsilon,$$

uniformly for all small nonnegative ε and all t in a suitable expanding interval. Here, E is a fixed positive constant, and the resulting inequality here yields (4.3.38). [See, for example, Birkhoff and Rota (1960, 1978) for a more general discussion on such *comparison results* for differential inequalities.]

Exercise 4.3.3: Consider two weakly inductively coupled LC circuits as discussed in Section 2.5, modelled by the system

$$Q_1''(t) + k^2 Q_1(t) = \varepsilon k^2 Q_2(t),$$
$$Q_2''(t) + k^2 Q_2(t) = \varepsilon k^2 Q_1(t) \quad \text{for } t > 0,$$

along with the initial conditions

$$Q_1(0) = q, \qquad Q_1' = 0,$$
$$Q_2(0) = 0, \qquad Q_2' = 0 \quad \text{at } t = 0.$$

Let $q_1(t, \varepsilon)$ and $q_2(t, \varepsilon)$ be the first-order Krylov/Bogoliubov approximations found in Section 2.5, given by (2.5.13). (These same functions q_1, q_2 can also be obtained by the two-variable method as in Exercise 3.5.1.) Derive the estimates [compare with (2.5.14)]

$$|Q_j(t, \varepsilon) - q_j(t, \varepsilon)| \le \tfrac{1}{2}(e^{kT} - 1)|q\varepsilon| \quad \text{for } 0 \le \varepsilon t \le T$$

for $j = 1, 2$ and for any fixed positive constant T. (Similar estimates can be obtained for the derivatives.) *Hint:* Use a suitable integral representation along with a Gronwall-type argument.

4.4 Global estimates for damped oscillators

The linearly damped problem

$$u'' + 2\varepsilon u' + u = 0 \quad \text{for } t > 0,$$
$$u(0) = a, \qquad u'(0) = b \quad \text{at } t = 0 \tag{4.4.1}$$

has a solution that decays toward zero with increasing t $(t \to \infty)$ for fixed $\varepsilon > 0$, whereas the solution becomes exponentially unbounded with increasing t for fixed $\varepsilon < 0$. For the case of positive damping ($\varepsilon > 0$), Reiss (1971) showed that the two-timing expansion

$$u(t, \varepsilon) \sim \sum_{k=0}^{\infty} u_k(t, \varepsilon t) \varepsilon^k \qquad (4.4.2)$$

obtained in Section 3.2 [see (3.2.11)] provides an asymptotic expansion for the exact solution that is valid for all $t \geq 0$. That is, there is a positive constant C_N such that

$$|u(t, \varepsilon) - U_N(t, \varepsilon)| \leq C_N \varepsilon^{N+1} \quad \text{for all } t \geq 0 \qquad (4.4.3)$$

and for all small nonnegative ε, where

$$U_N(t, \varepsilon) := \sum_{k=0}^{N} u_k(t, \varepsilon t) \varepsilon^k. \qquad (4.4.4)$$

We can similarly obtain an estimate of the type (4.4.3), with u and U_N replaced respectively by the derivatives u' and U_N', so that in this case the previous estimate (4.2.10) of Theorem 4.2.1 remains valid not only on an expanding (but compact) interval but also for all $t \geq 0$.

For the more general problem

$$u'' + u = \varepsilon f(\varepsilon, t, u, u') \qquad \text{for } t > 0, \qquad (4.4.5)$$

with

$$u(0) = a, \qquad u'(0) = b \quad \text{at } t = 0, \qquad (4.4.6)$$

Greenlee and Snow (1975) considered three subcases of (4.4.5)–(4.4.6) *with damping* and obtained results of the type

$$|u(t, \varepsilon) - U_N(t, \varepsilon)| + |u'(t, \varepsilon) - U_N'(t, \varepsilon)| \leq C_N \varepsilon^{N+1} \sigma(\varepsilon t) \qquad (4.4.7)$$

for all $t \geq 0$, where in each case $\sigma = \sigma(\tau)$ is a specified function that tends to zero at a known rate as $\tau = \varepsilon t \to \infty$. These results of Greenlee and Snow will be described briefly in the following.

First consider the special case of (4.4.5) with

$$f(\varepsilon, t, u, v) := -h(t)u - g(t)v, \qquad (4.4.8)$$

so that (4.4.5) becomes

$$u'' + \varepsilon g(t)u' + [1 + \varepsilon h(t)]u = 0 \quad \text{for } t > 0. \qquad (4.4.9)$$

The functions g and h are given 2π-periodic functions with g continuous and h continuously differentiable, subject to the additional assumptions

$$\int_0^{2\pi} g(t)\, dt > 0, \qquad (4.4.10)$$

and

$$(1 + \varepsilon h)^{-1} h' + 2g \geq 0 \qquad (4.4.11)$$

for all sufficiently small $\varepsilon > 0$. The condition (4.4.11) implies

$$h' + 2g \geq 0, \qquad (4.4.12)$$

while the stronger condition

$$h' + 2g > 0 \qquad (4.4.13)$$

implies also (4.4.11).

The periodicity of g, along with (4.4.10), guarantees that the linear equation (4.4.9) experiences *positive damping on the average*. This equation (4.4.9) includes the equation (4.4.1) as a special case.

Theorem 4.4.1 (Greenlee and Snow 1975): *Let u be the solution of (4.4.6) and (4.4.9), and let U_N be the $(N + 1)$st-order two-timing approximation (4.4.4) obtained as in Section 4.2, where g and h are 2π-periodic, with g continuous and h continuously differentiable. Then, for all sufficiently small $\varepsilon > 0$, (i) under assumptions (4.4.10)–(4.4.11) there is a constant C_N such that (4.4.7) holds for all $t \geq 0$, with*

$$\sigma(\tau): = 1 \quad \text{for all } \tau,$$

and (ii) under assumptions (4.4.10) and (4.4.13), and for each μ satisfying

$$0 < 4\mu < \min_{0 \leq t \leq 2\pi} \left[h'(t) + 2g(t) \right], \qquad (4.4.14)$$

there is a constant C_N such that (4.4.7) holds for all $t \geq 0$ with

$$\sigma(\tau): = e^{-\mu\tau}.$$

In both cases (i) and (ii) the approximate solution U_N decays exponentially to zero with increasing t, and then in case (ii) the same result holds also for the exact solution u.

An application of Theorem 4.4.1 is given in Example 4.4.1. We shall also discuss the proof of Theorem 4.4.1, but first we state two additional theorems of Greenlee and Snow involving two other subcases of (4.4.5)–(4.4.6).

As the next subcase of (4.4.5), consider the following autonomous nonlinear equation with at least linear damping:

$$u'' + u + \varepsilon[2cu' + h(\varepsilon, u, u')] = 0 \quad \text{for } t > 0, \qquad (4.4.15)$$

where, as usual, here $\varepsilon > 0$, and where in this case the following conditions of (4.4.16)–(4.4.20) are required to hold:

$$c \text{ is a fixed positive constant}, \qquad (4.4.16)$$

$$h(\varepsilon, u, v): = G(\varepsilon, u)u + H(\varepsilon, u, v)v, \qquad (4.4.17)$$

where $G(\cdot,\cdot)$ and $H(\cdot,\cdot,\cdot)$ are polynomials with $G(\cdot,0)=0$ and $H(\cdot,0,0)=0$,

$$H(\varepsilon,u,v)\geq 0 \quad \text{for all } (\varepsilon,u,v)\in[0,\varepsilon_0]\times\mathbb{R}^2, \quad (4.4.18)$$

$$\int_0^u sG(\varepsilon,s)\,ds\geq 0 \quad \text{for all } (\varepsilon,u)\in[0,\varepsilon_0]\times\mathbb{R}^1, \quad (4.4.19)$$

and

$$\int_0^u s^2\frac{d}{ds}G(\varepsilon,s)\,ds\geq 0 \quad \text{for all } (\varepsilon,u)\in[0,\varepsilon_0]\times\mathbb{R}^1. \quad (4.4.20)$$

Note that (4.4.18) implies that H has no terms of degree one in u and v, and (4.4.19) implies that G has no term of degree one in u. The polynomial dependence of h on ε is not essential.

This problem (4.4.15)–(4.4.20) occurs in nonlinear electric-circuit theory; an example taken from Greenlee and Snow (1975) is included as Example 4.4.2. A prototype of this problem (4.4.15)–(4.4.20) was studied by Baum (1972) without error estimates, in the case $G\equiv 0$ and $H=v^2$.

Theorem 4.4.2 (Greenlee and Snow 1975): *Let u be the solution of* (4.4.6) *and* (4.4.15) *subject to the conditions* (4.4.16)–(4.4.20), *and let U_N be the* $(N+1)$st*-order two-timing approximation* (4.4.4) *obtained as in Section 4.2. Then, for all sufficiently small $\varepsilon>0$, both u and U_N exist for $t\geq 0$, and there is a constant C_N such that* (4.4.7) *holds for $t\geq 0$ with*

$$\sigma(\tau):=(1+\tau)^{N+2}e^{-c\tau}.$$

As pointed out by Greenlee and Snow, analogues of Theorem 4.4.2 can be obtained under weaker conditions on h by restricting the initial values a and b in (4.4.6).

As the final subcase of (4.4.5), consider the following nonlinear autonomous equation with velocity damping:

$$u''+u+\varepsilon(u')^{2\nu+1}=0 \quad \text{for } t>0,\ \varepsilon>0, \quad (4.4.21)$$

where ν is a fixed positive integer. Here the damping is weaker than linear, and the solution has algebraic rather than exponential decay. Two-timing has already been used in Example 3.4.2 for a generalized version of (4.4.21) in the case of cubic damping, $\nu=1$.

Theorem 4.4.3 (Greenlee and Snow 1975): *Let u be the solution of* (4.4.6) *and* (4.4.21), *and let U_N be the* $(N+1)$st*-order two-timing approximation* (4.4.4) *obtained as in Section 3.4. (See also Example 4.3.2.) Then, for sufficiently small $\varepsilon>0$ and for each fixed μ satisfying*

$$0<\mu<\frac{1}{2\nu},$$

there is a constant C_N such that (4.4.7) holds for $t \geq 0$ with

$$\sigma(\tau): = \frac{1}{(1 + \tau)^\mu}.$$

In addition, for each such μ there is a positive constant C_N (different from the previous constant C_N) such that (4.4.7) holds for $t \geq 0$ with the function $\sigma(\tau)$ replaced by the following function σ of t and ε:

$$\sigma = \left(|u(t, \varepsilon)| + |u'(t, \varepsilon)|\right)(1 + \varepsilon t)^\mu. \tag{4.4.22}$$

Note that the last result in Theorem 4.4.3 with σ given by (4.4.22) implies that if $\gamma \geq 0$, then the relative error between (u, u') and (U_N, U_N') is order($\varepsilon^{N+1-\gamma}$) for times of order $\varepsilon^{-(\gamma+\mu)/\mu}$.

In each of these three subcases it is shown in Greenlee and Snow (1975) that the two-timing approximants U_N and the exact solution u exist for all $t \geq 0$. The stated estimate (4.4.7) is obtained with the aid of various auxiliary estimates, such as the energy estimate of the following lemma.

Lemma 4.4.4 (Greenlee and Snow 1975): *Let u be the solution of the linear problem*

$$u'' + \varphi(t)u' + [1 + \psi(t)]u = F(t) \quad \text{for } t > 0,$$
$$u(0) = a, \qquad u'(0) = b \qquad \text{at } t = 0, \tag{4.4.23}$$

where φ, ψ, ψ', and F are continuous on $[0, \infty)$, with $1 + \psi(t) > 0$ for $t \geq 0$, $F/\sqrt{1 + \psi} \in L^1[0, \infty)$, and

$$\frac{\psi'(t)}{1 + \psi(t)} + 2\varphi(t) \geq 0 \quad \text{for } t \geq 0. \tag{4.4.24}$$

Then

$$\left\{ [u(t)]^2 + \frac{[u'(t)]^2}{1 + \psi(t)} \right\}^{1/2}$$
$$\leq \left[a^2 + \frac{b^2}{1 + \psi(0)} \right]^{1/2} + \int_0^\infty \frac{|F(s)|}{\sqrt{1 + \psi(s)}} \, ds$$

holds for all $t \geq 0$.

Proof of Lemma 4.4.4: The function

$$E(t): = [u(t)]^2 + \frac{[u'(t)]^2}{1 + \psi(t)}$$

can be differentiated, and then (4.4.23)–(4.4.24) yield directly the differential inequality

$$E'(t) \le \frac{2u'(t)F(t)}{1 + \psi(t)} \le \frac{2\sqrt{E(t)}\,|F(t)|}{\sqrt{1 + \psi(t)}}.$$

This differential inequality can be integrated easily to yield the stated result of the lemma. ∎

Proof of Theorem 4.4.1: As in Example 4.1.2, the two-timing construction shows that U_N satisfies [see Greenlee and Snow (1975) for details]

$$U_N'' + \varepsilon g(t)U_N' + [1 + \varepsilon h(t)]U_N = -\rho_N(t, \varepsilon) \quad \text{for } t > 0,$$

$$(4.4.25)$$

and

$$U_N(0, \varepsilon) = a - \alpha_N(\varepsilon), \qquad U_N'(0, \varepsilon) = b - \beta_N(\varepsilon) \quad \text{at } t = 0,$$

$$(4.4.26)$$

with residuals α_N, β_N, and ρ_N such that

$$\alpha_N(\varepsilon) = 0, \qquad \beta_N(\varepsilon) = -u_{N,\tau}(0,0)\varepsilon^{N+1}, \qquad (4.4.27)$$

and

$$|\rho_N(t, \varepsilon)| \le C_N \varepsilon^{N+1}(1 + \varepsilon t)^{2N+1}\exp[\varepsilon t(\operatorname{Re}\lambda)] \qquad (4.4.28)$$

for a fixed positive constant C_N and a constant λ given as

$$\lambda := \frac{1}{4\pi}\left\{-\int_0^{2\pi}g(t)\,dt + \left[\left(\int_0^{2\pi}[g(t)\cos 2t + h(t)\sin 2t]\,dt\right)^2\right.\right.$$
$$\left.\left. + \left(\int_0^{2\pi}[h(t)\cos 2t - g(t)\sin 2t]\,dt\right)^2 - \left(\int_0^{2\pi}h(t)\,dt\right)^2\right]^{1/2}\right\}.$$

$$(4.4.29)$$

Hence, the remainder, or difference,

$$R(t, \varepsilon) := u(t, \varepsilon) - U_N(t, \varepsilon) \qquad (4.4.30)$$

satisfies

$$R'' + \varepsilon g(t)R' + [1 + \varepsilon h(t)]R = \rho_N(t, \varepsilon) \quad \text{for } t > 0 \qquad (4.4.31)$$

and

$$R(0, \varepsilon) = \alpha_N(\varepsilon), \qquad R'(0, \varepsilon) = \beta_N(\varepsilon) \quad \text{at } t = 0. \qquad (4.4.32)$$

The assumptions (4.4.10) and (4.4.11) can be shown to imply that the parameter λ of (4.4.29) satisfies [see Greenlee and Snow (1975)]

$$\text{Re } \lambda < 0, \qquad (4.4.33)$$

and we then see that Lemma 4.4.4 can be applied to the problem (4.4.31)–(4.4.32) with $\varphi(t) = \varepsilon g(t)$, $\psi(t) = \varepsilon h(t)$, $F(t) = \rho_N(t, \varepsilon)$, $a = \alpha_N(\varepsilon)$, and $b = \beta_N(\varepsilon)$. The resulting estimate given by Lemma 4.4.4 for the function $u = R(t, \varepsilon)$ can be used with (4.4.27) and (4.4.28) to give

$$\left\{ [R(t, \varepsilon)]^2 + \frac{[R'(t, \varepsilon)]^2}{1 + \varepsilon h(t)} \right\}^{1/2}$$

$$\leq \text{const. } \varepsilon^{N+1} \left\{ 1 + \int_0^\infty \frac{(1 + \varepsilon s)^{2N+1}}{\sqrt{1 + \varepsilon h(s)}} \exp[\varepsilon s (\text{Re } \lambda)] \, ds \right\}$$

$$\text{for } t \geq 0 \quad (4.4.34)$$

and for all sufficiently small $\varepsilon > 0$. These last two results lead directly to an estimate of the form

$$|R(t, \varepsilon)| + |R'(t, \varepsilon)| \leq \text{const. } \varepsilon^N \quad \text{for } t \geq 0 \qquad (4.4.35)$$

for all small $\varepsilon \geq 0$ and for a fixed constant depending on N. The previous argument used in going from (4.2.11) to (4.2.10) now allows us to replace ε^N with ε^{N+1} on the right side of (4.4.35) (with a different constant), and this completes the proof of part (i) of Theorem 4.4.1.

To prove part (ii), we put

$$w = w(t, \varepsilon) := e^{\mu \varepsilon t} R(t, \varepsilon) \qquad (4.4.36)$$

and find, with (4.4.31)–(4.4.32),

$$w'' + \varepsilon [g(t) - 2\mu] w' + \left\{ 1 + \varepsilon \left[h(t) + \varepsilon \mu^2 - \varepsilon \mu g(t) \right] \right\} w$$

$$= e^{\mu \varepsilon t} \rho_N(t, \varepsilon) \quad \text{for } t > 0 \qquad (4.4.37)$$

and

$$w(0) = 0, \qquad w'(0) = \varepsilon^{N+1} u_{N,\tau}(0,0) \quad \text{at } t = 0, \qquad (4.4.38)$$

where (4.4.27) has been used. A direct calculation shows that

$$\varphi(t) := \varepsilon [g(t) - 2\mu]$$

and

$$\psi(t) := \varepsilon \left[h(t) + \varepsilon \mu^2 - \varepsilon \mu g(t) \right]$$

satisfy (4.4.24) if μ satisfies (4.4.14). Using (4.4.13)–(4.4.14), we can also show the result [see Greenlee and Snow (1975)]

$$\mu + \operatorname{Re} \lambda < 0, \qquad (4.4.39)$$

which, with (4.4.28), yields

$$\int_0^\infty e^{\varepsilon\mu t} |\rho_N(t, \varepsilon)| \, dt \leq \text{const. } \varepsilon^N. \qquad (4.4.40)$$

Hence, Lemma 4.4.4 can be applied to the problem (4.4.37)–(4.4.38), and we find

$$\left\{ [w(t, \varepsilon)]^2 + \frac{[w'(t, \varepsilon)]^2}{1 + \psi(t)} \right\}^{1/2} \leq \text{const. } \varepsilon^N,$$

or, in terms of R, we find, with (4.4.36),

$$|R(t, \varepsilon)| + |R'(t, \varepsilon)| \leq \text{const. } \varepsilon^N e^{-\mu\varepsilon t} \quad \text{for } t \geq 0. \quad (4.4.41)$$

The usual device can now be used to replace ε^N on the right side of (4.4.41) with ε^{N+1}, and this completes the proof of part (ii) of Theorem 4.4.1. ∎

See Greenlee and Snow (1975) for the proofs of Theorems 4.4.2 and 4.4.3, which are proved much like Theorem 4.4.1.

Example 4.4.1 (Greenlee and Snow 1975): The initial-value problem

$$u'' + \varepsilon(2c + d \sin t)u' + (1 + 2\varepsilon C + \varepsilon D \cos t)u = 0 \quad (4.4.42)$$

for $t > 0$, with

$$u(0) = a, \qquad u'(0) = b \quad \text{at } t = 0, \qquad (4.4.43)$$

represents a model of a linear oscillator that is damped on the average by damping that is out of phase with the restoring force. The quantities a, b, c, d, C, and D are constants, with

$$2c > |d - \tfrac{1}{2}D|. \qquad (4.4.44)$$

This condition (4.4.44) permits the damping coefficient $\varepsilon(2c + d \sin t)$ to assume both positive and negative values for $t \geq 0$, but this damping coefficient is *positive on the average*, so that

$$g(t) := 2c + d \sin t$$

satisfies (4.4.10). Similarly, with $h(t) := 2C + D \cos t$, we find, with (4.4.44), that the previous condition (4.4.13) is satisfied. Hence, this problem (4.4.42)–(4.4.44) is of the type covered by Theorem 4.4.1.

The functions $u_k(t, \tau)$ in the two-variable expansion (4.4.2) are easily seen to satisfy [compare with (3.3.15)–(3.3.16), (3.4.17)–(3.4.18),

(3.4.35), etc.]

$$u_{k,tt} + u_k = G_k(t,\tau) \quad \text{for } k = 0,1,2,\ldots, \qquad (4.4.45)_k$$

with

$$G_k(t,\tau) = -u_{k-2,\tau\tau} - g(t)u_{k-2,\tau} - 2u_{k-1,t\tau}$$
$$- g(t)u_{k-1,t} - h(t)u_{k-1}, \qquad (4.4.46)_k$$

where g and h are given as before,

$$g(t):= 2c + d\sin t,$$
$$h(t):= 2C + D\cos t, \qquad (4.4.47)$$

and with $u_n := 0$ for negative n in the right side of $(4.4.46)_k$. In particular, $G_0 \equiv 0$, so that we find from $(4.4.45)_0$ the result

$$u_0(t,\tau) = A_0(\tau)\cos t + B_0(\tau)\sin t. \qquad (4.4.48)$$

Then $(4.4.46)_1$ and (4.4.48) imply

$$G_1(t,\tau) = -\big[2B_0'(\tau) + g(t)B_0(\tau) + h(t)A_0(\tau)\big]\cos t$$
$$+ \big[2A_0'(\tau) + g(t)A_0(\tau) - h(t)B_0(\tau)\big]\sin t,$$
$$(4.4.49)$$

and the Lindstedt/Poincaré condition (3.4.22) then leads directly to the system

$$A_0'(\tau) = -cA_0 + CB_0,$$
$$B_0'(\tau) = -CA_0 - cB_0, \qquad (4.4.50)$$

where the expressions from (4.4.47) have been used for $g(t)$ and $h(t)$. The solution of (4.4.50) subject to the appropriate initial conditions $[A_0(0) = a, B_0(0) = b]$ is given as

$$A_0(\tau) = e^{-c\tau}(a\cos C\tau + b\sin C\tau),$$
$$B_0(\tau) = e^{-c\tau}(b\cos C\tau - a\sin C\tau), \qquad (4.4.51)$$

and then the resulting first-order approximate solution is given, with (4.4.48), as

$$U_0(t,\varepsilon):= u_0(t,\varepsilon t)$$
$$= e^{-c\varepsilon t}\big[a\cos(1 + \varepsilon C)t + b\sin(1 + \varepsilon C)t\big]. \quad (4.4.52)$$

From Theorem 4.4.1 we have the estimate

$$|u(t,\varepsilon) - U_0(t,\varepsilon)| + |u'(t,\varepsilon) - U_0'(t,\varepsilon)| \le C_0\varepsilon e^{-\mu\varepsilon t} \qquad (4.4.53)$$

for a fixed positive constant C_0, uniformly for all $t \ge 0$ and all small

$\varepsilon \geq 0$, where μ can be any fixed positive constant satisfying [see (4.4.44)]

$$0 < \mu < c - \tfrac{1}{2}|d - \tfrac{1}{2}D|. \tag{4.4.54}$$

It follows from (4.4.53) that the function U_0 of (4.4.52) provides a useful approximation to the exact solution u. We see that the parameter C influences the phase shift. Moreover, the exact solution decays exponentially toward zero as $t \to \infty$, and the parameter c influences the decay constant. The constants d and D are relatively unimportant to first order, entering only into the condition (4.4.54). However, if we go to higher-order approximations U_N with $N > 1$, then d and D enter directly into the approximations.

Example 4.4.2 (Greenlee and Snow 1975): The initial-value problem

$$u'' + \varepsilon(2u' + du^3 + 6\varepsilon\,du^2u') + u = 0 \quad \text{for } t > 0,$$
$$u(0) = a, \qquad u'(0) = b \quad \text{at } t = 0, \tag{4.4.55}$$

arises as a nonlinear model in electric-circuit theory with ε a small positive parameter and d a given positive constant. The hypotheses of Theorem 4.4.2 are satisfied with

$$c := 1, \qquad d > 0, \tag{4.4.56}$$
$$G(\varepsilon, u) := du^2, \qquad H(\varepsilon, u, v) := 6\varepsilon\,du^2.$$

The leading term u_0 in the two-variable expansion can be obtained explicitly, though it need not be given here, and then Theorem 4.4.2 gives the result

$$|u(t, \varepsilon) - U_0(t, \varepsilon)| + |u'(t, \varepsilon) - U_0'(t, \varepsilon)|$$
$$\leq \text{const. } \varepsilon(1 + \varepsilon t)^2 e^{-2\varepsilon t} \tag{4.4.57}$$

uniformly for $t \geq 0$ and for all small $\varepsilon \geq 0$, where

$$U_0(t, \varepsilon) := u_0(t, \varepsilon t).$$

The general methods of Greenlee and Snow (1975) can be applied to related problems not contained in the foregoing theorems. For example, these techniques of Greenlee and Snow yield a result analogous to Theorem 4.4.3 for certain equations such as

$$u'' + u + \varepsilon p(\varepsilon t)(u')^{2\nu+1} = 0, \tag{4.4.58}$$

which are related to (4.4.21), where $p = p(\tau)$ is a given positive-valued function satisfying

$$\int_0^\infty p(\tau)\,d\tau = \infty.$$

This equation (4.4.58) has already been considered in Example 3.4.2 in the case $\nu = 1$.

Similarly, these techniques can be used to obtain a result analogous to Theorem 4.4.1 for the previous linear equation (4.4.9),

$$u'' + \varepsilon g(t)u' + \left[1 + \varepsilon h(t)\right]u = 0, \qquad (4.4.59)$$

without the previous requirement that g and h be 2π-periodic, provided that the previous condition (4.4.10) is replaced by some such condition as (4.4.62).

For example, suppose that g and h satisfy the conditions

$$\int_0^t g(s)e^{is}\sin(t - s)\, ds = t(g_1 \cdot \cos t + g_2 \cdot \sin t) + g_3(t),$$

$$\int_0^t h(s)e^{is}\sin(t - s)\, ds = t(h_1 \cdot \cos t + h_2 \cdot \sin t) + h_3(t),$$

$$(4.4.60)$$

for all $t \geq 0$, where $i = \sqrt{-1}$, and where (4.4.60) holds for suitable smooth functions $g_3(t)$ and $h_3(t)$ that (along with their derivatives) satisfy certain growth conditions as $t \to \infty$, and where g_1, g_2, h_1, and h_2 are fixed constants such that the matrix M defined as

$$M: = -\operatorname{Re}\begin{bmatrix} h_1 & g_1 \\ h_2 & g_2 \end{bmatrix} + \operatorname{Im}\begin{bmatrix} g_1 & -h_1 \\ g_2 & -h_2 \end{bmatrix} \qquad (4.4.61)$$

satisfies the following condition:

The eigenvalues of M have negative real parts.   (4.4.62)

Then we find a result analogous to Theorem 4.4.1. Rather than give a general result here, we shall be content to illustrate the situation by the following example.

Example 4.4.3: Consider the problem of Exercise 2.4.7:

$$u'' + \varepsilon(2c + 5de^{-t})u' + \left[1 + \varepsilon(2C + 5De^{-t})\right]u = 0,$$
$$u(0) = a, \qquad u'(0) = b \quad \text{at } t = 0, \qquad (4.4.63)$$

for given constants c, d, C, D with c positive,

$$c > 0. \qquad (4.4.64)$$

Hence, in (4.4.59),

$$g(t): = 2c + 5de^{-t},$$
$$h(t): = 2C + 5De^{-t}, \qquad (4.4.65)$$

and we find the conditions of (4.4.60), with

$$g_1 = -ic, \qquad g_2 = c,$$
$$h_1 = -iC, \qquad h_2 = C, \qquad (4.4.66)$$

and

$$g_3(t) = d\left[(-1 + e^{-t})\cos t + (3 - 2e^{-t})\sin t\right]$$
$$+ i\left\{c\sin t + d\left[(1 + e^{-t})\sin t + 2(-1 + e^{-t})\cos t\right]\right\}, \qquad (4.4.67)$$

with a similar result for $h_3(t)$ given by replacing c and d, respectively, with C and D in (4.4.67). The matrix M of (4.4.61) is given as

$$M = \begin{bmatrix} -c & C \\ -C & -c \end{bmatrix}, \qquad (4.4.68)$$

with eigenvalues λ given as

$$\lambda = -c \pm iC, \qquad (4.4.69)$$

so that the assumption (4.4.64) guarantees the validity of condition (4.4.62).

The terms u_k in the two-variable expansion (4.4.2) are seen to satisfy (4.4.45)–(4.4.46), with g and h given by (4.4.65). Hence, we have the result

$$u_k(t,\tau) = A_k(\tau)\cos t + B_k(\tau)\sin t + \int_0^t G_k(s,\tau)\sin(t - s)\, ds, \qquad (4.4.70)_k$$

with G_k given by $(4.4.46)_k$.

The Lindstedt/Poincaré requirement asks that A_k and B_k be chosen so as to eliminate spurious secular terms in the integral appearing on the right side of $(4.4.70)_{k+1}$. For example, (4.4.49) and (4.4.60), along with (3.2.25), give

$$\int_0^t G_1(s,\tau)\sin(t - s)\, ds$$
$$= t\cos t\left[-A_0'(\tau) - (\operatorname{Re} g_1 + \operatorname{Im} h_1)B_0(\tau)\right.$$
$$\left. - (\operatorname{Re} h_1 - \operatorname{Im} g_1)A_0(\tau)\right]$$
$$+ t\sin t\left[-B_0'(\tau) - (\operatorname{Re} g_2 + \operatorname{Im} h_2)B_0(\tau)\right.$$
$$\left. - (\operatorname{Re} h_2 - \operatorname{Im} g_2)A_0(\tau)\right]$$
$$- A_0(\tau)\left[\operatorname{Re} h_3(t) - \operatorname{Im} g_3(t)\right]$$
$$- B_0(\tau)\left[\operatorname{Re} g_3(t) + \operatorname{Im} h_3(t)\right]. \qquad (4.4.71)$$

Hence, A_0 and B_0 are taken to satisfy the system [compare (2.4.13)]

$$\frac{d}{d\tau}\begin{bmatrix} A_0 \\ B_0 \end{bmatrix} = M\begin{bmatrix} A_0 \\ B_0 \end{bmatrix}, \tag{4.4.72}$$

with the matrix M given by (4.4.61), where this system (4.4.72) is to be solved subject to the initial conditions $A_0(0) = a$, $B_0(0) = b$. In the present case it follows from (4.4.68) that this initial-value problem for A_0 and B_0 coincides with the previous problem involving the system (4.4.50), and so the solution is again given by (4.4.51). Hence, the first-order approximation is again given by (4.4.52), and then $u_1(t, \tau)$ has the form [see (4.4.70)$_1$–(4.4.72)]

$$\begin{aligned} u_1(t,\tau) = &\, A_1(\tau)\cos t + B_1(\tau)\sin t \\ &- A_0(\tau)[\operatorname{Re} h_3(t) - \operatorname{Im} g_3(t)] \\ &- B_0(\tau)[\operatorname{Re} g_3(t) + \operatorname{Im} h_3(t)], \end{aligned} \tag{4.4.73}$$

with A_0 and B_0 given explicitly by (4.4.51). The procedure can be continued recursively as usual, and the coefficients u_k are found to be well determined for $k = 0, 1, \ldots$. The functions A_k, B_k satisfy a nonhomogeneous version of (4.4.72) for $k \geq 1$.

The resulting function U_N of (4.4.4) satisfies (4.4.25)–(4.4.26), with g and h given by (4.4.65), and with residuals α_N, β_N given by (4.4.27), and residual ρ_N given as

$$\begin{aligned} \rho_N(t, \varepsilon) = &-\varepsilon^{N+1}[2u_{N,t\tau} + g(t)u_{N,t} + h(t)u_N \\ &+ \varepsilon(u_{N,\tau\tau} + g(t)u_{N,\tau}) + u_{N-1,\tau\tau} + g(t)u_{N-1,\tau}], \end{aligned} \tag{4.4.74}$$

where the right side here is evaluated at $\tau = \varepsilon t$. The construction of the u_k's leads directly, with (4.4.74), to an estimate of the type

$$|\rho_N(t, \varepsilon)| \leq C_N \varepsilon^{N+1}(1 + \varepsilon t)^N e^{-c\varepsilon t},$$

and the method of proof of Theorem 4.4.1 yields the estimate (4.4.7), with

$$\sigma(\tau) := e^{-\mu\tau}$$

for any $\mu < c$. [We choose t_1 large enough with (4.4.64) so that $4c + (10d - 5D)\exp(-t_1) > 0$ holds, and then a version of Theorem 4.4.1 can be applied for $t \geq t_1$. We use Theorem 4.2.1 for $0 \leq t \leq t_1$.]

In particular, for any positive $\mu < c$, we find again an estimate of the form (4.4.53) for all $t \geq 0$ and for all sufficiently small $\varepsilon \geq 0$, where $U_0(t, \varepsilon)$ is given again by (4.4.52).

Note that if g and h are both 2π-periodic, as in Theorem 4.4.1, then the conditions of (4.4.60) hold automatically, with

$$\begin{bmatrix} g_1 \\ g_2 \end{bmatrix} = \frac{1}{2\pi} \int_0^{2\pi} g(s) \begin{bmatrix} -\sin s \\ \cos s \end{bmatrix} e^{is} \, ds, \qquad (4.4.75)$$

and a similar result with g replaced by h, where the resulting functions g_3 and h_3 are then defined by (4.4.60) and (4.4.75). The conditions (4.4.10) and (4.4.11) then imply the validity of (4.4.62).

As a further illustration, the problem

$$u'' + u + \varepsilon \left[2\varepsilon + (u')^2 \right] u' = 0 \quad \text{for } t > 0,$$
$$u(0, \varepsilon) = a, \qquad u'(0, \varepsilon) = b \quad \text{at } t = 0 \ (\varepsilon \geq 0), \qquad (4.4.76)$$

is considered in the following exercises. This problem is not included in the results of Theorems 4.4.1, 4.4.2, and 4.4.3, but similar methods can be applied to (4.4.76).

Exercises

Exercise 4.4.1: Show that the solution of (4.4.76) exists and is bounded for $t \geq 0$. *Hint:* Differentiate $E(t) := \frac{1}{2}[u^2 + (u')^2]$, and obtain the inequality $E' \leq 0$, which can be integrated to give $E(t) \leq E(0)$. Hence, we find the a priori bounds

$$|u(t, \varepsilon)|, |u'(t, \varepsilon)| \leq (a^2 + b^2)^{1/2}$$

for the solution. It then follows from a basic continuation theorem in the theory of ordinary differential equations that the solution $u(t, \varepsilon)$, which is known to exist on some interval $0 \leq t < t_1$, must exist uniquely for all $t \geq 0$ [see p. 172 of Hirsch and Smale (1974)].

Exercise 4.4.2: Let $U = U(t, \varepsilon)$ be a given function that satisfies

$$U'' + U + \varepsilon \left[2\varepsilon + (U')^2 \right] U' = -\rho(t, \varepsilon) \quad \text{for } t > 0,$$
$$U(0, \varepsilon) = a - \alpha(\varepsilon), \qquad U'(0, \varepsilon) = b - \beta(\varepsilon) \quad \text{at } t = 0, \qquad (4.4.77)$$

for residuals satisfying

$$|\alpha(\varepsilon)|, |\beta(\varepsilon)| \leq \gamma_1 \varepsilon^n \qquad (4.4.78)$$

and

$$|\rho(t, \varepsilon)| \leq \gamma_2 \varepsilon^{n+2} (1 + \varepsilon^2 t)^m e^{-\varepsilon^2 t} \quad \text{for } t \geq 0 \qquad (4.4.79)$$

for fixed constants γ_1 and γ_2 and fixed nonnegative integers m and n.

For any positive constant μ satisfying

$$0 < \mu < 1, \tag{4.4.80}$$

show that there is a constant γ_3 such that

$$|u(t, \varepsilon) - U(t, \varepsilon)| + |u'(t, \varepsilon) - U'(t, \varepsilon)| \le \gamma_3 \varepsilon^n e^{-\varepsilon^2 \mu t} \tag{4.4.81}$$

for all $t \ge 0$ and for all sufficiently small ε, where u is the solution of (4.4.76). *Hint:* Put $R = R(t) = R(t, \varepsilon) := u(t, \varepsilon) - U(t, \varepsilon)$, and find

$$R'' + R + \varepsilon[2\varepsilon + g(t, \varepsilon)] R' = \rho(t, \varepsilon) \quad \text{for } t > 0,$$
$$R(0) = \alpha(\varepsilon), \qquad R'(0) = \beta(\varepsilon) \qquad \text{at } t = 0, \tag{4.4.82}$$

with

$$g(t, \varepsilon) := [u'(t, \varepsilon)]^2 + u'(t, \varepsilon)U'(t, \varepsilon) + [U'(t, \varepsilon)]^2, \tag{4.4.83}$$

where, following Greenlee and Snow (1975), we have

$$g(t, \varepsilon) \ge 0.$$

Lemma 4.4.4 can then be applied to (4.4.82), and we find

$$|R(t, \varepsilon)|, |R'(t, \varepsilon)| \le (\sqrt{2}\,\gamma_1 + e\gamma_2 m!)\varepsilon^n$$

for all $t \ge 0$ and for all sufficiently small $\varepsilon \ge 0$. To obtain the exponential decay indicated by (4.4.81), put $S(t, \varepsilon) := R(t, \varepsilon)\exp(\varepsilon^2 \mu t)$, and apply the preceding argument to S. In order to apply Lemma 4.4.4 to the problem for S, we require a uniform bound on g and g', which can be readily obtained. The question of the *construction* of a suitable approximant U is discussed in Rubenfeld (1978); see also Exercise 3.4.5.

PART II

Initial-value problems of overdamped type

5

Linear overdamped initial-value problems

This chapter concerns certain initial-value problems that depend on a small parameter in such a way that the perturbation is nonuniform near the initial time, taken here to be $t = 0$. Some typical problems of this type are electric-circuit problems with large resistance and/or small inductance, mechanical problems with small masses and/or large damping, and the propagation of radiation through a highly absorbing medium. The method of matched asymptotic approximation and the multivariable method are illustrated for these problems, and error estimates are obtained for the resulting approximations.

5.1 Introduction

The problems considered in this chapter depend on a parameter in such a way that solutions behave nonuniformly near the initial time for small values of the parameter. The small parameter enters into these problems in such a way that the reduced equation (obtained by putting the small parameter equal to zero in the differential equation) is of lower order or of different type than the original equation. Hence, for small values of the parameter, there is a small *boundary-layer region* adjoining $t = 0$, interior to which the solution function undergoes rapid changes. A suitable uniform approximation to the solution of such a problem can be written in an additive form, consisting of an *outer approximation*, which approximates the solution outside the boundary-layer region, plus a *boundary-layer correction term*, which is negligible outside the boundary-layer region, but which is required for uniform approximation of the solution inside the boundary-layer region.

A simple prototype problem of such a type can be given for the differential equation

$$\varepsilon \frac{d^2 u}{dt^2} + 2 \frac{du}{dt} + u = 0 \qquad \text{for } t > 0, \tag{5.1.1}$$

subject to the initial conditions

$$u = a \quad \text{and} \quad \frac{du}{dt} = b + \frac{c}{\varepsilon} \quad \text{at } t = 0, \tag{5.1.2}$$

where ε is a small positive parameter, and a, b, c are specified constants. This problem (5.1.1)–(5.1.2) can be interpreted as a mathematical model for a simple linear mechanical oscillator, in which case $u = u(t, \varepsilon)$ represents a displacement of the system from equilibrium, the constant a represents an initial displacement, b represents an initial velocity, and c represents an initial impulse [see pp. 4–8 of J. Cole (1968)]. The small parameter ε can be taken to be proportional to the mass or inversely proportional to the resistance.

The solution of (5.1.1)–(5.1.2) is

$$u(t, \varepsilon) = \left(\frac{a\mu_1 + c + \varepsilon b}{\mu_1 - \mu_2} \right) e^{-\mu_2 t / \varepsilon} - \left(\frac{a\mu_2 + c + \varepsilon b}{\mu_1 - \mu_2} \right) e^{-\mu_1 t / \varepsilon},$$

$$\tag{5.1.3}$$

where the quantities μ_1 and μ_2 are defined as

$$\mu_1 := 1 + \sqrt{1 - \varepsilon}, \qquad \mu_2 := 1 - \sqrt{1 - \varepsilon}. \tag{5.1.4}$$

Hence, we have

$$\frac{\mu_1}{\varepsilon} = \frac{2}{\varepsilon} - \frac{1}{2} + \text{order}(\varepsilon)$$

$$\frac{\mu_2}{\varepsilon} = \frac{1}{2} + \text{order}(\varepsilon) \quad \text{as } \varepsilon \to 0,$$

and (5.1.3) leads to a result of the type

$$u(t, \varepsilon) = e^{-t/2} \left[A_0(t) + \varepsilon A_1(t) + \cdots \right]$$
$$+ e^{-2t/\varepsilon} \left[B_0(t) + \varepsilon B_1(t) + \cdots \right] \tag{5.1.5}$$

for suitable functions A_k and B_k. From this last result we see that both the original slow time t and an associated fast time $\tau := t/\varepsilon$ play roles in describing the behavior of u for small ε.

The result (5.1.5) can be interpreted as giving a decomposition of the solution as the sum of an outer approximation plus a boundary-layer correction term that depends on the fast time. Hence, we have the decomposition

$$u(t, \varepsilon) = u_{\text{outer}} + u_{\text{BL}} \qquad (5.1.6)$$

where u_{BL} denotes a boundary-layer correction that is important only in a thin region near $t = 0$. At any fixed $t = t_1 > 0$, the outer solution gives a good approximation to $u(t, \varepsilon)$ as $\varepsilon \to 0+$.

There is a nonuniformity at $(t, \varepsilon) = (0, 0)$, with

$$\lim_{\substack{\varepsilon \to 0+ \\ \text{fixed } \varepsilon > 0}} \lim_{t \to 0+} u(t, \varepsilon) = a \qquad (5.1.7a)$$

and

$$\lim_{\substack{t \to 0+ \\ \text{fixed } t > 0}} \lim_{\varepsilon \to 0+} u(t, \varepsilon) = A_0(0), \qquad (5.1.7b)$$

where

$$A_0(0) = a + \tfrac{1}{2}c,$$

so that in general the two limits of (5.1.7) are different, $A_0(0) \neq a$.

If we let $v(t)$ denote the leading term in the outer solution, given as the limit

$$v(t) := \lim_{\substack{\varepsilon \to 0+ \\ \text{fixed } t > 0}} u(t, \varepsilon) = \lim_{\varepsilon \to 0+} u_{\text{outer}}, \qquad (5.1.8)$$

then

$$v(t) = \left(a + \frac{c}{2}\right)e^{-t/2} \quad \text{for } t > 0. \qquad (5.1.9)$$

Hence, the limit (5.1.8) satisfies the reduced equation

$$2\frac{dv}{dt} + v = 0 \qquad \text{for } t > 0, \qquad (5.1.10)$$

where this latter equation is obtained by putting $\varepsilon = 0$ in the original differential equation (5.1.1). The initial value $v(0)$ obtained from (5.1.9) differs from the value $u(0, \varepsilon) = a$ that is specified for the exact solution of the full problem (5.1.1)–(5.1.2).

A more general version of equation (5.1.1) is

$$\varepsilon\frac{d^2u}{dt^2} + \alpha(t)\frac{du}{dt} + \beta(t)u = f(t) \quad \text{for } t > 0, \qquad (5.1.11)$$

where ε is again a small positive parameter, and the functions α, β, and f are given smooth functions for $t \geq 0$. The given function α is always *positive-valued* in the applications we have in mind, and so we shall assume that there is a fixed constant $\kappa > 0$ such that

$$\alpha(t) \geq \kappa > 0 \quad \text{for all } t \geq 0. \qquad (5.1.12)$$

The Sturm transformation (0.3.4) transforms (5.1.11) into

$$\frac{d^2\bar{u}}{dt^2} - \left\{ \frac{\alpha(t)^2 + 2\varepsilon[\alpha'(t) - 2\beta(t)]}{4\varepsilon^2} \right\}\bar{u} = \frac{1}{\varepsilon}\bar{f}, \qquad (5.1.13)$$

and the condition (5.1.12), along with the smoothness of the data, implies that the homogeneous equation has *no oscillatory solutions* for small ε (see Exercise 5.1.1). Equation (5.1.11) subject to (5.1.12) is *overdamped* for small positive ε. The differential equation also has no oscillatory solutions for negative α with $\alpha(t) \le -\kappa < 0$, but the positivity of α, as in (5.1.12), is required for stability as $\varepsilon \to 0+$.

The differential equation (5.1.11) will be considered in Section 5.2 subject to the previous initial conditions of (5.1.2). A matching approach will be used to obtain an approximation to the solution of this initial-value problem for small ε.

As usual, the data of the problem, including the given functions α, β, and f, and the given initial quantities a, b, and c, can be permitted to depend regularly on ε. That is, the data may have asymptotic expansions in powers of ε. However, for simplicity we shall deal only with the case in which these quantities are independent of ε.

Exercise

Exercise 5.1.1: Prove that this homogeneous version of (5.1.13),

$$\frac{d^2\bar{u}}{dt^2} - \left\{ \frac{\alpha(t)^2 + 2\varepsilon[\alpha'(t) - 2\beta(t)]}{4\varepsilon^2} \right\}\bar{u} = 0, \qquad (5.1.14)$$

has *no oscillatory solutions* subject to the condition (5.1.12). *Hint:* Use the maximum principle (see Theorem 0.3.5) to prove that any solution $\bar{u}$ must vanish identically, with $\bar{u} \equiv 0$, if $\bar{u}$ has two zeros, say $\bar{u}(t_1) = 0$ and $\bar{u}(t_2) = 0$ for $t_1 < t_2$ (and for sufficiently small $\varepsilon > 0$).

5.2 A matching approach

In this section we use a matching approach to approximate the solution of the initial-value problem

$$\varepsilon\frac{d^2u}{dt^2} + \alpha(t)\frac{du}{dt} + \beta(t)u = f(t) \quad \text{for } t > 0, \qquad (5.2.1)$$

with

$$u = a \quad \text{and} \quad \frac{du}{dt} = b + \frac{c}{\varepsilon} \quad \text{at } t = 0. \qquad (5.2.2)$$

The data are as described near the end of Section 5.1, so that in particular we have

$$\alpha(t) \geq \kappa > 0 \quad \text{for } t \geq 0. \tag{5.2.3}$$

On the basis of the constant-coefficient example of Section 5.1, we expect the solution $u = u(t, \varepsilon)$ of (5.2.1)–(5.2.2) to exhibit a rapid transition near $t = 0$, characterized in terms of some suitable *fast variable* τ. Here a natural conjecture for the form of τ is [see (5.1.5)]

$$\tau := t/\varepsilon, \tag{5.2.4}$$

and we shall use (5.2.4) here. There is some flexibility in the choice of such a *stretched variable*. For example, a constant multiple of (5.2.4) would equally effective, and the quantity t on the right side of (5.2.4) could be replaced with an integral of α. The polyhedron algorithm of Nipp (1978, 1980) can be used in the selection of such stretched variables from within parametric families of possible choices; this matter is also discussed in Section 8.4 in the context of certain boundary-value problems. Following Prandtl (1905), the idea is that the stretched variable should be chosen so as to magnify the boundary-layer region and thereby eliminate any rapid variation that might be exhibited by the solution when the solution is considered as a function of the stretched variable.

For the preceding problem, we now consider the solution u as a function of τ, and for this purpose it is convenient to introduce the function $h = h(\tau, \varepsilon)$ defined as $h(\tau, \varepsilon) = u(t, \varepsilon)$, with $\tau = t/\varepsilon$, so that

$$h(\tau, \varepsilon) := u(\varepsilon\tau, \varepsilon) \quad \text{for } \tau \geq 0. \tag{5.2.5}$$

The differential equation (5.2.1) is easily rewritten in terms of the stretched variable τ and the function $h = h(\tau, \varepsilon)$, and we find

$$\frac{d^2h}{d\tau^2} + \alpha(\varepsilon\tau)\frac{dh}{d\tau} + \varepsilon\beta(\varepsilon\tau)h = \varepsilon f(\varepsilon\tau) \quad \text{for } \tau > 0. \tag{5.2.6}$$

Similarly, the initial conditions are rewritten as

$$h = a, \quad \frac{dh}{d\tau} = c + \varepsilon b \quad \text{at } \tau = 0. \tag{5.2.7}$$

Only the leading term $h_0 = h_0(\tau)$ in an asymptotic expansion for h will be considered here, where h_0 is defined as

$$h_0(\tau) := \lim_{\substack{\varepsilon \to 0+ \\ \text{fixed } \tau}} h(\tau, \varepsilon), \tag{5.2.8}$$

and where this limit is *assumed* to exist at this point in the analysis. The procedure will be justified later by suitable error estimates discussed in Section 5.3.

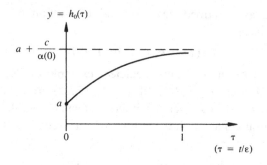

$y = h_0(\tau)$

$a + \dfrac{c}{\alpha(0)}$

a

0 1 τ

$(\tau = t/\varepsilon)$

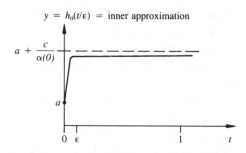

$y = h_0(t/\varepsilon) = $ inner approximation

$a + \dfrac{c}{\alpha(0)}$

a

$0\ \ \varepsilon$ 1 t

Figure 5.1
The inner approximation

From (5.2.6)–(5.2.8), we are led to impose on h_0 the conditions

$$h_0'' + \alpha(0)h_0' = 0 \quad \text{for } \tau > 0, \tag{5.2.9}$$

and

$$h_0(0) = a, \qquad h_0'(0) = c \quad \text{at } \tau = 0, \tag{5.2.10}$$

where the primes denote differentiation with respect to τ. This initial-value problem for h_0 yields directly the solution

$$h_0(\tau) = a + \frac{c}{\alpha(0)}\left(1 - e^{-\alpha(0)\tau}\right) \quad \text{for } \tau \geq 0. \tag{5.2.11}$$

The graph of this first-order inner solution is indicated in Figure 5.1.

We next consider an *outer solution* $g = g(t, \varepsilon)$, which is intended to approximate the exact solution u for values of t well away from $t = 0$, for small ε. This outer solution g is inserted into the original differential equation (5.2.1), giving

$$\alpha(t)g' + \beta(t)g = f(t) - \varepsilon g'' \quad \text{for } t > 0, \tag{5.2.12}$$

where the primes denote differentiation with respect to t. We seek an

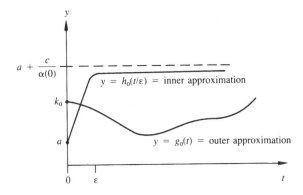

Figure 5.2

asymptotic power-series expansion for g in the form

$$g(t, \varepsilon) \sim \sum_{k=0}^{\infty} g_k(t)\varepsilon^k,$$

but again we shall obtain only the leading term g_0, where

$$g_0(t): = \lim_{\substack{\varepsilon \to 0 \\ \text{fixed } t > 0}} g(t, \varepsilon). \qquad (5.2.13)$$

Again, the entire procedure is justified by the error estimates to be obtained later.

From (5.2.12)–(5.2.13), we are led to impose on g_0 the condition

$$\alpha(t)g_0' + \beta(t)g_0 = f(t) \quad \text{for } t > 0. \qquad (5.2.14)$$

Hence, the leading outer approximation g_0 is determined as a solution of the reduced differential equation obtained by putting $\varepsilon = 0$ in the original second-order differential equation.

The most general solution of (5.2.14) involves a single constant of integration and can be given as

$$g_0(t) = k_0 \exp\left(-\int_0^t \frac{\beta}{\alpha}\right) + \int_0^t \frac{f(s)}{\alpha(s)}\exp\left(-\int_s^t \frac{\beta}{\alpha}\right) ds \qquad (5.2.15)$$

for some suitable constant k_0. The graph of g_0 is indicated in Figure 5.2 along with the graph of the previous inner approximation h_0.

We can now employ the *asymptotic matching principle* of Van Dyke (1964) to determine the remaining constant k_0 appearing in g_0. To this end we compute the leading term in the *inner expansion of the outer approximation* g_0 and equate this with the leading term in the *outer*

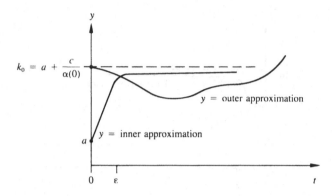

Figure 5.3

expansion of the inner approximation h_0. The former, inner expansion of g_0, is obtained by rewriting $g_0(t)$ as a function of the inner variable $\tau = t/\varepsilon$ and then passing to the limit $\varepsilon \to 0+$ to find

$$\lim_{\varepsilon \to 0} g_0(\varepsilon\tau) = k_0 = \begin{array}{l} \text{leading term in inner expansion} \\ \text{of outer approximation } g_0. \end{array} \quad (5.2.16)$$

Similarly, the outer expansion of h_0 is obtained by writing $h_0(\tau)$ as a function of t and passing to the limit $\varepsilon \to 0+$, to find

$$\lim_{\varepsilon \to 0} h_0(t/\varepsilon) = a + \frac{c}{\alpha(0)}$$

$$= \begin{array}{l} \text{leading term in outer expansion} \\ \text{of inner approximation } h_0. \end{array} \quad (5.2.17)$$

Hence, in this case the asymptotic matching principle of Van Dyke implies the matching condition

$$k_0 = a + \frac{c}{\alpha(0)}, \quad (5.2.18)$$

in agreement with what we would naturally expect, based on Figure 5.2. This result (5.2.18), along with (5.2.15), now completely determines the outer approximation $g_0(t)$, which is expected to provide a good approximation to the exact solution away from the initial boundary-layer region. The resulting inner and outer approximations are indicated in Figure 5.3.

It is convenient to form a single *composite approximation* U_0 defined as

$$U_0(t, \varepsilon) := g_0(t) + h_0\left(\frac{t}{\varepsilon}\right) - \left(a + \frac{c}{\alpha(0)}\right), \quad (5.2.19)$$

where the *common part* $a + c/\alpha(0)$ has been subtracted from the sum of the outer and inner approximations. The boundary-layer correction as described in (5.1.6) is given here by the last few terms in (5.2.19) as

$$\text{(first-order) boundary-layer correction} = h_0\left(\frac{t}{\varepsilon}\right) - \left(a + \frac{c}{\alpha(0)}\right).$$

$$(5.2.20)$$

It will be seen in Section 5.3 that (5.2.19) actually provides a uniformly valid first-order approximation to the exact solution. The asymptotic matching principle can also be used systematically to obtain additional, higher-order approximations, as in Vasil'eva (1963) [see the survey in Wasow (1965, 1976)] and Sibuya (1963b, 1964b), but we shall omit these developments here, because the O'Malley/Hoppensteadt two-variable approach seems preferable for these problems, as discussed in Section 5.4.

See O'Malley (1974a, pp. 15–17) for a brief history and references for matching techniques. The matching approach has been used successfully for difficult boundary-value problems involving ordinary and partial differential equations. The matching conditions become somewhat more intricate for this problems. A modern analysis of the foundations of matching techniques is given in Eckhaus (1979).

5.3 Error estimates

In this section, error estimates are given for the difference between the exact solution u and a conjectured approximate solution U for the linear problem

$$\mathscr{L}u = f(t,\varepsilon) \quad \text{for } t \geq 0,$$
$$u = a, \quad u' = b + \frac{c}{\varepsilon} \quad \text{at } t = 0,$$

$$(5.3.1)$$

where the differential operator $\mathscr{L}$ is here defined as

$$\mathscr{L} := \varepsilon\frac{d^2}{dt^2} + \alpha(t,\varepsilon)\frac{d}{dt} + \beta(t,\varepsilon),$$

$$(5.3.2)$$

and where the given functions $\alpha(t,\varepsilon)$, $\beta(t,\varepsilon)$, and $f(t,\varepsilon)$ are smooth, with

$$\alpha(t,\varepsilon) \geq \kappa > 0 \quad \text{for } t \geq 0, \varepsilon \geq 0$$

$$(5.3.3)$$

for a fixed positive constant κ.

The function U is assumed to be a given function that satisfies (5.3.1) approximately, with

$$\mathscr{L}U = f(t,\varepsilon) - \rho(t,\varepsilon) \quad \text{for } t \geq 0,$$
$$U = a - \gamma_1(\varepsilon), \quad U' = b + \frac{c}{\varepsilon} - \gamma_2(\varepsilon) \quad \text{at } t = 0,$$

$$(5.3.4)$$

for certain known residuals $\rho(t, \varepsilon)$, $\gamma_1(\varepsilon)$, and $\gamma_2(\varepsilon)$, where these residuals are known to be small.

The remainder, or difference between the exact solution u and approximate solution U, is denoted as R, where

$$R = R(t) = R(t, \varepsilon): = u(t, \varepsilon) - U(t, \varepsilon). \qquad (5.3.5)$$

Then (5.3.1), (5.3.4), and (5.3.5) yield

$$\mathscr{L}R = \rho(t, \varepsilon) \quad \text{for } t \geq 0,$$
$$R(0, \varepsilon) = \gamma_1(\varepsilon), \qquad R'(0, \varepsilon) = \gamma_2(\varepsilon) \qquad \text{at } t = 0, \qquad (5.3.6)$$

from which we find directly the following theorem.

Theorem 5.3.1: *The solution R of* (5.3.6) *satisfies the bounds*

$$|R(t, \varepsilon)|, \frac{\varepsilon}{\kappa}|R'(t, \varepsilon)|$$

$$\leq \left[|\gamma_1(\varepsilon)| + \frac{\varepsilon}{\kappa}|\gamma_2(\varepsilon)| + \frac{1}{\kappa} \int_0^t |\rho(s, \varepsilon)| \, ds \right] e^{\|\beta\| t / \kappa} \qquad (5.3.7)$$

for $t \geq 0$, where

$$\|\beta\|: = \max_{0 \leq s \leq t} |\beta(s, \varepsilon)|.$$

Proof: The estimates (5.3.7) follow directly from various previous results of Section 0.3, such as (0.3.18) or similar results from Theorem 0.3.2. In the notation of Section 0.3, we take $a(t) = \alpha(t, \varepsilon)/\varepsilon$, $b(t) = \beta(t, \varepsilon)/\varepsilon$, and $f(t) = \rho(t, \varepsilon)/\varepsilon$, with $a_0 = \kappa/\varepsilon$, and so forth. The condition (0.3.3) of Section 0.3 holds automatically in the present case for all small $\varepsilon > 0$ as a consequence of (5.3.3) and the assumed smoothness of the data. [We assume, for example, that $\alpha'(t, \varepsilon)$ and $\beta(t, \varepsilon)$ are bounded on any compact t interval, uniformly for all small ε.] Further details are omitted. ∎

If the function β is nonnegative, with

$$\beta(t, \varepsilon) \geq 0 \quad \text{for } t \geq 0, \varepsilon \geq 0, \qquad (5.3.8)$$

then we find similarly the following result.

Theorem 5.3.2: *If* (5.3.8) *holds along with* (5.3.3), *then the solution R of* (5.3.6) *satisfies*

$$|R(t, \varepsilon)| \leq |\gamma_1(\varepsilon)| + \frac{\varepsilon}{\kappa}|\gamma_2(\varepsilon)| + \frac{1}{\kappa} \int_0^t |\rho(s, \varepsilon)| \, ds \qquad (5.3.9)$$

for $t \geq 0$, and a related inequality for R' that need not be given here.

Proof: An application of Theorem 0.3.3 yields

$$|R(t,\varepsilon)| \le |\gamma_1(\varepsilon)| + |\gamma_2(\varepsilon)| \int_0^t \exp\left(-\int_0^s \frac{\alpha}{\varepsilon}\right) ds$$

$$+ \frac{1}{\varepsilon} \int_0^t |\rho(s,\varepsilon)| \left[\int_s^t \exp\left(-\int_s^\tau \frac{\alpha}{\varepsilon}\right) d\tau\right] ds,$$

from which (5.3.9) follows directly using the known result [see (5.3.3)]

$$\int_s^t \exp\left(-\int_s^\tau \frac{\alpha}{\varepsilon}\right) d\tau \le \frac{\varepsilon}{\kappa} \quad \text{for } t \ge s. \quad \blacksquare$$

The preceding theorems can be used to prove that the composite function U_0 constructed by matching in Section 5.2 actually provides a useful approximation to the solution u of the problem

$$\varepsilon u'' + \alpha(t)u' + \beta(t)u = f(t) \quad \text{for } t \ge 0, \tag{5.3.10}$$

$$u = a, \quad u' = b + \frac{c}{\varepsilon} \quad \text{at } t = 0,$$

where for simplicity, as in Section 5.2, we are taking the functions α, β, and f to be smooth functions of t independent of ε. Indeed, if we define $U = U(t,\varepsilon)$ as

$$U(t,\varepsilon) := U_0(t,\varepsilon), \tag{5.3.11}$$

with U_0 given by (5.2.19), then for this U we find the corresponding results of (5.3.4), with residuals given as

$$\gamma_1(\varepsilon) = 0,$$

$$\gamma_2(\varepsilon) = b + \frac{\beta(0)}{\alpha(0)}\left(a + \frac{c}{\alpha(0)}\right) - \frac{f(0)}{\alpha(0)}, \tag{5.3.12}$$

and

$$\rho(t,\varepsilon) = ce^{-\alpha(0)t/\varepsilon}\left\{\frac{\beta(t)}{\alpha(0)} + \frac{c}{\varepsilon}[\alpha(0) - \alpha(t)]\right\}$$

$$+ \varepsilon\left[\frac{\beta(t)f(t)}{\alpha(t)^2} - \frac{d}{dt}\left(\frac{f(t)}{\alpha(t)}\right)\right]$$

$$+ \varepsilon\left[\frac{d}{dt}\left(\frac{\beta(t)}{\alpha(t)}\right) - \left(\frac{\beta(t)}{\alpha(t)}\right)^2\right]$$

$$\times\left[\left(a + \frac{c}{\alpha(0)}\right)\exp\left(-\int_0^t \beta/\alpha\right) + \int_0^t \frac{f(s)}{\alpha(s)}\exp\left(-\int_s^t \beta/\alpha\right) ds\right]. \tag{5.3.13}$$

Using the smoothness of the data along with (5.3.3) and (5.3.13), it follows directly that there holds, for any fixed $T > 0$, an estimate of the form

$$\int_0^T |\rho(s,\varepsilon)|\, ds \le \text{const. } \varepsilon \quad \text{as } \varepsilon \to 0+, \tag{5.3.14}$$

for a fixed constant independent of ε. In obtaining (5.3.14), we use the results

$$\int_0^T e^{-\kappa s/\varepsilon}\, ds \le \frac{\varepsilon}{\kappa},$$

$$\int_0^T s e^{-\kappa s/\varepsilon}\, ds \le \frac{\varepsilon^2}{\kappa^2},$$

for any $T > 0$. Hence, we find the following theorem.

Theorem 5.3.3: *Let u be the exact solution of* (5.3.10), *and let U_0 be the composite function constructed by matching, given by* (5.2.19). *Then, for any fixed $T > 0$,*

$$|u(t,\varepsilon) - U_0(t,\varepsilon)|,\ \varepsilon|u'(t,\varepsilon) - U_0'(t,\varepsilon)| \le \text{const. } \varepsilon$$

for $0 \le t \le T$ as $\varepsilon \to 0+$, for a fixed constant independent of ε.

Proof: The stated estimates follow directly from Theorem 5.3.1 along with (5.3.5) and (5.3.11)–(5.3.14). ∎

   If the data α, β, f satisfy certain global integrability conditions, then (5.3.13) leads to a bound of the type (5.3.14), with $T = \infty$. In this case, if β is nonnegative, we can use Theorem 5.3.2 to obtain estimates of the type given in Theorem 5.3.3, *valid for all $t \ge 0$*. We leave the details to the reader.

   As mentioned earlier, matching can be used to construct higher-order approximations, but we do not pursue this here. Rather, we shall use a two-timing approach in Section 5.4 to obtain such higher-order approximations.

Exercise

Exercise 5.3.1: Let a, b, c be given smooth functions for $0 \le t \le T$, with

$$a(t) \ge \kappa > 0 \quad \text{for } 0 \le t \le T, \tag{5.3.15}$$

where κ is a fixed positive constant, and let the third-order differential operator $\mathcal{L}$ be defined as

$$\mathcal{L} := \varepsilon \frac{d^3}{dt^3} + a(t)\frac{d^2}{dt^2} + b(t)\frac{d}{dt} + c(t), \tag{5.3.16}$$

where ε is a small positive parameter. Show that the solution $R = R(t, \varepsilon)$ of the initial-value problem

$$\mathcal{L}R = \rho(t, \varepsilon) \quad \text{for } 0 \leq t \leq T,$$

$$R = \alpha(\varepsilon), \qquad R' = \beta(\varepsilon), \qquad R'' = \gamma(\varepsilon) \qquad \text{at } t = 0, \tag{5.3.17}$$

must satisfy a bound of the type

$$|R(t, \varepsilon)| \leq \text{const.} \left[|\alpha(\varepsilon)| + |\beta(\varepsilon)| + \varepsilon|\gamma(\varepsilon)| + \int_0^T |\rho| \right] \tag{5.3.18}$$

for $0 \leq t \leq T$, for a fixed constant independent of ε as $\varepsilon \to 0+$. Related bounds hold also for the derivatives of R, but we need not consider these here. *Hint*: Apply the result of Exercise 0.3.4.

5.4 A multivariable approach

In this section we use a multivariable approach to study the problem of Section 5.2,

$$\mathcal{L}u = f(t) \quad \text{for } t > 0,$$

$$u = a, \qquad u' = b + \frac{c}{\varepsilon} \quad \text{at } t = 0, \tag{5.4.1}$$

where $\mathcal{L}$ is the differential operator

$$\mathcal{L} := \varepsilon \frac{d^2}{dt^2} + \alpha(t) \frac{d}{dt} + \beta(t). \tag{5.4.2}$$

We could directly handle the case in which the data have asymptotic power series in ε, but for simplicity we consider only the case in which α, β, and f are smooth functions of t alone and a, b, and c are fixed constants.

The approach followed here is due to O'Malley (1971*a*, 1971*b*) and Hoppensteadt (1971*b*). This approach can be viewed as a particularly effective and happy repackaging of earlier techniques, such as the matching approach illustrated in Section 5.2 and the matching approaches of Vishik and Lyusternik (1957), Vasil'eva (1963), and Sibuya (1963*b*, 1964*b*). The resulting multivariable method of O'Malley and Hoppensteadt has proved to be considerably simpler than earlier techniques to apply in practice, and indeed the method easily handles more general nonlinear problems, as illustrated in Chapter 6.

Following O'Malley and Hoppensteadt, we seek an additive decomposition of the solution u of (5.4.1) in the form of an outer solution U plus a boundary-layer correction term U^*, where U^* is sought as a function of the stretched variable τ given as

$$\tau := t/\varepsilon. \tag{5.4.3}$$

Hence, we seek u in the form

$$u(t, \varepsilon) \sim U(t, \varepsilon) + U^*(\tau, \varepsilon), \quad \text{with } \tau = t/\varepsilon, \tag{5.4.4}$$

for suitable functions $U = U(t, \varepsilon)$ and $U^* = U^*(\tau, \varepsilon)$, which are to be obtained in the form of asymptotic expansions as

$$U(t, \varepsilon) \sim \sum_{k=0}^{\infty} U_k(t) \varepsilon^k \tag{5.4.5}$$

and

$$U^*(\tau, \varepsilon) \sim \sum_{k=0}^{\infty} U_k^*(\tau) \varepsilon^k. \tag{5.4.6}$$

The boundary-layer correction $U^*(\tau, \varepsilon)$, when evaluated at $\tau = t/\varepsilon$, must be negligible for any fixed positive $t > 0$ as $\varepsilon \to 0+$, and so we impose the matching conditions

$$\lim_{\tau \to \infty} U_k^*(\tau) = 0 \quad \text{for } k = 0, 1, 2, \dots . \tag{5.4.7}_k$$

The asymptotic relations of (5.4.4)–(5.4.6) are used in the efficient calculation of the coefficient functions U_k and U_k^*. After these coefficient functions are determined, we define the remainder $R_N = R_N(t, \varepsilon)$ by the relation

$$u(t, \varepsilon) = \sum_{k=0}^{N} \left[U_k(t) + U_k^*\!\left(\frac{t}{\varepsilon}\right) \right] \varepsilon^k + R_N(t, \varepsilon), \tag{5.4.8}$$

and we then use Theorem 5.3.1 or 5.3.2 to obtain suitable estimates on R_N.

In view of (5.4.3)–(5.4.7), we are led to require that the outer solution U satisfy the original equation $\mathscr{L}U = f(t)$, or

$$\alpha(t)\frac{dU}{dt} + \beta(t)U = f(t) - \varepsilon\frac{d^2U}{dt^2} \quad \text{for } t > 0. \tag{5.4.9}$$

Then (5.4.4) and (5.4.9), along with the differential equation of (5.4.1) for u, lead to the requirement that the boundary-layer correction U^* satisfy the homogeneous equation $\mathscr{L}U^* = 0$, or, in terms of the stretched variable $\tau = t/\varepsilon$ [note that $d/dt = (1/\varepsilon)(d/d\tau)$, etc.],

$$\frac{d^2U^*}{d\tau^2} + \alpha(\varepsilon\tau)\frac{dU^*}{d\tau} + \varepsilon\beta(\varepsilon\tau)U^* = 0 \quad \text{for } \tau > 0. \tag{5.4.10}$$

The expansion (5.4.5) for the outer solution is inserted into (5.4.9), and we are led to the conditions

$$\alpha(t)\frac{dU_k}{dt} + \beta(t)U_k = \begin{cases} f(t) & \text{for } k = 0, \\ -\dfrac{d^2U_{k-1}}{dt^2} & \text{for } k = 1, 2, \ldots . \end{cases}$$

$$(5.4.11)_k$$

Similarly, the expansion (5.4.6) for the boundary-layer correction is inserted into (5.4.10), and we are led to the conditions

$$\frac{d^2U_k^*}{d\tau^2} + \alpha(0)\frac{dU_k^*}{d\tau}$$

$$= \begin{cases} 0 & \text{for } k = 0, \\ -\displaystyle\sum_{n=1}^{k}\left[\dfrac{\alpha^{(n)}(0)\tau^n}{n!}\dfrac{dU_{k-n}^*}{d\tau} + \dfrac{\beta^{(n-1)}(0)\tau^{n-1}}{(n-1)!}U_{k-n}^*\right] & \\ & \text{for } k = 1, 2, \ldots . \end{cases}$$

$$(5.4.12)_k$$

In obtaining (5.4.12), we use the result

$$\alpha(\varepsilon\tau) \sim \sum_{k=0}^{\infty}\frac{\tau^k\alpha^{(k)}(0)}{k!}\varepsilon^k \tag{5.4.13}$$

and a similar result for $\beta(\varepsilon\tau)$, where (5.4.13) follows directly from the identity

$$\alpha(\varepsilon\tau) = \sum_{k=0}^{N}\frac{\tau^k\alpha^{(k)}(0)}{k!}\varepsilon^k + \frac{(\varepsilon\tau)^{N+1}}{N!}\int_0^1(1-s)^N\alpha^{(N+1)}(s\varepsilon\tau)\,ds$$

$$(5.4.14)$$

for any smooth function α, where $\alpha^{(k)}$ denotes the kth-order derivative of the function α with respect to its argument. This latter result follows in turn by taking $\Phi(s) := \alpha(s\varepsilon\tau)$, for fixed ε and fixed τ, in the Taylor/Cauchy identity

$$\Phi(1) = \sum_{k=0}^{N}\frac{\Phi^{(k)}(0)}{k!} + \int_0^1\frac{(1-s)^N}{N!}\Phi^{(N+1)}(s)\,ds.$$

Similarly, we insert (5.4.4)–(5.4.6) into the initial conditions of (5.4.1) and find

$$U_k(0) + U_k^*(0) = \begin{cases} a & \text{for } k = 0, \\ 0 & \text{otherwise,} \end{cases} \tag{$5.4.15)_k$}$$

and

$$\frac{dU_k^*(0)}{d\tau} = \begin{cases} c & \text{for } k = 0, \\ b - U_0'(0) & \text{for } k = 1, \\ -U_{k-1}'(0) & \text{for } k = 2, 3, \dots . \end{cases} \qquad (5.4.16)_k$$

We indicate briefly now how the required functions U_k and U_k^* can be determined recursively, using the conditions (5.4.11), (5.4.12), (5.4.15), and (5.4.16) along with the matching conditions of (5.4.7).

Equation $(5.4.12)_0$ yields

$$U_0^*(\tau) = c_1 + c_2 e^{-\alpha(0)\tau}$$

for suitable constants of integration c_1 and c_2, and then the conditions $(5.4.7)_0$ and $(5.4.16)_0$ can be used to determine these constants as [note that $\alpha(0)$ is positive]

$$c_1 = 0 \quad \text{and} \quad c_2 = -\frac{c}{\alpha(0)},$$

so that we have $U_0^*(\tau)$ determined uniquely as

$$U_0^*(\tau) = -\frac{c}{\alpha(0)} e^{-\alpha(0)\tau}. \qquad (5.4.17)$$

Using (5.4.17) with $(5.4.15)_0$, we now have the initial condition

$$U_0(0) = a + \frac{c}{\alpha(0)},$$

and this condition determines a unique solution for the first-order differential equation $(5.4.11)_0$, given as

$$U_0(t) = \left[a + \frac{c}{\alpha(0)} \right] \exp\left(-\int_0^t \frac{\beta}{\alpha} \right) + \int_0^t \frac{f(s)}{\alpha(s)} \exp\left(-\int_s^t \frac{\beta}{\alpha} \right) ds. \qquad (5.4.18)$$

We now continue recursively. After $U_n^*(\tau)$ and $U_n(t)$ have been determined for $n = 0, 1, 2, \dots, k - 1$, we then use $(5.4.7)_k$, $(5.4.12)_k$, and $(5.4.16)_k$ to determine $U_k^*(\tau)$ uniquely, and then $(5.4.11)_k$ and $(5.4.15)_k$ uniquely determine $U_k(t)$. We find directly that each boundary-layer correction-coefficient function $U_k^*(\tau)$ and its derivatives decay exponentially to zero as $\tau \to \infty$, with

$$|U_k^*(\tau)|, \left| \frac{dU_k^*(\tau)}{d\tau} \right| \le C_k \tau^{2k} e^{-\alpha(0)\tau} \quad \text{for } \tau \ge 0 \qquad (5.4.19)$$

for suitable constants C_k independent of τ [see (5.4.17) for the case $k = 0$].

The resulting expansions of (5.4.4)–(5.4.6) can now be truncated to provide useful approximations to the exact solution u. For example, the remainder R_N defined by (5.4.8) is found directly to satisfy the initial conditions [see (5.4.15) and (5.4.16) along with (5.4.1)]

$$R_N(0, \varepsilon) = 0,$$
$$R'_N(0, \varepsilon) = -U'_N(0)\varepsilon^N \quad \text{at } t = 0,$$

(5.4.20)

and similarly R_N satisfies the differential equation [see (5.4.11) and (5.4.12) along with (5.4.1)]

$$\mathscr{L}R_N = \rho(t, \varepsilon) \quad \text{for } t > 0,$$

(5.4.21)

with residual $\rho(t, \varepsilon)$ given as

$$\rho(t, \varepsilon) = \rho_1(t, \varepsilon) + \frac{1}{\varepsilon}\rho_2\left(\frac{t}{\varepsilon}, \varepsilon\right),$$

(5.4.22)

where

$$\rho_1(t, \varepsilon) := -U''_N(t)\varepsilon^{N+1}$$

(5.4.23)

and

$$\rho_2(\tau, \varepsilon) := -\sum_{k=0}^{N}\left[\frac{d^2U_k^*(\tau)}{d\tau^2} + \alpha(\varepsilon\tau)\frac{dU_k^*(\tau)}{d\tau} + \varepsilon\beta(\varepsilon\tau)U_k^*(\tau)\right]\varepsilon^k.$$

(5.4.24)

For any fixed $T > 0$, we find directly from (5.4.23) a bound of the type

$$\int_0^T |\rho_1(t, \varepsilon)| \, dt \leq \text{const. } \varepsilon^{N+1}$$

(5.4.25)

for a fixed constant independent of ε (as $\varepsilon \to 0+$). Indeed, we shall now see that the total residual ρ of (5.4.22) also satisfies a similar bound of the type (5.4.25).

To check this latter assertion, we need only note, following Fife (1973a), that the residual $\rho_2(\tau, \varepsilon)$ of (5.4.24) is a smooth function of ε that is small near $\varepsilon = 0$. Indeed, the functions $U_k^*(\tau)$ have been constructed with $(5.4.12)_k$ (for $k = 0, 1, \ldots, N$), so that (see Exercise 5.4.1)

$$\frac{\partial^n}{\partial\varepsilon^n}\rho_2(\tau, \varepsilon)\bigg|_{\varepsilon=0} = 0 \quad \text{for } n = 0, 1, \ldots, N,$$

(5.4.26)

and then the Taylor/Cauchy formula yields [compare with (5.4.14)]

$$\rho_2(\tau, \varepsilon) = \frac{\varepsilon^{N+1}}{N!}\int_0^1 (1 - s)^N \frac{\partial^{N+1}}{\partial\varepsilon^{N+1}}\rho_2(\tau, s\varepsilon) \, ds.$$

(5.4.27)

It follows directly from (5.4.12) and (5.4.24) that $(\partial^{N+1}/\partial\varepsilon^{N+1})[\rho_2(\tau, \varepsilon)]$ is a linear combination of $U_k^*(\tau)$ and $dU_k^*(\tau)/d\tau$ (for $k = 0, 1, \ldots, N$)

with coefficients involving $\alpha(\varepsilon\tau)$ and $\beta(\varepsilon\tau)$ and their derivatives through respective orders $N + 1$ and N, multiplied also by certain powers of τ. The terms involving $\alpha(\varepsilon\tau) = \alpha(t)$ and $\beta(\varepsilon\tau) = \beta(t)$ and their derivatives are uniformly bounded for $0 \le t \le T$ (any fixed $T > 0$), and taking into account the aforementioned powers of τ, we then find directly from (5.4.24) and (5.4.19) a bound of the type

$$\left| \frac{\partial^{N+1}}{\partial \varepsilon^{N+1}} p_2(\tau, s\varepsilon) \right|_{\tau=\frac{t}{\varepsilon}} \le \text{const.} \left[1 + \left(\frac{t}{\varepsilon} \right)^{3N+1} \right] e^{-\alpha(0)t/\varepsilon} \quad (5.4.28)$$

for all $0 \le t \le T$, for $0 \le s \le 1$, and for all small $\varepsilon > 0$. We can now use this last estimate in the right side of (5.4.27), along with the known result

$$\int_0^T \left(\frac{t}{\varepsilon} \right)^{3N+1} e^{-\alpha(0)t/\varepsilon}\, dt \le \frac{\varepsilon}{\alpha(0)^{3N+2}} \int_0^\infty t^{3N+1} e^{-t}\, dt = \text{const. } \varepsilon,$$

and we find directly the result

$$\frac{1}{\varepsilon} \int_0^T \left| p_2\left(\frac{t}{\varepsilon}, \varepsilon \right) \right| dt \le \text{const. } \varepsilon^{N+1}. \quad (5.4.29)$$

This derivation of (5.4.29) parallels the argument used by Fife (1973a) for a related problem involving a singularly perturbed boundary-value problem.

Combining (5.4.25) and (5.4.29), we find now, with (5.4.22), the desired result

$$\int_0^T |\rho(t, \varepsilon)|\, dt \le \text{const. } \varepsilon^{N+1} \quad (5.4.30)$$

for a fixed constant independent of ε as $\varepsilon \to 0+$.

These results yield directly the following theorem.

Theorem 5.4.1: *Let the given functions α, β, and f be of class $C^{N+1}[0, T]$, let $u(t, \varepsilon)$ be the exact solution of (5.4.1)–(5.4.2) subject to the stability condition (5.2.3), and let $U_k(t)$ and $U_k^*(\tau)$ be the outer terms and boundary-layer correction terms constructed using (5.4.7)$_k$, (5.4.11)$_k$, (5.4.12)$_k$, (5.4.15)$_k$, and (5.4.16)$_k$ for $k = 0, 1, \dots, N$. Then the remainder $R_N(t, \varepsilon)$ defined by (5.4.8) satisfies the bounds*

$$|R_N(t, \varepsilon)|, \varepsilon|R_N'(t, \varepsilon)| \le \text{const. } \varepsilon^{N+1} \quad (5.4.31)$$

uniformly for $0 \le t \le T$ and uniformly for all small $\varepsilon \ge 0$.

Proof: R_N satisfies (5.4.20)–(5.4.21), so that the stated bounds follow directly from Theorem 5.3.1 along with (5.4.30) and the results $\gamma_1(\varepsilon) = 0$

and $\gamma_2(\varepsilon) = -U_N'(0)\varepsilon^N$. The assumed regularity of the data α, β, f guarantees the existence and regularity of U_k and U_k^*, along with the validity of the bound (5.4.30). Further details are omitted. ∎

If the coefficient β is nonnegative and if the data α, β, f satisfy certain growth conditions at infinity, then we can use Theorem 5.3.2 to obtain bounds of the type (5.4.31) *valid for all* $t \geq 0$. Details are omitted.

We see from (5.4.17)–(5.4.18) that the first-order two-variable approximation

$$U_0(t) + U_0^*(t/\varepsilon)$$

agrees with the matching approximation (5.2.19) obtained in Section 5.2 [see (5.2.11), (5.2.15), (5.2.18), and (5.2.19)].

Slightly different forms are possible for the boundary-layer correction in (5.4.4). For example, we can replace $U^*(t/\varepsilon, \varepsilon)$ with a related $U^{**}(t, \varepsilon)$ of the form

$$U^{**}(t, \varepsilon) \sim \exp\left(-\frac{1}{\varepsilon} \int_0^t \alpha\right) \sum_{k=0}^\infty U_k^{**}(t)\varepsilon^k, \qquad (5.4.32)$$

and different forms are possible as well. Such boundary-layer corrections as (5.4.32) have been used by Smith and Palmer (1970), Smith (1971, 1975a), and others in the study of overdamped initial-value problems for ordinary and partial differential equations. These different forms of the boundary-layer correction are asymptotically equivalent. The foregoing approach of O'Malley and Hoppensteadt based on (5.4.3)–(5.4.7) provides a convenient formulation that generalizes nicely to nonlinear problems, as we shall see in Chapter 6. On the other hand, for linear problems, the slightly different packaging given by (5.4.32) sometimes has a quantitative advantage within the boundary layer, because the resulting remainder R_N may be slightly smaller. We illustrate the use of (5.4.32) in Section 5.5 for an overdamped problem involving a hyperbolic system of partial differential equations. See also Exercise 5.4.3, where (5.4.32) is used.

Related overdamped initial-value problems can be considered for certain higher-order equations for which the reduced equation is of lower order than the full equation, as illustrated in Exercise 5.4.4 [see Tschen (1935) and Vishik and Lyusternik (1957)]. Similarly, overdamped initial-value problems for ordinary differential equations in a Banach space can be considered, and the results can be used to study related singularly perturbed initial-boundary-value problems for partial differential equations, as in Trenogin (1963, 1970), Hoppensteadt (1971a, 1971b), and Krein (1971).

Exercise 5.4.1: Verify the result (5.4.26) explicitly in the cases $N = 0$ and $N = 1$.

Exercise 5.4.2: Let $u = u(t)$ be the solution of the initial-value problem
$$0.01u'' + (1 + t)u' + u = 1 \quad \text{for } t > 0,$$
$$u(0) = 0, \qquad u'(0) = 100 \quad \text{at } t = 0.$$
Give a numerical approximation for the value $u(0.01)$ at $t = 0.01$ with an error less than 0.001. Justify your approximation as regards the error. *Hint:* Take $a = b = 0$ and $c = 1$ in (5.4.1).

Exercise 5.4.3: Let $u = u(t, \varepsilon)$ be the solution of the problem
$$\varepsilon u'' + (1 + t)u' + u = 1 \quad \text{for } t > 0,$$
$$u = 0 \quad \text{and} \quad u' = \frac{1}{\varepsilon} \quad \text{at } t = 0,$$
and find the leading terms
$$U_0(t) + U_0^{**}(t)\exp\left[-\frac{1}{\varepsilon}\left(t + \tfrac{1}{2}t^2\right)\right]$$
in an expansion of the form [see (5.4.4), with U^* replaced there by U^{**}, as discussed for (5.4.32)]
$$u(t, \varepsilon) \sim \sum_{k=0}^{\infty} \left\{ U_k(t) + U_k^{**}(t)\exp\left[-\frac{1}{\varepsilon}\left(t + \tfrac{1}{2}t^2\right)\right] \right\} \varepsilon^k.$$
What can we say about the difference (error) between the exact solution and this leading approximation? [This is an extreme example of the quantitative advantage of the boundary-layer packaging as in (5.4.32).]

Exercise 5.4.4: Use a multivariable approach to study the problem
$$\mathscr{L}u = f(t) \quad \text{for } t > 0,$$
$$u = \alpha, \qquad u' = \beta, \qquad u'' = \gamma \qquad \text{at } t = 0,$$
where $\mathscr{L}$ is the third-order differential operator
$$\mathscr{L} := \varepsilon\frac{d^3}{dt^3} + a(t)\frac{d^2}{dt^2} + b(t)\frac{d}{dt} + c(t),$$
with ε a small positive parameter and with a, b, c given functions that for simplicity are taken here to be independent of ε. Assume also that the constants α, β, γ appearing in the preceding initial conditions are inde-

pendent of ε, and assume that the function a is uniformly positive, with

$$a(t) \geq \kappa > 0 \quad \text{for all } t \geq 0,$$

where κ is a fixed positive constant. Seek the solution u in the form (5.4.3)–(5.4.7). Show that the leading boundary-layer correction term vanishes, with $U_0^*(\tau) \equiv 0$, and use Exercise 5.3.1 to obtain an estimate of the type

$$|u(t, \varepsilon) - U_0(t)| \leq \text{const. } \varepsilon,$$

uniformly for $0 \leq t \leq T$ and for all sufficiently small $\varepsilon > 0$, for any fixed $T > 0$.

5.5 The telegraph equation with large resistance

In this section we shall discuss briefly an initial-value problem for the following hyperbolic system of partial differential equations:

$$\varepsilon \left(\frac{\partial}{\partial t} + \frac{\partial}{\partial x} \right) u + au + bv = f(t, x),$$

$$\varepsilon \left(\frac{\partial}{\partial t} - \frac{\partial}{\partial x} \right) v + cu + dv = g(t, x) \quad \text{for } t > 0, \ -\infty < x < \infty, \tag{5.5.1}$$

for small values of the positive parameter ε. The coefficients a, b, c, d are fixed constants, and the forcing terms f and g are specified given functions of x and t that are defined and smooth on the upper half-plane $t \geq 0$. We also specify the initial values of u and v as

$$u = u_0(x),$$

$$v = v_0(x) \quad \text{at } t = 0, \ -\infty < x < \infty, \tag{5.5.2}$$

where u_0 and v_0 are given smooth function of x. Finally, we shall assume that the coefficients satisfy the conditions

$$a \geq 0, \quad d \geq 0, \quad a + d > 0, \quad \text{and} \quad 0 \leq bc \leq ad. \tag{5.5.3}$$

These conditions are stability conditions that ensure that the solutions of (5.5.1) do not become exponentially unbounded as $\varepsilon \to 0+$ for $t > 0$.

We can easily handle this initial-value problem in the more general case in which the data $a, b, c, d, f, g, u_0, v_0$ are permitted to depend regularly also on ε, but for simplicity we consider here only the case in which these data are independent of ε.

We can eliminate either u or v from (5.5.1)–(5.5.2) so as to obtain an initial-value problem for a second-order wave equation for the remaining

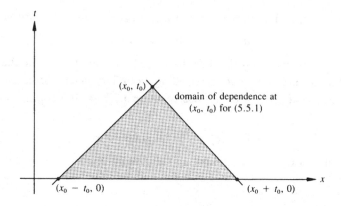

(x_0, t_0)

domain of dependence at
(x_0, t_0) for (5.5.1)

$(x_0 - t_0, 0)$

$(x_0 + t_0, 0)$

Figure 5.4

variable. For example, if we eliminate v, we find for u the equation

$$\varepsilon^2(u_{tt} - u_{xx}) + \varepsilon[(a + d)u_t - (a - d)u_x] + (ad - bc)u$$
$$= df - bg + \varepsilon(f_t - f_x) \quad \text{for } t > 0 \qquad (5.5.4)$$

subject to the initial conditions

$$u = u_0(x),$$

$$u_t = -\frac{du_0(x)}{dx} + \frac{1}{\varepsilon}\{f(x,0) - [au_0(x) + bv_0(x)]\} \quad \text{at } t = 0, \qquad (5.5.5)$$

where $u_t = \partial u/\partial t$, $u_{tt} = \partial^2 u/\partial t^2$, and so forth.

The problem (5.5.1)–(5.5.2) has a unique solution for each fixed value of $\varepsilon > 0$. The values at $(x, t) = (x_0, t_0)$ of the solution functions $u = u(x_0, t_0, \varepsilon)$ and $v = v(x_0, t_0, \varepsilon)$ depend on the data restricted to the fixed triangular domain of dependence D indicated in Figure 5.4, where

$$D := \{(x,t)|\,|x - x_0| \le x_0 + t_0 - t, \quad \text{for } 0 \le t \le t_0\}. \qquad (5.5.6)$$

Note that this domain of dependence is independent of ε.

By way of comparison, we shall see that *the limiting behavior of* $u(x_0, t_0, \varepsilon)$ *and* $v(x_0, t_0, \varepsilon)$ *as* $\varepsilon \to 0+$ *is governed entirely by the data restricted to* (*a neighborhood of*) *the subcharacteristic*

$$D_0 := \left\{(x,t)|x = x_0 + \frac{a - d}{a + d}(t_0 - t), \quad \text{for } 0 \le t \le t_0\right\}, \qquad (5.5.7)$$

as indicated in Figure 5.5. Note that the conditions of (5.5.3) ensure that D_0 is a subset of D.

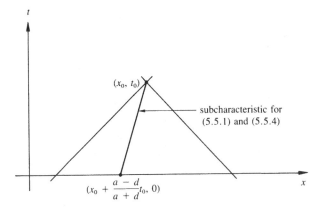

Figure 5.5

Various special cases of the problem (5.5.1)–(5.5.2) represent mathematical models for certain overdamped vibration problems, such as the motion of a vibrating string embedded in a highly viscous medium, the propagation of electrical signals along a conducting wire of large resistance, and the propagation of radiation through a highly absorbing medium. In these applications, the parameter ε is inversely proportional to the damping or absorption coefficient.

The problem (5.5.1)–(5.5.2) is linear, with constant coefficients, and may, of course, be solved in closed form by means of various techniques such as integral transforms (e.g., take the Laplace transform in t, and solve the resulting ordinary differential equations in x) or a Riemann function (integrating factor) approach. For example, if

$$ad = bc \quad \text{and} \quad f = g = 0, \tag{5.5.8}$$

then the solution can be written as

$$
\begin{aligned}
u(x,t,\varepsilon) = {}& e^{at/\varepsilon} u_0(x - t) \\
&+ \frac{t}{2\varepsilon} \exp\left[-\frac{(a+d)t}{2\varepsilon}\right] \int_{-\pi/2}^{\pi/2} \exp\left[\frac{(a-d)t\sin\theta}{2\varepsilon}\right] \\
&\times \left[\sqrt{ad}\,(1 - \sin\theta) I_1\!\left(\frac{\sqrt{ad}\,t\cos\theta}{\varepsilon}\right) u_0(x + t\sin\theta) \right. \\
&\left. \quad - b(\cos\theta) I_0\!\left(\frac{\sqrt{ad}\,t\cos\theta}{\varepsilon}\right) v_0(x + t\sin\theta) \right] d\theta, \tag{5.5.9}
\end{aligned}
$$

where $I_0(z) = J_0(iz)$ and $I_1(z) = -iJ_1(iz)$ are modified Bessel functions, and a similar formula can be given for v. The conditions (5.5.8) are assumed here only for convenience. The solution functions u and v can be similarly given in the general case, but the results are slightly more complicated without (5.5.8). In any case, we include (5.5.9) here only so as to indicate the flavor of these general expressions for the solution functions; we make no actual use of (5.5.9).

Representations such as (5.5.9) for the exact solution functions are not easy to interpret as $\varepsilon \to 0+$. Rather, it is often more convenient to use a perturbation approach that appeals directly to (5.5.1)–(5.5.2). We shall use a multivariable approach here for this purpose, as in Smith (1971).

The reduced system obtained by putting $\varepsilon = 0$ in (5.5.1) is

$$\begin{bmatrix} a & b \\ c & d \end{bmatrix} \begin{bmatrix} u \\ v \end{bmatrix} = \begin{bmatrix} f(x,t) \\ g(x,t) \end{bmatrix}, \tag{5.5.10}$$

which is no longer a differential equation at all, but rather a linear algebraic system that may have a unique solution, no solution, or infinitely many solutions, depending on the data a, b, c, d, f, and g. If the algebraic system (5.5.10) is singular, with $ad - bc = 0$, then the original singularly perturbed system (5.5.1) is said to be a *singular singularly perturbed system*, in the terminology of O'Malley and Flaherty (1977, 1980) and O'Malley (1978b). Such problems are discussed further in Chapter 7.

Turning now to a perturbation analysis of (5.5.1)–(5.5.2), we shall first consider only the special case

$$a = b = c = d > 0, \tag{5.5.11}$$
$$f(x,t) = g(x,t) \quad \text{for } -\infty < x < \infty, \, t \ge 0,$$

and then later we shall discuss the general case subject only to (5.5.3). Hence, we shall first consider the special case of (5.5.1) given as

$$\varepsilon\left(\frac{\partial}{\partial t} + \frac{\partial}{\partial x}\right)u + \tfrac{1}{2}(u + v) = f(x,t),$$
$$\varepsilon\left(\frac{\partial}{\partial t} - \frac{\partial}{\partial x}\right)v + \tfrac{1}{2}(u + v) = f(x,t), \tag{5.5.12}$$

where (5.5.11) holds and where, without loss, we have normalized the coefficients so that they all have the value $a = b = c = d = \tfrac{1}{2}$. Note in this special case that the reduced system (5.5.10) is singular and consistent, so that the reduced equation

$$\tfrac{1}{2}(u + v) = f(x,t) \tag{5.5.13}$$

has infinitely many solutions for u and v. We shall see that the solution functions $u(x, t, \varepsilon)$ and $v(x, t, \varepsilon)$ of the full problem (5.5.12) and (5.5.2) have well-defined limits as $\varepsilon \to 0+$, and these limiting values will satisfy the reduced equation (5.5.13) for $t > 0$. We shall show which of the infinitely many solutions of (5.5.13) is selected by the initial-value problem (5.5.12) and (5.5.2) as $\varepsilon \to 0+$.

The initial-value problem (5.5.12) and (5.5.2) can be easily solved and interpreted in the special case in which *the forcing terms and initial data are all independent of x*, because the solution functions u and v will also be independent of x in this case, and hence the partial differential equations will reduce to ordinary differential equations. In this case we can easily integrate (5.5.12) subject to (5.5.2), and we find

$$u(t, \varepsilon) = \tfrac{1}{2}(u_0 - v_0) + \tfrac{1}{2}e^{-t/\varepsilon}(u_0 + v_0)$$
$$+ \frac{1}{\varepsilon}\int_0^t \exp\left(-\frac{t-s}{\varepsilon}\right) f(s)\,ds, \qquad (5.5.14)$$

and a similar result for v. The integral on the right side of (5.5.14) can be integrated repeatedly by parts (provided f is smooth), and we find the formal expansion

$$u(t, \varepsilon) \sim \sum_{k=0}^{\infty}\left[A_k(t) + e^{-t/\varepsilon}B_k(t)\right]\varepsilon^k, \qquad (5.5.15)$$

with

$$A_k(t) = \begin{cases} \tfrac{1}{2}(u_0 - v_0) + f(t) & \text{for } k = 0, \\ (-1)^k f^{(k)}(t) & \text{for } k = 1, 2, \ldots, \end{cases}$$

and

$$B_k(t) = \begin{cases} \tfrac{1}{2}(u_0 + v_0) - f(0) & \text{for } k = 0, \\ (-1)^{k+1} f^{(k)}(0) & \text{for } k = 1, 2, \ldots, \end{cases}$$

where $f^{(k)} = d^k f/dt^k$. Hence, just as in Section 5.4 [see (5.4.3)], the solution functions depend in an essential way on the two variables t and $\tau := t/\varepsilon$. For small ε, the solution changes rapidly near $t = 0$, and the fast variable $\tau = t/\varepsilon$ plays an important role in describing the fast dynamics of this initial rapid variation.

In the general case in which the forcing terms and the initial data depend also on x, it is natural to seek multivariable expansions of the type

$$u(x, t, \varepsilon) \sim \sum_{k=0}^{\infty}\left[A_k(x, t) + e^{-\tau}B_k(x, t)\right]\varepsilon^k,$$
$$\qquad (5.5.16)$$
$$v(x, t, \varepsilon) \sim \sum_{k=0}^{\infty}\left[C_k(x, t) + e^{-\tau}D_k(x, t)\right]\varepsilon^k,$$

where the fast time τ is to be given here as

$$\tau: = \frac{h(x,t)}{\varepsilon} \qquad (5.5.17)$$

for some suitable function h that is to be determined later. We require that τ must vanish initially when $t = 0$, so that we impose on h the requirement

$$h(x,0) = 0 \quad \text{for all } x. \qquad (5.5.18)$$

The boundary-layer correction terms in the expansions of (5.5.16) are of the general type as in (5.4.32). Different forms are possible for the boundary-layer corrections. In particular, one could also use suitable O'Malley/Hoppensteadt boundary-layer correction terms here, similar to (5.4.6). These various forms for the boundary-layer correction terms are asymptotically equivalent.

If we insert the expansions (5.5.16) into the initial conditions (5.5.2) and use (5.5.17)–(5.5.18), we find formally

$$\sum_{k=0}^{\infty} \left[A_k(x,0) + B_k(x,0) \right] \varepsilon^k \sim u_0(x),$$

$$\sum_{k=0}^{\infty} \left[C_k(x,0) + D_k(x,0) \right] \varepsilon^k \sim v_0(x),$$

and these equations suggest that we impose the requirements

$$A_k(x,0) + B_k(x,0) = \begin{cases} u_0(x) & \text{for } k = 0, \\ 0 & \text{for } k = 1,2,\dots, \end{cases}$$

$$C_k(x,0) + D_k(x,0) = \begin{cases} v_0(x) & \text{for } k = 0, \\ 0 & \text{for } k = 1,2,\dots. \end{cases} \qquad (5.5.19)_k$$

Similarly, if we insert the expressions (5.5.16) into the differential equations (5.5.12), we find formally

$$\sum_{k=0}^{\infty} \left[D_+ A_{k-1} + \tfrac{1}{2} A_k + \tfrac{1}{2} C_k \right] \varepsilon^k$$

$$+ e^{-\tau} \sum_{k=0}^{\infty} \left[D_+ B_{k-1} + \left(\tfrac{1}{2} - D_+ h \right) B_k + \tfrac{1}{2} D_k \right] \varepsilon^k \sim f(x,t)$$

and

$$\sum_{k=0}^{\infty} \left[D_- C_{k-1} + \tfrac{1}{2} A_k + \tfrac{1}{2} C_k \right] \varepsilon^k$$

$$+ e^{-\tau} \sum_{k=0}^{\infty} \left[D_- D_{k-1} + \tfrac{1}{2} B_k + \left(\tfrac{1}{2} - D_- h \right) D_k \right] \varepsilon^k \sim f(x,t),$$

where we have put

$$A_{-1} \equiv B_{-1} \equiv C_{-1} \equiv D_{-1} \equiv 0, \tag{5.5.20}$$

and where we have introduced the differential operators

$$D_+ := \frac{\partial}{\partial t} + \frac{\partial}{\partial x}, \qquad D_- := \frac{\partial}{\partial t} - \frac{\partial}{\partial x}. \tag{5.5.21}$$

The foregoing asymptotic relations obtained from the differential equations of (5.5.12) will hold automatically for all x, t, and τ if we set to zero the expressions multiplying the exponentials and then equate coefficients of like powers of ε in the resulting equations. In this way we obtain the conditions

$$D_+ A_{k-1} + \tfrac{1}{2}(A_k + C_k) = \begin{cases} f(x,t) & \text{for } k = 0, \\ 0 & \text{for } k = 1, 2, \dots, \end{cases}$$

$$D_- C_{k-1} + \tfrac{1}{2}(A_k + C_k) = \begin{cases} f(x,t) & \text{for } k = 0, \\ 0 & \text{for } k = 1, 2, \dots, \end{cases}$$

$$\tag{5.5.22}_k$$

and

$$D_+ B_{k-1} + \left(\tfrac{1}{2} - D_+ h\right) B_k + \tfrac{1}{2} D_k = 0,$$

$$D_- D_{k-1} + \tfrac{1}{2} B_k + \left(\tfrac{1}{2} - D_- h\right) D_k = 0 \quad \text{for all } k = 0, 1, 2, \dots. \tag{5.5.23}_k$$

We now choose the function h so that the matrix

$$\begin{bmatrix} \tfrac{1}{2} - D_+ h & \tfrac{1}{2} \\ \tfrac{1}{2} & \tfrac{1}{2} - D_- h \end{bmatrix} \tag{5.5.24}$$

that occurs in (5.5.23) is singular, because otherwise (5.5.23) and (5.5.20) would imply

$$B_k \equiv D_k \equiv 0 \quad \text{for all } k,$$

and this would eliminate the required boundary-layer correction terms from (5.5.16). Hence, we must impose the requirement that the matrix (5.5.24) be singular with zero determinant, so that the function h must satisfy the following nonlinear first-order partial differential equation:

$$(D_+ h)(D_- h) - \tfrac{1}{2}(D_+ h + D_- h) = 0. \tag{5.5.25}$$

We can easily check that the unique nontrivial solution of (5.5.25) subject to the initial condition (5.5.18) is given as [see Garabedian (1964)]

$$h(x, t) = t. \tag{5.5.26}$$

With this choice of h, $(5.5.23)_k$ becomes

$$D_+ B_{k-1} - \tfrac{1}{2}(B_k - D_k) = 0,$$

$$D_- D_{k-1} + \tfrac{1}{2}(B_k - D_k) = 0 \quad \text{for } k = 0, 1, 2, \ldots.$$

$(5.5.27)_k$

Equations $(5.5.19)_k$, $(5.5.22)_k$, and $(5.5.27)_k$ can now be used recursively to determine the functions A_k, B_k, C_k, and D_k. For example, $(5.5.22)_0$ and $(5.5.27)_0$ imply [with (5.5.20)]

$$A_0(x, t) + C_0(x, t) = 2f(x, t),$$

$$B_0(x, t) - D_0(x, t) = 0,$$

$(5.5.28)$

and $(5.5.22)_1$ and $(5.5.27)_1$ imply

$$D_+ A_0(x, t) = D_- C_0(x, t),$$

$$D_+ B_0(x, t) = -D_- D_0(x, t).$$

$(5.5.29)$

The equations of (5.5.29) can be written as differential equations involving only A_0 and B_0 by using (5.5.28) to eliminate C_0 and D_0 in (5.5.29). The resulting differential equations can be solved for A_0 and B_0, giving all together [with (5.5.28)]

$$A_0(x, t) = A_0(x, 0) + f(x, t) - f(x, 0) - \int_0^t \frac{\partial f(x, s)}{\partial x} \, ds,$$

$$C_0(x, t) = -A_0(x, 0) + f(x, t) + f(x, 0) + \int_0^t \frac{\partial f(x, s)}{\partial x} \, ds, \quad (5.5.30)$$

$$B_0(x, t) = D_0(x, t) = B_0(x, 0).$$

These last equations can now be used with $(5.5.19)_0$ to determine the initial values

$$A_0(x, 0) = f(x, 0) + \tfrac{1}{2}[u_0(x) - v_0(x)],$$

$$B_0(x, 0) = -f(x, 0) + \tfrac{1}{2}[u_0(x) + v_0(x)],$$

$(5.5.31)$

so that (5.5.30) gives finally

$$A_0(x, t) = f(x, t) + \tfrac{1}{2}[u_0(x) - v_0(x)] - \int_0^t \frac{\partial f(x, s)}{\partial x} \, ds,$$

$$C_0(x, t) = f(x, t) - \tfrac{1}{2}[u_0(x) - v_0(x)] + \int_0^t \frac{\partial f(x, s)}{\partial x} \, ds,$$

$(5.5.32)$

and

$$B_0(x, t) = D_0(x, t) = -f(x, 0) + \tfrac{1}{2}[u_0(x) + v_0(x)] \quad (5.5.33)$$

This procedure can be continued recursively to give A_1, B_1, C_1, D_1, A_2, and so forth, and the resulting expansions of (5.5.16) can be shown to be asymptotically correct: *The truncated expansions provide accurate approximations to the exact solution functions* (to the expected orders in ε). The verification of the correctness of the procedure can be patterned after the analogous results of Section 5.3 and Section 5.4, based in this case on suitable comparison results for hyperbolic partial differential equations [see Weinstein and Smith (1976)]. Alternatively, we can give a direct Gronwall-type argument as in Smith and Palmer (1970) and Smith (1971), or we can use energy estimates as in de Jager (1975) and Geel (1981a, 1981b). In these latter references, de Jager and Geel also consider more general problems involving related quasilinear hyperbolic equations.

In the present case we find that the leading terms from (5.5.16) provide first-order approximations $u_0(x, t, \varepsilon)$ and $v_0(x, t, \varepsilon)$ given as [see (5.5.17) and (5.5.26)]

$$
\begin{aligned}
u_0(x, t, \varepsilon) &:= A_0(x, t) + e^{-t/\varepsilon} B_0(x, t), \\
v_0(x, t, \varepsilon) &:= C_0(x, t) + e^{-t/\varepsilon} D_0(x, t).
\end{aligned}
\tag{5.5.34}
$$

The remainders R and S defined by the relations

$$
\begin{aligned}
u(x, t, \varepsilon) &= u_0(x, t, \varepsilon) + R(x, t, \varepsilon), \\
v(x, t, \varepsilon) &= v_0(x, t, \varepsilon) + S(x, t, \varepsilon),
\end{aligned}
\tag{5.5.35}
$$

then satisfy estimates of the type

$$
|R(x, t, \varepsilon)|, |S(x, t, \varepsilon)| \leq \text{const. } \varepsilon
\tag{5.5.36}
$$

uniformly on compact subsets of the upper half-plane $t \geq 0$, as $\varepsilon \to 0+$.

In particular, we have, for any x and any fixed positive value of t, the result

$$
\lim_{\substack{\varepsilon \to 0+ \\ \text{fixed } t > 0}}
\begin{bmatrix} u(x, t, \varepsilon) \\ v(x, t, \varepsilon) \end{bmatrix}
=
\begin{bmatrix} A_0(x, t) \\ B_0(x, t) \end{bmatrix},
\tag{5.5.37}
$$

so that, from among the infinitely many solutions of the reduced equation (5.5.13), we see that the full initial-value problem (5.5.2) and (5.5.12) selects the particular solution

$$
u = A_0(x, t), \qquad v = C_0(x, t),
\tag{5.5.38}
$$

as $\varepsilon \to 0+$, where A_0 and C_0 are given by (5.5.32). It should be noted that this result is not at all obvious from the exact solution (5.5.9). [We see directly with (5.5.32) that the functions of (5.5.38) do in fact satisfy the reduced equation (5.5.13).]

Finally, we turn now to a brief discussion of the more general system (5.5.1) subject to the stability conditions of (5.5.3), but without the more restrictive conditions of (5.5.11). There are several cases to consider. First, if the reduced (algebraic) system (5.5.10) has infinitely many solutions along the subcharacteristic issuing down from (x, t) illustrated in Figure 5.5 [that is, if the coefficient matrix on the left side of (5.5.10) is singular and if the forcing vector on the right side is in the column space of this coefficient matrix along the subcharacteristic, as occurs for (5.5.11)], then the values of the solution functions $u(x, t, \varepsilon)$ and $v(x, t, \varepsilon)$ of (5.5.1)–(5.5.2) have well-defined (finite) limits as $\varepsilon \to 0+$, and these limiting values satisfy the reduced system (5.5.10) for $t > 0$. On the other hand, if the reduced system (5.5.10) has no solution along the previously mentioned subcharacteristic (that is, if the coefficient matrix is singular and the forcing vector is not in the column space of the coefficient matrix along the subcharacteristic), then the values $u(x, t, \varepsilon)$ and $v(x, t, \varepsilon)$ become unbounded like $1/\varepsilon$ as $\varepsilon \to 0+$ for fixed $t > 0$, where in this case the summations in (5.5.16) are modified to include also the summation index $k = -1$. Finally, if the reduced system (5.5.10) is nonsingular, then the values of the solution functions u and v of (5.5.1)–(5.5.2) tend to the unique solution of (5.5.10) as $\varepsilon \to 0+$, for fixed $t > 0$. In this latter case we use two new variables τ_1 and τ_2 in addition to x and t, because equation (5.5.25) now becomes

$$(D_+h)(D_-h) - (aD_-h + dD_+h) + (ad - bc) = 0, \quad (5.5.39)$$

which has two nontrivial solutions subject to the initial conditions (5.5.18). In all cases, the solution functions $u(x, t, \varepsilon)$ and $v(x, t, \varepsilon)$ exhibit boundary-layer behavior near $t = 0$ as $\varepsilon \to 0+$, and the multivariable technique leads to correct (uniformly valid) asymptotic expansions in all cases. We refer the reader to Smith (1971) for details.

As mentioned in Section 5.4, different forms are possible for the boundary-layer correction terms in multivariable expansions such as (5.5.16). In particular, the present boundary-layer terms in (5.5.16)–(5.5.17) can be replaced in every case with suitable O'Malley/Hoppensteadt boundary-layer terms similar to (5.4.6). This latter approach generalizes nicely to variable-coefficient and nonlinear problems, as shown in de Jager (1975) and Geel (1978, 1981a, 1981b).

Hsiao and Weinacht (1983) used a multivariable approach along with energy estimates to study initial-value problems for singularly perturbed hyperbolic equations for which the reduced equation was parabolic. Such problems have also been studied by Bobisud (1966) and others.

6

Nonlinear overdamped initial-value problems

In this chapter we consider certain nonlinear generalizations of the linear problems considered in Chapter 5. Such nonlinear problems occur in many areas, including biochemical kinetics, genetics, plasma physics, and mechanical and electrical systems involving large damping or resistance. The O'Malley/Hoppensteadt method is used here in the construction of approximate solutions for such problems. The use of such boundary-layer techniques for the *numerical* solution of overdamped initial-value problems is also discussed, with emphasis on the work of Miranker (1973).

6.1 Introduction

In this chapter we consider the initial-value problem

$$\frac{dx}{dt} = u(t, x, y, \varepsilon),$$

$$\varepsilon \frac{dy}{dt} = v(t, x, y, \varepsilon) \quad \text{for } t > 0,$$

(6.1.1)

and

$$x = \alpha(\varepsilon),$$

$$y = \beta(\varepsilon) \quad \text{at } t = 0.$$

(6.1.2)

We can handle the vector case for solution functions x and y that are, respectively, m-dimensional and n-dimensional real vector-valued functions, where the data quantities u, v, α, and β are suitable vector-valued functions. Indeed, the vector case is discussed near the end of this section and in later sections as well, and similar techniques are used in Chapter 7 for a related problem involving a vector system. However, for simplicity we shall consider first the scalar case in which all functions are real-

163

valued. Hence, in (6.1.1), the functions $u = u(t, x, y, \varepsilon)$ and $v = v(t, x, y, \varepsilon)$ are assumed for the moment to be given, smooth real-valued functions of four real variables, and the initial quantities $\alpha = \alpha(\varepsilon)$ and $\beta = \beta(\varepsilon)$ in (6.1.2) are given real-valued functions of the real parameter ε for all small $\varepsilon > 0$.

We assume that the given functions u and v are regular at $\varepsilon = 0$, by which we mean that these functions have asymptotic power-series expansions as

$$\begin{bmatrix} u(t, x, y, \varepsilon) \\ v(t, x, y, \varepsilon) \end{bmatrix} \sim \sum_{k=0}^{\infty} \begin{bmatrix} u_k(t, x, y) \\ v_k(t, x, y) \end{bmatrix} \varepsilon^k \qquad (6.1.3)$$

for suitable given smooth functions $u_k = u_k(t, x, y)$ and $v_k = v_k(t, x, y)$ for all required values of t, x, and y. Expansions such as (6.1.3) need only hold to some specified finite order for a suitable partial sum on the right side of (6.1.3), but this is not of primary interest here.

The initial quantities $\alpha(\varepsilon)$ and $\beta(\varepsilon)$ in (6.1.2) are assumed to satisfy either

$$\begin{bmatrix} \alpha(\varepsilon) \\ \beta(\varepsilon) \end{bmatrix} \sim \sum_{k=0}^{\infty} \begin{bmatrix} \alpha_k \\ \beta_k \end{bmatrix} \varepsilon^k \qquad (6.1.4)$$

or

$$\begin{bmatrix} \alpha(\varepsilon) \\ \beta(\varepsilon) \end{bmatrix} \sim \sum_{k=0}^{\infty} \begin{bmatrix} \alpha_k \\ \frac{1}{\varepsilon}\beta_k \end{bmatrix} \varepsilon^k, \quad \beta_0 \neq 0, \qquad (6.1.5)$$

as $\varepsilon \to 0+$, for given constants α_k and β_k. Note that the leading term in the expansion for $\beta(\varepsilon)$ in (6.1.5) is of the form $\beta_0 \varepsilon^{-1}$ as $\varepsilon \to 0+$. Each of these two possibilities (6.1.4) and (6.1.5) occurs in certain different applications, and other possibilities exist as well. Following O'Malley (1971a, 1971b) and Hoppensteadt (1971b), we shall consider mainly the case (6.1.4), although we also indicate the situation for (6.1.5) in Exercises 6.2.3 and 6.4.2. Roughly speaking, the difference between the two cases is that we first construct the outer solution in the case (6.1.4), as in Section 5.4 in the case $c = 0$ in (5.4.1), whereas we first construct the boundary-layer correction in the case (6.1.5), as in Section 5.4 in the case $c \neq 0$. Hence, as expected, we see that the detailed structure of solution functions $x = x(t, \varepsilon)$ and $y = y(t, \varepsilon)$ for (6.1.1)–(6.1.2) depends on the precise structure of the initial data at $\varepsilon = 0$. We have here yet another illustration of the richness of phenomena for singularly perturbed problems.

The data need not possess complete asymptotic expansions of these types; useful results are obtained if the data possess only partial expansions of some finite order.

Example 6.1.1: If the given functions u and v are defined as

$$u(t, x, y, \varepsilon) \equiv u_0(t, x, y): = y,$$

$$v(t, x, y, \varepsilon) \equiv v_0(t, x, y): = f(t) - B(t)x - A(t)y,$$

(6.1.6)

for given functions $A = A(t)$, $B = B(t)$, and $f = f(t)$, then the problem (6.1.1)–(6.1.2) is equivalent to the second-order initial-value problem

$$\varepsilon \frac{d^2 x}{dt^2} + A(t) \frac{dx}{dt} + B(t)x = f(t) \quad \text{for } t > 0,$$

(6.1.7)

$$x = \alpha(\varepsilon), \qquad \frac{dx}{dt} = \beta(\varepsilon) \quad \text{at } t = 0,$$

as considered earlier in Chapter 5, where the functions $A(t)$ and $B(t)$ were labelled as $A(t) = \alpha(t)$ and $B(t) = \beta(t)$ in (5.2.1) and (5.4.2), and where the earlier initial conditions of (5.2.2) and (5.4.1) are given here with

$$\alpha(\varepsilon) = a, \qquad \beta(\varepsilon) = b + c/\varepsilon.$$

(6.1.8)

Hence, depending on whether $c = 0$ or $c \neq 0$ in Chapter 5, we see that the initial quantities $\alpha(\varepsilon)$ and $\beta(\varepsilon)$ are here either of the form (6.1.4) or of the form (6.1.5). Initial values of the form (6.1.5) also appear in certain perturbation problems in optimal control theory; see O'Malley (1978a).

Example 6.1.2 (Heineken, Tsuchiya, and Aris 1967): The following system arises as a mathematical model in the kinetic theory of enzyme reactions:

$$\frac{dx}{dt} = -x + (x + \kappa - \lambda)y,$$

(6.1.9)

$$\varepsilon \frac{dy}{dt} = x - (x + \kappa)y \quad \text{for } t > 0,$$

with initial conditions

$$x = 1, \qquad y = 0 \quad \text{at } t = 0.$$

(6.1.10)

Here, κ and λ are given positive constants, x and y give the concentrations of two species in the reaction, and the positive parameter ε gives the ratio of the initial enzyme and substrate concentrations.

This problem (6.1.9)–(6.1.10) is of the form (6.1.1)–(6.1.4), with

$$\begin{bmatrix} u_0(t, x, y) \\ v_0(t, x, y) \end{bmatrix} := \begin{bmatrix} -x + (x + \kappa - \lambda) y \\ x - (x + \kappa) y \end{bmatrix}, \tag{6.1.11}$$

$$\begin{bmatrix} \alpha_0 \\ \beta_0 \end{bmatrix} := \begin{bmatrix} 1 \\ 0 \end{bmatrix}, \tag{6.1.12}$$

and all other u_k, v_k, α_k, and β_k zero for $k \geq 1$.

Before considering the general nonlinear problem (6.1.1)–(6.1.4), it is useful first to consider the special linear case in which the given functions u and v have the forms

$$\begin{aligned} u(t, x, y, \varepsilon) &:= A(t, \varepsilon)x + B(t, \varepsilon)y + f(t, \varepsilon), \\ v(t, x, y, \varepsilon) &:= C(t, \varepsilon)x + D(t, \varepsilon)y + g(t, \varepsilon), \end{aligned} \tag{6.1.13}$$

for given functions $A = A(t, \varepsilon)$, $B = B(t, \varepsilon)$, $C = C(t, \varepsilon)$, $D = D(t, \varepsilon)$, $f = f(t, \varepsilon)$ and $g = g(t, \varepsilon)$ defined on the region

$$0 \leq t \leq T, \qquad 0 < \varepsilon \leq \varepsilon_0, \tag{6.1.14}$$

for given positive constants T and ε_0. These functions are assumed to be uniformly continuous in t for $0 \leq t \leq T$, for each fixed positive ε in the given range $0 < \varepsilon \leq \varepsilon_0$. Moreover, the function D is assumed to be uniformly negative, with

$$D(t, \varepsilon) \leq -\kappa < 0 \tag{6.1.15}$$

for all (t, ε) satisfying (6.1.14), for a fixed positive constant $\kappa > 0$.

Hence, for this linear case we insert (6.1.13) into (6.1.1) and consider the initial-value problem

$$\begin{aligned} \frac{dx}{dt} &= A(t, \varepsilon)x + B(t, \varepsilon)y + f(t, \varepsilon), \\ \varepsilon \frac{dy}{dt} &= C(t, \varepsilon)x + D(t, \varepsilon)y + g(t, \varepsilon) \quad \text{for } 0 \leq t \leq T, \end{aligned} \tag{6.1.16}$$

with

$$x = \alpha(\varepsilon), \qquad y = \beta(\varepsilon) \quad \text{at } t = 0. \tag{6.1.17}$$

We shall study (6.1.16)–(6.1.17) using variation of parameters, and for this purpose it is convenient to consider also the corresponding homogeneous system

$$\begin{aligned} \frac{dx}{dt} &= A(t, \varepsilon)x + B(t, \varepsilon)y, \\ \varepsilon \frac{dy}{dt} &= C(t, \varepsilon)x + D(t, \varepsilon)y. \end{aligned} \tag{6.1.18}$$

We can integrate (6.1.18) and find the following equivalent integral equation for the slow variable x [see Exercise 7.1.1, and note that the integration order has been reversed in the last double integral on the right side of (6.1.19)],

$$x(t) = \exp\left(\int_0^t A\right)x(0) + \int_0^t \exp\left(\int_s^t A\right)B(s,\varepsilon)\exp\left(\frac{1}{\varepsilon}\int_0^s D\right)y(0)\,ds$$
$$+ \frac{1}{\varepsilon}\int_0^t\left[\int_s^t \exp\left(\int_\sigma^t A\right)B(\sigma,\varepsilon)\exp\left(\frac{1}{\varepsilon}\int_s^\sigma D\right)d\sigma\right]C(s,\varepsilon)x(s)\,ds,$$

(6.1.19)

along with the equation for the fast variable y,

$$y(t) = \exp\left(\frac{1}{\varepsilon}\int_0^t D\right)y(0) + \frac{1}{\varepsilon}\int_0^t \exp\left(\frac{1}{\varepsilon}\int_s^t D\right)C(s,\varepsilon)x(s)\,ds, \quad (6.1.20)$$

where the dependence of x and y on ε is being suppressed here.

Using (6.1.15), we have the bound

$$\int_s^t D \le -\kappa\cdot(t-s) \quad \text{for } s \le t, \tag{6.1.21}$$

and then (6.1.19) and (6.1.21) lead directly to the inequality

$$|x(t)| \le e^{\|A\|T}\left(|x(0)| + \frac{\varepsilon}{\kappa}|y(0)|\cdot\|B\| + \frac{\|B\|\cdot\|C\|}{\kappa}\int_0^t|x(s)|\,ds\right),$$

(6.1.22)

where the maximum norm is used here,

$$\|h\| := \max_{0\le t\le T}|h(t,\varepsilon)|, \tag{6.1.23}$$

for any suitable function $h = h(t,\varepsilon)$, and where we are suppressing the dependence of $\|h\|$ on ε. Gronwall's inequality can be used to resolve (6.1.22), and then the resulting estimate for $|x(t)|$ can be used in (6.1.20) to provide an estimate for $|y(t)|$. In this way we find directly for any solution of the homogeneous system (6.1.18) the estimates

$$|x(t,\varepsilon)| \le e^{\|A\|T}\left(|x(0)| + \frac{\varepsilon}{\kappa}|y(0)|\cdot\|B\|\right)\exp\left(\|B\|\cdot\|C\|\frac{T}{\kappa}e^{\|A\|T}\right),$$

$$|y(t,\varepsilon)| \le e^{-\kappa t/\varepsilon}|y(0)| + \frac{\|C\|\cdot\|x\|}{\kappa} \quad \text{for } 0\le t\le T, \tag{6.1.24}$$

where $\|x\|$ in the second inequality can be estimated here by the right side of the first inequality of (6.1.24).

We now let

$$\begin{bmatrix} x_1(t,\varepsilon) \\ y_1(t,\varepsilon) \end{bmatrix} \quad \text{and} \quad \begin{bmatrix} x_2(t,\varepsilon) \\ y_2(t,\varepsilon) \end{bmatrix}$$

denote the solutions of the homogeneous system (6.1.18) satisfying the initial conditions of

$$\begin{bmatrix} x_1(0,\varepsilon) \\ y_1(0,\varepsilon) \end{bmatrix} = \begin{bmatrix} 1 \\ 0 \end{bmatrix}, \quad \begin{bmatrix} x_2(0,\varepsilon) \\ y_2(0,\varepsilon) \end{bmatrix} = \begin{bmatrix} 0 \\ 1 \end{bmatrix}, \qquad (6.1.25)$$

respectively, and we let $Z = Z(t,\varepsilon)$ be the fundamental matrix with column vectors given by these independent solutions,

$$Z(t,\varepsilon) := \begin{bmatrix} x_1(t,\varepsilon) & x_2(t,\varepsilon) \\ y_1(t,\varepsilon) & y_2(t,\varepsilon) \end{bmatrix}. \qquad (6.1.26)$$

From (6.1.24)–(6.1.25) we have the bounds

$$|x_1(t,\varepsilon)| \le \gamma, \qquad |x_2(t,\varepsilon)| \le \frac{\varepsilon \|B\| \gamma}{\kappa},$$

$$|y_1(t,\varepsilon)| \le \frac{\|C\| \gamma}{\kappa}, \qquad |y_2(t,\varepsilon)| \le e^{-\kappa t/\varepsilon} + \frac{\varepsilon \|B\| \cdot \|C\| \gamma}{\kappa^2}, \qquad (6.1.27)$$

uniformly for $0 \le t \le T$, where

$$\gamma := \exp(\|A\| T) \exp\left(\|B\| \cdot \|C\| \frac{T}{\kappa} e^{\|A\| T}\right). \qquad (6.1.28)$$

From these developments we find the following theorem.

Theorem 6.1.1: *Let the real-valued data functions A, B, C, D, f, and g be continuous in t for each ε on the region (6.1.14), and assume that the function D satisfies the stability condition (6.1.15). Then the solution of the initial-value problem (6.1.16)–(6.1.17) can be represented as*

$$\begin{bmatrix} x(t,\varepsilon) \\ y(t,\varepsilon) \end{bmatrix} = K(t,0,\varepsilon) \begin{bmatrix} \alpha(\varepsilon) \\ \beta(\varepsilon) \end{bmatrix} + \int_0^t K(t,s,\varepsilon) \begin{bmatrix} f(s,\varepsilon) \\ \dfrac{1}{\varepsilon} g(s,\varepsilon) \end{bmatrix} ds, \quad (6.1.29)$$

where the resolvent matrix $K(t,s,\varepsilon)$ is given as

$$K(t,s,\varepsilon) = \begin{bmatrix} k_{11}(t,s,\varepsilon) & k_{12}(t,s,\varepsilon) \\ k_{21}(t,s,\varepsilon) & k_{22}(t,s,\varepsilon) \end{bmatrix} := Z(t,\varepsilon) Z(s,\varepsilon)^{-1}$$

$$(6.1.30)$$

and satisfies the estimates

$$|k_{11}(t, s, \varepsilon)| \leq \gamma, \qquad |k_{12}(t, s, \varepsilon)| \leq \frac{\varepsilon \|B\| \gamma}{\kappa}$$

$$|k_{21}(t, s, \varepsilon)| \leq \frac{\|C\| \gamma}{\kappa}, \qquad |k_{22}(t, s, \varepsilon)| \leq e^{-\kappa(t-s)/\varepsilon} + \frac{\varepsilon \|B\| \cdot \|C\| \gamma}{\kappa^2}$$

$$(6.1.31)$$

uniformly for $0 \leq s \leq t \leq T$, *with* γ *given by* (6.1.28).

Proof: The variation-of-parameters formula (6.1.29)–(6.1.30) is well known (see Exercise 0.1.1), so that we need only prove the estimates (6.1.31). To this end we note that the column vectors of the resolvent matrix $K(t, s, \varepsilon) := Z(t, \varepsilon)Z(s, \varepsilon)^{-1}$ satisfy the homogeneous system (6.1.18) as functions of t, with appropriate initial conditions at $t = s$ obtained from

$$K(t, s, \varepsilon) = \begin{bmatrix} 1 & 0 \\ 0 & 1 \end{bmatrix} \quad \text{at } t = s.$$

Then the argument leading to (6.1.27) can be repeated for these column vectors of $K(t, s, \varepsilon)$ with initial time $t = s$, and we find the estimates (6.1.31). ∎

An analogous version of Theorem 6.1.1 is valid also in the case of a *vector system* of the type (6.1.16) for solution functions $x = x(t, \varepsilon)$ and $y = y(t, \varepsilon)$, which are, respectively, m-dimensional and n-dimensional real vector-valued functions, where the given data functions A, B, C, D, f, and g are real matrix-valued and vector-valued functions with appropriate compatible orders, and where we assume always that these data functions are at least piecewise continuous in t, and uniformly bounded for all t and ε satisfying (6.1.14).

In this case we let $X = X(t, \varepsilon)$ and $Y = Y(t, \varepsilon)$ denote fundamental solution matrices for the respective homogeneous systems

$$\frac{dx}{dt} = A(t, \varepsilon)x \qquad (6.1.32)$$

and

$$\varepsilon \frac{dy}{dt} = D(t, \varepsilon)y, \qquad (6.1.33)$$

so that the corresponding resolvent (Green) matrices ξ and η for the initial-value problem for the systems (6.1.32) and (6.1.33) are given as

$$\xi(t, s, \varepsilon) := X(t, \varepsilon)X(s, \varepsilon)^{-1} \qquad (6.1.34)$$

and

$$\eta(t, s, \varepsilon) := Y(t, \varepsilon) Y(s, \varepsilon)^{-1}. \qquad (6.1.35)$$

Then $\xi(t, s, \varepsilon)$ satisfies

$$\frac{d\xi}{dt} = A(t, \varepsilon)\xi \quad \text{for } t > s,$$

$$\xi = I_m = m \times m \text{ identity matrix, at } t = s, \qquad (6.1.36)$$

and $\eta(t, s, \varepsilon)$ satisfies

$$\varepsilon \frac{d\eta}{dt} = D(t, \varepsilon)\eta \quad \text{for } t > s,$$

$$\eta = I_n = n \times n \text{ identity matrix, at } t = s. \qquad (6.1.37)$$

The previous scalar equations of (6.1.19)–(6.1.20) are replaced by the corresponding vector equations

$$x(t) = \xi(t, 0, \varepsilon)x(0) + \int_0^t \xi(t, s, \varepsilon) B(s, \varepsilon)\eta(s, 0, \varepsilon) y(0) \, ds$$

$$+ \frac{1}{\varepsilon} \int_0^t \left[\int_s^t \xi(t, \sigma, \varepsilon) B(\sigma, \varepsilon)\eta(\sigma, s, \varepsilon) \, d\sigma \right] C(s, \varepsilon)x(s) \, ds$$

$$(6.1.38)$$

and

$$y(t) = \eta(t, 0, \varepsilon) y(0) + \frac{1}{\varepsilon} \int_0^t \eta(t, s, \varepsilon) C(s, \varepsilon)x(s) \, ds, \qquad (6.1.39)$$

and these equations lead directly to an appropriate vector version of Theorem 6.1.1 *subject to the requirement that the resolvent function* $\eta(t, s, \varepsilon)$ *satisfies the stability condition*

$$|\eta(t, s, \varepsilon)| \le \kappa_0 \cdot \exp\left[-\kappa_1 \cdot (t - s)/\varepsilon\right] \qquad (6.1.40)$$

for fixed positive constants κ_0 and κ_1, uniformly for $0 \le s \le t \le T$, $0 < \varepsilon \le \varepsilon_0$, where for definiteness we shall use the matrix norm

$$|\eta| := \max_{i, j} |\eta_{ij}|$$

for any matrix $\eta = (\eta_{ij})$.

The condition (6.1.40) replaces the previous result

$$\exp\left(\frac{1}{\varepsilon} \int_s^t D\right) \le \exp\left[-\kappa \cdot (t - s)/\varepsilon\right] \quad \text{for } t \ge s, \qquad (6.1.41)$$

where (6.1.41) follows from the assumption (6.1.15) in the scalar case. A well-known result of Flatto and Levinson (1955) guarantees in the vector case that the analogous condition (6.1.40) holds if $D(t, \varepsilon)$ is continuous in t and uniformly bounded on the region of (6.1.14), and if all eigenvalues $\lambda = \lambda(t, \varepsilon)$ of the matrix-valued function $D(t, \varepsilon)$ have uniformly negative real parts,

$$\operatorname{Re}\lambda(t,\varepsilon) \le -\kappa_2 < 0, \tag{6.1.42}$$

for a fixed positive constant $\kappa_2 > \kappa_1$, uniformly for all t and ε satisfying (6.1.14). This condition (6.1.42) then replaces the previous condition (6.1.15). The proof of this result of Flatto and Levinson is discussed in Exercise 6.1.5; see also Ferguson (1975).

Let the $m \times m$ matrix-valued function $x_1(t,\varepsilon)$ and the $n \times m$ matrix-valued function $y_1(t,\varepsilon)$ be the solution functions of the homogeneous vector system (6.1.18) subject to the initial conditions

$$\begin{bmatrix} x_1 \\ y_1 \end{bmatrix} = \begin{bmatrix} I_m \\ 0 \end{bmatrix}_{n \times m} \quad \text{at } t = 0. \tag{6.1.43}$$

Then, just as in the scalar case, the boundedness of the data and the stability condition (6.1.40), along with the integral relations (6.1.38) and (6.1.39) and the initial conditions (6.1.43), lead directly to estimates of the type

$$|x_1(t,\varepsilon)|, |y_1(t,\varepsilon)| \le \text{const.} \tag{6.1.44}$$

uniformly for all t and ε satisfying (6.1.14). The derivation of this result actually provides an explicit representation for the constant on the right side of (6.1.44) in terms of the data, analogous to the right sides appearing in the estimates of (6.1.27), but this explicit representation for the constant need not be given here. Note also that the derivations of both (6.1.44) and the related following result (6.1.47) make use of several auxiliary estimates such as the uniform bound

$$|\xi(t,s,\varepsilon)| \le |I_m|e^{\|A\|T} \tag{6.1.45}$$

for $0 \le s \le t \le T$, all small $\varepsilon > 0$, where this bound follows easily from (6.1.36) by means of a Gronwall-type argument.

As in the derivation of (6.1.44), we now let the $m \times n$ matrix-valued function $x_2(t,\varepsilon)$ and the $n \times n$ matrix-valued function $y_2(t,\varepsilon)$ be the solution functions of the system (6.1.18) subject to the initial conditions

$$\begin{bmatrix} x_2 \\ y_2 \end{bmatrix} = \begin{bmatrix} 0 \\ I_n \end{bmatrix}^{m \times n} \quad \text{at } t = 0. \tag{6.1.46}$$

In this case, we easily find estimates of the type

$$\begin{aligned} |x_2(t,\varepsilon)| &\le \text{const. } \varepsilon, \\ |y_2(t,\varepsilon)| &\le \text{const.} \left(\varepsilon + e^{-\kappa_1 t/\varepsilon}\right), \end{aligned} \tag{6.1.47}$$

uniformly for all t and ε satisfying (6.1.14).

As in the scalar case, we now let $Z = Z(t, \varepsilon)$ be the particular fundamental solution matrix for the vector homogeneous system (6.1.18) given in block-partitioned form as

$$Z(t, \varepsilon): = \begin{bmatrix} x_1(t, \varepsilon) & \vdots & x_2(t, \varepsilon) \\ \cdots\cdots\cdots & \vdots & \cdots\cdots\cdots \\ y_1(t, \varepsilon) & \vdots & y_2(t, \varepsilon) \end{bmatrix}, \tag{6.1.48}$$

and then variation of parameters again gives the representation (6.1.29)–(6.1.30) for the solution functions x and y of the nonhomogeneous vector initial-value problem (6.1.16)–(6.1.17), with the $(m + n) \times (m + n)$ resolvent matrix K again given as

$$K(t, s, \varepsilon): = Z(t, \varepsilon) Z(s, \varepsilon)^{-1}. \tag{6.1.49}$$

As in the scalar case, we again partition K along the lines of the partition of Z in (6.1.48), as

$$K(t, s, \varepsilon) = \begin{bmatrix} k_{11}(t, s, \varepsilon) & \vdots & k_{12}(t, s, \varepsilon) \\ \cdots\cdots\cdots & \vdots & \cdots\cdots\cdots \\ k_{21}(t, s, \varepsilon) & \vdots & k_{22}(t, s, \varepsilon) \end{bmatrix}, \tag{6.1.50}$$

where k_{11} and k_{22} are square $m \times m$ and $n \times n$ matrix-valued functions, respectively, and k_{12} and k_{21} are $m \times n$ and $n \times m$ matrix-valued functions. We then have the following theorem.

Theorem 6.1.2: *Let the data functions A, B, C, D, f, g, α, and β be suitable matrix-valued and vector-valued functions of appropriate compatible orders as discussed earlier, and assume that the functions A, B, C, D, f, and g are piecewise continuous in t, with A, B, C, and D uniformly bounded for all t and ε satisfying (6.1.14), for fixed positive constants T and ε_0. Assume also that the matrix-valued function D is uniformly stable; i.e., assume that the resolvent function η of (6.1.37) satisfies (6.1.40) for fixed positive constants κ_0 and κ_1. Then the $(m + n) \times (m + n)$ resolvent matrix $K = K(t, s, \varepsilon)$ of (6.1.49)–(6.1.50) appearing in the variation-of-parameters formula (6.1.29)–(6.1.30) for the initial-value problem (6.1.16)–(6.1.17) satisfies estimates of the type [compare with (6.1.44) and (6.1.47)]*

$$|k_{11}(t, s, \varepsilon)| \le \gamma_{11}, \qquad |k_{12}(t, s, \varepsilon)| \le \gamma_{12} \cdot \varepsilon,$$

$$|k_{21}(t, s, \varepsilon)| \le \gamma_{21}, \qquad |k_{22}(t, s, \varepsilon)| \le \gamma_{22}(\varepsilon + e^{-\kappa_1(t-s)/\varepsilon}), \tag{6.1.51}$$

for suitable fixed positive constants γ_{ij} $(i, j = 1, 2)$, uniformly for all t and ε satisfying (6.1.14).

Proof: As in the scalar case, the various column vectors of $K(t, s, \varepsilon)$ satisfy the homogeneous system (6.1.18) as functions of t with appropriate initial conditions at $t = s$. The argument leading to the estimates (6.1.44) and (6.1.47) can be repeated for these column vectors of $K(t, s, \varepsilon)$ with initial time $t = s$, and we find the estimates of (6.1.51). ∎

Alternatively, the estimates of (6.1.51) can be derived using a Riccati transformation, as discussed in Sections 9.2 and 9.3.

The case in which the given matrices $A = A(\varepsilon)$, $B = B(\varepsilon)$, $C = C(\varepsilon)$, and $D = D(\varepsilon)$ are independent of t is studied in Campbell (1980) subject to the weaker assumption that D is only semistable, which permits D to be singular with all nonzero eigenvalues having negative real parts. Related nonlinear problems are considered in Campbell (1982). See also Coppel (1965, 1978) and Harris and Lutz (1977) for general theories of asymptotic integration.

Exercises

Exercise 6.1.1: Show that any solution of the homogeneous system (6.1.18) must satisfy the integral equations of (6.1.19)–(6.1.20) or (6.1.38)–(6.1.39) in the respective scalar and vector cases. *Hint:* Write the system (6.1.18) as

$$\frac{dx}{dt} - A(t, \varepsilon)x = B(t, \varepsilon)y,$$

$$\varepsilon \frac{dy}{dt} - D(t, \varepsilon)y = C(t, \varepsilon)x,$$

and integrate these last equations by variation of parameters, treating the right sides as given functions.

Exercise 6.1.2: Give the details in the derivation of the estimates (6.1.24).

Exercise 6.1.3: Consider the initial-value problem

$$\frac{dx}{dt} = A(t)x + B(t)y,$$

$$\frac{dy}{dt} = C(t)x + D(t)y \quad \text{for } t > 0,$$

(6.1.52)

with x and y specified at $t = 0$, for real-valued solution functions x and y, where the given real-valued functions A, B, C, and D are defined and continuous (or piecewise-continuous) on the half-line $[0, \infty)$. In this case it is well known (and follows from the discussion in Section 0.2) that

(6.1.52) has a unique global solution $x(t)$, $y(t)$ for all $t \geq 0$. Assume that the coefficients in (6.1.52) are known to satisfy the inequalities

$$|A(t)|, |C(t)| \leq \text{const. } e^{-\kappa t},$$

$$|B(t)|, |D(t)| \leq \text{constant}, \quad \text{for all } t \geq 0,$$

$$(6.1.53)$$

along with the condition

$$D(t) \leq -\kappa < 0 \quad \text{for } t \geq 0, \tag{6.1.54}$$

for a fixed positive constant κ. Prove that the solution function y decays exponentially toward zero with increasing t, with

$$|y(t)| \leq \text{const. } (1+t)e^{-\kappa t} \leq \text{const. } e^{-\kappa \mu t} \quad \text{for } t \geq 0, \quad (6.1.55)$$

for any fixed positive constant μ satisfying $0 < \mu < 1$. Prove also that the solution function x is bounded and moreover has a well-defined limit at infinity, so that

$$\lim_{t \to \infty} x(t) \text{ exists.} \tag{6.1.56}$$

Hint: Put $\varepsilon = 1$ in (6.1.19)–(6.1.20), and use (6.1.53)–(6.1.54) to obtain an integral inequality of the type

$$|x(t)| \leq \text{const. } \left(1 + \int_0^t e^{-\kappa s}|x(s)|\, ds\right) \quad \text{for } t \geq 0.$$

A Gronwall-type argument then gives a uniform bound on x of the type $|x(t)| \leq \text{constant}$, for all $t \geq 0$. This latter result can be used in (6.1.20) along with (6.1.53)–(6.1.54) to obtain the result (6.1.55). The result (6.1.56) then follows from the first equation of (6.1.52) on integration, treating $y = y(t)$ as a given function.

Exercise 6.1.4: Consider the nonhomogeneous problem

$$\frac{dx}{dt} = A(t)x + B(t)y + F(t),$$

$$\frac{dy}{dt} = C(t)x + D(t)y + G(t) \quad \text{for } t > 0,$$

$$(6.1.57)$$

with x and y specified at $t = 0$, where the coefficient functions A, B, C, and D are as in Exercise 6.1.3, and the given continuous (or piecewise-continuous) functions F and G satisfy

$$|F(t)|, |G(t)| \leq \text{const. } e^{-\kappa \mu t} \quad \text{for all } t \geq 0, \tag{6.1.58}$$

where κ is the constant of (6.1.53)–(6.1.54) and μ is a fixed constant as in (6.1.55), with $0 < \mu < 1$. Prove that the solution function y decays exponentially as

$$|y(t)| \leq \text{const. } e^{-\kappa \mu t} \quad \text{for } t \geq 0, \tag{6.1.59}$$

and prove that the solution function x is uniformly bounded and has a well-defined limit at infinity,

$$\lim_{t \to \infty} x(t) \text{ exists.} \tag{6.1.60}$$

Hint: Use an appropriate representation as (6.1.29) for a suitable matrix $K = K(t, s) = [k_{ij}(t, s)]_{i, j=1,2}$, with

$$|k_{1j}(t, s)| \le \text{constant}, \quad \text{for } j = 1, 2,$$

$$|k_{2j}(t, s)| \le \text{const.} \left[1 + (t - s)\right] e^{-\kappa(t-s)} \quad \text{for } j = 1, 2,$$

uniformly for $t \ge s$. (Corresponding versions of Exercises 6.1.3 and 6.1.4 are valid also for analogous vector systems, as can be proved with methods like those used earlier.)

Exercise 6.1.5: Let $D = D(t, \varepsilon)$ be a given $n \times n$ matrix-valued function defined on the region

$$0 \le t \le T, \quad 0 < \varepsilon \le \varepsilon_0, \tag{6.1.61}$$

for fixed positive constants T and ε_0, and assume that D is uniformly continuous in t (for each small $\varepsilon > 0$) and uniformly bounded for all t, ε in the region (6.1.61). Assume also that all eigenvalues $\lambda = \lambda(t, \varepsilon)$ of $D(t, \varepsilon)$ have uniformly negative real parts, with

$$\text{Re}\,\lambda(t, \varepsilon) \le -\kappa_2 < 0 \tag{6.1.62}$$

for a fixed positive constant κ_2, uniformly for all t, ε satisfying (6.1.61). Then a result of Levin and Levinson (1954) guarantees that for any fixed positive κ_3 less than κ_2,

$$|e^{D(t, \varepsilon)s}| \le \text{const.}\, e^{-\kappa_3 s} \quad \text{for all } s \ge 0, \text{ any } 0 < \kappa_3 < \kappa_2, \tag{6.1.63}$$

uniformly for all t, ε satisfying (6.1.61). [This result is well known for a *constant* matrix D with stable eigenvalues having negative real parts; see Chapter 7 of Hirsch and Smale (1974).] Assume the validity of (6.1.63), and for simplicity assume also that D is Hölder-continuous in t, uniformly for all t and ε satisfying (6.1.61), with

$$|D(t_1, \varepsilon) - D(t_2, \varepsilon)| \le \text{const.}\, |t_1 - t_2|^{\delta} \tag{6.1.64}$$

for some fixed positive constant δ, uniformly for all t_1, t_2, and ε satisfying (6.1.61). Given these assumptions (6.1.63) and (6.1.64), derive the stability condition (6.1.40) for the resolvent function η of (6.1.37) for any fixed positive $\kappa_1 < \kappa_2$. *Hint (Flatto and Levinson 1955):* For any fixed τ, write the differential equation of (6.1.37) as

$$\varepsilon \frac{d\eta}{dt} - D(\tau, \varepsilon)\eta = \left[D(t, \varepsilon) - D(\tau, \varepsilon)\right]\eta,$$

and then use variation of parameters to obtain the result

$$\eta(t, s, \varepsilon) = e^{D(\tau, \varepsilon)(t-s)/\varepsilon}$$
$$+ \frac{1}{\varepsilon} \int_s^t e^{D(\tau, \varepsilon)(t-\sigma)/\varepsilon} [D(\sigma, \varepsilon) - D(\tau, \varepsilon)] \eta(\sigma, s, \varepsilon) \, d\sigma$$

for the solution of (6.1.37). Now put $\tau = t$ in this last result and obtain

$$\eta(t, s, \varepsilon) = e^{D(t, \varepsilon)(t-s)/\varepsilon}$$
$$+ \frac{1}{\varepsilon} \int_s^t e^{D(t, \varepsilon)(t-\sigma)/\varepsilon} [D(\sigma, \varepsilon) - D(t, \varepsilon)] \eta(\sigma, s, \varepsilon) \, d\sigma,$$

$$(6.1.65)$$

which, with (6.1.63) and (6.1.64), yields

$$|\eta(t, s, \varepsilon)| \le \text{const.} \left[e^{-\kappa_3(t-s)/\varepsilon} \right.$$
$$\left. + \frac{1}{\varepsilon} \int_s^t e^{-\kappa_3(t-\sigma)/\varepsilon} (t - \sigma)^\delta |\eta(\sigma, s, \varepsilon)| \, d\sigma \right]$$

$$(6.1.66)$$

uniformly for $0 \le s \le t \le T$, $0 < \varepsilon \le \varepsilon_0$. Now put

$$\Phi(t, s, \varepsilon) := |\eta(t, s, \varepsilon)| e^{\kappa_4(t-s)/\varepsilon} \qquad (6.1.67)$$

for any fixed positive $\kappa_4 < \kappa_3$, and find

$$0 \le \Phi(t, s, \varepsilon) \le \text{const.} \left[1 + \frac{1}{\varepsilon} \int_s^t e^{-(\kappa_3-\kappa_4)(t-\sigma)/\varepsilon} (t - \sigma)^\delta \Phi(\sigma, s, \varepsilon) \, d\sigma \right]$$
$$\le \text{const.} \left[1 + \|\Phi\| \varepsilon^\delta \right], \qquad (6.1.68)$$

where

$$\|\Phi\| := \max_{0 \le s \le t \le T} |\Phi(t, s, \varepsilon)|.$$

Use (6.1.68) to conclude $\|\Phi\| \le$ constant, which with (6.1.67) leads to the desired result

$$|\eta(t, s, \varepsilon)| \le \text{const.} \, e^{-\kappa_4(t-s)/\varepsilon} \qquad (6.1.69)$$

uniformly for $0 \le s \le t \le T$ as $\varepsilon \to 0+$, for any fixed positive κ_4 less than κ_3. Because κ_3 can in turn be any fixed positive constant less than κ_2, it follows from (6.1.69) that the desired result (6.1.40) is valid for any $\kappa_1 := \kappa_4$ less than κ_2, which was to be shown. The requirement (6.1.64) that D be Hölder-continuous can be weakened, as in Flatto and Levinson (1955). The 2×2 example

$$D(\varepsilon) = \begin{bmatrix} -1 & \varepsilon^{-1} \\ 0 & -1 \end{bmatrix}, \quad \text{with } e^{Ds} = e^{-s} \begin{bmatrix} 1 & s/\varepsilon \\ 0 & 1 \end{bmatrix},$$

shows that (6.1.63) need not hold if D is not suitably bounded [see Kreiss and Nichols (1975)].

6.2 The O'Malley/Hoppensteadt construction

In this section we return to the problem (6.1.1)–(6.1.2), which we rewrite here for convenience:

$$\frac{dx}{dt} = u(t, x, y, \varepsilon),$$
$$\varepsilon \frac{dy}{dt} = v(t, x, y, \varepsilon) \quad \text{for } t > 0, \tag{6.2.1}$$

and

$$x = \alpha(\varepsilon),$$
$$y = \beta(\varepsilon) \quad \text{at } t = 0, \tag{6.2.2}$$

where we shall assume that the data u, v, α, and β possess expansions of the type (6.1.3) and (6.1.4). For simplicity we shall think of the scalar case in which x and y (and hence also u and v) are scalar-valued functions, although the same techniques suffice to handle the vector case, with the obvious modifications.

For the construction of approximate solutions to this problem, we use the approach of O'Malley (1971*a*, 1971*b*) and Hoppensteadt (1971*b*), already used for the linear problem of Section 5.4. Hence, we introduce the stretched variable τ given as

$$\tau := \frac{t}{\varepsilon}, \tag{6.2.3}$$

and we seek asymptotic representations of the solution functions x and y in the form [compare with (5.4.4)–(5.4.7)]

$$x(t, \varepsilon) \sim X(t, \varepsilon) + \varepsilon X^*(\tau, \varepsilon),$$
$$y(t, \varepsilon) \sim Y(t, \varepsilon) + Y^*(\tau, \varepsilon), \tag{6.2.4}$$

for suitable outer functions $X = X(t, \varepsilon)$, $Y = Y(t, \varepsilon)$ and suitable boundary-layer correction functions $X^* = X^*(\tau, \varepsilon)$, $Y^* = Y^*(\tau, \varepsilon)$, which are to be obtained in the form of asymptotic expansions as

$$\begin{bmatrix} X(t, \varepsilon) \\ Y(t, \varepsilon) \end{bmatrix} \sim \sum_{k=0}^{\infty} \begin{bmatrix} X_k(t) \\ Y_k(t) \end{bmatrix} \varepsilon^k \tag{6.2.5}$$

and

$$\begin{bmatrix} X^*(\tau, \varepsilon) \\ Y^*(\tau, \varepsilon) \end{bmatrix} \sim \sum_{k=0}^{\infty} \begin{bmatrix} X_k^*(\tau) \\ Y_k^*(\tau) \end{bmatrix} \varepsilon^k. \tag{6.2.6}$$

The boundary-layer correction functions $X^*(\tau, \varepsilon)$ and $Y^*(\tau, \varepsilon)$, when evaluated at $\tau = t/\varepsilon$, must be negligible for any fixed positive $t > 0$ as $\varepsilon \to 0+$, and so we impose the matching conditions [see (5.4.7)]

$$\lim_{\tau \to \infty} \begin{bmatrix} X_k^*(\tau) \\ Y_k^*(\tau) \end{bmatrix} = \begin{bmatrix} 0 \\ 0 \end{bmatrix} \quad \text{for } k = 0, 1, 2, \dots . \qquad (6.2.7)_k$$

The explicit factor of ε that multiplies X^* in the right side of (6.2.4) has been conveniently included for bookkeeping purposes only. This is related to the fact that we have taken the initial data in (6.2.2) to be regular at $\varepsilon = 0$, in accordance with (6.1.4). By way of comparison, if the initial data are given as in (6.1.5), then (6.2.4) will be replaced with

$$\begin{aligned} x(t, \varepsilon) &\sim X(t, \varepsilon) + X^*(\tau, \varepsilon), \\ y(t, \varepsilon) &\sim Y(t, \varepsilon) + \frac{1}{\varepsilon} Y^*(\tau, \varepsilon), \end{aligned} \qquad (6.2.8)$$

and then we again seek expansions of the forms (6.2.5) and (6.2.6) (see Exercise 6.2.3).

As in Section 5.4, we require that the outer solution functions X, Y satisfy the full system (6.2.1) for $t > 0$,

$$\begin{aligned} \frac{dX}{dt} &= u(t, X, Y, \varepsilon), \\ \varepsilon \frac{dY}{dt} &= v(t, X, Y, \varepsilon) \quad \text{for } t > 0. \end{aligned} \qquad (6.2.9)$$

From (6.2.1), (6.2.3), (6.2.8), and (6.2.9), we are then led to require that the boundary-layer correction functions X^*, Y^* satisfy the system

$$\begin{aligned} \frac{dX^*}{d\tau} &= u\big[\varepsilon\tau, X(\varepsilon\tau, \varepsilon) + \varepsilon X^*, Y(\varepsilon\tau, \varepsilon) + Y^*, \varepsilon\big] \\ &\quad - u\big[\varepsilon\tau, X(\varepsilon\tau, \varepsilon), Y(\varepsilon\tau, \varepsilon), \varepsilon\big], \\ \frac{dY^*}{d\tau} &= v\big[\varepsilon\tau, X(\varepsilon\tau, \varepsilon) + \varepsilon X^*, Y(\varepsilon\tau, \varepsilon) + Y^*, \varepsilon\big] \\ &\quad - v\big[\varepsilon\tau, X(\varepsilon\tau, \varepsilon), Y(\varepsilon\tau, \varepsilon), \varepsilon\big], \qquad (6.2.10) \end{aligned}$$

for $\tau > 0$.

The expansions of (6.1.3) and (6.2.5) are used for X, Y, u, and v in the outer system (6.2.9), and we expand the resulting expressions in Taylor expansions about $\varepsilon = 0$ {e.g., $u(t, X, Y, \varepsilon) \sim u(t, X_0 + \varepsilon X_1 + \dots, Y_0 + \varepsilon Y_1 + \dots, \varepsilon) \sim u_0(t, X_0, Y_0) + \varepsilon[u_{0,x}(t, X_0, Y_0)X_1 + u_{0,y}(t, X_0, Y_0)Y_1 + u_1(t, X_0, Y_0)] + \dots$, etc.}, from which we are led to impose the following conditions on the outer coefficients X_k, Y_k:

$$\frac{dX_0}{dt} = u_0(t, X_0, Y_0),$$

$$0 = v_0(t, X_0, Y_0),$$

$(6.2.11)_0$

$$\frac{dX_1}{dt} = u_{0,x}(t, X_0, Y_0)X_1 + u_{0,y}(t, X_0, Y_0)Y_1 + u_1(t, X_0, Y_0),$$

$$\frac{dY_0}{dt} = v_{0,x}(t, X_0, Y_0)X_1 + v_{0,y}(t, X_0, Y_0)Y_1 + v_1(t, X_0, Y_0),$$

$(6.2.11)_1$

and, in general,

$$\frac{dX_k}{dt} = u_{0,x}(t, X_0, Y_0)X_k + u_{0,y}(t, X_0, Y_0)Y_k + A_{k-1}(t),$$

$$0 = v_{0,x}(t, X_0, Y_0)X_k + v_{0,y}(t, X_0, Y_0)Y_k + B_{k-1}(t),$$

$(6.2.11)_k$

for $k = 1, 2, \ldots$, with

$$A_0(t) = u_1[t, X_0(t), Y_0(t)],$$

$$B_0(t) = v_1[t, X_0(t), Y_0(t)] - \frac{dY_0(t)}{dt},$$

$(6.2.12)$

and, in general, with $A_{k-1}(t)$ and $B_{k-1}(t)$ determined recursively in terms of X_j, Y_j for $j \le k - 1$.

Similarly, the expansions of (6.2.5) and (6.2.6) are inserted into the boundary-layer correction system (6.2.10), and we use (6.2.3) along with suitable Taylor expansions about $\varepsilon = 0$ [e.g., $u[\varepsilon\tau, X_0(\varepsilon\tau) + \varepsilon X_1(\varepsilon\tau) + \cdots + \varepsilon X_0^*(\tau) + \cdots, Y_0(\varepsilon\tau) + \varepsilon Y_1(\varepsilon\tau) + \cdots + Y_0^*(\tau) + \varepsilon Y_1^*(\tau) + \ldots, \varepsilon] \sim u_0[0, X_0(0), Y_0(0) + Y_0^*(\tau)] + \varepsilon\{\tau u_{0,t}[0, X_0(0), Y_0(0) + Y_0^*(\tau)] + [\tau X_0'(0) + X_1(0) + X_0^*(\tau)]u_{0,x}[0, X_0(0), Y_0(0) + Y_0^*(\tau)] + [\tau Y_0'(0) + Y_1(0) + Y_1^*(\tau)]u_{0,y}[0, X_0(0), Y_0(0) + Y_0^*(\tau)] + u_1[0, X_0(0), Y_0(0) + Y_0^*(\tau)]\} + \cdots$] to obtain the following conditions on the boundary-layer correction coefficients X_k^*, Y_k^*:

$$\frac{dX_0^*}{d\tau} = u_0[0, X_0(0), Y_0(0) + Y_0^*] - u_0[0, X_0(0), Y_0(0)],$$

$$\frac{dY_0^*}{d\tau} = v_0[0, X_0(0), Y_0(0) + Y_0^*] - v_0[0, X_0(0), Y_0(0)],$$

$(6.2.13)_0$

and for $k = 1, 2, \ldots,$

$$\frac{dX_k^*}{d\tau} = u_{0,y}\left[0, X_0(0), Y_0(0) + Y_0^*\right]Y_k^* + A_{k-1}^*(\tau),$$

$$\frac{dY_k^*}{d\tau} = v_{0,y}\left[0, X_0(0), Y_0(0) + Y_0^*\right]Y_k^* + B_{k-1}^*(\tau),$$

$$(6.2.13)_k$$

where $A_{k-1}^*(\tau)$ and $B_{k-1}^*(\tau)$ are determined recursively in terms of X_{j+1}, Y_{j+1}, X_j^*, and Y_j^* for $j \leq k - 1$. For example, we have

$$A_0^*(\tau) := \tau\{u_{0,t}\left[0, X_0(0), Y_0(0) + Y_0^*(\tau)\right] - u_{0,t}\left[0, X_0(0), Y_0(0)\right]\}$$
$$+ \{u_{0,x}\left[0, X_0(0), Y_0(0) + Y_0^*(\tau)\right]$$
$$- u_{0,x}\left[0, X_0(0), Y_0(0)\right]\}\left[\tau X_0'(0) + X_1(0)\right]$$
$$+ \{u_{0,y}\left[0, X_0(0), Y_0(0) + Y_0^*(\tau)\right]$$
$$- u_{0,y}\left[0, X_0(0), Y_0(0)\right]\}\left[\tau Y_0'(0) + Y_1(0)\right]$$
$$+ u_{0,x}\left[0, X_0(0), Y_0(0) + Y_0^*(\tau)\right]X_0^*(\tau)$$
$$+ \{u_1\left[0, X_0(0), Y_0(0) + Y_0^*(\tau)\right] - u_1\left[0, X_0(0), Y_0(0)\right]\}$$

$$(6.2.14)$$

and a similar expression for $B_0^*(\tau)$.

Similarly, we insert (6.2.3)–(6.2.6) into the initial conditions of (6.2.2) and find, with (6.1.4),

$$X_k(0) + X_{k-1}^*(0) = \alpha_k,$$
$$Y_k(0) + Y_k^*(0) = \beta_k \quad \text{for } k = 0, 1, 2, \ldots,$$

$$(6.2.15)_k$$

where $X_{-1}^* := 0$.

We wish to use the conditions (6.2.7), (6.2.11), (6.2.13), and (6.2.15) to determine recursively the required functions X_k, Y_k, X_k^*, and Y_k^*. For this purpose, following O'Malley (1971a, 1971b), we use assumptions A.1 and A.2:

A.1: There is a continuously differentiable function $\phi = \phi(t, x)$ such that [compare with (6.2.1) with $\varepsilon = 0$]

$$v_0\left[t, x, \phi(t, x)\right] = 0 \quad \text{for all suitable } t \text{ and } x, \qquad (6.2.16)$$

and such that the initial-value problem

$$\frac{dX_0}{dt} = u_0\left[t, X_0, \phi(t, X_0)\right] \quad \text{for } t > 0, \qquad (6.2.17)$$

$$X_0 = \alpha_0 \quad \text{at } t = 0,$$

has a solution $X_0 = X_0(t)$ on some compact interval, say $0 \leq t \leq T$, with [see (6.1.15) in the linear case]

$$v_{0,y}\left[t, X_0(t), Y_0(t)\right] \leq -\kappa < 0 \quad \text{for } 0 \leq t \leq T \qquad (6.2.18)$$

for some fixed positive constant $\kappa > 0$, where

$$Y_0(t) := \phi\left[t, X_0(t)\right] \quad \text{for } 0 \leq t \leq T. \qquad (6.2.19)$$

The condition (6.2.16) need hold only for all (t, x) near $[t, X_0(t)]$ for $0 \leq t \leq T$.

A.2: With the same constant κ of (6.2.18), there holds

$$v_{0,y}(0, \alpha_0, \zeta) \leq -\kappa < 0 \qquad (6.2.20)$$

for all values of ζ between β_0 and $Y_0(0) = \phi(0, \alpha_0)$.

In the case of a vector system with x, y, u, v, α, and β suitable vector-valued quantities in (6.2.1)–(6.2.2), then the assumptions (6.2.18) and (6.2.20) are replaced with related assumptions requiring that all eigenvalues of the matrix $v_{0,y}$ have negative real parts, at least for all suitable values of the variables involved; see (6.1.42) in the linear case. The assumption (6.2.20) can be weakened; see Exercise 6.2.7. These assumptions A.1 and A.2 and their vector generalizations are related to conditions introduced by Tikhonov (1950, 1952) and Levin and Levinson (1954) in early studies of certain initial-value problems for (6.2.1). Related *conditionally stable problems* can also be considered for the case in which the matrix $v_{0,y}$ has some eigenvalues with negative real parts and other eigenvalues with positive real parts, provided that the initial value for y is restricted to a suitable stable manifold. We discuss such conditionally stable problems briefly in Chapter 7, with emphasis there on a particular illustrative example.

Example 6.2.1: If the functions u and v are given by (6.1.6), as in Example 6.1.1, with

$$u \equiv u_0(t, x, y) := y,$$

$$v \equiv v_0(t, x, y) := f(t) - B(t)x - A(t)y,$$

for given functions A, B, and f, then (6.2.16) holds with

$$\phi(t, x) := A(t)^{-1}\left[f(t) - B(t)x\right],$$

and (6.2.19) gives

$$Y_0(t) = A(t)^{-1}\left[f(t) - B(t)X_0(t)\right]$$

where X_0 is the solution of the reduced problem

$$\frac{dX_0}{dt} = A(t)^{-1}[f(t) - B(t)X_0] \quad \text{for } t > 0,$$

$$X_0 = \alpha_0 \quad \text{at } t = 0.$$

As in Chapter 5, we assume that the given function $A = A(t)$ is positive, with

$$A(t) \geq \kappa > 0 \quad \text{for } t \geq 0,$$

so that X_0 and Y_0 exist for $t > 0$. In this case we find $v_{0, y} = -A(t) \leq -\kappa$, so that assumptions A.1 and A.2 are valid. In the present case we require $c = 0$ in (6.1.8), so that (6.1.4) holds. Also, in the vector case, the foregoing positivity condition on $A(t)$ is replaced with a corresponding positivity condition on all eigenvalues of the square matrix-valued function $A = A(t)$, so that the eigenvalues $\lambda = \lambda(t)$ of $-A(t)$ satisfy (6.1.42).

Example 6.2.2: If u and v are given by (6.1.11), as in Example 6.1.2, with

$$u \equiv u_0(t, x, y) := -x + (x + \kappa - \lambda)y,$$

$$v \equiv v_0(t, x, y) := x - (x + \kappa)y,$$

for given positive constants κ and λ, then (6.2.16) holds with

$$\phi(t, x) := \frac{x}{\kappa + x} \quad \text{for all } t, \text{ all } x \neq -\kappa. \tag{6.2.21}$$

Hence, (6.2.19) gives

$$Y_0(t) = \frac{X_0(t)}{\kappa + X_0(t)}, \tag{6.2.22}$$

where X_0 is the solution of the initial-value problem [see (6.2.17)]

$$\frac{dX_0}{dt} = -\frac{\lambda X_0}{\kappa + X_0} \quad \text{for } t > 0,$$

$$X_0 = 1 \quad \text{at } t = 0, \tag{6.2.23}$$

where the initial condition is taken from Example 6.1.2.

Because $X_0 \equiv 0$ is a solution of the differential equation of (6.2.23), we see, with the uniqueness theorem from differential-equation theory, that the solution of the initial-value problem (6.2.23) is positive-valued for $t \geq 0$. In fact, we see that the solution X_0 of (6.2.23) exists for all $t \geq 0$ and satisfies the uniform bounds $0 < X_0(t) \leq 1$. Indeed, (6.2.23) yields directly the differential inequalities

$$-\frac{\lambda}{\kappa} X_0 \leq \frac{dX_0}{dt} \leq 0,$$

from which we find easily, on integration,

$$e^{-\lambda t/\kappa} \le X_0(t) \le 1 \quad \text{for all } t \ge 0. \tag{6.2.24}$$

The function X_0 can be determined implicitly with the equation

$$X_0 + \kappa \log X_0 = 1 - \lambda t, \tag{6.2.25}$$

obtained by direct integration from (6.2.23).

In this case we find $v_{0,y}(t, x, y) = -(\kappa + x)$, and then (6.2.24) implies

$$v_{0,y}[t, X_0(t), Y_0(t)] \le -\kappa < 0 \tag{6.2.26}$$

for all $t \ge 0$, so that assumption A.1 is satisfied in terms of the same positive constant κ appearing in the data of the present initial-value problem. Similarly, with $\alpha_0 = 1$ and $\beta_0 = 0$, as in Example 6.1.2, we find

$$v_{0,y}(0, \alpha_0, \zeta) = -(\kappa + 1) < -\kappa < 0 \quad \text{for all } \zeta, \tag{6.2.27}$$

so that assumption A.2 is also satisfied with the same constant κ as in A.1. See Exercise 6.2.6 for a further consideration of this problem.

We now return to the original problem (6.2.1)–(6.2.4) subject to the assumptions A.1 and A.2. We wish to show that conditions (6.2.7), (6.2.11), (6.2.13), and (6.2.15) can be used recursively to determine the required functions X_k, Y_k, X_k^*, and Y_k^* for the solution expansions (6.2.3)–(6.2.6).

First, note that A.1 already provides functions $X_0(t)$ and $Y_0(t)$ satisfying $(6.2.11)_0$ [see (6.2.16)–(6.2.19)] and also satisfying the first initial condition of $(6.2.15)_0$, with $X_{-1}^* \equiv 0$. Moreover, because $Y_0(t)$ is now known, we find, from the second initial condition of $(6.2.15)_0$, the result

$$Y_0^*(0) = \beta_0 - Y_0(0) = \beta_0 - \phi(0, \alpha_0) = \text{known quantity}, \tag{6.2.28}$$

and the second equation of $(6.2.13)_0$ gives the following differential equation for Y_0^*:

$$\frac{dY_0^*}{d\tau} = v_0\big[0, \alpha_0, \phi(0, \alpha_0) + Y_0^*\big] - v_0\big[0, \alpha_0, \phi(0, \alpha_0)\big]$$
$$= v_0\big[0, \alpha_0, \phi(0, \alpha_0) + Y_0^*\big] \quad \text{for } \tau > 0, \tag{6.2.29}$$

where we have used the second equation of $(6.2.11)_0$ at $t = 0$. The identity

$$f(1) - f(0) = \int_0^1 f'(s)\, ds \tag{6.2.30}$$

can be applied, with $f(s) := v_0[0, \alpha_0, \phi(0, \alpha_0) + s Y_0^*(\tau)]$, and we find the result

$$v_0\big[0, \alpha_0, \phi(0, \alpha_0) + Y_0^*(\tau)\big]$$

$$= \left\{ \int_0^1 v_{0,y}\big[0, \alpha_0, \phi(0, \alpha_0) + s Y_0^*(\tau)\big]\, ds \right\} Y_0^*(\tau). \quad (6.2.31)$$

Then condition A.2, along with (6.2.28), (6.2.29), and (6.2.31), yields the differential inequality

$$\frac{d}{d\tau}|Y_0^*(\tau)|^2 \le -2\kappa|Y_0^*(\tau)|^2, \quad (6.2.32)$$

which can be integrated to give

$$|Y_0^*(\tau)| \le |Y_0^*(0)|e^{-\kappa\tau} \quad \text{for } \tau \ge 0. \quad (6.2.33)$$

In particular, it follows that the solution Y_0^* of the initial-value problem (6.2.28)–(6.2.29) exists for all $\tau \ge 0$, and Y_0^* satisfies the estimate (6.2.33).

Now that Y_0^* is determined, we can integrate the first equation of $(6.2.13)_0$ along with the matching condition $(6.2.7)_0$ to find

$$X_0^*(\tau) = -\int_\tau^\infty \big\{ u_0\big[0, \alpha_0, \phi(0, \alpha_0) + Y_0^*(\sigma)\big]$$

$$- u_0\big[0, \alpha_0, \phi(0, \alpha_0)\big] \big\}\, d\sigma. \quad (6.2.34)$$

If we take $f(s) := u_0[0, \alpha_0, \phi(0, \alpha_0) + s Y_0^*(\sigma)]$ in (6.2.30), we are led to an inequality of the type (note that $u_{0,y}$ is bounded on any fixed compact set)

$$\big| u_0\big[0, \alpha_0, \phi(0, \alpha_0) + Y_0^*(\sigma)\big] - u_0\big[0, \alpha_0, \phi(0, \alpha_0)\big] \big|$$

$$\le \text{const.}\, |Y_0^*(\sigma)| \le \text{const.}\, e^{-\kappa\sigma}$$

for $\sigma \ge 0$, where (6.2.33) has also been used. This last result and (6.2.34) imply the inequality

$$|X_0^*(\tau)| \le \text{const.}\, e^{-\kappa\tau} \quad \text{for } \tau \ge 0. \quad (6.2.35)$$

We now continue recursively. After $X_j(t)$, $Y_j(t)$, $X_j^*(\tau)$, and $Y_j^*(\tau)$ have been determined for $j = 0, 1, \ldots, k-1$, then the second equation of $(6.2.11)_k$ gives

$$Y_k(t) = -\big\{ v_{0,y}\big[t, X_0(t), Y_0(t)\big] \big\}^{-1}$$

$$\times \big\{ B_{k-1}(t) + v_{0,x}\big[t, X_0(t), Y_0(t)\big] X_k(t) \big\}, \quad (6.2.36)_k$$

where (6.2.18) guarantees that $v_{0,y}[t, X_0(t), Y_0(t)]$ is nonzero (or, in the vector case, the corresponding assumption on the eigenvalues guarantees that this matrix is nonsingular). This result (6.2.36) can be used back in the first equation of $(6.2.11)_k$, which then provides a given linear first-order differential equation for $X_k(t)$. This latter equation can be solved subject to the initial condition provided by the first equation of $(6.2.15)_k$, so that $X_k(t)$ is now determined, and then (6.2.36) gives $Y_k(t)$, which now is also determined. The initial value of $Y_k^*(\tau)$ is then given by the second equation of $(6.2.15)_k$, and this initial value is used with the linear differential equation provided by the second equation of $(6.2.13)_k$ to determine $Y_k^*(\tau)$ for $\tau \geq 0$. After $Y_k^*(\tau)$ is known, we integrate the first equation of $(6.2.13)_k$ along with the appropriate matching condition from $(6.2.7)_k$ so as to determine $X_k^*(\tau)$ for $\tau \geq 0$. Taking into account the structure of the terms $A_{k-1}^*(\tau)$ and $B_{k-1}^*(\tau)$ on the right side of $(6.2.13)_k$, and using the condition A.1, we find estimates of the type [see pp. 84–5 of O'Malley (1974*a*)]

$$|X_k^*(\tau)|, |Y_k^*(\tau)| \leq C_k e^{-\kappa \mu \tau} \quad \text{for } \tau \geq 0 \qquad (6.2.37)$$

for any fixed positive constant $\mu < 1$.

This discussion shows that the coefficient functions in the expansions of (6.2.3)–(6.2.7) can be determined recursively beginning with a suitable fixed function ϕ satisfying (6.2.16), provided that the conditions A.1 and A.2 hold. We can then consider the truncated expansions, and for this purpose we define functions x^N and y^N by the formulas

$$x^N(t, \varepsilon) := \sum_{k=0}^{N} \left[X_k(t) + X_{k-1}^* \left(\frac{t}{\varepsilon} \right) \right] \varepsilon^k,$$

$$y^N(t, \varepsilon) := \sum_{k=0}^{N} \left[Y_k(t) + Y_k^* \left(\frac{t}{\varepsilon} \right) \right] \varepsilon^k \quad \text{for } t \geq 0, \qquad (6.2.38)$$

where $X_{-1}^*(\tau) \equiv 0$.

In Section 6.3 we show that, subject to the foregoing assumptions and for all sufficiently small $\varepsilon > 0$, the original initial-value problem (6.2.1)–(6.2.2) has a unique solution of the form

$$x(t, \varepsilon) = x^N(t, \varepsilon) + R_N(t, \varepsilon),$$

$$y(t, \varepsilon) = y^N(t, \varepsilon) + S_N(t, \varepsilon), \qquad (6.2.39)$$

with x^N and y^N given by (6.2.38), and for suitable remainders R_N and S_N that satisfy estimates of the type

$$|R_N(t, \varepsilon)|, |S_N(t, \varepsilon)| \leq \text{const.} \ \varepsilon^{N+1} \qquad (6.2.40)$$

uniformly for all t, ε satisfying

$$0 \leq t \leq T, \quad 0 < \varepsilon \leq \varepsilon_0, \qquad (6.2.41)$$

for suitable fixed positive constants T and ε_0. Hence, the functions x^N, y^N given by (6.2.38) and provided by the O'Malley/Hoppensteadt construction do in fact provide suitable, useful approximate solutions to the original initial-value problem.

The construction remains valid for suitable vector systems, with obvious modifications as discussed briefly earlier.

Note that the nonlinear equation (6.2.16) may have more than one solution for $\phi = \phi(t, x)$. In such a case the specific numerical values of the initial data α, β can be crucial in determining the admissibility of any particular solution ϕ in accordance with the conditions A.1 and A.2; see Exercise 6.2.2.

Example 6.2.3: Consider the problem

$$x' = y, \qquad\qquad x = 1,$$
$$\varepsilon y' = -x^2 - y \quad \text{for } t > 0, \qquad y = 0 \quad \text{at } t = 0. \tag{6.2.42}$$

We seek the lowest-order approximations x^0, y^0 of (6.2.38) with $N = 0$. Note that this problem (6.2.42) is of the type (6.2.1)–(6.2.2), with

$$u \equiv u_0(t, x, y) := y, \qquad\qquad \alpha \equiv \alpha_0 := 1,$$
$$v \equiv v_0(t, x, y) := -x^2 - y, \qquad \beta \equiv \beta_0 := 0. \tag{6.2.43}$$

The problem is equivalent to the second-order initial-value problem

$$\varepsilon x'' + x' + x^2 = 0 \quad \text{for } t > 0,$$
$$x = 1 \quad \text{and} \quad x' = 0 \quad \text{at } t = 0.$$

The function ϕ of (6.2.16) is easily seen to be given as

$$\phi(x) = -x^2,$$

independent of t in this case, so that the initial-value problem (6.2.17) is

$$\frac{dX_0}{dt} = -(X_0)^2 \quad \text{for } t > 0,$$
$$X_0 = 1 \qquad\qquad \text{at } t = 0,$$

with solution

$$X_0(t) = \frac{1}{1 + t} \qquad \text{for } t \geq 0.$$

Then (6.2.19) gives

$$Y_0(t) = \phi[X_0(t)] = -\frac{1}{(1 + t)^2}.$$

In this case, $v_{0, y} \equiv -1$, so that the assumptions A.1 and A.2 both hold.

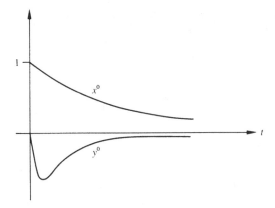

Figure 6.1

The initial-value problem for the boundary-layer correction term $Y_0^*(\tau)$ is given by $(6.2.13)_0$ and $(6.2.15)_0$ as

$$\frac{dY_0^*}{d\tau} = -Y_0^* \quad \text{for } \tau > 0,$$

$$Y_0^* = 1 \quad \text{at } \tau = 0,$$

with solution $Y_0^*(\tau) = e^{-\tau}$. Hence, we have the lowest-order approximations

$$x^0(t,\varepsilon) = \frac{1}{1+t},$$

$$y^0(t,\varepsilon) = -\frac{1}{(1+t)^2} + e^{-t/\varepsilon}, \tag{6.2.44}$$

as illustrated in Figure 6.1.

The general estimates given in the next section will be seen to imply [see (6.2.39)–(6.2.40)]

$$|x(t,\varepsilon) - x^0(t,\varepsilon)|, |y(t,\varepsilon) - y^0(t,\varepsilon)| \le \text{const. } \varepsilon \tag{6.2.45}$$

uniformly for all t, ε satisfying (6.2.41). However, we give here a direct derivation of (6.2.45) so as to assess the magnitude of the constant on the right side of (6.2.45). Moreover, the present derivation shows that (6.2.45) actually remains valid here on an expanding t interval of the type $0 \le t \le T/\varepsilon$ as $\varepsilon \to 0+$.

Hence, we introduce the remainders $R(t)$, $S(t)$ as

$$R(t) := x(t,\varepsilon) - x^0(t,\varepsilon),$$

$$S(t) := y(t,\varepsilon) - y^0(t,\varepsilon), \tag{6.2.46}$$

where we suppress the dependence of R and S on ε. It follows from (6.2.42), (6.2.44), and (6.2.46) that

$$\frac{dR}{dt} = S + e^{-t/\varepsilon},$$

$$\varepsilon\frac{dS}{dt} = -\frac{2}{1+t}R - S - \frac{2\varepsilon}{(1+t)^3} - R^2 \quad \text{for } t > 0, \quad (6.2.47)$$

and

$$R(0) = S(0) = 0 \quad \text{at } t = 0. \tag{6.2.48}$$

We can treat (6.2.47)–(6.2.48) directly, but we prefer to consider instead the equivalent second-order initial-value problem for R, found to be

$$\varepsilon R'' + R' + \frac{2}{1+t}R = -\frac{2\varepsilon}{(1+t)^3} - R^2 \quad \text{for } t > 0,$$

$$R(0)=0, \qquad R'(0) = 1 \quad \text{at } t = 0. \tag{6.2.49}$$

It follows now directly from (6.2.49) and Theorem 5.3.2 that R satisfies the inequality

$$|R(t)| \le \varepsilon|R'(0)| + \int_0^t\left[\frac{2\varepsilon}{(1+s)^3} + R(s)^2\right]ds$$

$$\le 2\varepsilon + \int_0^t|R|^2 \tag{6.2.50}$$

for all $t \ge 0$, and this latter inequality can be resolved by a Gronwall-type argument so as to obtain

$$|R(t)| \le \frac{2\varepsilon}{1 - 2\varepsilon t} \quad \text{for } 0 \le t < \frac{1}{2\varepsilon}. \tag{6.2.51}$$

In particular, if we restrict t to the interval

$$0 \le t \le \frac{1}{4\varepsilon}, \tag{6.2.52}$$

then we find the estimates

$$|R(t)| \le 4\varepsilon, \qquad |S(t)| \le 11\varepsilon, \tag{6.2.53}$$

uniformly as $\varepsilon \to 0+$ and uniformly for all t satisfying (6.2.52), where the inequality for S is obtained by integrating the appropriate differential inequality obtained from the second equation of (6.2.47) along with the homogeneous initial condition on S and the known estimate for R; see Exercise 6.2.1.

Note that (6.2.46) and (6.2.53) provide specific versions of the inequalities of (6.2.45), and these estimates are uniformly valid on the expanding interval (6.2.52) as $\varepsilon \to 0+$.

We can similarly obtain higher-order approximations, such as

$$x^1(t, \varepsilon): = X_0(t) + \varepsilon\left[X_1(t) + X_0^*\left(\frac{t}{\varepsilon}\right)\right]$$

$$= \frac{1}{1+t} + \varepsilon\left[\frac{1 - \log(1 + t)^2}{(1 + t)^2} - e^{-t/\varepsilon}\right]$$

and

$$y^1(t, \varepsilon): = Y_0(t) + Y_0^*\left(\frac{t}{\varepsilon}\right) + \varepsilon\left[Y_1(t) + Y_1^*\left(\frac{t}{\varepsilon}\right)\right]$$

$$= -\frac{1}{(1 + t)^2} + e^{-t/\varepsilon}$$

$$+ 2\varepsilon\left[\frac{-2 + \log(1 + t)^2}{(1 + t)^3} + \left(2 + \frac{t}{\varepsilon}\right)e^{-t/\varepsilon}\right].$$

In this case we find

$$|x(t, \varepsilon) - x^1(t, \varepsilon)|, |y(t, \varepsilon) - y^1(t, \varepsilon)| \leq \text{const. } \varepsilon^2$$

uniformly as $\varepsilon \to 0+$ and uniformly for all t on a larger expanding interval of length proportional to ε^{-2}.

The functions x^N, y^N provide useful approximations to the exact solution functions x, y inside the boundary layer for small t even if ε is not small; see Exercises 6.2.5 and 6.3.3.

The results of this section have been extended to related problems involving several small parameters by O'Malley (1971b, 1974a).

Example 6.2.4: The problem

$$\begin{aligned} x' &= y, \\ \varepsilon y' &= -x^2 - y \qquad \text{for } t > 0, \end{aligned}$$

(6.2.54)

and

$$x = 0, \qquad y = \frac{1}{\varepsilon} \quad \text{at } t = 0,$$

(6.2.55)

resembles the problem (6.2.42) of the previous Example 6.2.3. However, the initial condition for y is here proportional to $1/\varepsilon$, so that (6.1.4) does not hold, but rather (6.1.5) is valid. The present construction of this Section 6.2 cannot be used to provide approximate solutions for (6.2.54)–(6.2.55), although similar ideas suffice; see Exercises 6.2.3 and 6.2.4.

Exercises

Exercise 6.2.1: Show that (6.2.47)–(6.2.48) imply

$$S(t) = -\frac{1}{\varepsilon}\int_0^t e^{-(t-s)/\varepsilon}\left[\frac{2\varepsilon}{(1+s)^3} + \frac{2}{1+s}R(s) + R(s)^2\right]ds.$$

Use this integral equation along with the first inequality of (6.2.53) to derive the estimate

$$|S(t)| \le 11\varepsilon \quad \text{for } 0 \le t \le \frac{1}{4\varepsilon}$$

for all small enough $\varepsilon > 0$.

Exercise 6.2.2 (O'Malley 1974a, pp. 86–92): The initial-value problem

$$\frac{dx}{dt} = xy,$$

$$\varepsilon\frac{dy}{dt} = -y^3 + y \quad \text{for } t > 0,$$

(6.2.56)

and

$$\begin{bmatrix} x(0,\varepsilon) \\ y(0,\varepsilon) \end{bmatrix} = \begin{bmatrix} \alpha(\varepsilon) \\ \beta(\varepsilon) \end{bmatrix} \sim \sum_{k=0}^{\infty} \begin{bmatrix} \alpha_k \\ \beta_k \end{bmatrix}\varepsilon^k$$

(6.2.57)

corresponds to (6.2.1)–(6.2.2) with

$$u \equiv u_0(t, x, y) := xy,$$

$$v \equiv v_0(t, x, y) := -y^3 + y.$$

In this case there are three possibilities $\phi^{(j)}$ ($j = 1, 2, 3$) for the function ϕ of (6.2.16), given as

$$\phi^{(j)}(t, x) := \begin{cases} 1 & \text{for } j = 1, \\ -1 & \text{for } j = 2, \\ 0 & \text{for } j = 3, \end{cases}$$

for all t and all x. Discuss the problem (6.2.56)–(6.2.57) for small $\varepsilon > 0$ using the approach discussed in Section 6.2. Note that the exact solution can be easily given in this case, and we find, in particular,

$$\lim_{\substack{\varepsilon \to 0+ \\ \text{fixed } t > 0}} y(t, \varepsilon) = \begin{cases} +1 & \text{if } \beta(0) > 0, \\ -1 & \text{if } \beta(0) < 0. \end{cases}$$

Exercise 6.2.3: Consider the problem

$$\frac{dx}{dt} = u(t, x, y, \varepsilon),$$

$$\varepsilon \frac{dy}{dt} = v(t, x, y, \varepsilon) \quad \text{for } t > 0,$$
(6.2.58)

and

$$\begin{bmatrix} x(0, \varepsilon) \\ y(0, \varepsilon) \end{bmatrix} = \begin{bmatrix} \alpha(\varepsilon) \\ \beta(\varepsilon) \end{bmatrix} \sim \sum_{k=0}^{\infty} \begin{bmatrix} \alpha_k \\ \frac{1}{\varepsilon} \beta_k \end{bmatrix} \varepsilon^k, \quad \beta_0 \neq 0, \quad (6.2.59)$$

so that (6.1.5) holds, rather than (6.1.4). Assume that u and v are of the form

$$u(t, x, y, \varepsilon) := f(t, x, \varepsilon) + g(t, x, \varepsilon) y,$$

$$v(t, x, y, \varepsilon) := h(t, x, \varepsilon) + k(t, x, \varepsilon) y,$$
(6.2.60)

for suitable given functions f, g, h, and k depending on t, x, and ε, where these functions have asymptotic power-series expansions of the form

$$\begin{bmatrix} f(t, x, \varepsilon) \\ g(t, x, \varepsilon) \\ h(t, x, \varepsilon) \\ k(t, x, \varepsilon) \end{bmatrix} \sim \sum_{j=0}^{\infty} \begin{bmatrix} f_j(t, x) \\ g_j(t, x) \\ h_j(t, x) \\ k_j(t, x) \end{bmatrix} \varepsilon^j. \quad (6.2.61)$$

Hence, the system (6.2.58) is quasi-linear, as

$$\frac{dx}{dt} = f(t, x, \varepsilon) + g(t, x, \varepsilon) y$$

$$\varepsilon \frac{dy}{dt} = h(t, x, \varepsilon) + k(t, x, \varepsilon) y.$$
(6.2.62)

This latter system includes certain cases in practice, such as the system of Example 6.2.1 and that of Example 6.2.2, but not others, such as the system of Exercise 6.2.2. Write $u_{0,y}(t, x)$ and $v_{0,y}(t, x)$, respectively, for [see (6.2.60)]

$$u_{0,y}(t, x) = g_0(t, x) \quad \text{and} \quad v_{0,y}(t, x) = k_0(t, x),$$

and study the problem (6.2.59)–(6.2.60) subject to the following assumptions B.1, B.2, and B.3.

B.1: The problem

$$\frac{d\xi}{d\tau} = u_{0,y}(0, \xi) Y_0^*,$$

$$(6.2.63)$$

$$\frac{dY_0^*}{d\tau} = v_{0,y}(0, \xi) Y_0^* \qquad \text{for } \tau > 0,$$

and

$$\xi = \alpha_0, \qquad Y_0^* = \beta_0 \qquad \text{at } \tau = 0, \qquad (6.2.64)$$

has a solution $\xi = \xi(\tau)$, $Y_0^* = Y_0^*(\tau)$ for $\tau \geq 0$ such that the following limit exists, denoted as $X_0(0)$:

$$X_0(0) := \lim_{\tau \to \infty} \xi(\tau). \qquad (6.2.65)$$

B.2: There are positive constants κ_0, κ_1, and κ_2 such that

$$v_{0,y}(0, x) \leq -\kappa_0 < 0 \quad \text{and} \quad 0 < \kappa_1 \leq |u_{0,y}(0, x)| \leq \kappa_2 \quad (6.2.66)$$

hold for all numbers x between α_0 and $X_0(0)$, where $X_0(0)$ is defined by (6.2.65).

B.3: The problem

$$\frac{dX_0}{dt} = f_0(t, X_0) - g_0(t, X_0) k_0(t, X_0)^{-1} h_0(t, X_0) \quad \text{for } t > 0,$$

$$X_0 = \text{value defined by (6.2.65)}, \quad \text{at } t = 0, \qquad (6.2.67)$$

has a solution $X_0 = X_0(t)$ on some interval $0 \leq t \leq T$, such that

$$k_0[t, X_0(t)] = v_{0,y}[t, X_0(t)] \leq -\kappa_0 < 0 \qquad (6.2.68)$$

holds for $0 \leq t \leq T$.

Given the validity of these assumptions, show that asymptotic expansions of the form (6.2.5)–(6.2.8) can be constructed for the solution of

(6.2.58)–(6.2.62), with $X_0^*(\tau) := \xi(\tau) - X_0(0)$. Verify the exponential decay of $X_0^*(\tau)$ and $Y_0^*(\tau)$ as $\tau \to \infty$.
Hint: For the latter result, we can use the relations

$$\frac{d}{d\tau}(Y_0^*)^2 = 2v_{0,y}[0, \xi(\tau)](Y_0^*)^2 \leq -2\kappa_0(Y_0^*)^2,$$

which can be integrated to yield $|Y_0^*(\tau)| \leq |\beta_0| e^{-\kappa_0 \tau}$ for $\tau \geq 0$. We can show that Y_0^* and ξ are monotonic, and then the first equation of (6.2.63) can be integrated to give

$$X_0^*(\tau) = -\int_\tau^\infty u_{0,y}[0, X_0(0) + X_0^*(\sigma)] Y_0^*(\sigma) \, d\sigma,$$

from which the exponential decay of X_0^* follows. See Exercises 6.4.2 and 6.4.3 for further details.

Exercise 6.2.4: Use the technique of Exercise 6.2.3 to study the problem [see (6.2.54)–(6.2.55)]

$$x' = y, \qquad\qquad x = 0,$$

$$\varepsilon y' = -x^2 - y \quad \text{for } t > 0, \qquad y = \frac{1}{\varepsilon} \quad \text{at } t = 0.$$

In particular, obtain the approximate solution functions [see (6.2.8)]

$$x^0(t, \varepsilon) := X_0(t) + X_0^*(t/\varepsilon) = (1 + t)^{-1} - e^{-t/\varepsilon}$$

$$y^0(t, \varepsilon) := (1/\varepsilon) Y_0^*(t/\varepsilon) + Y_0(t) + Y_1^*(t/\varepsilon)$$

$$= (1/\varepsilon)e^{-t/\varepsilon} - (1 + t)^{-2} + (2t/\varepsilon)e^{-t/\varepsilon} + e^{-2t/\varepsilon}.$$

Error estimates are given in Exercise 6.3.4. The graphs of the solution functions are indicated in Figure 6.2 and can be compared with the graphs of Figure 6.1 in Example 6.2.3.

Exercise 6.2.5: Let x^0, y^0 be the first-order approximations for the problem (6.2.42) of Example 6.2.3, given by (6.2.44). Prove that x^0 and y^0 provide useful approximations to the exact solution functions x and y for small values of t (near the initial time) even if ε is not small. Specifically, if δ is a given positive number, prove the estimate

$$|x(t, \varepsilon) - x^0(t, \varepsilon)| \leq \tfrac{3}{2}(1 + 2\varepsilon)\varepsilon\delta \qquad\qquad (6.2.69)$$

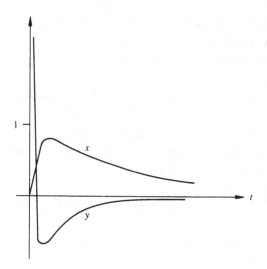

Figure 6.2

for $0 \leq t \leq \varepsilon\delta$, provided that $(1 + 2\varepsilon)(\varepsilon\delta)^2 \leq \frac{1}{3}$. (A similar estimate holds also for $y - y^0$, but we need not obtain it here.) Hence, if ε is not small, say, for example, $\varepsilon = 1$, then we find the estimate

$$|x(t,1) - x^0(t,1)| \leq \tfrac{9}{2}\delta \quad \text{for } 0 \leq t \leq \delta,$$

provided that $\delta \leq \frac{1}{3}$. *Hint*: Return to the proof of Theorem 5.3.2, and improve on the inequality (6.2.50). For example, we can obtain the inequality

$$|R(t)| \leq \left[2t + (1 - e^{-t/\varepsilon})\right]\varepsilon + \int_0^t |R|^2 \quad \text{for } t \geq 0.$$

Also, $1 - e^{-t/\varepsilon} \leq \delta$ for $t/\varepsilon \leq \delta$, so that we find

$$|R(t)| \leq (1 + 2\varepsilon)\varepsilon\delta + \int_0^t |R|^2 \quad \text{for } 0 \leq t \leq \varepsilon\delta.$$

A Gronwall-type argument can be applied to this last inequality so as to yield the stated result.

Exercise 6.2.6 (Heineken, Tsuchiya, and Aris 1967): The following problem arises as a mathematical model in the kinetic theory of enzyme reactions (see Examples 6.1.2 and 6.2.2):

$$\frac{dx}{dt} = -x + (x + \kappa - \lambda)y, \qquad\qquad x = 1,$$

$$\varepsilon\frac{dy}{dt} = x - (x + \kappa)y \quad \text{for } t > 0, \qquad y = 0 \quad \text{at } t = 0. \tag{6.2.70}$$

For this problem, discuss the construction of the functions x^1, y^1 given by (6.2.38) as

$$x^1(t, \varepsilon) := \left[X_0(t) + \varepsilon X_1(t)\right] + \varepsilon X_0^*\left(\frac{t}{\varepsilon}\right)$$

$$y^1(t, \varepsilon) := \left[Y_0(t) + \varepsilon Y_1(t)\right] + \left[Y_0^*\left(\frac{t}{\varepsilon}\right) + \varepsilon Y_1^*\left(\frac{t}{\varepsilon}\right)\right]. \tag{6.2.71}$$

In particular, derive explicit formulas for $X_0^*(\tau)$ and $Y_0^*(\tau)$, and show that Y_0 and Y_1 can be given as

$$Y_0(t) = \frac{X_0(t)}{\kappa + X_0(t)},$$

$$Y_1(t) = \frac{\kappa\lambda X_0(t)}{\left[\kappa + X_0(t)\right]^4} + \frac{\kappa X_1(t)}{\left[\kappa + X_0(t)\right]^2}, \tag{6.2.72}$$

where X_0 and X_1 are determined by the initial-value problem

$$\frac{dX_0}{dt} = f_0(X_0), \qquad\qquad X_0 = 1,$$

$$\frac{dX_1}{dt} = f_1(X_0, X_1) \quad \text{for } t > 0, \qquad X_1 = \frac{\lambda + \kappa(1 + \kappa)}{(1 + \kappa)^2} \quad \text{at } t = 0, \tag{6.2.73}$$

with the functions $f_0 = f_0(x_0)$ and $f_1 = f_1(x_0, x_1)$ defined by the formulas

$$f_0(x_0) := -\frac{\lambda x_0}{\kappa + x_0}$$

$$f_1(x_0, x_1) := -\frac{\kappa\lambda x_1}{(\kappa + x_0)^2} + \frac{\kappa\lambda x_0(x_0 + \kappa - \lambda)}{(\kappa + x_0)^4}. \tag{6.2.74}$$

Finally, show that $Y_1^*(\tau)$ can be determined as the solution of the initial-value problem

$$\frac{dY_1^*}{d\tau} = -(1 + \kappa)Y_1^* + B_0^*(\tau) \quad \text{for } \tau > 0,$$

$$Y_1^* = -\kappa\frac{2\lambda + \kappa(1 + \kappa)}{(1 + \kappa)^4} \qquad \text{at } \tau = 0, \tag{6.2.75}$$

where the function B_0^* is given as

$$B_0^*(\tau) := \frac{2\kappa(1 + \kappa) + \lambda(1 - \kappa) - \lambda(1 + \kappa)\tau}{(1 + \kappa)^3}e^{-(1+\kappa)\tau}$$

$$+ \frac{1 + \kappa - \lambda}{(1 + \kappa)^3}e^{-2(1+\kappa)\tau}.$$

This latter problem (6.2.75) can be solved in closed form, but we need not do this here. In practice, the problem (6.2.73) must be solved numerically using any of the standard algorithms for regular initial-value problems, such as (the convenient, though nonoptimal) Euler's method with extrapolation, or a suitable Runge/Kutta method. The functions x^1, y^1 of (6.2.71) give uniformly valid approximations for the exact solution functions, with error of order(ε^2); see Heineken, Tsuchiya, and Aris (1967) for related numerical results.

Exercise 6.2.7: Use the O'Malley/Hoppensteadt construction and find the leading approximation (6.2.38) in the case $N = 0$ for the problem

$$x' = y,$$

$$\varepsilon y' = xy + y^2 \qquad \text{for } 0 < t < 1, \tag{6.2.76}$$

and

$$x = 4, \qquad y = -1 \quad \text{at } t = 0. \tag{6.2.77}$$

Hint: We have $v(t, x, y, \varepsilon) \equiv v(x, y) := xy + y^2$ in (6.2.1), and the stable root of $v(x, y) = 0$ is found to be given as [see (6.2.16) and (6.2.18)]

$$y = \phi(t, x) := -x. \tag{6.2.78}$$

Note that the condition (6.2.20) of assumption A.2 is not satisfied in this case. However, the weakened assumption

$$y \cdot \left[\int_0^1 v_{0,y}[0, \alpha_0, \phi(0, \alpha_0) + sy] \, ds \right] y \le -\kappa |y|^2, \qquad (6.2.79)$$

for all points (or vectors) y between 0 and $\beta_0 - \phi(0, \alpha_0)$, is seen to hold (with $\alpha_0 = 4$, $\beta_0 = -1$) for a fixed positive constant κ, and this latter condition suffices for the O'Malley/Hoppensteadt construction. The dot product in (6.2.79) is interpreted as an inner product in the vector case.

6.3 Existence, uniqueness, and error estimates

In this section we consider questions of existence, uniqueness, and error estimates for the previous initial-value problem (6.2.1)–(6.2.2), which we rewrite here for convenience as

$$\frac{dx}{dt} = u(t, x, y, \varepsilon),$$

$$\varepsilon \frac{dy}{dt} = v(t, x, y, \varepsilon) \qquad \text{for } t > 0, \qquad (6.3.1)$$

and

$$x = \alpha(\varepsilon), \qquad y = \beta(\varepsilon) \quad \text{at } t = 0, \qquad (6.3.2)$$

where the data u, v, α, and β are assumed to possess expansions of the type (6.1.3)–(6.1.4) of Section 6.1, and where the assumptions A.1 and A.2 of Section 6.2 are assumed to hold. The same techniques can be used for other related problems, such as the initial-value problem of Exercise 6.2.3 subject to the assumptions B.1, B.2, and B.3 in the case that the initial data satisfy expansions of the type (6.1.5) (see Exercise 6.3.4), but we consider here only initial data satisfying (6.1.4).

Of course, the basic result from the theory of ordinary differential equations guarantees that this initial-value problem (6.3.1)–(6.3.2) has precisely one solution, at least locally near $t = 0$, and for any fixed $\varepsilon \ne 0$. However, the interval of existence might conceivably shrink with decreasing ε. In fact, we give a direct proof of existence for all small enough $\varepsilon > 0$, and we find that the solution exists on a fixed interval $0 \le t \le T$ independent of ε as $\varepsilon \to 0+$, where T is as in the assumption A.1 [see (6.2.17)–(6.2.18)]. Moreover, we prove that the O'Malley/

Hoppensteadt functions $x^N = x^N(t, \varepsilon)$, $y^N = y^N(t, \varepsilon)$ constructed in Section 6.2 [see (6.2.38)] provide useful approximations to the exact solution functions, with

$$|x(t, \varepsilon) - x^N(t, \varepsilon)|, |y(t, \varepsilon) - y^N(t, \varepsilon)| \leq C_N \varepsilon^{N+1} \qquad (6.3.3)$$

for a fixed constant C_N depending on the data, uniformly for $0 \leq t \leq T$ and uniformly for all small $\varepsilon > 0$ ($\varepsilon \to 0+$).

We begin with the observation that these functions x^N, y^N have been constructed so that they satisfy the initial-value problem (6.3.1)–(6.3.2) approximately, with

$$\frac{dx^N}{dt} = u(t, x^N, y^N, \varepsilon) - \rho_1(t, \varepsilon),$$

$$\varepsilon \frac{dy^N}{dt} = v(t, x^N, y^N, \varepsilon) - \rho_2(t, \varepsilon) \quad \text{for } 0 \leq t \leq T, \qquad (6.3.4)$$

and

$$x^N = \sum_{k=0}^{N} \alpha_k \varepsilon^k = \alpha(\varepsilon) - \sigma_1(\varepsilon),$$

$$y^N = \sum_{k=0}^{N} \beta_k \varepsilon^k = \beta(\varepsilon) - \sigma_2(\varepsilon) \quad \text{at } t = 0, \qquad (6.3.5)$$

where the residuals ρ_j, σ_j ($j = 1, 2$) are defined by (6.3.4)–(6.3.5) and can be shown directly to satisfy the estimates

$$|\rho_j(t, \varepsilon)| \leq \text{const. } \varepsilon^N,$$

$$|\sigma_j(\varepsilon)| \leq \text{const. } \varepsilon^{N+1} \quad \text{for } j = 1, 2, \qquad (6.3.6)$$

for fixed constants that depend on N, uniformly as $\varepsilon \to 0+$, and where the inequality for ρ_j is uniformly valid for $0 \leq t \leq T$. We are suppressing the obvious dependence of these residuals ρ_j, σ_j on N. The proofs of the inequalities of (6.3.6) are based on the construction of x^N, y^N given in Section 6.2; see also Section 5.4 for an analogous proof for a related linear problem. For example, the inequality of (6.3.6) for σ_j follows directly from (6.2.2), (6.2.15), (6.2.38), and (6.1.4). We omit the proof of the inequality for ρ_j, although the proof is indicated in Exercise 6.3.1 for the case $N = 0$, where it is seen that a slightly better estimate than that of (6.3.6) is actually valid for ρ_j. However, (6.3.6) suffices for our purpose.

Remainders $R = R_N$ and $S = S_N$ are now defined as [see (6.2.39)]

$$R \equiv R_N(t, \varepsilon) := x(t, \varepsilon) - x^N(t, \varepsilon),$$
$$S \equiv S_N(t, \varepsilon) := y(t, \varepsilon) - y^N(t, \varepsilon), \tag{6.3.7}$$

and are found to satisfy

$$\frac{dR}{dt} = u(t, x^N + R, y^N + S, \varepsilon) - u(t, x^N, y^N, \varepsilon) + \rho_1(t, \varepsilon),$$
$$\varepsilon \frac{dS}{dt} = v(t, x^N + R, y^N + S, \varepsilon) - v(t, x^N, y^N, \varepsilon) + \rho_2(t, \varepsilon), \tag{6.3.8}$$

for $t > 0$, subject to the initial conditions

$$R = \sigma_1(\varepsilon), \qquad S = \sigma_2(\varepsilon) \quad \text{at } t = 0. \tag{6.3.9}$$

As in (4.1.39)–(4.1.41), the Taylor/Cauchy formula can be used to write

$$u\big[t, x^N(t, \varepsilon) + R, y^N(t, \varepsilon) + S, \varepsilon\big] - u\big[t, x^N(t, \varepsilon), y^N(t, \varepsilon), \varepsilon\big]$$
$$= A(t, \varepsilon)R + B(t, \varepsilon)S + E(t, R, S, \varepsilon),$$
$$v\big[t, x^N(t, \varepsilon) + R, y^N(t, \varepsilon) + S, \varepsilon\big] - v\big[t, x^N(t, \varepsilon), y^N(t, \varepsilon), \varepsilon\big]$$
$$= C(t, \varepsilon)R + D(t, \varepsilon)S + F(t, R, S, \varepsilon), \tag{6.3.10}$$

with

$$A(t, \varepsilon) := u_x\big[t, x^N(t, \varepsilon), y^N(t, \varepsilon), \varepsilon\big],$$
$$B(t, \varepsilon) := u_y\big[t, x^N(t, \varepsilon), y^N(t, \varepsilon), \varepsilon\big],$$
$$C(t, \varepsilon) := v_x\big[t, x^N(t, \varepsilon), y^N(t, \varepsilon), \varepsilon\big],$$
$$D(t, \varepsilon) := v_y\big[t, x^N(t, \varepsilon), y^N(t, \varepsilon), \varepsilon\big], \tag{6.3.11}$$

and

$$E(t, R, S, \varepsilon) := \int_0^1 (1 - s)\big\{ u_{xx}[Z(s)] R^2 + 2u_{xy}[Z(s)] RS$$
$$+ u_{yy}[Z(s)] S^2 \big\}\, ds,$$
$$F(t, R, S, \varepsilon) := \int_0^1 (1 - s)\big\{ v_{xx}[Z(s)] R^2 + 2v_{xy}[Z(s)] RS$$
$$+ v_{yy}[Z(s)] S^2 \big\}\, ds. \tag{6.3.12}$$

where for brevity we are writing $Z(s)$ for the argument of u and v, with

$$Z(s) \equiv Z(s, t, R, S, \varepsilon) := \left[t, x^N(t, \varepsilon) + sR, y^N(t, \varepsilon) + sS, \varepsilon \right].$$

$$(6.3.13)$$

Note that we are suppressing the dependence of these quantities A, B, C, D, E, and F on N, and we are similarly suppressing the dependence of $Z(s)$ on t, R, S, ε, and N. The subscripts on u_x, u_y, u_{xx}, u_{xy}, and so forth, denote partial derivatives, such as

$$u_x = \frac{\partial u(t, x, y, \varepsilon)}{\partial x} \quad \text{and} \quad u_{xy} = \frac{\partial^2 u(t, x, y, \varepsilon)}{\partial x \, \partial y}.$$

We now put $R = R(t, \varepsilon)$ and $S = S(t, \varepsilon)$ in (6.3.10), and then the differential equations of (6.3.8) can be rewritten as

$$\frac{dR}{dt} = A(t, \varepsilon)R + B(t, \varepsilon)S + \rho_1(t, \varepsilon) + E(t, R, S, \varepsilon),$$

$$(6.3.14)$$

$$\varepsilon \frac{dS}{dt} = C(t, \varepsilon)R + D(t, \varepsilon)S + \rho_2(t, \varepsilon) + F(t, R, S, \varepsilon),$$

for $t > 0$.

For each fixed N and each fixed $\varepsilon > 0$, the preceding functions A, B, C, and D defined by (6.3.11) are smooth functions of t for $0 \le t \le T$, as follows from the regularity of u and v along with the construction of x^N and y^N. Moreover, these functions are uniformly bounded as $\varepsilon \to 0+$, with

$$\|A\|, \|B\|, \|C\|, \|D\| \le \text{constant}, \qquad (6.3.15)$$

where the norm here is the maximum norm as in (6.1.23), and the constant depends on N but is independent of ε as $\varepsilon \to 0$. Finally, the assumptions A.1 and A.2 of Section 6.2, along with the construction of x^N and y^N, guarantee that the function $D = D(t, \varepsilon)$ of (6.3.11) is uniformly negative, with [see (6.1.3), (6.2.18), (6.2.20), (6.2.37), and (6.2.38)]

$$D(t, \varepsilon) \le -\tfrac{1}{2}\kappa < 0 \quad \text{for } 0 \le t \le T, 0 < \varepsilon \le \varepsilon_0, \qquad (6.3.16)$$

for some suitably small $\varepsilon_0 > 0$, where κ is the constant appearing in A.1 and A.2.

We can apply Theorem 6.1.1 to the initial-value problem (6.3.9) and (6.3.14) (with κ replaced in Theorem 6.1.1 by $\kappa/2$, with $f = \rho_1 + E$,

$g = \rho_2 + F$, etc.), and we obtain the equivalent integral equation

$$\begin{bmatrix} R(t) \\ S(t) \end{bmatrix} = \begin{bmatrix} \xi(t, \varepsilon) \\ \eta(t, \varepsilon) \end{bmatrix} + \int_0^t K(t, s, \varepsilon) \begin{bmatrix} E[s, R(s), S(s), \varepsilon] \\ \dfrac{1}{\varepsilon} F[s, R(s), S(s), \varepsilon] \end{bmatrix} ds,$$

(6.3.17)

where we define

$$\begin{bmatrix} \xi(t, \varepsilon) \\ \eta(t, \varepsilon) \end{bmatrix} := K(t, 0, \varepsilon) \begin{bmatrix} \sigma_1(\varepsilon) \\ \sigma_2(\varepsilon) \end{bmatrix} + \int_0^t K(t, s, \varepsilon) \begin{bmatrix} \rho_1(s, \varepsilon) \\ \dfrac{1}{\varepsilon} \rho_2(s, \varepsilon) \end{bmatrix} ds,$$

(6.3.18)

and where the kernel $K = K(t, s, \varepsilon)$ is the fundamental matrix given by (6.1.30) that satisfies the estimates of (6.1.31). We are suppressing the dependence of R and S on the parameter ε in (6.3.17).

We now use the Banach/Picard fixed-point theorem to prove that the integral equation (6.3.17) has a suitable solution. For this purpose we use the Banach space $\mathcal{V}$ of continuous vector-valued functions $\begin{bmatrix} R \\ S \end{bmatrix}$, where $R = R(t)$ and $S = S(t)$ are continuous functions on $0 \le t \le T$, with norm

$$\left\| \begin{bmatrix} R \\ S \end{bmatrix} \right\| := \|R\| + \|S\|, \tag{6.3.19}$$

where $\|R\|$ and $\|S\|$ denote the maximum norms over $0 \le t \le T$ as in (6.1.23).

Define the operator M for suitable vectors $\begin{bmatrix} R \\ S \end{bmatrix}$ in $\mathcal{V}$ by the formula

$$M \begin{bmatrix} R \\ S \end{bmatrix}(t) := \text{right side of equation (6.3.17)}$$

$$= \begin{bmatrix} \xi(t, \varepsilon) \\ \eta(t, \varepsilon) \end{bmatrix} + \int_0^t K(t, s, \varepsilon) \begin{bmatrix} E[s, R(s), S(s), \varepsilon] \\ \dfrac{1}{\varepsilon} F[s, R(s), S(s), \varepsilon] \end{bmatrix} ds,$$

(6.3.20)

where this operator M is well defined by (6.3.20), at least for all vectors $\begin{bmatrix} R \\ S \end{bmatrix}$ in some fixed ball $B_{\hat{r}}$ centered at the origin in the space $\mathcal{V}$,

$$B_{\hat{r}} := \left\{ \begin{bmatrix} R \\ S \end{bmatrix} \middle| \left\| \begin{bmatrix} R \\ S \end{bmatrix} \right\| \le \hat{r} \right\},$$

for some fixed positive radius $\hat{r}$. Equation (6.3.17) can be written now as the fixed-point equation

$$\begin{bmatrix} R \\ S \end{bmatrix} = M \begin{bmatrix} R \\ S \end{bmatrix}. \tag{6.3.21}$$

In the following we initially restrict M to a fixed ball $B_{\hat{r}}$ of radius $\hat{r}$ as indicated earlier.

It follows from the definition of $\begin{bmatrix} \xi \\ \eta \end{bmatrix}$ given by (6.3.18), along with (6.1.31), (6.3.6), and (6.3.15), that the given vector $\begin{bmatrix} \xi \\ \eta \end{bmatrix}$ has small norm, of order ε^N, with

$$\left\| \begin{bmatrix} \xi \\ \eta \end{bmatrix} \right\| \leq \gamma_1 \varepsilon^N \tag{6.3.22}$$

for a fixed positive constant γ_1 and for all small positive ε, say for $0 < \varepsilon \leq \varepsilon_1$, for some fixed ε_1 that can be taken to satisfy

$$\gamma_1 \cdot \varepsilon_1^N \leq \tfrac{1}{2}\hat{r}. \tag{6.3.23}$$

Note that, for the moment, we are assuming the condition $N \geq 1$, so that (6.3.23) will automatically hold for small enough ε_1.

We now find easily that for all small $\varepsilon > 0$, *the operator M maps the ball*

$$B_r := \left\{ \begin{bmatrix} R \\ S \end{bmatrix} \middle| \left\| \begin{bmatrix} R \\ S \end{bmatrix} \right\| \leq r \right\},$$

into itself if the radius r is chosen as

$$r := 2\gamma_1 \cdot \varepsilon^N. \tag{6.3.24}$$

Indeed, for all vectors in $B_{\hat{r}}$, it follows from (6.3.12)–(6.3.13) that estimates of the form

$$\left| E\left[t, R(t), S(t), \varepsilon\right] \right|, \left| F\left[t, R(t), S(t), \varepsilon\right] \right| \leq \gamma_2 \cdot \left\| \begin{bmatrix} R \\ S \end{bmatrix} \right\|^2 \tag{6.3.25}$$

hold for a fixed constant γ_2, uniformly for $0 \leq t \leq T$ and for all small $\varepsilon > 0$. This result and (6.1.31) yield the inequality

$$\left\| \int_0^t K(t, s, \varepsilon) \begin{bmatrix} E\left[s, R(s), S(s), \varepsilon\right] \\ \dfrac{1}{\varepsilon} F\left[s, R(s), S(s), \varepsilon\right] \end{bmatrix} ds \right\| \leq \gamma_3 \cdot r^2 \tag{6.3.26}$$

for a fixed constant γ_3 and for all vectors $\begin{bmatrix} R \\ S \end{bmatrix}$ in B_r, for $0 < r \leq \hat{r}$. These results, along with (6.3.20), (6.3.22), and (6.3.24), then imply

$$\left\| M \begin{bmatrix} R \\ S \end{bmatrix} \right\| \leq \tfrac{1}{2}r + \gamma_3 \cdot r^2 \quad \text{for all } \begin{bmatrix} R \\ S \end{bmatrix} \text{ in } B_r, \tag{6.3.27}$$

for $0 < \varepsilon \leq \varepsilon_1$. In addition to the previous requirement (6.3.23) on ε_1, we also require that ε_1 satisfy the inequality

$$2\gamma_1\gamma_3 \cdot \varepsilon_1^N \leq \tfrac{1}{2}, \tag{6.3.28}$$

and then (6.3.24) and (6.3.27) yield

$$\left\| M\begin{bmatrix} R \\ S \end{bmatrix} \right\| \leq r \quad \text{for all vectors } \begin{bmatrix} R \\ S \end{bmatrix} \text{ in } B_r, \tag{6.3.29}$$

for $0 < \varepsilon \leq \varepsilon_1$.

Hence, M maps any such ball B_r into itself, so that the Banach/Picard fixed-point theorem can be applied to the equation (6.3.21) on such a ball *if we can show that M is a contraction operator on B_r*. In order to show this latter result, we use (6.3.20) to compute, for any two elements $\begin{bmatrix} R_1 \\ S_1 \end{bmatrix}$ and $\begin{bmatrix} R_2 \\ S_2 \end{bmatrix}$ in B_r, the difference

$$M\begin{bmatrix} R_1 \\ S_1 \end{bmatrix}(t) - M\begin{bmatrix} R_2 \\ S_2 \end{bmatrix}(t)$$

$$= \int_0^t K(t,s,\varepsilon) \begin{bmatrix} E[s,R_1(s),S_1(s),\varepsilon] - E[s,R_2(s),S_2(s),\varepsilon] \\ \dfrac{1}{\varepsilon}\{F[s,R_1(s),S_1(s),\varepsilon] - F[s,R_2(s),S_2(s),\varepsilon]\} \end{bmatrix} ds.$$

$$(6.3.30)$$

Taylor's theorem, along with (6.3.12)–(6.3.13), gives [see (4.1.61)]

$$\left| E[t,R_1(t),S_1(t),\varepsilon] - E[t,R_2(t),S_2(t),\varepsilon] \right|$$

$$\leq \gamma_4 \cdot r \left\| \begin{bmatrix} R_1 \\ S_1 \end{bmatrix} - \begin{bmatrix} R_2 \\ S_2 \end{bmatrix} \right\| \tag{6.3.31}$$

for all vectors $\begin{bmatrix} R_1 \\ S_1 \end{bmatrix}$ and $\begin{bmatrix} R_2 \\ S_2 \end{bmatrix}$ in B_r, for all $0 \leq t \leq T$ and all small $\varepsilon > 0$, for some fixed constant γ_4, and the same inequality holds also for F. Then (6.1.31), along with these last two results (6.3.30) and (6.3.31), gives

$$\left\| M\begin{bmatrix} R_1 \\ S_1 \end{bmatrix} - M\begin{bmatrix} R_2 \\ S_2 \end{bmatrix} \right\| \leq \gamma_5 \cdot r \left\| \begin{bmatrix} R_1 \\ S_1 \end{bmatrix} - \begin{bmatrix} R_2 \\ S_2 \end{bmatrix} \right\| \tag{6.3.32}$$

for a fixed constant γ_5 and for all vectors in the ball B_r, with r given by (6.3.24). This result (6.3.32) is uniformly valid for $0 < \varepsilon \leq \varepsilon_1$, for any fixed ε_1 satisfying (6.3.23) and (6.3.28). If we now impose the further condition on ε_1 that

$$2\gamma_1\gamma_5 \cdot \varepsilon_1^N \leq \tfrac{1}{2}, \tag{6.3.33}$$

then (6.3.32) yields

$$\left\| M\begin{bmatrix} R_1 \\ S_1 \end{bmatrix} - M\begin{bmatrix} R_2 \\ S_2 \end{bmatrix} \right\| \leq \tfrac{1}{2} \left\| \begin{bmatrix} R_1 \\ S_1 \end{bmatrix} - \begin{bmatrix} R_2 \\ S_2 \end{bmatrix} \right\| \qquad (6.3.34)$$

everywhere in the ball B_r uniformly for all $0 < \varepsilon \leq \varepsilon_1$. Hence, M is contracting. Putting these results together, we find the following theorem.

Theorem 6.3.1: *Let the initial-value problem* (6.1.1)–(6.1.4) *satisfy the assumptions of Section 6.2, including the assumptions A.1 and A.2 [see* (6.2.16)–(6.2.20)], *and let* x^N, y^N *be the O'Malley/Hoppensteadt approximants given by* (6.2.38), *as constructed in Section 6.2. Then there are fixed constants* C_N, T, *and* ε_1 *so that the given initial-value problem has a unique solution for t and* ε *satisfying*

$$0 \leq t \leq T, \qquad 0 < \varepsilon \leq \varepsilon_1, \qquad (6.3.35)$$

and, moreover, the exact solution functions x, y *satisfy the uniform estimates*

$$|x(t, \varepsilon) - x^N(t, \varepsilon)|, |y(t, \varepsilon) - y^N(t, \varepsilon)| \leq C_N \cdot \varepsilon^{N+1} \quad (6.3.36)$$

uniformly for all t, ε *satisfying* (6.3.35).

Proof: We take ε_1 to be any fixed positive number that satisfies simultaneously all three of the conditions (6.3.23), (6.3.28), and (6.3.33). Then the foregoing discussion, along with the Banach/Picard fixed-point theorem (Theorem 0.4.1), gives the stated result of the present theorem, but with ε^{N+1} replaced by ε^N on the right side of (6.3.36) [see (6.3.7), (6.3.21), (6.3.24), and (6.3.29)]. That is, we have the result

$$|x(t, \varepsilon) - x^N(t, \varepsilon)|, |y(t, \varepsilon) - y^N(t, \varepsilon)| \leq C_N \cdot \varepsilon^N \quad (6.3.37)_N$$

for any fixed positive integer $N \geq 1$. In addition, we have the identity

$$\begin{bmatrix} x - x^N \\ y - y^N \end{bmatrix} = \begin{bmatrix} x - x^{N+1} \\ y - y^{N+1} \end{bmatrix} + \begin{bmatrix} x^{N+1} - x^N \\ y^{N+1} - y^N \end{bmatrix}, \qquad (6.3.38)$$

from which the actual result (6.3.36) follows, with a different constant C_N. Indeed, the result $(6.3.37)_{N+1}$ (with N replaced by $N + 1$) shows that the first term on the right side of (6.3.38) is of order ε^{N+1}, and (6.2.38) shows that the last term on the right side of (6.3.38) is also of order ε^{N+1}, uniformly for all t and ε satisfying (6.3.35). Note that this last argument holds also for the case $N = 0$, so that the stated result (6.3.36) is seen to be valid for any nonnegative integer $N \geq 0$. ∎

An analogous theorem is valid also for the corresponding initial-value problem (6.1.1)–(6.1.4) in the vector case, in which the data and solution functions are suitable vector-valued quantities, subject to suitably modified versions of the foregoing assumptions, as discussed briefly in Sections 6.1 and 6.2. The appropriate integral equation (6.3.17) is obtained in the vector case from Theorem 6.1.2.

Results related to Theorem 6.3.1 have been obtained by many authors, including Tikhonov (1950, 1952), Levin and Levinson (1954), Butuzov (1965), Chang and Coppel (1969), and Chang (1969a); see also the references listed in these works and additional references given in O'Malley (1974a). A generalization of Theorem 6.3.1 on an infinite interval with $T = \infty$ has been given by Hoppensteadt (1966, 1971b). The conclusions of Theorem 6.3.1 remain true if the condition (6.2.20) of assumption A.2 is replaced by the weaker assumption (6.2.79). In this latter case the previous estimates of (6.1.31) must be replaced by certain more delicate estimates; see Smith (1984c), where such estimates are given for a related boundary-value problem.

Exercises

Exercise 6.3.1: Let $\rho_1(t, \varepsilon), \rho_2(t, \varepsilon)$ be defined by (6.3.4) in the case $N = 0$. Prove directly the estimates

$$|\rho_1(t, \varepsilon)| \leq \text{const.} \left(\varepsilon + e^{-\kappa t / \varepsilon} \right),$$
$$|\rho_2(t, \varepsilon)| \leq \text{const.} \ \varepsilon, \tag{6.3.39}$$

for $0 \leq t \leq T$ and for all small enough $\varepsilon > 0$. [Note that the estimates of (6.3.39) are stronger than those of (6.3.6) for ρ_1 and ρ_2 in the case $N = 0$. For example, in the case $N = 0$, (6.3.6) implies that the following integrals ($j = 1, 2$) are bounded:

$$\int_0^T |\rho_j(t, \varepsilon)| \, dt \leq \text{constant},$$

whereas (6.3.39) implies that these integrals are actually smaller, of order ε.] *Hint*: The result for ρ_2 follows in part from the estimates

$$\left| v_0 \left[t, X_0(t), Y_0(t) + Y_0^*(\tau) \right] - v_0 \left[0, X_0(0), Y_0(0) + Y_0^*(\tau) \right] \right|$$
$$\leq \text{const.} \ |t| e^{-\kappa t / \varepsilon} \leq \text{const.} \ \varepsilon,$$

where we have put $\tau = t/\varepsilon$, and where the first estimate here can be shown to follow from the identity

$$f(t, r) - f(0, r) = \int_0^1 \int_0^1 \frac{\partial^2 f(s_1 t, s_2 r)}{\partial s_1 \, \partial s_2} \, ds_1 \, ds_2 \tag{6.3.40}$$

for the function

$$f(t, r) := v_0[t, X_0(t), Y_0(t) + r]$$

evaluated at $r = Y_0^*(\tau)$. Note that the function f satisfies $f(t, 0) \equiv 0$ for all t, as follows from the second equation of $(6.2.11)_0$, and hence also $\partial f(t, 0)/\partial t \equiv 0$.

Exercise 6.3.2: Use the estimate (6.3.39) to give a direct proof of the result (6.3.36) of Theorem 6.3.1 in the case $N = 0$. *Hint:* Show that (6.3.39) permits an improved version of the estimate (6.3.22) in the case $N = 0$.

Exercise 6.3.3: Let the initial values $\alpha \equiv \alpha_0$, $\beta \equiv \beta_0$ be independent of ε in (6.1.2), so that the residuals σ_j vanish for $j = 1, 2$ in (6.3.5) and (6.3.9). In the case $N = 0$, use the estimates of (6.3.39) to derive the result (see Exercise 6.2.5)

$$|R(t, \varepsilon)|, |S(t, \varepsilon)| \leq \text{const. } \varepsilon\delta \quad \text{for } 0 \leq t \leq \varepsilon\delta \qquad (6.3.41)$$

if $\varepsilon\delta$ is small enough. [This proves that the functions x^0, y^0 provide useful approximations to x, y for small t even if ε is not small. If the initial values $\alpha(\varepsilon), \beta(\varepsilon)$ depend on ε as in (6.1.4), then we still obtain the result (6.3.41) in the case $N = 0$ if the construction of x^0, y^0 is modified so that $x^0(0, \varepsilon) = \alpha(\varepsilon)$, $y^0(0, \varepsilon) = \beta(\varepsilon)$ hold. We need only replace the previous initial conditions of $(6.2.15)_0$ with the modified initial conditions $X_0(0) = \alpha(\varepsilon)$, $Y_0(0) + Y_0^*(0) = \beta(\varepsilon)$. In this case, $X_0(t)$, $Y_0(t)$, and $Y_0^*(\tau)$ will depend explicitly on ε in addition to t and τ, as $X_0 = X_0(t, \varepsilon)$, and so forth, but this causes no difficulties.]

Exercise 6.3.4: Let x^0, y^0 be the approximate solution functions given in Exercise 6.2.4 for the problem

$$x' = y, \qquad\qquad x = 0,$$

$$\varepsilon y' = -x^2 - y \quad \text{for } t > 0, \qquad y = \frac{1}{\varepsilon} \quad \text{at } t = 0,$$

and prove the estimates

$$|R(t, \varepsilon)|, |S(t, \varepsilon)| \leq \text{const. } \varepsilon \quad \text{for } 0 \leq t \leq 1$$

and for all small $\varepsilon > 0$, where $R = x - x^0$, $S = y - y^0$. *Hint:* Use Theorem 6.1.1 to obtain a suitable integral representation of the form

$$\begin{bmatrix} R(t) \\ S(t) \end{bmatrix} = \int_0^t K(t, s, \varepsilon) \begin{bmatrix} \rho_1(s, \varepsilon) \\ \frac{1}{\varepsilon}\rho_2(s, \varepsilon) \end{bmatrix} ds + \int_0^t K(t, s, \varepsilon) \begin{bmatrix} 0 \\ -\frac{1}{\varepsilon}R(s)^2 \end{bmatrix} ds,$$

where the dependence of R and S on ε is suppressed here, and where K is the fundamental matrix given by (6.1.30) that satisfies the estimates of (6.1.31) with $A = 0$, $B = 1$, $C = -2x^0$, $D = -1$. The residuals ρ_j can be examined, and we can then use a Gronwall-type argument to obtain estimates for R and S from the integral representation, or we can use the Banach/Picard fixed-point theorem. In this way we can obtain, for example,

$$|R(t, \varepsilon)| \leq 60\varepsilon \quad \text{for } 0 \leq t \leq 1$$

and for all small $\varepsilon > 0$, with a similar result for S. (An analogue of Exercise 6.3.3 can also be obtained in this case.)

6.4 Numerical methods

In this section we consider the numerical solution of the following problem from Section 6.2 and Section 6.3:

$$\frac{dx}{dt} = u(t, x, y, \varepsilon),$$

$$\varepsilon\frac{dy}{dt} = v(t, x, y, \varepsilon) \quad \text{for } t > 0, \tag{6.4.1}$$

and

$$x = \alpha(\varepsilon), \qquad y = \beta(\varepsilon) \quad \text{at } t = 0, \tag{6.4.2}$$

where the data u, v, α, and β are assumed to possess expansions of the type (6.1.3)–(6.1.4) of Section 6.1, and where the assumptions A.1 and A.2 of Section 6.2 are assumed to hold. As usual, we can consider the case of vector systems, but we restrict consideration here to the scalar case in which x, y, u, v, α, and β are all suitable real-valued quantities. Also, the same techniques can be used for other related problems such as the initial-value problem of Exercise 6.2.3 subject to the assumptions B.1, B.2, and B.3 [see (6.2.63)–(6.2.68)] in the case that the initial data satisfy expansions of the type (6.1.5) (see Exercises 6.4.2, 6.4.3, and 6.4.4).

We have seen in previous sections that the fast variable y for the solution functions $x(t, \varepsilon)$, $y(t, \varepsilon)$ for (6.4.1)–(6.4.2) undergoes rapid change within boundary layers, for small $\varepsilon > 0$. For this reason the problem is difficult to handle numerically with classical numerical algorithms based on discretization of the differential equations using finite-difference approximations. The use of various such finite-difference techniques for (6.4.1)–(6.4.2) when ε is small requires an excessively small step size in the discretization if we wish to follow the rapid

Table 6.1

ε	1	10^{-1}	10^{-2}	10^{-3}
N	14	30	510	524,286

N = number of function evaluations required to obtain 2-digit accuracy for solution of (6.4.3) at $t = 0.1$ using Euler's method with extrapolation (calculations performed using 16-digit precision).

variation through the thin boundary layer. The initial-value problem (6.4.1)–(6.4.2) falls within the class of problems described as *stiff*.

For example, suppose that we wish to use *Euler's method with extrapolation* to obtain numerical approximations for the solution values $x(0.1, \varepsilon)$, $y(0.1, \varepsilon)$ at $t = 0.1$ for the problem

$$\frac{dx}{dt} = 1 + 6y^2, \qquad\qquad x = 0,$$

$$\varepsilon \frac{dy}{dt} = -y + t \quad \text{for } t > 0, \qquad y = 1 \quad \text{at } t = 0. \qquad (6.4.3)$$

Table 6.1 shows the number of function evaluations [total combined number of evaluations for the functions u and v in (6.4.1)] required in the case (6.4.3) to obtain just 2-digit accuracy, where the accuracy is measured in this case against the following exact solution values:

$$x(t, \varepsilon) = t + 2\left[\varepsilon^3 + (t - \varepsilon)^3\right] - 12\varepsilon(1 + \varepsilon)te^{-t/\varepsilon}$$
$$\qquad\qquad + 3\varepsilon(1 + \varepsilon)^2 (1 - e^{-2t/\varepsilon})$$

$$y(t, \varepsilon) = t - \varepsilon + (1 + \varepsilon)e^{-t/\varepsilon}.$$

Table 6.1 shows that the computational cost, as measured by the number of function evaluations, eventually increases precipitously with decreasing ε. Euler's method with extrapolation is actually capable of yielding high numerical accuracy here, say 8-digit accuracy for the values of ε listed in Table 6.1 (using 16-digit arithmetic), but the computational cost is relatively high for the smaller values of ε. For even smaller ε, the resulting excessively large number of function evaluations eventually causes the accuracy of the numerical calculation to be destroyed by roundoff errors. This phenomenon, as ε decreases, of precipitous increase in computational cost and decrease in numerical accuracy is not limited to Euler's method with extrapolation, but is shared with other classical techniques such as the standard Runge/Kutta methods.

Miranker (1973, 1981) has shown that the problem (6.4.1)–(6.4.2) can be handled nicely numerically for small ε with boundary-layer methods based on the developments of Sections 6.2 and 6.3. These methods, with the boundary-layer correction terms included, are uniformly accurate for all small t when ε is small and also when ε is not small (see Exercises 6.2.5 and 6.3.3), and, moreover, the methods remain accurate for larger t as well, as $\varepsilon \to 0+$. Indeed, the method of Miranker is intended primarily as a means of integrating *across* boundary layers using a step size h that is large compared with ε. The method is most accurate in this regard for small ε, and indeed the method *improves* with decreasing ε, whereas many other methods fail precipitously for these problems as $\varepsilon \to 0+$. Hence, the method of Miranker complements other methods nicely.

Miranker's approach is based directly on results such as those contained in the previous two sections. Specifically, Miranker constructs suitable numerical approximations to the functions x^1, y^1 of (6.2.38) that give order(ε^2) uniform approximations to the exact solution functions x, y [cf. (6.2.39)–(6.2.40) with $N = 1$]. Hence, following the developments of Section 6.2, the method of Miranker is based on obtaining suitable numerical solutions for the following four subproblems, from which suitable approximate values are constructed for x^1 and y^1, at some chosen $t = h$ for some given step size h that is ordinarily taken to be much larger than the small parameter ε.

Subproblem 1:

$$\frac{dX_0}{dt} = u_0(t, X_0, Y_0), \qquad 0 = v_0(t, X_0, Y_0) \quad \text{for } t > 0,$$

$$X_0(0) = \alpha_0 \quad \text{at } t = 0.$$

Subproblem 2:

$$\frac{dY_0^*}{d\tau} = v_0\big[0, X_0(0), Y_0(0) + Y_0^*(\tau)\big] \quad \text{for } \tau > 0,$$

$$Y_0^*(0) = \beta_0 - Y_0(0) \quad \text{at } \tau = 0.$$

Subproblem 3:

$$X_0^*(\tau) = -\int_\tau^\infty \big\{ u_0\big[0, X_0(0), Y_0(0) + Y_0^*(\sigma)\big]$$

$$- u_0\big[0, X_0(0), Y_0(0)\big]\big\} \, d\sigma.$$

Subproblem 4:

$$\frac{dX_1}{dt} = u_{0,x}(t, X_0, Y_0)X_1 + u_{0,y}(t, X_0, Y_0)Y_1 + u_1(t, X_0, Y_0),$$

$$\frac{dY_0}{dt} = v_{0,x}(t, X_0, Y_0)X_1 + v_{0,y}(t, X_0, Y_0)Y_1 + v_1(t, X_0, Y_0) \quad \text{for } t > 0,$$

$$X_1(0) = \alpha_1 - X_0^*(0) \quad \text{at } t = 0,$$

where dY_0/dt can be given on the left side of the second equation in Subproblem 4 as

$$\frac{dY_0}{dt} = -(v_{0,y})^{-1}\left(v_{0,t} + v_{0,x}\frac{dX_0}{dt}\right)\Bigg|_{\substack{x=X_0(t)\\y=Y_0(t)}}$$

$$= -(v_{0,y})^{-1}(v_{0,t} + v_{0,x}u_0)\Bigg|_{\substack{x=X_0(t)\\y=Y_0(t)}},$$

as obtained from Subproblem 1 on differentiation of the second equation there ($0 = v_0$) with respect to t.

Finally, if we wish to obtain approximate solution values *inside* the boundary layer (with step size h that is small compared with ε), then it is useful also to consider the following fifth subproblem (Aiken and Lapidus 1974):

Subproblem 5:

$$\frac{dY_1^*}{d\tau} = v_{0,y}[0, X_0(0), Y_0(0) + Y_0^*(\tau)]Y_1^* + B_0^*(\tau) \quad \text{for } \tau > 0,$$

$$Y_1^*(0) = \beta_1 - Y_1(0) \quad \text{at } \tau = 0,$$

where the quantity B_0^* is defined as [compare with (6.2.14)]

$$\begin{aligned}
B_0^*(\tau) := &\ \tau\{v_{0,t}[0, X_0(0), Y_0(0) + Y_0^*(\tau)] - v_{0,t}[0, X_0(0), Y_0(0)]\}\\
&+ \{v_{0,x}[0, X_0(0), Y_0(0) + Y_0^*(\tau)]\\
&\quad - v_{0,x}[0, X_0(0), Y_0(0)]\}[\tau X_0'(0) + X_1(0)]\\
&+ \{v_{0,y}[0, X_0(0), Y_0(0) + Y_0^*(\tau)]\\
&\quad - v_{0,y}[0, X_0(0), Y_0(0)]\}[\tau Y_0'(0) + Y_1(0)]\\
&+ v_{0,x}[0, X_0(0), Y_0(0) + Y_0^*(\tau)]X_0^*(\tau)\\
&+ \{v_1[0, X_0(0), Y_0(0) + Y_0^*(\tau)] - v_1[0, X_0(0), Y_0(0)]\},
\end{aligned}$$

with the quantities on the right side of this last equation for B_0^* determined by the earlier subproblems.

Each of the foregoing subproblems is regular, with no stiffness occurring, and these subproblems can be solved using standard algorithms. As mentioned earlier, Miranker gives the method primarily as a means of integrating across boundary layers directly with a step size h that is large compared with the small parameter ε.

We first solve Subproblem 1 numerically to get approximations for

$$Y_0(0), \quad X_0(h), \quad \text{and} \quad Y_0(h), \tag{6.4.4}$$

using any suitable self-starting method such as Euler's method with extrapolation or a Runge/Kutta method (for the differential equation $dX_0/dt = u_0$), along with Newton's method (for the equation $0 = v_0$). More specifically, using the given values $t = 0$ and $X_0(0) = \alpha_0$, we first solve the algebraic equation $0 = v_0$ for a suitable value of $Y_0 = Y_0(0)$, and then this value for $Y_0(0)$ is used in the differential equation $dX_0/dt = u_0$ in the numerical integration for $X_0(h)$. We generally need to solve the algebraic equation for $Y_0(t)$ for several different values of t and $X_0(t)$.

The known values of $X_0(0)$ and $Y_0(0)$ are then used in Subproblem 2, and this latter subproblem is solved numerically (using any suitable initial-value solver) to provide approximations for the values

$$Y_0^*(\eta), Y_0^*(2\eta), \ldots, Y_0^*(M\eta) \tag{6.4.5}$$

for a chosen step size η and a fixed positive integer M.

These values (6.4.5) of Y_0^* are then used to obtain an approximate value for

$$X_0^*(0) \tag{6.4.6}$$

by performing a numerical quadrature for the integral in Subproblem 3 at $\tau = 0$, using perhaps the midpoint rule with extrapolation, or any other suitable numerical quadrature rule. Note that $Y_0^*(\sigma)$, and hence also the integrand in Subproblem 3, decays exponentially toward zero with increasing σ [see (6.2.33)], so that the major contribution in the integral comes from "small" values of the integration variable.

Finally, we use the approximate value $X_0^*(0)$ from the previous subproblem as an input value in the initial condition in Subproblem 4, and this latter subproblem is solved after the fashion of Subproblem 1, so as to yield

$$Y_1(0), \quad X_1(h), \quad \text{and} \quad Y_1(h). \tag{6.4.7}$$

We use the second equation of Subproblem 4 to provide Y_1 directly in terms of X_1, and this result for Y_1 is substituted back into the first equation, which then gives a differential equation for the determination of X_1 subject to the given initial condition. This latter initial-value problem for X_1 can be solved numerically with any of several appropriate initial-value solvers.

These computed values are now used to form approximations to x^1 and y^1 from (6.2.38), evaluated at $t = h$ as

$$x^1(h, \varepsilon) := X_0(h) + \varepsilon X_1(h) + \varepsilon X_0^* \left(\frac{h}{\varepsilon} \right)$$

$$y^1(h, \varepsilon) := Y_0(h) + \varepsilon Y_1(h) + \left[Y_0^* \left(\frac{h}{\varepsilon} \right) + \varepsilon Y_1^* \left(\frac{h}{\varepsilon} \right) \right].$$

(6.4.8)

The boundary-layer correction terms on the right side of (6.4.8) are exponentially small for large values of their arguments [see (6.2.37)], so that in particular the terms $X_0^*(h/\varepsilon)$, $Y_0^*(h/\varepsilon)$, and $Y_1^*(h/\varepsilon)$ are exponentially small if h is large compared with ε,

$$\varepsilon \ll h. \qquad (6.4.9)$$

Miranker chooses h large compared with ε, so that (6.4.9) holds, and then the boundary-layer correction terms can be neglected in (6.4.8) to give the approximate values

$$x^1(h, \varepsilon) \cong X_0(h) + \varepsilon X_1(h),$$

$$y^1(h, \varepsilon) \cong Y_0(h) + \varepsilon Y_1(h).$$

(6.4.10)

These approximate values (6.4.10), which involve certain of the previously computed quantities from (6.4.4) and (6.4.7), are now used as the *Miranker approximations* for the exact values $x(h, \varepsilon)$, $y(h, \varepsilon)$. Note that, for fixed step size h, the Miranker approximation *improves* as $\varepsilon \to 0+$.

If h is not large compared with ε, so that (6.4.9) fails to hold, then we include the boundary-layer correction terms in the approximations, so that we use (6.4.8) in this case to give approximations for the exact solution values. In this case we must also solve Subproblem 5 numerically for $Y_1^*(h/\varepsilon)$. Moreover, we choose the step size η used in the solution of Subproblem 2 [see (6.4.5)] so that the quantity h/ε is an integral multiple of η, and we also solve Subproblem 3 for $X_0^*(\tau)$ at $\tau = h/\varepsilon$.

The entire procedure can be performed again starting at the new initial time $t = h$, with the current approximate solution values as the new initial data, so as to advance the numerical solution to $t = 2h$. We can continue the procedure repeatedly as desired, so as to advance the numerical solution to $t = h, 2h, 3h, \ldots$. We can also incorporate a step-size control so as to permit the selection of a new step size h for each new step. Note that because of numerical roundoff error, we must allow for the possibility of the occurrence of (numerical) boundary layers at each step. See Miranker (1973, 1981) for further details, along with numerical examples.

Example 6.4.1: We consider again the problem (6.4.3), rewritten here as

$$\frac{dx}{dt} = 1 + 6y^2, \qquad\qquad x = 0,$$

$$\varepsilon\frac{dy}{dt} = -y + t \quad \text{for } t > 0, \qquad y = 1 \quad \text{at } t = 0. \qquad (6.4.11)$$

Hence, the data of (6.4.1)–(6.4.2) are given here as

$$u(t, x, y, \varepsilon) \equiv u_0(t, x, y): \quad = 1 + 6y^2,$$
$$v(t, x, y, \varepsilon) \equiv v_0(t, x, y): \quad = -y + t, \qquad\qquad (6.4.12)$$
$$\alpha(\varepsilon) \equiv \alpha_0: = 0 \quad \text{and} \quad \beta(\varepsilon) \equiv \beta_0: = 1,$$

with $u_1 \equiv v_1 \equiv \alpha_1 \equiv \beta_1 \equiv 0$. The subproblems of Miranker can be solved exactly in this case, with infinite precision, and we find the following solutions:

Subproblem 1: $X_0(t) = t + 2t^3, \qquad Y_0(t) = t$

Subproblem 2: $Y_0^*(\tau) = e^{-\tau}$

Subproblem 3: $X_0^*(\tau) = -3e^{-2\tau}$

Subproblem 4: $X_1(t) = 3 - 6t^2, \qquad Y_1(t) = -1$

Subproblem 5: $Y_1^*(\tau) = e^{-\tau}$.

We take step size $h = 0.1$, and the resulting approximations at $t = h$ are listed in Table 6.2, as computed by (6.4.8) and also by (6.4.10), along with the exact solution values at $t = h$, for different values of ε. The values in Table 6.2 are given with all digits correct as shown, with the exception of the last digit, which is rounded to the nearest integer.

We see from the values in Table 6.2 that both (6.4.8) and (6.4.10) yield accurate results that agree closely with each other, *for small values of ε.* Of course, this is not surprising, because the boundary-layer correction terms in (6.4.8) are negligible as $\varepsilon \to 0+$. Hence, in such cases, when (6.4.9) holds, we are justified in using the Miranker values (6.4.10). On the other hand, if ε is not small compared with h, so that (6.4.9) does not hold, then the approximate values of (6.4.10) are not expected to be accurate, and indeed in the example they are then *not accurate*, as illustrated in Table 6.2. In this latter situation the approximate values of (6.4.8) may still be sufficiently accurate so as to provide some useful information.

Table 6.2. Tabulated values for (6.4.11) at $t = h = 0.1$.

	ε	1.0	0.1	0.01	0.001	0.0001
(6.4.10)	$x^1(0.1, \varepsilon)$	3.0420000	0.3960000	0.1314000	0.1049400	0.1022940
	$y^1(0.1, \varepsilon)$	-0.9000000	0.0000000	0.0900000	0.0990000	0.0999000
(6.4.8)	$x^1(0.1, \varepsilon)$	0.5858077	0.3553994	0.1314000	0.1049400	0.1022940
	$y^1(0.1, \varepsilon)$	0.9096748	0.4046674	0.0900459	0.0990000	0.0999000
Exact values	$x(0.1, \varepsilon)$	0.6456212	0.3673132	0.1320624	0.1049466	0.1022941
	$y(0.1, \varepsilon)$	0.9096748	0.4046674	0.0900459	0.0990000	0.0999000

Note that the approximate values given here by the boundary-layer techniques improve with decreasing values of ε, whereas many classical methods fail catastrophically for small ε, as indicated earlier in Table 6.1.

Miranker (1973, 1981) also develops the foregoing boundary-layer method for related problems in which there may be no easily identifiable small parameter. In practice, we must exercise good judgment in using these (as any other) techniques, as illustrated in the following example.

Example 6.4.2: Consider the system of differential equations (6.2.42) of Example 6.2.3 with $\varepsilon = 0.01$,

$$\frac{dx}{dt} = y,$$

$$0.01 \frac{dy}{dt} = -x^2 - y \quad \text{for } t > 0, \tag{6.4.13}$$

and consider the initial conditions

$$x = 1 \quad \text{and} \quad y = 50 \quad \text{at } t = 0. \tag{6.4.14}$$

This problem can be modelled as (6.4.1)–(6.4.2) with

$$u(t, x, y, \varepsilon) \equiv u_0(t, x, y) := y,$$

$$v(t, x, y, \varepsilon) \equiv v_0(t, x, y) := -x^2 - y, \tag{6.4.15}$$

$$\alpha(\varepsilon) \equiv \alpha_0 := 1, \qquad \beta(\varepsilon) \equiv \beta_0 := 50, \qquad \varepsilon: = 0.01,$$

and with $u_k \equiv v_k \equiv \alpha_k \equiv \beta_k \equiv 0$ for $k \geq 1$. The subproblems of Miranker can be solved exactly, and we find the following results:

Subproblem 1:

$$X_0(t) = \frac{1}{1 + t}, \qquad Y_0(t) = -\frac{1}{(1 + t)^2},$$

Subproblem 2:

$$Y_0^*(\tau) = 51e^{-\tau},$$

Subproblem 3:

$$X_0^*(\tau) = -51e^{-\tau},$$

Subproblem 4:

$$X_1(t) = \frac{51 + \log(1 + t)^{-2}}{(1 + t)^2}, \qquad Y_1(t) = \frac{-104 + \log(1 + t)^4}{(1 + t)^3}.$$

We take $h = 0.1$, and this value can be considered to be large compared with $\varepsilon = 0.01$ [see (6.4.9)]. Hence, we use (6.4.10) to obtain approximate solution values at $t = h = 0.1$, and we find

$$x^1 \cong X_0(0.1) + (0.01) X_1(0.1) \cong 1.329,$$
$$y^1 \cong Y_0(0.1) + (0.01) Y_1(0.1) \cong -1.605, \tag{6.4.16}$$

based on (6.4.10), at $t = 0.1$.

The problem (6.4.13)–(6.4.14) actually falls within the range of solvability of various standard numerical algorithms, such as a suitable Runge/Kutta method or Euler's method with extrapolation, and using such methods we can determine the solution values accurately by a direct numerical integration. In this way we find

$$x = 1.3288\ldots,$$
$$y = -1.8132\ldots \quad \text{at } t = 0.1, \tag{6.4.17}$$

where the digits shown in (6.4.17) are correct.

A comparison of (6.4.16) and (6.4.17) shows that the approximation (6.4.10) yields an acceptable approximate value x^1 for $x(0.1)$, but the approximation y^1 based on (6.4.10) is rather poor (although even this approximation for y may be adequate in some situations).

Of course, the present *initial value* for the fast variable y specified in (6.4.14) is rather large, and we can expect only limited accuracy from (6.4.10) for the problem (6.4.13)–(6.4.14). Indeed, the initial value $y(0) = 50$ of (6.4.14) can be considered to be proportional to the reciprocal of the small parameter $\varepsilon = 1/100$, with

$$y(0) = 0.5/\varepsilon.$$

Hence, we should consider applying the general approach of Miranker in this case, starting with the analysis of Exercise 6.2.3 for a problem with initial data satisfying (6.2.59). Following the procedure to be outlined later in Exercise 6.4.2, we find in the present case (see also Exercise 6.4.4)

$$x^1 \cong X_0(0.1) + (0.01) X_1(0.1) \cong 1.327,$$
$$y^1 \cong Y_0(0.1) + (0.01) Y_1(0.1) \cong -1.805, \tag{6.4.18}$$

based on Exercise 6.4.2, where the functions X_0, X_1, Y_0, and Y_1 are, of course, different here than in (6.4.16).

We see that both approximations (6.4.16) and (6.4.18) are acceptable for the slow variable x [cf. (6.4.17)], but (6.4.18) gives a more accurate value for the fast variable y than does (6.4.16).

For more general problems involving systems of differential equations with specific numerical coefficients along with specific numerical initial data, it is sometimes difficult in practice to predict the expected accuracy for a numerical algorithm.

Boundary-layer methods can also be used in the development of numerical algorithms for singularly perturbed boundary-value problems, as discussed in Chapter 10.

Exercises

Exercise 6.4.1: Solve the subproblems of Miranker for the problem (6.4.13)–(6.4.14), and verify the results of (6.4.16).

Exercise 6.4.2: Consider the quasi-linear problem of Exercise 6.2.3 subject to the assumptions B.1, B.2, and B.3 [see (6.2.63)–(6.2.68)], so that we have asymptotic representations of the form

$$x(t, \varepsilon) \sim X(t, \varepsilon) + X^*(\tau, \varepsilon),$$
$$y(t, \varepsilon) \sim Y(t, \varepsilon) + \frac{1}{\varepsilon} Y^*(\tau, \varepsilon) \quad \text{with } \tau := \frac{t}{\varepsilon}, \tag{6.4.19}$$

where the functions X, Y, X^*, and Y^* have expansions of the form (6.2.5) and (6.2.6), subject also to the matching conditions of (6.2.7). Let x^1, y^1 be defined as

$$x^1(t, \varepsilon): = \left[X_0(t) + \varepsilon X_1(t) \right] + \left[X_0^*\left(\frac{t}{\varepsilon}\right) + \varepsilon X_1^*\left(\frac{t}{\varepsilon}\right) \right],$$
$$y^1(t, \varepsilon): = \left[Y_0(t) + \varepsilon Y_1(t) \right] + \left[\frac{1}{\varepsilon} Y_0^*\left(\frac{t}{\varepsilon}\right) + Y_1^*\left(\frac{t}{\varepsilon}\right) \right], \tag{6.4.20}$$

and show that x^1 and y^1 can be determined by the following six subproblems:

Subproblem 1:

$$\frac{d\xi_0(\tau)}{d\tau} = g_0[0, \xi_0(\tau)] Y_0^*(\tau),$$

$$\frac{dY_0^*(\tau)}{d\tau} = k_0[0, \xi_0(\tau)] Y_0^*(\tau) \quad \text{for } \tau > 0,$$

$$\xi_0 = \alpha_0 \quad \text{and} \quad Y_0^* = \beta_0 \quad \text{at } \tau = 0.$$

Subproblem 2:

$$X_0(0) := \lim_{\tau \to \infty} \xi_0(\tau),$$

$$X_0^*(\tau) := \xi_0(\tau) - X_0(0) \quad \text{for } \tau \ge 0.$$

Subproblem 3:

$$\frac{dX_0}{dt} = f_0(t, X_0) - g_0(t, X_0) k_0(t, X_0)^{-1} h_0(t, X_0) \quad \text{for } t > 0,$$

$$X_0 = \text{value given in Subproblem 2, at } t = 0,$$

$$Y_0(t) := -k_0[t, X_0(t)]^{-1} h_0[t, X_0(t)] \quad \text{for } t \ge 0.$$

Subproblem 4:

$$\frac{d\xi_1(\tau)}{d\tau} = g_{0,x}[0, \xi_0(\tau)] Y_0^*(\tau)\xi_1 + g_0[0, \xi_0(\tau)] Y_1^* + A_0^*(\tau),$$

$$\frac{dY_1^*(\tau)}{d\tau} = k_{0,x}[0, \xi_0(\tau)] Y_0^*(\tau)\xi_1 + k_0[0, \xi_0(\tau)] Y_1^* + B_0^*(\tau)$$
$$\text{for } \tau > 0,$$

$$\xi_1 = \alpha_1 \quad \text{and} \quad Y_1^* = \beta_1 - Y_0(0) \quad \text{at } \tau = 0,$$

with

$$A_0^*(\tau) := \left\{ \tau g_{0,t}[0, \xi_0(\tau)] + \tau g_{0,x}[0, \xi_0(\tau)] X_0'(0) \right.$$
$$\left. + g_1[0, \xi_0(\tau)] \right\} Y_0^*(\tau)$$
$$+ \left\{ f_0[0, \xi_0(\tau)] - f_0[0, X_0(0)] \right\}$$
$$+ \left\{ g_0[0, \xi_0(\tau)] - g_0[0, X_0(0)] \right\} Y_0(0),$$

and a similar formula for $B_0^*(\tau)$ obtained by replacing g and f in this formula for A_0^* with k and h, respectively.

Subproblem 5:

$$X_1(0) := \lim_{\tau \to \infty} \xi_1(\tau),$$

$$X_1^*(\tau) := \xi_1(\tau) - X_1(0) \quad \text{for } \tau \ge 0.$$

Subproblem 6:

$$\frac{dX_1}{dt} = \left[f_{0,x}(t, X_0) + g_{0,x}(t, X_0)Y_0 \right] X_1 + g_0(t, X_0)Y_1$$

$$+ f_1(t, X_0) + g_1(t, X_0)Y_0 \quad \text{for } t > 0,$$

$$X_1 = \text{value given in Subproblem 5, at } t = 0,$$

$$Y_1(t) := k_0(t, X_0)^{-1}\left(\frac{dY_0}{dt} - \{ \left[h_{0,x}(t, X_0) + k_{0,x}(t, X_0)Y_0 \right] X_1 \right.$$

$$\left. + h_1(t, X_0) + k_1(t, X_0)Y_0 \} \right) \quad \text{for } t \geq 0.$$

Exercise 6.4.3: Use the conditions of Exercise 6.2.3 to prove that the solution function $Y_0^* = Y_0^*(\tau)$ of Subproblem 1 of the previous Exercise 6.4.2 decays exponentially, with

$$|Y_0^*(\tau)| \leq \text{const. } e^{-\kappa\tau} \quad \text{for } \tau \geq 0,$$

where $\kappa = \kappa_0$ is the constant appearing in (6.2.66). Show also that the limit defining $X_0(0)$ given in Subproblem 2 exists and is determined by the equation

$$\int_{\alpha_0}^{X_0(0)} g_0(0, \xi)^{-1}\, d\xi = \int_0^\infty Y_0^*. \tag{6.4.21}$$

Finally, show that Subproblem 4 of Exercise 6.4.2 yields a solution function Y_1^* that decays exponentially with

$$|Y_1^*(\tau)| \leq \text{const. } e^{-\kappa\mu\tau} \quad \text{for } \tau \geq 0,$$

for any fixed μ satisfying $0 < \mu < 1$, and show that the solution ξ_1 is bounded and that the limit defining $X_1(0)$ exists in Subproblem 5. *Hint:* The last assertions follow directly from Exercise 6.1.4, with t replaced by τ in Exercise 6.1.4.

Exercise 6.4.4: Solve the subproblems of Exercise 6.4.2 for the problem

$$\frac{dx}{dt} = y, \qquad\qquad x = 1,$$

$$\varepsilon \frac{dy}{dt} = -x^2 - y \quad \text{for } t > 0, \qquad y = \frac{0.5}{\varepsilon} \quad \text{at } t = 0.$$

Evaluate the results at $t = 0.1$, and verify the results of (6.4.18).

7

Conditionally stable problems

This chapter introduces the concept of conditional stability for singularly perturbed initial-value problems. The ideas involved are discussed, for the most part, within the context of a specific linear system that is considered in detail as an illustrative example. The system is conditionally stable in the sense that it has both exponentially increasing and exponentially decreasing solutions (along with other bounded solutions of nonexponential type). Detailed information is obtained on a fundamental solution for the system, with results that are related to the classical Liouville/Green approximation. The linear system considered includes as a special case the second-order vector system

$$\varepsilon^2 \frac{d^2x}{dt^2} = C(t,\varepsilon)x + \varepsilon^2 D(t,\varepsilon)\frac{dx}{dt} + g(t,\varepsilon)$$

for nonsingular C subject to a certain condition on the eigenpairs of C. In yet a different special case, the system considered here also exhibits the property of being a *singular* singularly perturbed system, much like the earlier system of Section 5.5 involving the telegraph equation.

7.1 Introduction

Consider the initial-value problem for the system

$$\frac{dx}{dt} = A(t,\varepsilon)x + B(t,\varepsilon)y + f(t,\varepsilon),$$

$$\varepsilon^2 \frac{dy}{dt} = C(t,\varepsilon)x + \varepsilon^2 D(t,\varepsilon)y + g(t,\varepsilon) \quad (\text{small } \varepsilon > 0),$$

(7.1.1)

for solution functions $x = x(t,\varepsilon)$ and $y = y(t,\varepsilon)$ that are respectively

219

m-dimensional and n-dimensional real vector-valued functions, with

$$1 \le m \le n, \tag{7.1.2}$$

and where the given data functions A, B, C, D, f, and g are real matrix-valued or vector-valued functions with appropriate compatible orders. These data functions are assumed to be smooth and to have asymptotic power-series expansions in terms of the small nonnegative parameter ε, with coefficients that are smooth functions of t. The system (7.1.1) is considered on a real t interval containing $t = 0$, subject to the conditions

$$x(0, \varepsilon) = \alpha(\varepsilon), \qquad y(0, \varepsilon) = \beta(\varepsilon), \tag{7.1.3}$$

for given vectors $\alpha(\varepsilon)$ and $\beta(\varepsilon)$ that have asymptotic expansions in ε. The t interval is taken to be $[-T, 0]$, $[0, T]$, or $[-T, T]$ for a given positive number T. [Similar results can be obtained for unbounded intervals with $T = \infty$. Also, the data and solution functions can be complex-valued, but the independent variable t is real. With minor changes, the asymptotic expansions for the initial data $\alpha(\varepsilon), \beta(\varepsilon)$ can be permitted to include some negative powers of ε.]

Note that the system (7.1.1) can be considered to be in the same form as the earlier system (6.2.1) if we relabel the small quantity ε^2 in (7.1.1) as ε and if we take the functions u and v in (6.2.1) to be defined as

$$u(t, x, y, \varepsilon) := A(t, \varepsilon)x + B(t, \varepsilon)y + f(t, \varepsilon),$$
$$v(t, x, y, \varepsilon) := C(t, \varepsilon)x + \varepsilon D(t, \varepsilon)y + g(t, \varepsilon).$$

Hence, in this case the matrix $v_{0,y}$ *vanishes identically* at $\varepsilon = 0$, and so the stability condition of Section 6.2 [see (6.2.18) and the remarks in the paragraph following (6.2.20)] is not satisfied for (7.1.1). The results of Chapter 6 do not apply to the present problem.

Along with (7.1.1) we have the corresponding homogeneous system $(f = 0, g = 0)$

$$\frac{dx}{dt} = A(t, \varepsilon)x + B(t, \varepsilon)y,$$
$$\varepsilon^2 \frac{dy}{dt} = C(t, \varepsilon)x + \varepsilon^2 D(t, \varepsilon)y. \tag{7.1.4}$$

The system (7.1.1) can be rewritten as

$$\varepsilon \frac{d}{dt} \begin{bmatrix} \hat{x} \\ \hat{y} \end{bmatrix} = \begin{bmatrix} \varepsilon A & B \\ C & \varepsilon D \end{bmatrix} \begin{bmatrix} \hat{x} \\ \hat{y} \end{bmatrix} + \begin{bmatrix} \varepsilon f \\ g \end{bmatrix}, \tag{7.1.5}$$

with

$$\hat{x} := x, \qquad \hat{y} := \varepsilon y, \tag{7.1.6}$$

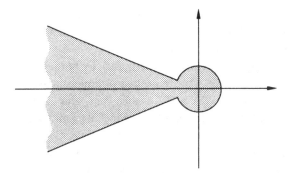

Figure 7.1

and (7.1.5) indicates that the matrix

$$\begin{bmatrix} 0 & B_0(t) \\ C_0(t) & 0 \end{bmatrix} \quad [B_0(t):=B(t,0),\text{ etc.}] \qquad (7.1.7)$$

should play a crucial role in the study of (7.1.1) for small ε. And, indeed, our key assumption here will imply that the matrix (7.1.7) has constant rank $2m \le m+n$, with an m-dimensional eigenspace corresponding to eigenvalues with positive real parts, an m-dimensional eigenspace corresponding to eigenvalues with negative real parts, and an $(n-m)$-dimensional eigenspace corresponding to the zero eigenvalue.

This key assumption consists of the requirement that *the $m \times m$ matrix $B_0(t)C_0(t)$ has a complete set of m linearly independent eigenvectors $z = z(t)$,*

$$z_j = z_j(t) \quad \text{for } j = 1, 2, \ldots, m, \qquad (7.1.8)$$

which can be taken to be continuously differentiable functions of t, *with all eigenvalues $\lambda = \lambda_j(t)$ excluded from the shaded region of* Figure 7.1. That is, there is a fixed positive constant δ, with $\delta < \pi$, such that all eigenvalues $\lambda = \lambda(t) = |\lambda|\exp[i(\arg \lambda)]$ of $B_0(t)C_0(t)$ satisfy

$$|\lambda(t)| \ge \delta > 0 \quad \text{and} \quad |\arg \lambda(t)| \le \pi - \delta, \qquad (7.1.9)$$

uniformly in t. No assumption is made regarding the multiplicities of the eigenvalues $\lambda = \lambda_j(t)$, where

$$B_0(t)C_0(t)z_j = \lambda_j z_j \quad (\text{no sum on } j) \qquad (7.1.10)$$

for $j = 1, 2, \ldots, m$.

This assumption on the eigenvectors and eigenvalues of $B_0 C_0$ will imply the existence of m independent solutions of (7.1.4) that are exponentially increasing for $t > 0$ but not for $t < 0$ as $\varepsilon \to 0+$, along with m additional independent solutions of (7.1.4) that are exponentially increasing for $t < 0$ but not for $t > 0$ as $\varepsilon \to 0+$, and also $n - m$ independent solutions that remain bounded for both positive and negative values of t as $\varepsilon \to 0+$. There are two n-dimensional affine manifolds $\mu_+(\varepsilon)$ and $\mu_-(\varepsilon)$ in $\mathbb{R}^m \times \mathbb{R}^n$ such that the solution $(x, y) = [x(t, \varepsilon), y(t, \varepsilon)]$ of the initial-value problem (7.1.1) and (7.1.3) contains no exponentially increasing component on the respective interval $[0, T]$ or $[-T, 0]$ if the initial point $[\alpha(\varepsilon), \beta(\varepsilon)]$ lies on $\mu_+(\varepsilon)$ or $\mu_-(\varepsilon)$, respectively. We shall see in Section 7.2 that these manifolds are of the form

$$\mu_\pm(\varepsilon): \alpha = \varepsilon M_\pm(\varepsilon)\beta + b_\pm(f, g; \varepsilon) \qquad (7.1.11)$$

for suitable $m \times n$ matrices $M_\pm(\varepsilon)$ and suitable m-vectors $b_\pm(f, g; \varepsilon)$ that depend smoothly on the data A, B, C, D, f, and g in (7.1.1), with $b_\pm(f, g; \varepsilon)$ depending linearly on f and g, but with $M_\pm(\varepsilon)$ independent of f and g. There is a fixed positive constant K such that the estimates

$$|M_\pm(\varepsilon)\beta| \leq K(|B_0(0)\beta| + \varepsilon|P(0)\beta|) \quad \text{for any } n\text{-vector } \beta,$$
$$(7.1.12)$$
$$|b_\pm(f, g; \varepsilon)| \leq K(\varepsilon\|f\| + \|B_0 g\| + \varepsilon\|Pg\|),$$

hold as $\varepsilon \to 0+$, where $|\cdot|$ denotes the l_∞ vector norm and $\|\cdot\|$ denotes the maximum norm over the appropriate t interval, and where $P = P(t): \mathbb{R}^n \to \mathbb{R}^n$ is the projection on the null space of $B_0(t)$ along the image of $C_0(t)$, given by

$$P(t) := I_n - C_0(t)[B_0(t)C_0(t)]^{-1}B_0(t), \qquad (7.1.13)$$

with $B_0 P = 0$ and $PC_0 = 0$. [Note that $\mathbb{R}^n$ is a direct sum of the respective null spaces of $B_0(t): \mathbb{R}^n \to \mathbb{R}^m$ and $P(t): \mathbb{R}^n \to \mathbb{R}^n$, as follows from the equivalence of the relations $P(t)\beta = \beta$ and $B_0(t)\beta = 0$ for n-vectors β.] Moreover, $M_\pm(\varepsilon)$ and $b_\pm(f, g; \varepsilon)$ have asymptotic expansions $M_\pm(\varepsilon) \sim \sum_{k=0}^\infty M_\pm^{(k)} \cdot \varepsilon^k$ and $b_\pm(f, g; \varepsilon) \sim \sum_{k=0}^\infty b_\pm^{(k)}(f, g)\varepsilon^k$, with leading coefficients given as

$$M_\pm^{(0)} = \mp Z(0)\Omega(0)^{-1}Z(0)^{-1}B_0(0),$$
$$(7.1.14)$$
$$b_\pm^{(0)}(f, g) = -[B_0(0)C_0(0)]^{-1}B_0(0)g_0(0),$$

where Z is the matrix with columns consisting of the eigenvectors of $B_0 C_0$, and Ω is the diagonal matrix with diagonal entries given by the positive square roots of the corresponding eigenvalues [see (7.2.1)–(7.2.4)]. Note that $M_+^{(0)} = -M_-^{(0)}$, whereas $b_+^{(0)}(f, g) = b_-^{(0)}(f, g)$.

The reduced system for (7.1.1) has the form

$$\frac{dx}{dt} = A_0(t)x + B_0(t)y + f_0(t),$$
$$0 = C_0(t)x + g_0(t),$$
(7.1.15)

where (7.1.15) fails to have a solution in the case $m < n$ if the given vector $g_0 := g(t, 0)$ is not in the image of C_0, that is, if $g_0(t)$ is not in the column space of $C_0(t)$. On the other hand, if $m < n$ and if g_0 is in the image of C_0, with

$$g_0(t) = C_0(t)v_0(t)$$
(7.1.16)

for some suitable m-vector v_0, then (7.1.15) has infinitely many solutions, given as

$$x = -(B_0 C_0)^{-1} B_0 g_0 = -v_0(t),$$
$$y = y^* - C_0 (B_0 C_0)^{-1} \left(\frac{dv_0}{dt} - A_0 v_0 + f_0 \right),$$
(7.1.17)

where $y^* = y^*(t)$ can be any n-vector in the $(n - m)$-dimensional null space of $B_0(t)$: $\mathbb{R}^n \rightarrow \mathbb{R}^m$. The reduced system (7.1.15) always has a unique solution, given by (7.1.17) with $y^* = 0$, in the case $m = n$. In this latter case the $m \times m$ matrices B_0 and C_0 are nonsingular as a result of the assumed nonsingularity of $B_0 C_0$ [cf. (7.1.9)]. The problem (7.1.1) and (7.1.3) is conditionally stable in every case here regardless of the status of (7.1.15).

Note that the reduced system for (7.1.5) is

$$B_0(t)\hat{y} = 0, \qquad C_0(t)\hat{x} = -g_0(t),$$

which differs from (7.1.15). Also, the reduced system for (7.1.5)–(7.1.6) is simply

$$C_0(t)x = -g_0,$$

which again differs from (7.1.15).

Example 7.1.1: Consider the system

$$\frac{dx}{dt} = x + y_1 + f(t),$$
(7.1.18)

$$\varepsilon^2 \frac{d}{dt} \begin{bmatrix} y_1 \\ y_2 \end{bmatrix} = \begin{bmatrix} 1 \\ 0 \end{bmatrix} x + \varepsilon^2 \begin{bmatrix} 1 & 1 \\ 0 & 1 \end{bmatrix} \begin{bmatrix} y_1 \\ y_2 \end{bmatrix} + \begin{bmatrix} g_1(t) \\ g_2(t) \end{bmatrix},$$

for real-valued functions x, y_1, and y_2 on $[0, T]$, subject to the initial conditions

$$x = \alpha, \qquad y_1 = \beta_1, \qquad y_2 = \beta_2 \quad \text{at } t = 0.$$
(7.1.19)

This problem is of the type (7.1.1)–(7.1.3), with $m = 1$, $n = 2$, and

$$A := [1], \qquad B := [1, 0],$$

$$C := \begin{bmatrix} 1 \\ 0 \end{bmatrix}, \qquad D := \begin{bmatrix} 1 & 1 \\ 0 & 1 \end{bmatrix}. \tag{7.1.20}$$

The projection P of (7.1.13) is given as

$$P = \begin{bmatrix} 0 & 0 \\ 0 & 1 \end{bmatrix}. \tag{7.1.21}$$

The solution of (7.1.18)–(7.1.19) is given as

$$
\begin{bmatrix} x(t, \varepsilon) \\ y_1(t, \varepsilon) \end{bmatrix} = \frac{1}{2\varepsilon}
\begin{bmatrix} \varepsilon(e^{\lambda_1 t} + e^{\lambda_2 t}) & \varepsilon^2(e^{\lambda_1 t} - e^{\lambda_2 t}) \\ (e^{\lambda_1 t} - e^{\lambda_2 t}) & \varepsilon(e^{\lambda_1 t} + e^{\lambda_2 t}) \end{bmatrix}
\begin{bmatrix} \alpha \\ \beta_1 \end{bmatrix}
$$

$$
+ \frac{1}{2\varepsilon} \int_0^t
\begin{bmatrix} \varepsilon(e^{\lambda_1(t-s)} + e^{\lambda_2(t-s)}) & \varepsilon^2(e^{\lambda_1(t-s)} - e^{\lambda_2(t-s)}) \\ e^{\lambda_1(t-s)} - e^{\lambda_2(t-s)} & \varepsilon(e^{\lambda_1(t-s)} + e^{\lambda_2(t-s)}) \end{bmatrix}
$$

$$
\times \begin{bmatrix} f(s) \\ y_2(s) + \dfrac{1}{\varepsilon^2} g_1(s) \end{bmatrix} ds, \tag{7.1.22}
$$

with

$$y_2(t) = y_2(t, \varepsilon) = \beta_2 e^t + \frac{1}{\varepsilon^2} \int_0^t e^{t-s} g_2(s)\, ds, \tag{7.1.23}$$

where the values λ_1 and λ_2 are given as

$$\lambda_1 := 1 + \frac{1}{\varepsilon}, \qquad \lambda_2 := 1 - \frac{1}{\varepsilon}. \tag{7.1.24}$$

If we evaluate the solution functions of (7.1.22)–(7.1.23) at $t = T$, then we find directly that the exponentially *increasing* solution components can be eliminated for $0 \le t \le T$ (as $\varepsilon \to 0+$) through imposition of the condition

$$\alpha + \varepsilon\beta_1 + \int_0^T e^{-\lambda_1 t}\left[f(t) + \varepsilon\beta_2 e^t + \frac{1}{\varepsilon} g_1(t) + \frac{1}{\varepsilon} \int_0^t e^{t-s} g_2(s)\, ds \right] dt = 0. \tag{7.1.25}$$

This equation (7.1.25) characterizes the stable manifold $\mu_+(\varepsilon)$ of (7.1.11), and we find directly from (7.1.25) that the quantities $M_+(\varepsilon)$ and

$b_+(f, g; \varepsilon)$ of (7.1.11) are given here as

$$M_+(\varepsilon) = -\left[1, \int_0^T e^{-t/\varepsilon}\, dt\right] = -\left[1, \varepsilon(1 - e^{-T/\varepsilon})\right],$$

(7.1.26)

$$b_+(f, g; \varepsilon) = -\int_0^T e^{-\lambda_1 t}\left[f(t) + \frac{1}{\varepsilon}g_1(t) + \frac{1}{\varepsilon}\int_0^t e^{t-s}g_2(s)\, ds\right] dt.$$

Hence, we find directly in this case the inequality (see Exercise 7.1.1)

$$\left|M_+(\varepsilon)\beta\right| \le |\beta_1| + \varepsilon|\beta_2|$$

$$= |B\beta| + \varepsilon|P\beta| \quad \text{for any } \beta = \begin{bmatrix} \beta_1 \\ \beta_2 \end{bmatrix}, \quad (7.1.27)$$

and, similarly,

$$\left|b_+(f, g; \varepsilon)\right| \le \varepsilon\|f\| + \|Bg\| + \varepsilon\|Pg\|. \tag{7.1.28}$$

These inequalities give explicit versions of the estimates of (7.1.12) in this case. We also verify the results of (7.1.14) directly in this case.

We shall see in Section 7.2 that the stable manifolds are not, strictly speaking, uniquely determined, because they can be modified in various ways by inclusion or deletion of terms that are asymptotically zero (see Exercise 7.2.7).

The nature of the general problem considered here for (7.1.1) and (7.1.3) is qualitatively insensitive to each of the transformations $t \mapsto -t$ and $\varepsilon \mapsto -\varepsilon$. However, for definiteness, the small parameter is taken to be positive, $\varepsilon > 0$. The full problem (7.1.1) and (7.1.3) always has a unique solution for each nonzero value of ε, and the behavior of this solution is studied as ε tends toward zero through positive values. The stable initial manifolds $\mu_+(\varepsilon)$ of (7.1.11) are obtained in Section 7.2, and then in Section 7.3 a suitable asymptotic expansion is constructed for the solution $(x, y) = [x(t, \varepsilon), y(t, \varepsilon)]$ of (7.1.1) and (7.1.3) for initial values $(\alpha, \beta) = [\alpha(\varepsilon), \beta(\varepsilon)]$ on a stable initial manifold. In general, for $t \ne 0$, the y component of the exact solution is seen to become unbounded like ε^{-2} (as $\varepsilon \to 0$) if (7.1.16) fails to hold with g_0 not in the image of $C_0(t): \mathbb{R}^m \to \mathbb{R}^n$. On the other hand, if (7.1.16) holds, then the solution of the full problem converges, again for $t \ne 0$, to a particular one of the reduced solutions (7.1.17). Error estimates are obtained in Section 7.4 for the difference between the exact and asymptotic solutions. The *unstable* exponentially increasing (as $\varepsilon \to 0+$) solutions are discussed in Section 7.5.

The results given here for the linear system (7.1.1) can be generalized to the semilinear system

$$\frac{dx}{dt} = A(t, x, \varepsilon) + B(t, \varepsilon)y,$$

$$\varepsilon^2 \frac{dy}{dt} = C(t, x, \varepsilon) + \varepsilon^2 D(t, \varepsilon)y. \tag{7.1.29}$$

In the special case $A = 0$, $B = I_m$, with $m = n$, the system (7.1.29) is equivalent to the second-order vector system

$$\varepsilon^2 \frac{d^2 x}{dt^2} = C(t, x, \varepsilon) + \varepsilon^2 D(t, \varepsilon) \frac{dx}{dt}. \tag{7.1.30}$$

The Dirichlet problem for a related nonlinear system is studied in Kelley (1979) using differential inequality techniques.

We can replace the vector

$$\begin{bmatrix} \hat{x} \\ \hat{y} \end{bmatrix}$$

in (7.1.5) with

$$z := \begin{bmatrix} \hat{x} \\ \hat{y} \end{bmatrix},$$

and then the system (7.1.5) can be written in the form

$$\varepsilon \frac{dz}{dt} = v(t, z, \varepsilon), \tag{7.1.31}$$

with

$$v(t, z, \varepsilon) := \begin{bmatrix} \varepsilon A(t, \varepsilon) & B(t, \varepsilon) \\ C(t, \varepsilon) & \varepsilon D(t, \varepsilon) \end{bmatrix} z + \begin{bmatrix} \varepsilon f(t, \varepsilon) \\ g(t, \varepsilon) \end{bmatrix}. \tag{7.1.32}$$

[It will be clear from the context whether z denotes an $(n + m)$-vector solution of (7.1.31) or an eigenvector, as in (7.1.8)–(7.1.10).]

The initial-value problem for (7.1.31),

$$\varepsilon \frac{dz}{dt} = v(t, z, \varepsilon) \quad \text{for } 0 \le t \le T,$$

$$z(0, \varepsilon) = z_0(\varepsilon), \tag{7.1.33}$$

is a special case of the initial-value problem considered in Chapter 6 for (6.2.1)–(6.2.2), with $u \equiv x \equiv \alpha \equiv 0$ and with y relabelled as z in (6.2.1). According to a classical result of Tikhonov (1950, 1952) and Levin and Levinson (1954), if the reduced equation

$$0 = v(t, z, 0) \tag{7.1.34}$$

is solvable for $z = Z_0(t)$, if the matrix $v_z(t, z, 0) = \partial v(t, z, 0)/\partial z$ is stable with all eigenvalues having negative real parts for all (t, z) near $[t, Z_0(t)]$ (uniformly for $0 \le t \le T$), and if the "boundary-layer jump" $z_0(0) - Z_0(0)$ is not too great, then the solution Z_0 of (7.1.34) is locally unique, and this function Z_0 provides a useful approximation to the solution of (7.1.33) away from an initial boundary-layer region. The initial value $z_0(\varepsilon)$ need not be restricted to any lower-dimensional manifold analogous to the manifolds of (7.1.11). {If (7.1.34) has two or more distinct solutions, e.g., $Z_1(t)$ and $Z_2(t)$, with $v_z(t, z, 0)$ locally stable in respective neighborhoods of $[t, Z_1(t)]$ and $[t, Z_2(t)]$ for $0 \le t \le T$, then $Z_j(t)$ provides a useful approximation to the solution of (7.1.33) outside an initial boundary-layer region, provided that the initial point $z_0(\varepsilon)$ lies in a suitable $(n + m)$-dimensional region containing $Z_j(0)$, for $j = 1, 2$.} Note that the results of Chapter 6 [see (6.2.16)–(6.2.20), along with Theorem 6.3.1] provide a result closely related to this result of Tikhonov/Levin/Levinson.

This classical result of Tikhonov/Levin/Levinson is generalized in Butuzov and Vasil'eva (1970), Vasil'eva (1975a, 1975b, 1976), and Vasil'eva and Butuzov (1980) for the case in which the matrix $v_z(t, z, 0)$ is *singular* [as occurs for (7.1.32) in the case $m < n$], with constant deficient rank $k = \text{rank}[v_z(t, z, 0)] < N$ ($z \in \mathbb{R}^N$), and where $v_z(t, z, 0)$ has a k-dimensional eigenspace corresponding to stable eigenvalues. In particular, no unstable eigenvalues are permitted. Related results are obtained in Gordon (1975), O'Malley and Flaherty (1977, 1980), O'Malley (1978b), Campbell and Rose (1979), Campbell (1980), and Mika (1982), where in all of these references it is required that there be no unstable eigenvalues (with positive real part).

O'Malley and Flaherty call such problems *singular* singularly perturbed problems when $v_z(t, z, 0)$ is singular. The reduced equation (7.1.34) may then have no solution or many solutions, as illustrated here by the function v of (7.1.32). Particular examples of such singularly perturbed problems involving ordinary and partial differential equations for which the reduced system is singular are studied in Smith (1971) and Fife (1972) and in other references cited in O'Malley and Flaherty (1980). The problem of Section 5.5 for the telegraph equation is of this type.

Related conditionally stable problems for systems including (7.1.33) are considered in Levin (1956, 1957), Chang and Coppel (1969), and Hoppensteadt (1971b) for the case in which $v_z(t, z, 0)$ is nonsingular, but with some stable and some unstable eigenvalues.

These results are generalized in a certain direction in Sacker and Sell (1980) so as to include both the singular and nonsingular cases for v_z, and including also the possibilities of stable and/or unstable eigenvalues. The results of Sacker and Sell (1980), which do not apply to the present problem of this chapter, are based on their spectral theory for linear differential equations. Geometric presentations are given in Fenichel (1979) and in other references cited in O'Malley (1984). See also Harris and Lutz (1977) for a general theory of asymptotic integration.

The foregoing problem (7.1.1) and (7.1.3) and the equivalent problem (7.1.3) and (7.1.5)–(7.1.6) are not included directly in these earlier works. However, standard techniques suffice for study of the present problem. The construction of the stable initial manifolds given in Section 7.2 is based directly on the approach of Levin (1956), which in turn is based on the study of an auxiliary system of integral equations, where this latter system is suitably chosen so as to exclude any unwanted solution components, following an approach that has long been used in the study of asymptotic behavior of solutions of differential equations; see Cotton (1910), Perron (1929), Levinson (1948), Bellman (1953), Atkinson (1954), Hartman and Wintner (1955), Coppel (1965, 1978), Harris and Lutz (1974, 1977), and other references cited by these authors. The O'Malley/ Hoppensteadt multivariable method is used in Section 7.3 for construction of uniformly valid approximations to the stable solutions. The previously mentioned method of Cotton, Perron, and Levinson is used in Section 7.5 for construction of the remaining components of the fundamental solution matrix. Alternatively, a Riccati transformation, as discussed in Sections 9.2 and 9.3, can be used in the construction of a fundamental solution.

Exercises

Exercise 7.1.1: In Example 7.1.1, show that $M_+(\varepsilon)$ and $b_+(f, g; \varepsilon)$ from (7.1.26) satisfy the inequalities (7.1.27) and (7.1.28). [The quantity $b_+(f, g; \varepsilon)$ actually depends primarily on the restrictions of f and g to any fixed subinterval adjoining $t = 0$.]

Exercise 7.1.2: In Example 7.1.1, show that the solution functions satisfy estimates of the type

$$|x(t, \varepsilon)| \le \varepsilon \text{ const.} \left(|B\beta| + \varepsilon|P\beta| + \|f\| + \frac{1}{\varepsilon}\|g\| \right),$$

$$|y(t, \varepsilon)| \le \text{ const.} \left(|\beta| + \|f\| + \frac{1}{\varepsilon}\|Bg\| + \frac{1}{\varepsilon^2}\|Pg\| \right),$$

$$(7.1.35)$$

provided that the initial values satisfy the stability condition (7.1.25).

7.2 The stable initial manifolds

Let $\Omega = \Omega(t) = \mathrm{diag}\{\omega_j(t)\}$ be the $m \times m$ diagonal matrix with jth diagonal entry $\omega_j(t)$ given by the following "positive" square root of the eigenvalue $\lambda_j(t)$ of $B_0 C_0$ [see (7.1.8)–(7.1.10)]:

$$\omega_j(t) := \left|\lambda_j(t)\right|^{1/2} \exp\left\{\tfrac{1}{2}\left[\arg\lambda_j(t)\right]i\right\} \quad \text{for } j = 1, 2, \ldots, m$$

$$\left(i = (-1)^{1/2}\right), \quad (7.2.1)$$

where $\left|\lambda_j(t)\right|^{1/2}$ denotes the positive square root. (No assumption is made regarding the multiplicities of the λ_j's and corresponding ω_j's.) It follows from (7.1.9) that the ω_j's have positive real parts, with

$$\mathrm{Re}\,\omega_j(t) \geq \sqrt{\delta}\,\sin(\delta/2) > 0 \quad \text{for } j = 1, 2, \ldots, m, \quad (7.2.2)$$

and for all t in the interval under consideration. From (7.2.1) we have the result

$$\Omega(t)^2 = \mathrm{diag}\{\lambda_1(t), \lambda_2(t), \ldots, \lambda_m(t)\}, \quad (7.2.3)$$

and then (7.1.10) can be rewritten in the matrix form

$$B_0(t)C_0(t)Z(t) = Z(t)\Omega(t)^2, \quad (7.2.4)$$

where $Z = Z(t)$ is the $m \times m$ nonsingular matrix with jth column given by the jth eigenvector $z_j = z_j(t)$ of $B_0 C_0$.

It follows from (7.1.2), along with the nonsingularity of $B_0 C_0$, that the $m \times n$ matrix $B_0(t)$ has rank m:

$$\mathrm{rank}\,B_0(t) = m, \quad \text{with } \dim\{\text{null space of } B_0\} = n - m. \quad (7.2.5)$$

Let the n-vectors $h_1(t), h_2(t), \ldots, h_{n-m}(t)$ form an orthonormal basis for the null space of $B_0(t) \colon \mathbb{R}^n \to \mathbb{R}^m$, and let $H = H(t)$ be the $n \times (n-m)$ matrix having its jth column given by the n-vector $h_j(t)$ for $j = 1, 2, \ldots, n - m$, so that

$$B_0(t)H(t) = 0. \quad (7.2.6)$$

The matrix H has rank $n - m$ and satisfies the normalization condition

$$H(t)^{\mathrm{T}}H(t) = I_{n-m}, \quad (7.2.7)$$

where H^{T} denotes the transpose of H, and I_{n-m} is the identity matrix with $n - m$ rows and columns. In view of the assumed smoothness of $B_0(t)$ as a function of t, there is no loss in assuming that $H = H(t)$ depends smoothly on t, and this assumption is made.

Let the square $(m + n) \times (m + n)$ matrix $S = S(t)$ be defined in block form as

$$S(t) := \begin{bmatrix} \frac{1}{2}Z(t)^{-1} & \frac{1}{2}\Omega(t)^{-1}Z(t)^{-1}B_0(t) \\ \frac{1}{2}Z(t)^{-1} & -\frac{1}{2}\Omega(t)^{-1}Z(t)^{-1}B_0(t) \\ \overset{(n-m)\times m}{0} & H(t)^{\mathsf{T}}P(t) \end{bmatrix}, \qquad (7.2.8)$$

where the projection $P(t)$ is given by (7.1.13). Then

$$S(t)^{-1} = \begin{bmatrix} Z(t) & Z(t) & 0 \\ C_0(t)Z(t)\Omega(t)^{-1} & -C_0(t)Z(t)\Omega(t)^{-1} & H(t) \end{bmatrix}, \qquad (7.2.9)$$

where the validity of the inverse S^{-1} given by (7.2.9) can be verified directly by computing SS^{-1} using (7.2.4)–(7.2.9). A simple calculation using these same results also yields

$$S(t)\begin{bmatrix} 0 & B_0(t) \\ C_0(t) & 0 \end{bmatrix}S(t)^{-1} = \begin{bmatrix} \Omega(t) & 0 & 0 \\ 0 & -\Omega(t) & 0 \\ 0 & 0 & 0 \end{bmatrix}. \qquad (7.2.10)$$

We rewrite the system (7.1.1) of (7.1.5)–(7.1.6) in terms of the variables ξ, η, and ζ defined as

$$\begin{bmatrix} \xi \\ \eta \\ \zeta \end{bmatrix} = S(t)\begin{bmatrix} x \\ \varepsilon y \end{bmatrix}, \qquad (7.2.11)$$

with

$$\begin{bmatrix} x \\ \varepsilon y \end{bmatrix} = S(t)^{-1}\begin{bmatrix} \xi \\ \eta \\ \zeta \end{bmatrix}. \qquad (7.2.12)$$

In terms of these new variables, (7.1.1) becomes

$$\varepsilon \frac{d}{dt}\begin{bmatrix} \xi \\ \eta \\ \zeta \end{bmatrix} = \begin{bmatrix} \Omega(t) & 0 & 0 \\ 0 & -\Omega(t) & 0 \\ 0 & 0 & 0 \end{bmatrix}\begin{bmatrix} \xi \\ \eta \\ \zeta \end{bmatrix}$$

$$+ \varepsilon E(t, \varepsilon)\begin{bmatrix} \xi \\ \eta \\ \zeta \end{bmatrix} + S(t)\begin{bmatrix} \varepsilon f(t, \varepsilon) \\ g(t, \varepsilon) \end{bmatrix}, \qquad (7.2.13)$$

where $E = E(t, \varepsilon)$ is defined as

$$E(t, \varepsilon) := \dot{S}(t) S(t)^{-1}$$

$$+ S(t) \begin{bmatrix} A(t, \varepsilon) & \varepsilon^{-1}[B(t, \varepsilon) - B_0(t)] \\ \varepsilon^{-1}[C(t, \varepsilon) - C_0(t)] & D(t, \varepsilon) \end{bmatrix}$$

$$\times S(t)^{-1}, \qquad (7.2.14)$$

and where $\dot{S} = (d/dt)S(t)$. Our assumptions on the regularity of the data yield, with (7.2.14), the uniform bound

$$|E(t, \varepsilon)| \leq \text{constant}, \quad \text{as } \varepsilon \to 0, \qquad (7.2.15)$$

uniformly in t, where for definiteness $|E|$ refers here to the maximum absolute value of all components of E.

The system (7.2.13), with ξ and ζ interchanged, corresponds to the system (1.4) of Levin (1956) with the function $H(t, \xi, \eta, \zeta, \varepsilon)$ of Levin replaced in (7.2.13) with $\varepsilon^{-1}H(t)^T P(t) g(t, \varepsilon)$ + regular terms. Hence, the regularity condition (1.6) of Levin (1956) fails to hold (as $\varepsilon \to 0+$) for the present system (7.2.13) if $g_0(t)$ is not in the image of $C_0(t) : \mathbb{R}^m \to \mathbb{R}^n$. Nevertheless, the method of Levin (1956) suffices for (7.2.13), as will now be shown.

The initial conditions of (7.1.3) can be written in terms of the new variables (7.2.11) as

$$\xi(0, \varepsilon) = \xi_0(\varepsilon), \qquad \eta(0, \varepsilon) = \eta_0(\varepsilon), \qquad \zeta(0, \varepsilon) = \zeta_0(\varepsilon), \qquad (7.2.16)$$

where ξ_0, η_0, and ζ_0 are defined as

$$\xi_0(\varepsilon) := \tfrac{1}{2} Z(0)^{-1} \alpha(\varepsilon) + \frac{\varepsilon}{2} \Omega(0)^{-1} Z(0)^{-1} B_0(0) \beta(\varepsilon),$$

$$\eta_0(\varepsilon) := \tfrac{1}{2} Z(0)^{-1} \alpha(\varepsilon) - \frac{\varepsilon}{2} \Omega(0)^{-1} Z(0)^{-1} B_0(0) \beta(\varepsilon), \qquad (7.2.17)$$

$$\zeta_0(\varepsilon) := \varepsilon H(0)^T P(0) \beta(\varepsilon).$$

It is convenient to partition the matrix E of (7.2.13)–(7.2.14) into blocks $E_{\alpha\beta}$ ($\alpha, \beta = \xi, \eta, \zeta$) of appropriate dimensions as

$$E(t, \varepsilon) = \begin{bmatrix} E_{\xi\xi}(t, \varepsilon) & E_{\xi\eta}(t, \varepsilon) & E_{\xi\zeta}(t, \varepsilon) \\ E_{\eta\xi}(t, \varepsilon) & E_{\eta\eta}(t, \varepsilon) & E_{\eta\zeta}(t, \varepsilon) \\ E_{\zeta\xi}(t, \varepsilon) & E_{\zeta\eta}(t, \varepsilon) & E_{\zeta\zeta}(t, \varepsilon) \end{bmatrix}, \qquad (7.2.18)$$

where

$$E(t, \varepsilon) \begin{bmatrix} \xi \\ \eta \\ \zeta \end{bmatrix} = \begin{bmatrix} E_{\xi\xi}\xi + E_{\xi\eta}\eta + E_{\xi\zeta}\zeta \\ E_{\eta\xi}\xi + E_{\eta\eta}\eta + E_{\eta\zeta}\zeta \\ E_{\zeta\xi}\xi + E_{\zeta\eta}\eta + E_{\zeta\zeta}\zeta \end{bmatrix}. \tag{7.2.19}$$

Let $\gamma(t, \varepsilon)$ be a fundamental solution matrix for the system

$$\frac{d\zeta}{dt} = E_{\zeta\zeta}(t, \varepsilon)\zeta, \tag{7.2.20}$$

and put

$$\Gamma(t, s, \varepsilon) := \gamma(t, \varepsilon)\gamma(s, \varepsilon)^{-1}, \tag{7.2.21}$$

so that Γ is just the resolvent (Green) matrix for the initial-value problem for (7.2.20). As is well known, the uniform boundedness of $E_{\zeta\zeta}$ [see (7.2.15)] implies also the uniform boundedness of Γ,

$$|\Gamma(t, s, \varepsilon)| \le \text{constant}, \tag{7.2.22}$$

uniformly for t, s on any compact interval and uniformly in ε ($\varepsilon \to 0+$). As in (7.2.15), $|\Gamma|$ here denotes a maximum absolute value for all components of Γ.

We consider first the initial-value problem for (7.2.13) on the positive interval $[0, T]$. Following the approach of Levin (1956), the system of differential equations (7.2.13) is replaced with the following integral equations [see Cotton (1910), Perron (1929), Levinson (1948), Bellman (1953), and Coppel (1965), where related integral equations are used in studying the asymptotic behavior of solutions of differential equations as the independent variable $t \to \infty$]:

$$\xi(t) = \xi^*(t, \varepsilon) - \int_t^T \exp\left[-\frac{1}{\varepsilon}\int_t^s \Omega\right]\left\{E_{\xi\xi}(s, \varepsilon)\xi(s) + E_{\xi\eta}(s, \varepsilon)\eta(s)\right.$$

$$\left. + E_{\xi\zeta}(s, \varepsilon)\zeta(s)\right\} ds,$$

$$\eta(t) = \eta^*(t, \varepsilon) + \int_0^t \exp\left[-\frac{1}{\varepsilon}\int_s^t \Omega\right]\left\{E_{\eta\xi}(s, \varepsilon)\xi(s) + E_{\eta\eta}(s, \varepsilon)\eta(s)\right.$$

$$\left. + E_{\eta\zeta}(s, \varepsilon)\zeta(s)\right\} ds, \tag{7.2.23}$$

$$\zeta(t) = \zeta^*(t, \varepsilon) + \int_0^t \Gamma(t, s, \varepsilon)\left\{E_{\zeta\xi}(s, \varepsilon)\xi(s) + E_{\zeta\eta}(s, \varepsilon)\eta(s)\right\} ds,$$

where the functions ξ^*, η^*, and ζ^* are defined as

$$\xi^*(t, \varepsilon) := \exp\left[-\frac{1}{\varepsilon}\int_t^T \Omega\right]\xi_1$$

$$-\frac{1}{2}\int_t^T \exp\left[-\frac{1}{\varepsilon}\int_t^s \Omega\right]\left\{Z(s)^{-1}f(s, \varepsilon)\right.$$

$$\left.+\frac{1}{\varepsilon}\Omega(s)^{-1}Z(s)^{-1}B_0(s)g(s, \varepsilon)\right\}ds,$$

$$(7.2.24)_\xi$$

$$\eta^*(t, \varepsilon) := \exp\left[-\frac{1}{\varepsilon}\int_0^t \Omega\right]\eta_1$$

$$+\frac{1}{2}\int_0^t \exp\left[-\frac{1}{\varepsilon}\int_s^t \Omega\right]\left\{Z(s)^{-1}f(s, \varepsilon)\right.$$

$$\left.-\frac{1}{\varepsilon}\Omega(s)^{-1}Z(s)^{-1}B_0(s)g(s, \varepsilon)\right\}ds,$$

$$(7.2.24)_\eta$$

and

$$\zeta^*(t, \varepsilon) := \Gamma(t, 0, \varepsilon)\zeta_1 + \varepsilon^{-1}\int_0^t \Gamma(t, s, \varepsilon)H(s)^TP(s)g(s, \varepsilon)\,ds,$$

$$(7.2.24)_\zeta$$

and where $\xi_1 = \xi_1(\varepsilon)$, $\eta_1 = \eta_1(\varepsilon)$, and $\zeta_1 = \zeta_1(\varepsilon)$ are fixed (constant) vectors that will be specified later. The definition of S given by (7.2.8) has been used in the last term on the right side of (7.2.13) in obtaining (7.2.24). Note that the dependence on ε of $\xi(t) = \xi(t, \varepsilon)$, $\eta(t) = \eta(t, \varepsilon)$, and $\zeta(t) = \zeta(t, \varepsilon)$ is being suppressed in (7.2.23). We see by a direct calculation that any solution of (7.2.23)–(7.2.24) is also a solution of (7.2.13), and this is true for any choice of the constant vectors ξ_1, η_1, and ζ_1 in (7.2.24). Following the usual approach, the system (7.2.23) is now solved using the Banach/Picard fixed-point theorem.

The last equation of (7.2.23) is used to eliminate ζ on the right side of the first two equations there, and in this way we find the following system of equations for ξ and η:

$$\xi(t) = \xi^{**}(t, \varepsilon) + Q_1\begin{bmatrix}\xi \\ \eta\end{bmatrix}(t),$$

$$\eta(t) = \eta^{**}(t, \varepsilon) + Q_2\begin{bmatrix}\xi \\ \eta\end{bmatrix}(t),$$

$$(7.2.25)$$

where the operators Q_1 and Q_2 take vector-valued functions

$$\begin{bmatrix} \xi \\ \eta \end{bmatrix} = \begin{bmatrix} \xi(t) \\ \eta(t) \end{bmatrix}$$

into the functions

$$Q_1 \begin{bmatrix} \xi \\ \eta \end{bmatrix} = Q_1 \begin{bmatrix} \xi \\ \eta \end{bmatrix}(t) \quad \text{and} \quad Q_2 \begin{bmatrix} \xi \\ \eta \end{bmatrix} = Q_2 \begin{bmatrix} \xi \\ \eta \end{bmatrix}(t),$$

defined as

$$
Q_1 \begin{bmatrix} \xi \\ \eta \end{bmatrix}(t) := -\int_t^T \exp\left[-\frac{1}{\varepsilon}\int_t^s \Omega\right] \Big\{ E_{\xi\xi}(s,\varepsilon)\xi(s) + E_{\xi\zeta}(s,\varepsilon)
$$
$$
\times \int_0^s \Gamma(s,\sigma,\varepsilon) E_{\zeta\xi}(\sigma,\varepsilon)\xi(\sigma)\,d\sigma \Big\}\,ds
$$
$$
-\int_t^T \exp\left[-\frac{1}{\varepsilon}\int_t^s \Omega\right] \Big\{ E_{\xi\eta}(s,\varepsilon)\eta(s) + E_{\xi\zeta}(s,\varepsilon)
$$
$$
\times \int_0^s \Gamma(s,\sigma,\varepsilon) E_{\zeta\eta}(\sigma,\varepsilon)\eta(\sigma)\,d\sigma \Big\}\,ds
$$

$$(7.2.26)_1$$

and

$$
Q_2 \begin{bmatrix} \xi \\ \eta \end{bmatrix}(t) := \int_0^t \exp\left[-\frac{1}{\varepsilon}\int_s^t \Omega\right] \Big\{ E_{\eta\xi}(s,\varepsilon)\xi(s) + E_{\eta\zeta}(s,\varepsilon)
$$
$$
\times \int_0^s \Gamma(s,\sigma,\varepsilon) E_{\zeta\xi}(\sigma,\varepsilon)\xi(\sigma)\,d\sigma \Big\}\,ds
$$
$$
+\int_0^t \exp\left[-\frac{1}{\varepsilon}\int_s^t \Omega\right] \Big\{ E_{\eta\eta}(s,\varepsilon)\eta(s) + E_{\eta\zeta}(s,\varepsilon)
$$
$$
\times \int_0^s \Gamma(s,\sigma,\varepsilon) E_{\zeta\eta}(\sigma,\varepsilon)\eta(\sigma)\,d\sigma \Big\}\,ds,
$$

$$(7.2.26)_2$$

and where the vector-valued functions ξ^{**} and η^{**} are defined as

$$
\xi^{**}(t,\varepsilon) := \xi^*(t,\varepsilon) - \int_t^T \exp\left[-\frac{1}{\varepsilon}\int_t^s \Omega\right] E_{\xi\zeta}(s,\varepsilon)\zeta^*(s,\varepsilon)\,ds,
$$

$$(7.2.27)$$

$$
\eta^{**}(t,\varepsilon) := \eta^*(t,\varepsilon) + \int_0^t \exp\left[-\frac{1}{\varepsilon}\int_s^t \Omega\right] E_{\eta\zeta}(s,\varepsilon)\zeta^*(s,\varepsilon)\,ds,
$$

with ξ^*, η^*, and ζ^* given by (7.2.24). The dependence of Q_1 and Q_2 on ε is suppressed in the left side of (7.2.26).

Finally, we put

$$Q = \begin{bmatrix} Q_1 \\ Q_2 \end{bmatrix}, \quad \text{with } Q\begin{bmatrix} \xi \\ \eta \end{bmatrix} := \begin{bmatrix} Q_1\begin{bmatrix} \xi \\ \eta \end{bmatrix} \\ Q_2\begin{bmatrix} \xi \\ \eta \end{bmatrix} \end{bmatrix},$$

and then the equations of (7.2.25) can be written as the single operator equation

$$\begin{bmatrix} \xi \\ \eta \end{bmatrix} = \begin{bmatrix} \xi^{**} \\ \eta^{**} \end{bmatrix} + Q\begin{bmatrix} \xi \\ \eta \end{bmatrix}, \tag{7.2.28}$$

where the right side of (7.2.28) defines a mapping of functions $\begin{bmatrix} \xi \\ \eta \end{bmatrix}$ onto functions

$$\begin{bmatrix} \xi^{**} \\ \eta^{**} \end{bmatrix} + Q\begin{bmatrix} \xi \\ \eta \end{bmatrix}.$$

Equation (7.2.28) asks for a fixed point of this latter mapping.

Let $\mathscr{B}$ be the Banach space of continuous vector-valued functions $\begin{bmatrix} \xi \\ \eta \end{bmatrix}$, where $\xi = \xi(t)$ and $\eta = \eta(t)$ are continuous m-vector-valued functions of t, with norm

$$\left\| \begin{bmatrix} \xi \\ \eta \end{bmatrix} \right\| = \|\xi\| + \|\eta\|, \quad \text{where } \|\xi\| := \max_{0 \le t \le T} |\xi(t)|,$$

and similarly for $\|\eta\|$, and where $|\xi(t)|$ denotes the maximum absolute value of all components of $\xi(t)$. For any linear operator $Q : \mathscr{B} \to \mathscr{B}$ as before, we use the operator norm $\|Q\|$ corresponding to the norm of $\mathscr{B}$.

Lemma 7.2.1: *The operator*

$$Q = \begin{bmatrix} Q_1 \\ Q_2 \end{bmatrix}$$

defined on the Banach space $\mathscr{B}$ by (7.2.26) and appearing in (7.2.28) has small norm of order ε as $\varepsilon \to 0+$; that is, there is a fixed constant $\rho > 0$ such that

$$\|Q\| \le \rho \cdot \varepsilon \quad \text{as } \varepsilon \to 0+. \tag{7.2.29}$$

Hence, the equation (7.2.28) has a unique solution that can be given by the Neumann series

$$\begin{bmatrix} \xi(t, \varepsilon) \\ \eta(t, \varepsilon) \end{bmatrix} = \begin{bmatrix} \xi^{**}(t, \varepsilon) \\ \eta^{**}(t, \varepsilon) \end{bmatrix} + \sum_{k=1}^{\infty} Q^k \begin{bmatrix} \xi^{**} \\ \eta^{**} \end{bmatrix}(t, \varepsilon), \tag{7.2.30}$$

where this series converges for each positive $\varepsilon < \rho^{-1}$. The resulting solution (7.2.30) *satisfies the bound*

$$\left\| \begin{bmatrix} \xi \\ \eta \end{bmatrix} \right\| \le 2 \cdot \left\| \begin{bmatrix} \xi^{**} \\ \eta^{**} \end{bmatrix} \right\| \quad \text{for } 0 < \varepsilon < \tfrac{1}{2}\rho^{-1}. \tag{7.2.31}$$

Proof: The matrix

$$\exp\left[-\frac{1}{\varepsilon} \int_s^t \Omega \right]$$

satisfies the well-known inequality

$$\left| \exp\left(-\frac{1}{\varepsilon} \int_s^t \Omega \right) \right| \le \text{const. } \exp\left[-\frac{1}{\varepsilon}\sqrt{\delta} \cdot \sin\left(\frac{\delta}{2}\right) \cdot (t-s) \right]$$

$$\text{for } s \le t,\ \varepsilon > 0, \quad (7.2.32)$$

as follows easily by a direct calculation for the diagonal matrix Ω (see Exercise 7.2.3). From (7.2.32), along with (7.2.15), (7.2.22), and (7.2.26), follows

$$\left\| Q_1 \begin{bmatrix} \xi \\ \eta \end{bmatrix} \right\| + \left\| Q_2 \begin{bmatrix} \xi \\ \eta \end{bmatrix} \right\| \le \text{const. } \int_0^T \exp\left[-\sqrt{\delta} \cdot \sin\left(\frac{\delta}{2}\right) \cdot \frac{t}{\varepsilon} \right] dt \cdot \left\| \begin{bmatrix} \xi \\ \eta \end{bmatrix} \right\|$$

$$\le \text{const. } \frac{\varepsilon}{\sqrt{\delta} \cdot \sin(\delta/2)} \cdot \left\| \begin{bmatrix} \xi \\ \eta \end{bmatrix} \right\|, \tag{7.2.33}$$

from which (7.2.29) follows, with $\rho := \text{constant}/[\sqrt{\delta} \sin(\delta/2)]$. It follows now from (7.2.29) that the operator $I - Q$ is invertible, with

$$(I - Q)^{-1} = \sum_{k=0}^{\infty} Q^k \quad \text{for } 0 < \varepsilon < \rho^{-1}, \tag{7.2.34}$$

and the result (7.2.30) then follows directly. Finally, (7.2.31) follows directly, as usual, from (7.2.29)–(7.2.30). ∎

The solution functions for ξ and η given by (7.2.30) can be inserted into the right side of the last equation of (7.2.23) to give ζ, and the resulting three functions ξ, η, and ζ provide the unique solution to the system of differential equations (7.2.13) satisfying the boundary conditions

$$\xi(T, \varepsilon) = \xi_1 \quad \text{at } t = T, \tag{7.2.35}$$

$$\eta(0, \varepsilon) = \eta_1 \quad \text{and} \quad \zeta(0, \varepsilon) = \zeta_1 \quad \text{at } t = 0,$$

where the vectors ξ_1, η_1, and ζ_1 are the constant vectors appearing in (7.2.24).

In order to eliminate any exponentially *increasing* component in the solution for $0 \le t \le T$ as $\varepsilon \to 0+$, we impose the condition

$$\xi_1 = 0 \qquad (7.2.36)$$

in (7.2.24) and (7.2.35). The vectors η_1 and ζ_1 of (7.2.24) and (7.2.35) are determined by the last two initial conditions of (7.2.16) as

$$\eta_1 = \eta_0(\varepsilon) \quad \text{and} \quad \zeta_1 = \zeta_0(\varepsilon), \qquad (7.2.37)$$

where $\eta_0(\varepsilon)$ and $\zeta_0(\varepsilon)$ are given by the last two equations of (7.2.17). The choices (7.2.36) and (7.2.37) suffice for a unique specification of the solution functions ξ, η, and ζ for the system of differential equations (7.2.13) subject to the auxiliary conditions (7.2.35), where these solution functions are given by the foregoing construction based on (7.2.23) and (7.2.30).

Of course, the resulting solution functions will generally not satisfy the remaining, first initial condition of (7.2.16)–(7.2.17) for ξ. Indeed, the imposition now of this remaining initial condition for ξ amounts to the following restriction on the original initial values α and β of (7.1.3): The vectors $\alpha = \alpha(\varepsilon)$ and $\beta = \beta(\varepsilon)$ must lie on an n-dimensional affine manifold $\mu_+(\varepsilon)$ in $\mathbb{R}^m \times \mathbb{R}^n$ characterized by the condition [see (7.2.16) and (7.2.30)]

$$\xi_0(\varepsilon)|_{(7.2.17)} = \xi \text{ component of the right side of } (7.2.30),$$

$$\text{evaluated at } t = 0, \text{ and subject to } (7.2.36)–(7.2.37).$$
$$(7.2.38)$$

We find directly from (7.2.17), (7.2.24), (7.2.26), (7.2.27), (7.2.30), (7.2.36), and (7.2.37) that (7.2.38) can be rewritten in terms of $\alpha = \alpha(\varepsilon)$ and $\beta = \beta(\varepsilon)$ in the form [note that ξ^{**} and η^{**} depend *linearly* on the data $\alpha(\varepsilon)$, $\beta(\varepsilon)$, f, and g, whereas the operator Q of (7.2.26) is *independent* of these particular data]

$$\mu_+(\varepsilon): \alpha = \varepsilon M_+(\varepsilon)\beta + b_+(f, g; \varepsilon) \qquad (7.2.39)$$

for a suitable $m \times m$ matrix $M_+(\varepsilon)$ depending (regularly) on the data A, B, C, and D (but independent of f and g), and a suitable m-vector $b_+(f, g; \varepsilon)$ depending on A, B, C, D, f, and g in (7.1.1). A straightforward analysis shows that $M_+(\varepsilon)$ and $b_+(f, g; \varepsilon)$ have asymptotic power-series expansions in ε, with leading terms [see (7.1.14)]

$$M_+^{(0)} = -Z(0)\Omega(0)^{-1}Z(0)^{-1}B_0(0),$$
$$b_+^{(0)}(f, g) = -[B_0(0)C_0(0)]^{-1}B_0(0)g_0(0),$$
$$(7.2.40)$$

where, in deriving (7.2.40), we use such results as

$$\frac{1}{\varepsilon}\int_0^T \exp\left[-\frac{1}{\varepsilon}\int_0^t \Omega\right] v(t)\, dt = \Omega(0)^{-1} v(0) + \text{order}(\varepsilon)$$

for smooth vector-valued functions $v = v(t)$. Details are omitted. Finally, we see that the dependence of ξ^{**} and η^{**} on the data leads to the estimate

$$\left\|\begin{bmatrix} \xi^{**} \\ \eta^{**} \end{bmatrix}\right\| \le K\left[|\alpha(\varepsilon)| + \varepsilon|B_0(0)\beta(\varepsilon)| + \varepsilon^2|P(0)\beta(\varepsilon)| + \varepsilon\|f\| + \|g\|\right]$$

$$(7.2.41)$$

for a fixed positive constant K independent of ε as $\varepsilon \to 0+$. Similarly, from (7.2.29), (7.2.30), (7.2.38), (7.2.39), and (7.2.41), we find directly the estimates

$$|M_+(\varepsilon)\beta| \le K\left[|B_0(0)\beta| + \varepsilon|P(0)\beta|\right] \quad \text{for any } n\text{-vector } \beta,$$
$$|b_+(f, g; \varepsilon)| \le K(\varepsilon\|f\| + \|B_0 g\| + \varepsilon\|Pg\|),$$

$$(7.2.42)$$

uniformly as $\varepsilon \to 0+$, for a fixed positive constant K that is not the same here as in (7.2.41). Note that the projection $P(t)$ of (7.1.13) is zero, $P(t) = 0$, in the case $m = n$.

Combining these results, we have the following theorems for the original initial-value problem (7.1.1) and (7.1.3) subject to the stated assumptions (7.1.8)–(7.1.9) on the eigenvectors and eigenvalues of $B_0(t)C_0(t)$. Theorems 7.2.2 and 7.2.3 deal with the homogeneous system (7.1.4) with $f = 0$ and $g = 0$ in (7.1.1), and Theorem 7.2.4 deals with the inhomogeneous system.

Theorem 7.2.2: *The solution functions* $x = x(t, \varepsilon)$ *and* $y = y(t, \varepsilon)$ *for the homogeneous system* (7.1.4) *subject to the initial conditions* $x(0, \varepsilon) = \alpha(\varepsilon)$ *and* $y(0, \varepsilon) = \beta(\varepsilon)$ *of* (7.1.3) *satisfy the bounds*

$$|x(t, \varepsilon)| \le K_1\left[|\alpha(\varepsilon)| + \varepsilon|B_0(0)\beta(\varepsilon)| + \varepsilon^2|P(0)\beta(\varepsilon)|\right]$$
$$\le \varepsilon K_2\left[|B_0(0)\beta(\varepsilon)| + \varepsilon|P(0)\beta(\varepsilon)|\right] \qquad (7.2.43)$$

and

$$|y(t, \varepsilon)| \le K_1\left[\frac{1}{\varepsilon}|\alpha(\varepsilon)| + |B_0(0)\beta(\varepsilon)| + |P(0)\beta(\varepsilon)|\right] \le K_2|\beta(\varepsilon)|$$

$$(7.2.44)$$

for $0 \le t \le T$ *and for fixed positive constants* K_1 *and* K_2, *uniformly as* $\varepsilon \to 0+$, *provided that the initial vectors* $\alpha(\varepsilon)$ *and* $\beta(\varepsilon)$ *satisfy the stability relation* [*see* (7.2.39)]

$$\alpha = \varepsilon M_+(\varepsilon)\beta, \tag{7.2.45}$$

where the vector $b_+(f, g; \varepsilon)$ *in* (7.2.39) *vanishes for the homogeneous system* (7.1.4) *with* $b_+(0, 0; \varepsilon) = 0$. *If* $\beta(\varepsilon)$ *lies in the null space of* $B_0(0)$, *then the solution function* x *satisfies*

$$|x(t, \varepsilon)| \le \varepsilon^2 \text{ const. } |\beta(\varepsilon)| \quad \text{for } 0 \le t \le T, \ \varepsilon \to 0+,$$

$$\text{if } B_0(0)\beta(\varepsilon) = 0. \tag{7.2.46}$$

Proof: The solution functions ξ and η given by (7.2.30) subject to (7.2.36)–(7.2.38) satisfy the bounds [see (7.2.31) and (7.2.41)]

$$|\xi(t, \varepsilon)|, |\eta(t, \varepsilon)| \le \text{const.} \left[|\alpha(\varepsilon)| + \varepsilon|B_0(0)\beta(\varepsilon)| + \varepsilon^2|P(0)\beta(\varepsilon)| \right]$$

$$\tag{7.2.47}$$

for $0 \le t \le T$ [note that $f = 0$ and $g = 0$ in (7.2.41)], and then these bounds, along with (7.2.17), (7.2.23), (7.2.24), and (7.2.37), yield for ζ the bound

$$|\zeta(t, \varepsilon)| \le \text{const.} \left[|\alpha(\varepsilon)| + \varepsilon|B_0(0)\beta(\varepsilon)| + \varepsilon|P(0)\beta(\varepsilon)| \right], \tag{7.2.48}$$

which differs from (7.2.47) only in the power of ε appearing in the last term. The first inequalities of both (7.2.43) and (7.2.44) follow now directly from (7.2.9) and (7.2.12), along with (7.2.47)–(7.2.48), and the second inequalities of (7.2.43) and (7.2.44) then follow from (7.2.45), along with the previous inequality for M_+ given by (7.2.42). Finally, (7.2.46) follows directly from (7.2.43) in the case $B_0(0)\beta(\varepsilon) = 0$, where in this case $P(0)\beta(\varepsilon) = \beta(\varepsilon)$. ∎

Note in the foregoing that the *existence* of the solution functions $x(t, \varepsilon)$ and $y(t, \varepsilon)$ for (7.1.3)–(7.1.4) subject to the stability condition (7.2.36) requires that the initial data $\alpha(\varepsilon)$ and $\beta(\varepsilon)$ satisfy (7.2.45).

In the case $m = n$ there is no ζ component in the preceding construction. For example, in this case the matrix S of (7.2.8) is replaced with the $2m \times 2m$ matrix

$$S(t) := \begin{bmatrix} \frac{1}{2}Z(t)^{-1} & \frac{1}{2}\Omega(t)^{-1}Z(t)^{-1}B_0(t) \\ \frac{1}{2}Z(t)^{-1} & -\frac{1}{2}\Omega(t)^{-1}Z(t)^{-1}B_0(t) \end{bmatrix},$$

and the projection P of (7.1.13) vanishes, $P(t) = 0$. We simply take

$\zeta = 0$ in the foregoing construction if $m = n$. The system (7.2.23) reduces to

$$\xi(t) = \xi^*(t,\varepsilon) - \int_t^T \exp\left[-\frac{1}{\varepsilon}\int_t^s \Omega\right]\{E_{\xi\xi}(s,\varepsilon)\xi(s) + E_{\xi\eta}(s,\varepsilon)\eta(s)\}\,ds,$$

(7.2.49)

$$\eta(t) = \eta^*(t,\varepsilon) + \int_0^t \exp\left[-\frac{1}{\varepsilon}\int_s^t \Omega\right]\{E_{\eta\xi}(s,\varepsilon)\xi(s) + E_{\eta\eta}(s,\varepsilon)\eta(s)\}\,ds,$$

with ξ^* and η^* given by (7.2.24). The system (7.2.49) replaces (7.2.25) in this case. As before, the conditions [see (7.2.36)–(7.2.37)] $\xi_1 = 0$ and $\eta_1 = \eta_0(\varepsilon)$ lead to a unique solution of (7.2.49), where in this case (7.2.24) yields (with $f = 0$ and $g = 0$)

$$\xi^*(t,\varepsilon) = 0, \quad \eta^*(t,\varepsilon) = \exp\left[-\frac{1}{\varepsilon}\int_0^t \Omega\right]\eta_0(\varepsilon). \qquad (7.2.50)$$

We find that the solution functions $\xi(t) = \xi(t,\varepsilon)$ and $\eta(t) = \eta(t,\varepsilon)$ of (7.2.49) decay exponentially for $t > 0$ provided that the initial vectors $\alpha(\varepsilon)$ and $\beta(\varepsilon)$ satisfy the stability relation (7.2.45). Specifically, we find estimates of the form [compare with (7.2.47)]

$$|\xi(t,\varepsilon)|, |\eta(t,\varepsilon)| \leq \text{const.}\,[|\alpha(\varepsilon)| + \varepsilon|\beta(\varepsilon)|]\,e^{-\kappa t/\varepsilon} \qquad (7.2.51)$$

for $0 \leq t \leq T$, $\varepsilon \to 0+$, and where κ can be any fixed positive constant satisfying [cf. (7.2.32)]

$$0 < \kappa < \sqrt{\delta}\,\sin(\delta/2). \qquad (7.2.52)$$

The proof of (7.2.51) follows along the lines of the proof of Lemma 7.2.1, but with a modified norm on the Banach space $\mathscr{B}$ so as to take into account the expected exponential decay. For example, the quantity $\|\xi\|$ is now defined as

$$\|\xi\| := \max_{0 \leq t \leq T} e^{\kappa t/\varepsilon}|\xi(t)|.$$

Details are omitted. We then have the following theorem.

Theorem 7.2.3: *In the case $m = n$, the solution functions $x = x(t,\varepsilon)$ and $y = y(t,\varepsilon)$ for the homogeneous system (7.1.4) subject to the initial conditions (7.1.3) satisfy the bounds*

$$|x(t,\varepsilon)| + \varepsilon|y(t,\varepsilon)| \leq K_1[|\alpha(\varepsilon)| + \varepsilon|\beta(\varepsilon)|]\,e^{-\kappa t/\varepsilon}$$

$$\leq \varepsilon K_2|\beta(\varepsilon)|\,e^{-\kappa t/\varepsilon} \qquad (7.2.53)$$

for $0 \leq t \leq T$, for fixed positive constants K_1 and K_2, and for any fixed constant κ satisfying (7.2.52), all uniformly as $\varepsilon \to 0+$, provided that the initial vectors $\alpha(\varepsilon)$ and $\beta(\varepsilon)$ satisfy the stability relation (7.2.45).

For the general nonhomogeneous system (7.1.1), we find related theorems such as the following.

Theorem 7.2.4: *The solution functions $x = x(t, \varepsilon)$ and $y = y(t, \varepsilon)$ for the initial-value problem (7.1.1) and (7.1.3) satisfy the bounds*

$$|x(t, \varepsilon)| \leq K_1\left[|\alpha(\varepsilon)| + \varepsilon|B_0(0)\beta(\varepsilon)| + \varepsilon^2|P(0)\beta(\varepsilon)| + \varepsilon\|f\| + \|g\|\right]$$

$$\leq \varepsilon K_2\left[|B_0(0)\beta(\varepsilon)| + \varepsilon|P(0)\beta(\varepsilon)| + \|f\| + \varepsilon^{-1}\|g\|\right]$$

$$(7.2.54)$$

and

$$\varepsilon|y(t, \varepsilon)| \leq K_1\left[|\alpha(\varepsilon)| + \varepsilon|\beta(\varepsilon)| + \varepsilon\|f\| + \|B_0 g\| + \varepsilon^{-1}\|Pg\|\right]$$

$$\leq \varepsilon K_2\left[|\beta(\varepsilon)| + \|f\| + \varepsilon^{-1}\|B_0 g\| + \varepsilon^{-2}\|Pg\|\right] \qquad (7.2.55)$$

for $0 \leq t \leq T$ and for fixed positive constants K_1 and K_2, uniformly for $\varepsilon \to 0+$, provided that the initial vectors $\alpha(\varepsilon)$ and $\beta(\varepsilon)$ satisfy the stability relation (7.2.39).

Proof: In the present case the estimates (7.2.47)–(7.2.48) become

$$|\xi(t, \varepsilon)|, |\eta(t, \varepsilon)|$$

$$\leq \text{const.} \left[|\alpha(\varepsilon)| + \varepsilon|B_0(0)\beta(\varepsilon)| + \varepsilon^2|P(0)\beta(\varepsilon)| + \varepsilon\|f\| + \|g\|\right]$$

$$(7.2.56)$$

and

$$|\zeta(t, \varepsilon)| \leq \text{const.} \left[|\alpha(\varepsilon)| + \varepsilon|\beta(\varepsilon)| + \varepsilon\|f\| + \|B_0 g\| + \varepsilon^{-1}\|Pg\|\right]$$

$$(7.2.57)$$

for $0 \leq t \leq T$ as $\varepsilon \to 0+$. The first inequalities of (7.2.54) and (7.2.55) follow directly from (7.2.56)–(7.2.57) along with (7.2.9) and (7.2.12), and the remaining inequalities of (7.2.54)–(7.2.55) follow then with (7.2.39) and (7.2.42). ■

The consideration of the terminal-value problem on the negative interval $[-T, 0]$ follows the same lines as the foregoing study for the initial-value problem on $[0, T]$. On the interval $[-T, 0]$ we replace the integral equations (7.2.23) with

$$\xi(t) = \xi^*(t, \varepsilon) - \int_t^0 \exp\left[-\frac{1}{\varepsilon}\int_t^s \Omega\right]\left\{E_{\xi\xi}(s, \varepsilon)\xi(s) + E_{\xi\eta}(s, \varepsilon)\eta(s)\right.$$

$$\left. + E_{\xi\zeta}(s, \varepsilon)\zeta(s)\right\} ds,$$

$$\eta(t) = \eta^*(t, \varepsilon) + \int_{-T}^t \exp\left[-\frac{1}{\varepsilon}\int_s^t \Omega\right]\left\{E_{\eta\xi}(s, \varepsilon)\xi(s) + E_{\eta\eta}(s, \varepsilon)\eta(s)\right.$$

$$\left. + E_{\eta\zeta}(s, \varepsilon)\zeta(s)\right\} ds, \qquad (7.2.58)$$

$$\zeta(t) = \zeta^*(t, \varepsilon) - \int_t^0 \Gamma(t, s, \varepsilon)\left\{E_{\zeta\xi}(s, \varepsilon)\xi(s) + E_{\zeta\eta}(s, \varepsilon)\eta(s)\right\} ds,$$

for $-T \le t \le 0$, where the functions ξ^*, η^*, and ζ^* are now defined as [cf. (7.2.24)]

$$\xi^*(t,\varepsilon) := \exp\left[-\frac{1}{\varepsilon}\int_t^0 \Omega\right]\xi_1 - \frac{1}{2}\int_t^0 \exp\left[-\frac{1}{\varepsilon}\int_t^s \Omega\right]$$

$$\times \left\{ Z(s)^{-1}f(s,\varepsilon) + \frac{1}{\varepsilon}\Omega(s)^{-1}Z(s)^{-1}B_0(s)g(s,\varepsilon)\right\} ds,$$

$$\tag{7.2.59}_\xi$$

$$\eta^*(t,\varepsilon) := \exp\left[-\frac{1}{\varepsilon}\int_{-T}^t \Omega\right]\eta_1 + \frac{1}{2}\int_{-T}^t \exp\left[-\frac{1}{\varepsilon}\int_s^t \Omega\right]$$

$$\times \left\{ Z(s)^{-1}f(s,\varepsilon) - \frac{1}{\varepsilon}\Omega(s)^{-1}Z(s)^{-1}B_0(s)g(s,\varepsilon)\right\} ds,$$

$$\tag{7.2.59}_\eta$$

and

$$\zeta^*(t,\varepsilon) := \Gamma(t,0,\varepsilon)\zeta_1$$

$$- \frac{1}{\varepsilon}\int_t^0 \Gamma(t,s,\varepsilon)H(s)^{\mathrm{T}}P(s)g(s,\varepsilon)\,ds. \tag{7.2.59}_\zeta$$

We omit details and give only a statement of a typical result. In the following, the norm of a (vector-valued) function $h = h(t)$ is taken to be

$$\|h\| := \max_{-T \le t \le 0}|h(t)|.$$

The stable manifold $\mu_-(\varepsilon)$ for the terminal data (7.1.3) for the problem on $[-T,0]$ can be written as

$$\mu_-(\varepsilon): \alpha = \varepsilon M_-(\varepsilon)\beta + b_-(f,g;\varepsilon), \tag{7.2.60}$$

where $M_-(\varepsilon)$ and $b_-(f,g;\varepsilon)$ satisfy the appropriate results of (7.1.12) and (7.1.14).

Theorem 7.2.5: *The solution functions $x = x(t,\varepsilon)$ and $y = y(t,\varepsilon)$ for the terminal-value problem (7.1.1) and (7.1.3) on $[-T,0]$ satisfy bounds of the form (7.2.54) and (7.2.55) for $-T \le t \le 0$, provided that the terminal vectors $\alpha(\varepsilon)$ and $\beta(\varepsilon)$ satisfy the stability relation (7.2.60).*

If the data functions A, B, C, D, f, and g are given on an interval containing $t = 0$ as an interior point, such as the interval $[-T,T]$, and if we consider the problem (7.1.1) and (7.1.3) on $[-T,T]$, then we find that the stable manifolds $\mu_-(\varepsilon)$ and $\mu_+(\varepsilon)$, each of dimension n, intersect along an $(n-m)$-dimensional submanifold determined by the relation

$$[M_+(\varepsilon) - M_-(\varepsilon)]\beta = -\frac{1}{\varepsilon}[b_+(f,g;\varepsilon) - b_-(f,g;\varepsilon)]. \tag{7.2.61}$$

We obtain an $(n - m)$-dimensional solution space for the homogeneous system (7.1.4) corresponding to solutions that satisfy bounds of the type (7.2.43)–(7.2.44) for $-T \leq t \leq T$, along with an m-dimensional solution space corresponding to solutions that become exponentially unbounded for $t > 0$ but not for $t < 0$ [with data α, β on $\mu_-(\varepsilon)$], and also along with an m-dimensional solution space corresponding to solutions that become exponentially unbounded for $t < 0$ but not for $t > 0$ [with data α, β on $\mu_+(\varepsilon)$], as $\varepsilon \to 0+$. Details are omitted.

Exercises

Exercise 7.2.1: Consider the plane autonomous homogeneous system

$$\frac{dx}{dt} = a \cdot x + b \cdot y,$$
$$\varepsilon^2 \frac{dy}{dt} = c \cdot x + \varepsilon^2 d \cdot y, \tag{7.2.62}$$

for real-valued solution functions x and y, where a, b, c, and d are given, fixed constants with

$$bc > 0. \tag{7.2.63}$$

Show that the origin in the phase plane is a saddle for small ε. Show also that the stable initial manifold μ_+ of (7.2.45) coincides with the collection of points obtainable by translations from the origin through scalar multiples of an eigenvector corresponding to the stable eigenvalue having negative real part, for the following matrix of the system (7.2.62):

$$\begin{bmatrix} a & b \\ \dfrac{c}{\varepsilon^2} & d \end{bmatrix}.$$

Finally, verify the results of Theorem 7.2.3 for the initial-value problem for (7.2.62) for $t \geq 0$.

Exercise 7.2.2: Carry out the transformation (7.2.11) for the system (7.2.62), where the matrix S of (7.2.11) can be given in this case as the following 2×2 constant matrix:

$$S = \frac{1}{2} \begin{bmatrix} 1 & \sqrt{b/c} \\ 1 & -\sqrt{b/c} \end{bmatrix}. \tag{7.2.64}$$

Show that the transformed functions ξ, η satisfy the system

$$\varepsilon \frac{d}{dt} \begin{bmatrix} \xi \\ \eta \end{bmatrix} = \left[\begin{bmatrix} \sqrt{bc} & 0 \\ 0 & -\sqrt{bc} \end{bmatrix} + \frac{\varepsilon}{2} \begin{bmatrix} a+d & a-d \\ a-d & a+d \end{bmatrix} \right] \begin{bmatrix} \xi \\ \eta \end{bmatrix}. \tag{7.2.65}$$

Exercise 7.2.3: Let the diagonal matrix

$$\Omega = \Omega(t) = \text{diag}\{\omega_1(t), \omega_2(t), \ldots, \omega_m(t)\}$$

be integrable in t, and assume that (7.2.2) is satisfied so that the (possibly complex-valued) components $\omega_j = \omega_j(t)$ have uniformly positive real parts. Derive the estimate (7.2.32). *Hint:*

$$\exp\left[-\frac{1}{\varepsilon}\int_s^t \Omega\right] = \text{diag}\left\{\exp\left[-\frac{1}{\varepsilon}\int_s^t \omega_j\right]\right\}_{j=1}^m,$$

with

$$\left|\exp\left[-\frac{1}{\varepsilon}\int_s^t \omega_j\right]\right| \le \exp\left[-\frac{1}{\varepsilon}\sqrt{\delta}\cdot\sin\left(\frac{\delta}{2}\right)\cdot(t-s)\right] \quad \text{for } s \le t, \varepsilon > 0.$$

Exercise 7.2.4: Verify directly the results of (7.2.40) for the problem (7.1.18)–(7.1.19) of Example 7.1.1.

Exercise 7.2.5: For the homogeneous system (cf. Example 7.1.1)

$$\frac{dx}{dt} = x + y_1,$$

$$\varepsilon^2 \frac{d}{dt}\begin{bmatrix} y_1 \\ y_2 \end{bmatrix} = \begin{bmatrix} 1 \\ 0 \end{bmatrix} x + \varepsilon^2 \begin{bmatrix} 1 & 1 \\ 0 & 1 \end{bmatrix}\begin{bmatrix} y_1 \\ y_2 \end{bmatrix}, \tag{7.2.66}$$

for real-valued functions x, y_1, and y_2, *show directly* that the stable manifolds μ_+ and μ_- for the respective intervals $[0, T]$ and $[-T, 0]$ $(T > 0)$ are given as

$$\mu_+: \alpha + \varepsilon\beta_1 + \varepsilon^2\beta_2(1 - e^{-T/\varepsilon}) = 0 \tag{7.2.67}$$

and

$$\mu_-: \alpha - \varepsilon\beta_1 + \varepsilon^2\beta_2(1 - e^{-T/\varepsilon}) = 0. \tag{7.2.68}$$

Give an explicit representation for a solution of (7.2.66) that remains uniformly bounded on $[-T, T]$ as $\varepsilon \to 0+$. Show that this solution is uniformly bounded. *Hint:* Take $\beta_1 = 0$ along with any fixed α and β_2 satisfying

$$\alpha + \varepsilon^2\beta_2(1 - e^{-T/\varepsilon}) = 0.$$

Exercise 7.2.6: Give an explicit representation for a solution of (7.2.66) that remains bounded on $[0, T]$ but becomes exponentially unbounded on $[-T, 0]$ as $\varepsilon \to 0+$. Hence, this solution must be linearly independent of the solution of Exercise 7.2.5. *Hint:* Take $\beta_1 = 1$ and $\beta_2 = 0$ along with a suitable value for α so that the initial point is on the manifold μ_+. (We could find a third independent solution that remains bounded on $[-T, 0]$ but not on $[0, T]$, but we need not do this here.)

Exercise 7.2.7: Solve the homogeneous system (7.2.66) for $t > 0$ subject to the initial conditions

$$
\begin{aligned}
x &= \alpha := -\varepsilon^2, \\
y_1 &= \beta_1 := 0, \\
y_2 &= \beta_2 := 1 \quad \text{at } t = 0.
\end{aligned}
\tag{7.2.69}
$$

Show that the resulting solution functions are uniformly bounded on $[0, T]$ (as $\varepsilon \to 0+$) even though the initial data of (7.2.69) do not satisfy the stability condition of (7.2.67). Rather, the data of (7.2.69) satisfy the relation

$$
\alpha + \varepsilon \beta_1 + \varepsilon^2 \beta_2 = 0,
\tag{7.2.70}
$$

which is asymptotically equivalent to (7.2.67). This illustrates the fact that the stability conditions introduced earlier can be modified in various ways by inclusion or deletion of terms that are asymptotically zero.

7.3 The multivariable expansion

As indicated in Section 7.1, the data of (7.1.1) and (7.1.3) are assumed to have asymptotic expansions of the form

$$
\begin{bmatrix}
A(t, \varepsilon) \\
B(t, \varepsilon) \\
C(t, \varepsilon) \\
D(t, \varepsilon)
\end{bmatrix}
\sim
\sum_{k=0}^{\infty}
\begin{bmatrix}
A_k(t) \\
B_k(t) \\
C_k(t) \\
D_k(t)
\end{bmatrix}
\varepsilon^k
\tag{7.3.1}
$$

and

$$
\begin{bmatrix}
f(t, \varepsilon) \\
g(t, \varepsilon) \\
\alpha(\varepsilon) \\
\beta(\varepsilon)
\end{bmatrix}
\sim
\sum_{k=0}^{\infty}
\begin{bmatrix}
f_k(t) \\
g_k(t) \\
\alpha_k \\
\beta_k
\end{bmatrix}
\varepsilon^k
\tag{7.3.2}
$$

for suitable matrix- and vector-valued coefficients A_k, B_k, C_k, D_k, f_k, g_k, α_k, and β_k of appropriate orders, where A_k, B_k, C_k, D_k, f_k, and g_k are smooth functions of t on the interval $[0, T]$, and where the initial data are assumed here to satisfy the stability condition (7.2.39). (The same analysis with obvious modifications applies also to the interval $[-T, 0]$.)

On the basis of Example 7.1.1 [see (7.1.22)–(7.1.23)] and the results of Theorem 7.2.4, along with known asymptotic results for various related problems as in Smith (1971), Fife (1972), and O'Malley and Flaherty (1980), we are led to seek representations for the (stable) solution functions x and y of the initial-value problem (7.1.1) and (7.1.3) for small $\varepsilon > 0$ in the asymptotic forms

$$x(t, \varepsilon) \sim X(t, \varepsilon) + X^*(\tau, \varepsilon),$$

$$y(t, \varepsilon) \sim \frac{1}{\varepsilon^2} Y(t, \varepsilon) + \frac{1}{\varepsilon} Y^*(\tau, \varepsilon), \qquad (7.3.3)$$

for suitable outer functions X and Y and suitable initial-layer corrections X^* and Y^*, where these functions are to have asymptotic expansions of the forms

$$\begin{bmatrix} X(t, \varepsilon) \\ Y(t, \varepsilon) \end{bmatrix} \sim \sum_{k=0}^{\infty} \begin{bmatrix} X_k(t) \\ Y_k(t) \end{bmatrix} \varepsilon^k \qquad (7.3.4)$$

and

$$\begin{bmatrix} X^*(\tau, \varepsilon) \\ Y^*(\tau, \varepsilon) \end{bmatrix} \sim \sum_{k=0}^{\infty} \begin{bmatrix} X_k^*(\tau) \\ Y_k^*(\tau) \end{bmatrix} \varepsilon^k, \qquad (7.3.5)$$

where the boundary-layer variable τ is given as

$$\tau := t/\varepsilon. \qquad (7.3.6)$$

The boundary-layer correction terms are required to satisfy the matching conditions

$$\lim_{\tau \to \infty} X_k^*(\tau) = 0,$$

$$\lim_{\tau \to \infty} Y_k^*(\tau) = 0 \quad \text{for } k = 0, 1, \ldots, \qquad (7.3.7)_k$$

and we also have the initial conditions [see (7.1.3) and (7.3.2)–(7.3.6)]

$$X_k^*(0) = \alpha_k - X_k(0) \qquad \text{for } k = 0, 1, \ldots \qquad (7.3.8)_k$$

and

$$Y_k(0) = \begin{cases} 0 & \text{for } k = 0 \\ -Y_0^*(0) & \text{for } k = 1 \\ -Y_{k-1}^*(0) + \beta_{k-2} & \text{for } k = 2, 3, \ldots. \end{cases} \qquad (7.3.9)_k$$

The outer solution functions $x = X$ and $y = \varepsilon^{-2}Y$ of (7.3.3) are required to satisfy the full system (7.1.1), from which we find, with (7.3.1), (7.3.2), and (7.3.4), the equations

$$B_0(t)Y_k(t) = \begin{cases} 0 & \text{for } k = 0, \\ -B_1(t)Y_0(t) & \text{for } k = 1, \\ -\sum_{i=0}^{k-1} B_{k-i}Y_i - \sum_{i=0}^{k-2} A_{k-2-i}X_i - f_{k-2} + \frac{d}{dt}X_{k-2} \end{cases}$$

$$\text{for } k = 2, 3, \ldots,$$

$$(7.3.10)_k$$

and

$$\frac{dY_k}{dt} = C_0(t)X_k + D_0(t)Y_k + g_k(t) + \sum_{i=0}^{k-1}(C_{k-i}X_i + D_{k-i}Y_i)$$

$$\text{for } k = 0,1,2,\ldots, \quad (7.3.11)_k$$

where the final sum on the right side of $(7.3.11)_k$ is understood to be *defined to be zero* in the case $k = 0$.

The initial-layer corrections $x = X^*$ and $y = \varepsilon^{-1}Y^*$ of (7.3.3) are required to satisfy the homogeneous system (7.1.4), which in terms of the variable τ of (7.3.6) becomes [compare with the homogeneous version of (7.1.5)]

$$\frac{dX^*}{d\tau} = \varepsilon A(\varepsilon\tau, \varepsilon)X^* + B(\varepsilon\tau, \varepsilon)Y^*,$$

$$\frac{dY^*}{d\tau} = C(\varepsilon\tau, \varepsilon)X^* + \varepsilon D(\varepsilon\tau, \varepsilon)Y^* \quad \text{for } \tau > 0. \tag{7.3.12}$$

The expansions (7.3.1) and (7.3.5) with (7.3.6) and (7.3.12) then lead to the equations

$$\frac{dX_k^*}{d\tau} = B_0(0)Y_k^*(\tau) + P_{k-1}^*(\tau),$$

$$\frac{dY_k^*}{d\tau} = C_0(0)X_k^*(\tau) + Q_{k-1}^*(\tau) \quad \text{for } k = 0,1,2,\ldots, \tag{7.3.13}_k$$

for suitable functions P_k^* and Q_k^* that are determined successively in terms of the coefficients X_i^* and Y_i^* for $i = 0,1,\ldots, k$. In particular,

$$P_{-1}^*(\tau) = 0, \quad Q_{-1}^*(\tau) = 0 \quad \text{for } \tau \geq 0, \tag{7.3.14}$$

and

$$P_0^*(\tau) = \big[B_1(0) + \tau B_0'(0)\big]Y_0^*(\tau) + A_0(0)X_0^*(\tau),$$

$$Q_0^*(\tau) = \big[C_1(0) + \tau C_0'(0)\big]X_0^*(\tau) + D_0(0)Y_0^*(\tau) \quad \text{for } \tau \geq 0. \tag{7.3.15}$$

We first solve $(7.3.10)_0$–$(7.3.11)_0$ for the leading outer functions X_0 and Y_0 subject to the initial condition $(7.3.9)_0$ and then solve $(7.3.13)_0$–$(7.3.14)$ for the boundary-layer corrections X_0^* and Y_0^* subject to the auxiliary conditions $(7.3.7)_0$–$(7.3.8)_0$. We turn next to the outer functions X_1 and Y_1, followed by the boundary-layer functions X_1^* and Y_1^*, and so forth, continuing the procedure iteratively for $k = 0,1,2,\ldots,$ as described in the following.

In order to solve the outer equations (7.3.10) and (7.3.11), we differentiate $(7.3.10)_k$ with respect to t to find

$$B_0 \frac{dY_k}{dt} = -\frac{dB_0}{dt} Y_k + Q_{k-1}(t) \quad \text{for } k = 0, 1, 2, \ldots, \qquad (7.3.16)_k$$

where

$$Q_k(t) := \begin{cases} 0 & \text{for } k = -1, \\ -(d/dt)\{B_1(t) Y_0(t)\} & \text{for } k = 0, \\ \dfrac{d}{dt}\left\{ -\displaystyle\sum_{i=0}^{k} B_{k+1-i} Y_i - \sum_{i=0}^{k-1} A_{k-1-i} X_i - f_{k-1} + \frac{d}{dt} X_{k-1} \right\} & \\ & \text{for } k = 1, 2, \ldots. \end{cases}$$

$$(7.3.17)_k$$

We now multiply $(7.3.11)_k$ on the left by $B_0(t)$ and use $(7.3.16)_k$ to find

$$B_0 C_0 X_k = -\left(B_0' + B_0 D_0 \right) Y_k - B_0 g_k + Q_{k-1}$$

$$- B_0 \sum_{i=0}^{k-1} (C_{k-i} X_i + D_{k-i} Y_i) \qquad (7.3.18)_k$$

for $k = 0, 1, 2, \ldots$, where $B_0' = (d/dt) B_0$, with the independent variable t generally suppressed in (7.3.18), and where the final sum on the right side of $(7.3.18)_k$ is understood to be zero in the case $k = 0$. The matrix $B_0 C_0$ is nonsingular in view of (7.1.9), and so $(7.3.18)_k$ determines X_k in terms of Y_k and earlier X_i and Y_i for $i = 0, 1, \ldots, k - 1$. The resulting expression for X_k can be inserted back into the right side of $(7.3.11)_k$ so as to eliminate X_k there, and we find the following system for Y_k:

$$\frac{dY_k}{dt} = \left\{ D_0 - \left[C_0 (B_0 C_0)^{-1} (B_0' + B_0 D_0) \right] \right\} Y_k + P(t) g_k$$

$$+ C_0 (B_0 C_0)^{-1} Q_{k-1} + P(t) \sum_{i=0}^{k-1} (C_{k-i} X_i + D_{k-i} Y_i), \qquad (7.3.19)_k$$

for $k = 0, 1, 2, \ldots$, where $P = P(t)$ is the projection given by (7.1.13), and where, as usual, the summation on i (from $i = 0$ to $i = k - 1$) is put equal to zero in the case $k = 0$.

The function Y_0 is now determined uniquely as the solution of the initial-value problem [see $(7.3.9)_0$ and $(7.3.19)_0$ along with $(7.3.17)_{-1}$]

$$\frac{dY_0}{dt} = \left\{ D_0 - \left[C_0 (B_0 C_0)^{-1} (B_0' + B_0 D_0) \right] \right\} Y_0 + P(t) g_0$$

$$\text{for } 0 < t < T, \quad \text{with } Y_0(0) = 0, \quad (7.3.20)$$

and then X_0 is determined by $(7.3.18)_0$ as

$$X_0(t) = -\left[B_0(t)C_0(t)\right]^{-1}\left[(B_0' + B_0D_0)Y_0(t) + B_0(t)g_0(t)\right]$$
(7.3.21)

for $0 \le t < T$, with Y_0 determined by (7.3.20).

From (7.3.20), note that $Y_0 \equiv 0$ if and only if g_0 is in the image of C_0, as in (7.1.16), with $g_0 = C_0\nu_0$, and in this case (7.3.21) gives $X_0 = -\nu_0$; that is, we have

$$Y_0(t) \equiv 0 \quad \text{with } X_0(t) = -\nu_0(t)$$
(7.3.22)

if and only if g_0 is in the column space of C_0, with $g_0(t) = C_0(t)\nu_0(t)$. This situation always obtains in the case $m = n$.

Now that $X_0(t)$ and $Y_0(t)$ are uniquely determined for $0 \le t < T$, we turn to the determination of the leading terms in the initial-layer correction functions, namely X_0^* and Y_0^*. From $(7.3.13)_0$–(7.3.14) we find for X_0^* the second-order vector system

$$\frac{d^2X_0^*}{d\tau^2} = B_0(0)C_0(0)X_0^* \quad \text{for } \tau > 0,$$
(7.3.23)

where this system is to be solved subject to the initial condition [see $(7.3.8)_0$]

$$X_0^*(0) = \alpha_0 - X_0(0)$$
(7.3.24)

and the matching condition [see $(7.3.7)_0$]

$$\lim_{\tau \to \infty} X_0^*(\tau) = 0.$$
(7.3.25)

The general solution of (7.3.23) can be given in the form

$$X_0^*(\tau) = Z(0)\left(e^{+\Omega(0)\tau}x_+ + e^{-\Omega(0)\tau}x_-\right),$$
(7.3.26)

where Ω and Z are as introduced at the beginning of Section 7.2, and where x_+ and x_- denote arbitrary (constant) m-vectors that are determined in the present case by (7.3.24)–(7.3.25) as

$$x_+ = 0, \qquad x_- = Z^{-1}(0)\left[\alpha_0 - X_0(0)\right].$$
(7.3.27)

Note that the matrix $e^{-\Omega(0)\tau}$ decays exponentially to zero as $\tau \to \infty$, as follows from Exercise 7.2.3. The resulting solution $X_0^*(\tau)$ then leads directly also to Y_0^* using $(7.3.13)_0$ along with the matching condition for Y_0^* from $(7.3.7)_0$. In this way we have the unique solutions (note that $\Omega e^{-\Omega\tau}\Omega^{-1} = e^{-\Omega\tau}$)

$$X_0^*(\tau) = Z(0)e^{-\Omega(0)\tau}Z(0)^{-1}\left[\alpha_0 - X_0(0)\right],$$

$$Y_0^*(\tau) = -C_0(0)Z(0)e^{-\Omega(0)\tau}\Omega(0)^{-1}Z(0)^{-1}\left[\alpha_0 - X_0(0)\right],$$
(7.3.28)

where $X_0(0)$ is given as in (7.3.21), with $X_0(0) = -[B_0(0)C_0(0)]^{-1}$ $B_0(0)g_0(0)$.

As usual, we now continue recursively. After $X_i(t), Y_i(t)$ and $X_i^*(\tau)$, $Y_i^*(\tau)$ have been determined for $i = 0, 1, \ldots, k-1$, we use $(7.3.9)_k$ and $(7.3.18)_k-(7.3.19)_k$ to determine $X_k(t), Y_k(t)$. Then $(7.3.7)_k-(7.3.8)_k$, along with $(7.3.13)_k$, determine $X_k^*(\tau), Y_k^*(\tau)$, where $(7.3.13)_k$ implies for X_k^* the second-order vector system

$$\frac{d^2 X_k^*}{d\tau^2} = B_0(0)C_0(0) X_k^* + F_k(\tau) \quad \text{for } \tau > 0, \qquad (7.3.29)$$

with

$$F_k(\tau) := B_0(0)Q_{k-1}^* + \frac{dP_{k-1}^*(\tau)}{d\tau}. \qquad (7.3.30)$$

We find (for $i = 0, 1, \ldots, k-1$) that each $X_i^*(\tau), Y_i^*(\tau)$ decays exponentially to zero as $\tau \to \infty$, and the same result then holds also for F_k, with

$$|F_k(\tau)| \leq \text{const. } e^{-\kappa\tau} \quad \text{for } \tau \geq 0, \qquad (7.3.31)$$

for any fixed positive constant κ satisfying (7.2.52). The most general solution of (7.3.29) can be given as

$$X_k^*(\tau) = Z(0)\big(e^{\Omega(0)\tau}x_+ + e^{-\Omega(0)\tau}x_-\big)$$
$$+ Z(0)\int_0^\tau \tfrac{1}{2}\big(e^{-\Omega(0)(\tau-\sigma)} - e^{-\Omega(0)(\tau-\sigma)}\big)\Omega(0)^{-1}Z(0)^{-1}F_k(\sigma)\,d\sigma$$

$$(7.3.32)$$

for suitable constants x_+ and x_- that are determined uniquely in the present case by the appropriate conditions of $(7.3.7)_k-(7.3.8)_k$. The resulting solution for X_k^* is given as

$$X_k^*(\tau) = Z(0)e^{-\Omega(0)\tau}Z(0)^{-1}[\alpha_k - X_k(0)]$$
$$+ \tfrac{1}{2}Z(0)e^{-\Omega(0)\tau}\int_0^\infty e^{-\Omega(0)\sigma}\Omega(0)^{-1}Z(0)^{-1}F_k(\sigma)\,d\sigma$$
$$- \tfrac{1}{2}Z(0)\int_\tau^\infty e^{-\Omega(0)(\sigma-\tau)}\Omega(0)^{-1}Z(0)^{-1}F_k(\sigma)\,d\sigma$$
$$- \tfrac{1}{2}Z(0)\int_0^\tau e^{-\Omega(0)(\tau-\sigma)}\Omega(0)^{-1}Z(0)^{-1}F_k(\sigma)\,d\sigma, \qquad (7.3.33)$$

and then Y_k^* is given by $(7.3.7)_k$ and $(7.3.13)_k$ as

$$Y_k^*(\tau) = -\int_\tau^\infty \big[C_0(0)X_k^*(\sigma) + Q_{k-1}^*(\sigma)\big]\,d\sigma. \qquad (7.3.34)$$

We find directly with (7.3.31), along with the exponential decay of

$e^{-\Omega(0)\tau}$, that the resulting functions X_k^* and Y_k^* also decay exponentially, with

$$\left|X_k^*(\tau)\right|, \left|Y_k^*(\tau)\right| \le \text{const.} \; e^{-\kappa\tau} \quad \text{for } \tau \ge 0, \qquad (7.3.35)$$

again for any fixed positive constant κ satisfying (7.2.52).

This completes the formal construction of the asymptotic representation (7.3.3)–(7.3.5).

Example 7.3.1: In the special case in which the data functions $A(t)$, $B(t)$, $C(t)$, $D(t)$, $f(t)$, and $g(t)$ are independent of ε, with, moreover [see (7.1.16)],

$$g(t) = C(t)\nu(t), \qquad (7.3.36)$$

then the foregoing construction on $[0, T]$ yields, for the first few terms,

$$X_0(t) = -\nu(t), \qquad X_0^*(\tau) = 0,$$
$$Y_0(t) = 0, \qquad Y_0^*(\tau) = 0,$$
$$X_1(t) = 0, \qquad X_1^*(\tau) = Z(0)e^{-\Omega(0)\tau}Z(0)^{-1}\alpha_1,$$
$$Y_1(t) = 0, \qquad Y_1^*(\tau) = -C(0)Z(0)e^{-\Omega(0)\tau}\Omega(0)^{-1}Z(0)^{-1}\alpha_1,$$
$$(7.3.37)$$

provided that the initial point $[\alpha(\varepsilon), \beta(\varepsilon)]$ lies on the stable manifold $\mu_+(\varepsilon)$:

$$[\alpha(\varepsilon), \beta(\varepsilon)] \in \mu_+(\varepsilon). \qquad (7.3.38)$$

Assuming that condition (7.3.38) holds for all small $\varepsilon \to 0+$, then the coefficients α_k and β_k in the expansions for $\alpha(\varepsilon)$ and $\beta(\varepsilon)$ [see (7.3.2)] must satisfy suitable stability conditions obtained from (7.1.11), the first two of which are

$$\alpha_0 = -\nu(0),$$
$$B(0)\beta_0 + Z(0)\Omega(0)Z(0)^{-1}\alpha_1 = -\nu'(0) + A(0)\nu(0) - f(0).$$
$$(7.3.39)$$

These latter conditions have been used in obtaining (7.3.37), where, for example, we find initially the result

$$X_0^*(\tau) = Z(0)e^{-\Omega(0)\tau}Z(0)^{-1}[\alpha_0 + \nu(0)], \qquad (7.3.40)$$

and then the first condition of (7.3.39) implies $X_0^* \equiv 0$, as listed in (7.3.37).

Along with the results of (7.3.37), we find also that $Y_2(t)$ is determined as the unique solution of the initial-value problem

$$\frac{dY_2}{dt} = \left\{ D - \left[C(BC)^{-1}(B' + BD) \right] \right\} Y_2$$

$$+ C(BC)^{-1} \frac{d}{dt} \left\{ -v'(t) + A(t)v(t) - f(t) \right\} \quad \text{for } t > 0,$$

with $Y_2(0) = \beta_0 + C(0)Z(0)\Omega(0)^{-1}Z(0)^{-1}\alpha_1.$ (7.3.41)

On differentiation of the quantity

$$B\left[Y_2 + C(BC)^{-1}(v' - Av + f) \right]$$

$$\equiv B(t)Y_2(t) + v'(t) - A(t)v(t) + f(t), \quad (7.3.42)$$

we find, with the differential equation of (7.3.41), that this latter quantity is constant in t. Moreover, the initial condition of (7.3.41), along with the second condition of (7.3.39), shows that this quantity is zero at $t = 0$, and hence the expression in (7.3.42) vanishes for all t, so that the function y^* defined as

$$y^*(t) := Y_2(t) + C(t)(BC)^{-1}\left[v' - A(t)v + f(t) \right]   (7.3.43)$$

is in the null space of $B(t)$.

If we now truncate the representation (7.3.3)–(7.3.5) and use the estimates of Theorem 7.4.1, we have for the exact solution functions $x(t, \varepsilon)$, $y(t, \varepsilon)$ the results

$$x(t, \varepsilon) = X_0(t) + \varepsilon X_1^*\left(\frac{t}{\varepsilon}\right) + O(\varepsilon^2),$$

$$y(t, \varepsilon) = Y_2(t) + Y_1^*\left(\frac{t}{\varepsilon}\right) + O(\varepsilon),$$ (7.3.44)

where the remainder terms here are respectively $O(\varepsilon^2)$ and $O(\varepsilon)$ *uniformly* for $t \in [0, T]$ as $\varepsilon \to 0+$. In particular, we have

$$\lim_{\substack{\varepsilon \to 0+ \\ \text{fixed } t > 0}} \begin{bmatrix} x(t, \varepsilon) \\ y(t, \varepsilon) \end{bmatrix} = \begin{bmatrix} X_0(t) \\ Y_2(t) \end{bmatrix},   (7.3.45)$$

with $X_0 = -v$ [see (7.3.36)] and with Y_2 determined by (7.3.41). The right side of (7.3.45) provides a particular solution of the reduced system (7.1.15), where this particular limiting solution is seen to be of the form (7.1.17), with the function y^* of (7.1.17) given here by (7.3.43). In the case $m < n$, there are infinitely many functions y^* in the null space of B, but in this case the original initial-value problem "chooses" the particular y^* of (7.3.43) in the limit $\varepsilon \to 0+$. In the case $m = n$, we see directly that (7.3.43) yields $y^* = 0$, which is the unique vector in the null space of B.

Example 7.3.2: In the case in which the data functions $A(t)$, $B(t)$, $C(t)$, $D(t)$, $f(t)$, and $g(t)$ are independent of ε, and with $g(t)$ not in the image of $C(t)$, so that (7.3.36) fails to hold, then the foregoing construction, along with the estimates of Theorem 7.4.1, yields [compare with (7.3.44)]

$$x(t, \varepsilon) = X_0(t) + \varepsilon X_1^*\left(\frac{t}{\varepsilon}\right) + O(\varepsilon^2),$$

$$y(t, \varepsilon) = \varepsilon^{-2} Y_0(t) + \left[Y_2(t) + Y_1^*\left(\frac{t}{\varepsilon}\right)\right] + O(\varepsilon), \tag{7.3.46}$$

where $Y_0(t)$ is the solution of the initial-value problem (7.3.20), and $X_0(t)$ is given by (7.3.21), and where the stability condition (7.3.38) is assumed to hold. We again have $\alpha_0 = X_0(0)$, and $X_0^*(\tau) \equiv X_1(t) \equiv 0$ and $Y_0^*(\tau) \equiv Y_1(t) \equiv 0$, and $X_1^*(\tau)$ and $Y_1^*(\tau)$ are again given as in (7.3.37).

In this case the reduced system (7.1.15) has no solution, and we find, with (7.3.46), that the solution function $y(t, \varepsilon)$ of the full problem becomes unbounded like ε^{-2} as $\varepsilon \to 0+$. On the other hand, the solution function $x(t, \varepsilon)$ has a well-defined (finite) limit as $\varepsilon \to 0+$ for fixed $t \in [0, T]$. For $t \geq 0$, this limiting function $X_0(t)$ does not satisfy the reduced relation $Cx + g = 0$.

Exercises

Exercise 7.3.1: Consider the initial-value problem (see Example 7.1.1)

$$\frac{dx}{dt} = x + y_1 + f(t),$$

$$\varepsilon^2 \frac{d}{dt}\begin{bmatrix} y_1 \\ y_2 \end{bmatrix} = \begin{bmatrix} 1 \\ 0 \end{bmatrix} x + \varepsilon^2 \begin{bmatrix} 1 & 1 \\ 0 & 1 \end{bmatrix} \begin{bmatrix} y_1 \\ y_2 \end{bmatrix} + \begin{bmatrix} g_1(t) \\ 0 \end{bmatrix}, \tag{7.3.47}$$

for real-valued functions x, y_1, and y_2 on $[0, T]$ ($T > 0$), subject to the initial conditions

$$x = \alpha(\varepsilon), \qquad y_1 = 0, \qquad y_2 = 1 \quad \text{at } t = 0, \tag{7.3.48}$$

for given smooth functions $f(t)$ and $g_1(t)$ independent of ε, and where $\alpha(\varepsilon)$ is given as

$$\alpha(\varepsilon) := -\varepsilon^2(1 - e^{-T/\varepsilon}) - \int_0^T \exp\left[-\left(1 + \frac{1}{\varepsilon}\right)t\right]\left[f(t) + \frac{1}{\varepsilon}g_1(t)\right] dt. \tag{7.3.49}$$

(i) Show that the real-valued quantity $\alpha(\varepsilon)$ of (7.3.49) has an asymptotic expansion of suitable type, as indicated in (7.3.2), and compute explicitly the first three coefficients α_0, α_1, and α_2 of this expansion. Also, show that the initial values of (7.3.48)–(7.3.49) satisfy the stability relation (7.1.25).

(ii) Compute explicitly the first few terms in the expansions of (7.3.3)–(7.3.6) for the solution functions x, y_1, and y_2, and use these results in the right side of (7.3.45) to obtain the limiting values of $x(t, \varepsilon)$, $y_1(t, \varepsilon)$, and $y_2(t, \varepsilon)$ as $\varepsilon \to 0+$, for fixed $t > 0$. Could we reasonably have predicted these limit functions in this case?

Exercise 7.3.2: Consider the initial-value problem

$$\frac{dx}{dt} = x + y_1 + f(t),$$

$$\varepsilon^2 \frac{d}{dt}\begin{bmatrix} y_1 \\ y_2 \end{bmatrix} = \begin{bmatrix} 1 \\ 0 \end{bmatrix} x + \varepsilon^2 \begin{bmatrix} 1 & 1 \\ 0 & 1 \end{bmatrix}\begin{bmatrix} y_1 \\ y_2 \end{bmatrix} + \begin{bmatrix} 0 \\ 1 \end{bmatrix}, \qquad (7.3.50)$$

for real-valued functions x, y_1, and y_2 on $[0, T]$ $(T > 0)$, subject to the initial conditions

$$x = \alpha(\varepsilon), \qquad y_1 = 0, \qquad y_2 = 1 \quad \text{at } t = 0, \qquad (7.3.51)$$

where $\alpha(\varepsilon)$ is given as

$$\alpha(\varepsilon) := -\left\{ \frac{1}{1 + \varepsilon} + \varepsilon^2 + \int_0^T \exp\left[-\left(1 + \frac{1}{\varepsilon}\right)t \right] f(t)\, dt \right\}$$

$$+ (1 + \varepsilon^2) e^{-T/\varepsilon} - \frac{1}{1 + \varepsilon} \exp\left[-\left(1 + \frac{1}{\varepsilon}\right)T \right]. \qquad (7.3.52)$$

Compute the quantities $X_0(t)$ and $Y_0(t)$ in (7.3.46), and thereby determine the limits

$$\lim_{\substack{\varepsilon \to 0+ \\ \text{fixed } t > 0}} x(t, \varepsilon) \quad \text{and} \quad \lim_{\substack{\varepsilon \to 0+ \\ \text{fixed } t > 0}} \varepsilon^2 y(t, \varepsilon).$$

7.4 Error estimates

Let the functions $x^n = x^n(t, \varepsilon)$ and $y^n = y^n(t, \varepsilon)$ be defined as

$$x^n(t, \varepsilon) := \sum_{k=0}^n \left[X_k(t) + X_k^*\left(\frac{t}{\varepsilon}\right) \right] \varepsilon^k,$$

$$y^n(t, \varepsilon) := \sum_{k=0}^{n+1} Y_k(t) \varepsilon^{k-2} + \sum_{k=0}^n Y_k^*\left(\frac{t}{\varepsilon}\right) \varepsilon^{k-1}, \qquad (7.4.1)$$

where for definiteness the functions $X_k(t)$, $Y_k(t)$, $X_k^*(\tau)$, and $Y_k^*(\tau)$ are those constructed in Section 7.3 for the initial-value problem on $[0, T]$ with initial data satisfying the stability condition (7.2.39). A direct calculation shows that x^n, y^n satisfy the original initial-value problem *approximately*, in the sense that

$$\frac{dx^n}{dt} = A(t, \varepsilon)x^n + B(t, \varepsilon)y^n + f(t, \varepsilon) - \rho_1(t, \varepsilon),$$

$$\varepsilon^2 \frac{dy^n}{dt} = C(t, \varepsilon)x^n + \varepsilon^2 D(t, \varepsilon)y^n + g(t, \varepsilon) - \rho_2(t, \varepsilon), \tag{7.4.2}$$

and

$$x^n(0, \varepsilon) = \alpha(\varepsilon) - \rho_3(\varepsilon), \tag{7.4.3}$$
$$y^n(0, \varepsilon) = \beta(\varepsilon) - \rho_4(\varepsilon),$$

for residuals $\rho_1(t, \varepsilon)$, $\rho_2(t, \varepsilon)$, $\rho_3(\varepsilon)$, and $\rho_4(\varepsilon)$ that are small. In fact, we see directly that these residuals, which are defined by (7.4.1)–(7.4.3), satisfy the estimates (see Exercise 7.4.1)

$$\rho_1(t, \varepsilon) = O(\varepsilon^n),$$
$$\rho_2(t, \varepsilon) = O(\varepsilon^{n+1}), \tag{7.4.4}$$

uniformly for $0 \leq t \leq T$ as $\varepsilon \to 0+$, and

$$\rho_3(\varepsilon) = O(\varepsilon^{n+1}),$$
$$\rho_4(\varepsilon) = O(\varepsilon^n), \tag{7.4.5}$$

as $\varepsilon \to 0+$.

The functions $\hat{x}$ and $\hat{y}$ defined as

$$\hat{x}(t, \varepsilon) := x(t, \varepsilon) - x^n(t, \varepsilon),$$
$$\hat{y}(t, \varepsilon) := y(t, \varepsilon) - y^n(t, \varepsilon), \tag{7.4.6}$$

then satisfy the system

$$\frac{d\hat{x}}{dt} = A(t, \varepsilon)\hat{x} + B(t, \varepsilon)\hat{y} + \rho_1(t, \varepsilon),$$

$$\varepsilon^2 \frac{d\hat{y}}{dt} = C(t, \varepsilon)\hat{x} + \varepsilon^2 D(t, \varepsilon)\hat{y} + \rho_2(t, \varepsilon), \tag{7.4.7}$$

along with the initial conditions

$$\hat{x}(0, \varepsilon) = \rho_3(\varepsilon), \qquad \hat{y}(0, \varepsilon) = \rho_4(\varepsilon). \tag{7.4.8}$$

Moreover, the construction of Section 7.3, along with the *linear* dependence of $b_+(f, g; \varepsilon)$ on f and g in (7.2.39), implies that the initial data ρ_3 and ρ_4 in (7.4.8) satisfy the stability condition (7.2.39), with f and g replaced by ρ_1 and ρ_2, respectively,

$$\rho_3 = \varepsilon M_+(\varepsilon)\rho_4 + b_+(\rho_1, \rho_2; \varepsilon). \tag{7.4.9}$$

The transformation (7.2.11) can be applied to the problem (7.4.7)–(7.4.8), and the resulting functions $\hat{\xi}$, $\hat{\eta}$, and $\hat{\zeta}$ satisfy the integral equations given by (7.2.23)–(7.2.28), with f, g replaced by ρ_1, ρ_2 and with α, β replaced by ρ_3, ρ_4. Hence, Lemma 7.2.1, along with the argument following Lemma 7.2.1, can be applied to the problem (7.4.7)–(7.4.9), and we find the result of Theorem 7.2.4 for the present functions $\hat{x}$, $\hat{y}$, with the data f, g, α, and β replaced by ρ_1, ρ_2, ρ_3, and ρ_4, respectively. The resulting inequalities of (7.2.54)–(7.2.55), along with the estimates of (7.4.4)–(7.4.5), yield

$$\begin{aligned} |\hat{x}(t, \varepsilon)| &\le \text{const. } \varepsilon^{n+1}, \\ |\hat{y}(t, \varepsilon)| &\le \text{const. } \varepsilon^{n-1}, \end{aligned} \qquad (7.4.10)$$

uniformly for $0 \le t \le T$ as $\varepsilon \to 0+$, and in this way we find the following result.

Theorem 7.4.1: *The functions $x^n = x^n(t, \varepsilon)$ and $y^n = y^n(t, \varepsilon)$ of (7.4.1) obtained by the two-variable procedure of Section 7.3 for the initial-value problem (7.1.1)–(7.1.3) on the interval $[0, T]$, subject to the stability relation (7.2.39), provide uniform approximations for the exact solution functions $x = x(t, \varepsilon)$ and $y = y(t, \varepsilon)$. Specifically, there is a positive constant K_n such that the estimates*

$$\begin{aligned} |x(t, \varepsilon) - x^n(t, \varepsilon)| &\le K_n \varepsilon^{n+1}, \\ |y(t, \varepsilon) - y^n(t, \varepsilon)| &\le K_n \varepsilon^n, \end{aligned} \qquad (7.4.11)$$

hold uniformly for t on $[0, T]$ as $\varepsilon \to 0+$.

Proof: The foregoing discussion with n replaced by $n + 1$ yields directly the results [see (7.4.6) and (7.4.10), with n replaced by $n + 1$]

$$\begin{aligned} |x(t, \varepsilon) - x^{n+1}(t, \varepsilon)| &\le K\varepsilon^{n+2}, \\ |y(t, \varepsilon) - y^{n+1}(t, \varepsilon)| &\le K\varepsilon^n, \end{aligned} \qquad (7.4.12)$$

for some suitable constant K. But (7.4.1) yields directly the uniform estimates

$$x^{n+1} = x^n + O(\varepsilon^{n+1}), \qquad y^{n+1} = y^n + O(\varepsilon^n), \qquad (7.4.13)$$

and then the stated results of (7.4.11) follow directly from (7.4.12)–(7.4.13). ∎

A similar result is valid for the terminal-value problem on $[-T, 0]$ subject to the corresponding stability relation (7.2.60).

Exercise

Exercise 7.4.1: Give a direct verification of the estimates of (7.4.4) and (7.4.5) in the case $n = 0$.

7.5 The fundamental solution

The construction of Section 7.2 yields $n - m$ independent solutions $(x, y) = [x(t, \varepsilon), y(t, \varepsilon)]$ of the homogeneous system (7.1.4) satisfying

$$|x(t, \varepsilon)| \leq \text{const. } \varepsilon^2$$
$$|y(t, \varepsilon)| \leq \text{constant}, \quad \text{for } t \in [0, T] \tag{7.5.1}$$

as $\varepsilon \to 0+$. Indeed, we need only take the initial vector $\beta(\varepsilon) = \beta_0$ to lie in the null space of $B_0(0)$, with corresponding $\alpha(\varepsilon)$ given as $\alpha(\varepsilon) := \varepsilon M_+(\varepsilon)\beta_0$, so that the initial point (α, β) lies on the stable manifold $\mu_+(\varepsilon)$. Theorem 7.2.2 then shows that the resulting solution of the homogeneous system satisfies the bounds of (7.5.1).

Because the null space of $B_0(0)$ has dimension $n - m$ [see (7.2.5)], it follows that there are $n - m$ independent solutions satisfying (7.5.1), say

$$\begin{bmatrix} x \\ y \end{bmatrix} = \begin{bmatrix} \varepsilon^2 \hat{x}_j(t, \varepsilon) \\ \hat{y}_j(t, \varepsilon) \end{bmatrix} \quad \text{for } j = 1, 2, \ldots, n - m,$$

for suitable vector-valued functions $\hat{x}_j, \hat{y}_j$ that are uniformly bounded for $0 \leq t \leq T$ as $\varepsilon \to 0+$. Let $\hat{X} = \hat{X}(t, \varepsilon)$ and $\hat{Y} = \hat{Y}(t, \varepsilon)$ be the respective $m \times (n - m)$ and $n \times (n - m)$ matrix-valued functions with jth columns given respectively by $\hat{x}_j(t, \varepsilon)$ and $\hat{y}_j(t, \varepsilon)$, for $j = 1, 2, \ldots, n - m$. Then the $(m + n) \times (n - m)$ matrix-valued function

$$\begin{bmatrix} X \\ Y \end{bmatrix} := \frac{1}{\varepsilon} \begin{bmatrix} \varepsilon^2 \hat{X} \\ \hat{Y} \end{bmatrix} = \begin{bmatrix} \varepsilon \hat{X} \\ \frac{1}{\varepsilon} \hat{Y} \end{bmatrix} \tag{7.5.2}$$

has rank $(n - m)$ and satisfies the homogeneous system [see (7.1.4)]

$$\frac{d}{dt} \begin{bmatrix} X \\ Y \end{bmatrix} = \begin{bmatrix} A & B \\ \varepsilon^{-2}C & D \end{bmatrix} \begin{bmatrix} X \\ Y \end{bmatrix}. \tag{7.5.3}$$

The solution (7.5.2) can now be used along with suitable additional solutions of exponential type in the construction of a fundamental solution matrix for the homogeneous system (7.1.4), as described in Theorem 7.5.1.

Condition (7.5.4) is used in the construction of the solutions of exponential type. If the matrix $B_0(t)C_0(t)$ is independent of t, then

(7.5.4) is not a restriction, because it can be achieved simply by a suitable ordering of the eigenvalues of $B_0 C_0$. However, if this matrix is a nonconstant function of t, then (7.5.4) represents an additional restriction that is required here for technical reasons; see Perron (1929), Levinson (1948), Bellman (1953), and Coppel (1965), where conditions related to (7.5.4) are used in the study of the asymptotic behavior of linear systems. Condition (7.5.4) can be weakened in several ways. For example, it suffices for the eigenvalues to satisfy an ordering of the type (7.5.4) on each subinterval of a finite partition of $[0, T]$, where the ordering can change from subinterval to subinterval.

Theorem 7.5.1: *In addition to the hypothesis of (7.1.8)–(7.1.10), let the eigenvalues* $\lambda_j = \lambda_j(t)$ *of* $B_0(t) C_0(t)$ *[see (7.1.8)–(7.1.10)] satisfy*

$$\operatorname{Re} \lambda_j(t) \geq \operatorname{Re} \lambda_k(t) \quad \text{for } j \leq k, \tag{7.5.4}$$

uniformly for all $t \in [0, T]$. *Then the homogeneous system (7.1.4) has a fundamental solution matrix* $F = F(t, \varepsilon)$ *of the form*

$$F(t, \varepsilon)$$

$$= \begin{bmatrix} X_+(t, \varepsilon) \exp\left(\dfrac{1}{\varepsilon} \displaystyle\int_0^t \Omega\right) & X_-(t, \varepsilon) \exp\left(-\dfrac{1}{\varepsilon} \displaystyle\int_0^t \Omega\right) & \varepsilon \hat{X}(t, \varepsilon) \\[3mm] \dfrac{1}{\varepsilon} Y_+(t, \varepsilon) \exp\left(\dfrac{1}{\varepsilon} \displaystyle\int_0^t \Omega\right) & \dfrac{1}{\varepsilon} Y_-(t, \varepsilon) \exp\left(-\dfrac{1}{\varepsilon} \displaystyle\int_0^t \Omega\right) & \dfrac{1}{\varepsilon} \hat{Y}(t, \varepsilon) \end{bmatrix}$$

$$\tag{7.5.5}$$

for suitable matrix-valued functions $X_+(t, \varepsilon)$, $Y_+(t, \varepsilon)$, $X_-(t, \varepsilon)$, $Y_-(t, \varepsilon)$, $\hat{X}(t, \varepsilon)$, *and* $\hat{Y}(t, \varepsilon)$, *all of which are uniformly bounded for* $t \in [0, T]$ *as* $\varepsilon \to 0+$. *[The functions* $\hat{X}$, $\hat{Y}$ *may be taken to be those appearing in (7.5.2), and* Ω *is the diagonal matrix introduced in Section 7.2.]*

Proof: In view of (7.5.1)–(7.5.3), there remains only to prove the existence of suitable exponential-type solutions as in the first two block-columns of (7.5.5), where these columns constitute the first $2m$ column vectors of (7.5.5). In obtaining any one of these solution vectors for (7.1.4) corresponding to any particular one of these first $2m$ column vectors of (7.5.5), we pattern the proof after the earlier proof of Lemma 7.2.1 for the (homogeneous version of the) system (7.2.13); however, we replace the integral equations of (7.2.23) with a different related system depending on the particular solution vector sought, where the particular system of integral equations chosen makes use of the ordering (7.5.4).

The technique is patterned after a technique used in Perron (1929), Levinson (1948), and Bellman (1953) in the asymptotic study of solutions of linear systems for large values of the independent variable; details are omitted. (Alternatively, a Riccati transformation, as discussed in Sections 9.2 and 9.3, can be used here.) ■

The functions $\hat{X}, \hat{Y}$ of (7.5.5) can be taken to be given as in (7.5.2), and in this case the analysis of Sections 7.3 and 7.4 yields

$$\hat{Y}(t, \varepsilon) = \hat{Y}(t, 0) + \text{order}(\varepsilon), \tag{7.5.6}$$

where $\hat{Y}(t, 0)$ is the solution of the initial-value problem [see $(7.3.19)_2$, or (7.3.41) with $\nu = 0$, $\alpha_1 = 0$]

$$\frac{d\hat{Y}}{dt} = \left\{ D_0 - \left[C_0 (B_0 C_0)^{-1} (B_0' + B_0 D_0) \right] \right\} \hat{Y} \quad \text{for } t > 0,$$

$$\text{with } \hat{Y} = H(0) \text{ at } t = 0, \tag{7.5.7}$$

where $H = H(t)$ is an $n \times (n - m)$ matrix whose columns span the null space of $B_0(t)$, as in (7.2.6).

A perturbation analysis similar to that of Sections 7.3 and 7.4 can be used to study the behavior of $X_{\pm}(t, \varepsilon)$ and $Y_{\pm}(t, \varepsilon)$ for small ε, and we find (up to inessential scalar multiples)

$$X_{\pm}(t, \varepsilon) = Z(t) + \text{order}(\varepsilon), \tag{7.5.8}$$

$$Y_{\pm}(t, \varepsilon) = \pm C_0(t) Z(t) \Omega(t)^{-1} + \text{order}(\varepsilon),$$

where Z is the $m \times m$ matrix of eigenvectors of $B_0 C_0$ as in (7.2.4).

From (7.5.5)–(7.5.8) we have for the preceding fundamental matrix the result

$$F(t, \varepsilon)$$

$$= \begin{bmatrix} [Z(t) + O(\varepsilon)] e^{\Psi} & [Z(t) + O(\varepsilon)] e^{-\Psi} & O(\varepsilon) \\ \frac{1}{\varepsilon} [C_0 Z \Omega^{-1} + O(\varepsilon)] e^{\Psi} & \frac{1}{\varepsilon} [-C_0 Z \Omega^{-1} + O(\varepsilon)] e^{-\Psi} & \frac{1}{\varepsilon} [\hat{Y}(t, 0) + O(\varepsilon)] \end{bmatrix},$$

$$\tag{7.5.9}$$

with

$$\Psi := \frac{1}{\varepsilon} \int_0^t \Omega$$

where $\hat{Y}(t, 0)$ is the solution of (7.5.7). The estimates of the first two block-columns on the right side of (7.5.9) for the exponential-type solutions correspond to the lowest-order Liouville/Green approximation; see Olver (1961, 1974). Higher-order approximations can be obtained here for $F(t, \varepsilon)$ using standard perturbation techniques.

From (7.5.9) we find for the inverse of F the related result

$$F(t, \varepsilon)^{-1} = \begin{bmatrix} e^{-\Psi}\left[\tfrac{1}{2}Z^{-1} + O(\varepsilon)\right] & \varepsilon e^{-\Psi}\left[\tfrac{1}{2}\Omega^{-1}Z^{-1}B_0 + O(\varepsilon)\right] \\ e^{\Psi}\left[\tfrac{1}{2}Z^{-1} + O(\varepsilon)\right] & \varepsilon e^{\Psi}\left[-\tfrac{1}{2}\Omega^{-1}Z^{-1}B_0 + O(\varepsilon)\right] \\ O(\varepsilon) & \varepsilon\left\{\left[\hat{Y}(t,0)^{+}\right]P(t) + O(\varepsilon)\right\} \end{bmatrix}$$

$$(7.5.10)$$

where $\hat{Y}(t, 0)^{+}$ denotes the Moore/Penrose generalized inverse,

$$\hat{Y}(t,0)^{+} := \left[\hat{Y}(t,0)^{\mathsf{T}}\hat{Y}(t,0)\right]^{-1}\hat{Y}(t,0)^{\mathsf{T}}, \qquad (7.5.11)$$

and $P(t)$ denotes the projection on the null space of $B_0(t)$ along the image of $C_0(t)$, as in (7.1.13).

PART III

Boundary-value problems

8

Linear scalar problems

In this chapter we shall consider certain linear two-point boundary-value problems for a single scalar equation subject to certain restrictions that eliminate consideration of turning points, except in Section 8.5, where examples involving turning points (interior layers) are given. Such linear problems and related nonlinear problems with solutions of boundary-layer type occur in many areas of applications, including chemical-reactor theory, fluid mechanics, elasticity theory, and the physical theory of semiconductors and transistors. A multivariable approach is used here to obtain uniformly valid approximations for solutions to these problems. Error estimates can be obtained by any of several different methods.

8.1 Introduction

Consider the second-order differential equation

$$\varepsilon \frac{d^2 x}{dt^2} + a(t, \varepsilon) \frac{dx}{dt} + b(t, \varepsilon) x = f(t, \varepsilon) \quad \text{for } t \in [t_1, t_2] \quad (8.1.1)$$

on a given interval $[t_1, t_2]$, for smooth data a, b, and f subject to various restrictions to be discussed later, and where the real-valued solution function $x = x(t) = x(t, \varepsilon)$ is also subject to the following separated boundary conditions:

$$\xi_j(\varepsilon) x(t, \varepsilon) + \eta_j(\varepsilon) \frac{dx(t, \varepsilon)}{dt} = \alpha_j(\varepsilon) \quad \text{at } t = t_j \quad \text{for } j = 1, 2.$$

$$(8.1.2)$$

The quantities ξ_j and η_j are assumed to be normalized as

$$\xi_j(\varepsilon)^2 + \eta_j(\varepsilon)^2 = 1 \quad \text{for } j = 1, 2. \quad (8.1.3)$$

263

The special case

$$\xi_j = 1 \quad \text{and} \quad \eta_j = 0 \quad \text{for } j = 1,2 \tag{8.1.4}$$

corresponds to the following boundary conditions of Dirichlet type:

$$x(t_j, \varepsilon) = \alpha_j(\varepsilon) \qquad \text{for } j = 1,2. \tag{8.1.5}$$

The boundary conditions (8.1.2) are separated in the sense that each of the two boundary conditions there, for $j = 1$ and also for $j = 2$, involves boundary values of x and dx/dt at only a single endpoint, either t_1 or t_2. Related problems are considered in Chapter 9 for coupled boundary conditions that mix the boundary values at the two different endpoints. Nonlinear boundary conditions can also be considered; see van Harten (1978*b*).

The problem (8.1.1)–(8.1.2) is a special case of linear boundary-value problems studied by Tschen (1935), Wasow (1941, 1944, 1965, 1976), Latta (1951), Vishik and Lyusternik (1957), Harris (1960, 1962*a*), O'Malley (1968*b*, 1969*b*, 1974*b*), and others. Nonlinear generalizations of this problem have been studied by many authors and will be discussed in Chapter 10.

The Sturm transformation (0.3.4) transforms the differential equation (8.1.1) into the corresponding equation*

$$\varepsilon^2 \frac{d^2 \bar{x}}{dt^2} - c(t, \varepsilon)\bar{x} = \varepsilon \bar{f}(t, \varepsilon) \qquad \text{for } t \in [t_1, t_2], \tag{8.1.6}$$

with

$$\bar{x}(t, \varepsilon) := \left[\exp\left(\frac{1}{2\varepsilon} \int_{t_0}^{t} a \right) \right] x(t, \varepsilon) \quad \text{for } t \in [t_1, t_2], \tag{8.1.7}$$

where t_0 is any fixed number in $[t_1, t_2]$, and with a similar relation giving $\bar{f}$ in terms of f. The function c is defined as

$$c(t, \varepsilon) := \tfrac{1}{4}\left[a(t, \varepsilon)^2 + 2\varepsilon\left(\frac{da(t, \varepsilon)}{dt} - 2b(t, \varepsilon) \right) \right]. \tag{8.1.8}$$

The behavior of solutions of (8.1.6) depends critically on the *sign* and *magnitude* of the coefficient function c. For example, if $f = \bar{f} = 0$, then solutions of the related homogeneous equation

$$\varepsilon^2 \frac{d^2 \bar{x}}{dt^2} - c(t, \varepsilon)\bar{x} = 0 \tag{8.1.9}$$

are expected to be oscillatory or nonoscillatory depending on whether c

---

* The remaining part of this section can be skimmed lightly during a first reading.

is everywhere negative or everywhere positive, whereas solutions may change from oscillatory to nonoscillatory across transition points or turning points, where the values of c change from negative to positive. And, indeed, the *maximum principle* implies that each nontrivial solution of (8.1.9) can have at most one zero, and hence is nonoscillatory, if $c \geq 0$ (see Exercise 8.1.1), whereas the *Sturm comparison theorem* [see Birkhoff and Rota (1960, 1978)] implies that any solution must have a zero in every interval of length $\varepsilon\pi/\kappa$ if $c \leq -\kappa^2 < 0$ for some fixed positive constant κ, so that nontrivial solutions are oscillatory in this latter case. The magnitude of c can be crucial in this latter case of negative c, as illustrated by the example [see Olver (1974, p. 190)]

$$c(t, \varepsilon) := \frac{\varepsilon^2 m(m-1)}{t^2} \quad \text{for } t > 0. \qquad (8.1.10)$$

If m is a fixed constant, with $0 < m < 1$, then c is everywhere negative, and we may expect to have the oscillatory case. However, solutions are in fact nonoscillatory, as is seen from the following expressions for the general solution of (8.1.9)–(8.1.10):

$$\bar{x}(t) = \begin{cases} (A + B \log t)t^{1/2} & \text{if } m = \frac{1}{2}, \\ At^m + Bt^{1-m} & \text{otherwise,} \end{cases} \qquad (8.1.11)$$

where A and B are constants of integration. The Sturm comparison theorem can be applied to (8.1.9)–(8.1.10) on any bounded interval such as $(0, t_2)$, and we conclude that any solution must have a zero in any subinterval of length $\pi t_2 / \sqrt{m(1-m)}$. But, of course, no such subinterval exists!

The following examples illustrate the situation for the Dirichlet problem for (8.1.9) on $[t_1, t_2]$ if the coefficient function c is either everywhere negative or everywhere positive.

Example 8.1.1: Let $c := -1$ for all t. Then the most general solution of (8.1.9) can be given as

$$\bar{x}(t, \varepsilon) = A \sin\frac{t}{\varepsilon} + B \cos\frac{t}{\varepsilon} \qquad (8.1.12)$$

for suitable constants A and B. The Dirichlet boundary conditions $\bar{x}(t_j, \varepsilon) = \bar{\alpha}_j(\varepsilon)$ $(j = 1, 2)$, along with (8.1.12), lead to the system

$$\begin{bmatrix} \sin\dfrac{t_1}{\varepsilon} & \cos\dfrac{t_1}{\varepsilon} \\ \sin\dfrac{t_2}{\varepsilon} & \cos\dfrac{t_2}{\varepsilon} \end{bmatrix} \begin{bmatrix} A \\ B \end{bmatrix} = \begin{bmatrix} \bar{\alpha}_1(\varepsilon) \\ \bar{\alpha}_2(\varepsilon) \end{bmatrix},$$

and this system has a unique solution for A and B if and only if

$$\sin \frac{t_2 - t_1}{\varepsilon} \neq 0. \qquad (8.1.13)$$

If (8.1.13) holds, that is, if $(t_2 - t_1)/\pi\varepsilon$ takes on a nonintegral value, then the given Dirichlet problem is uniquely solvable. However, if

$$\varepsilon = \frac{t_2 - t_1}{\pi k}$$

for some positive integer k, then the Dirichlet problem has no solution unless the boundary values $\bar{\alpha}_1$ and $\bar{\alpha}_2$ satisfy the compatibility condition

$$\bar{\alpha}_1 \cos \frac{t_2}{\varepsilon} - \bar{\alpha}_2 \sin \frac{t_1}{\varepsilon} = 0,$$

and in this latter case the Dirichlet problem has infinitely many solutions. We have here an elementary example of the *Fredholm alternative* in this, the oscillatory, case.

Example 8.1.2: Let $c := +1$ for all t. Then the most general solution of (8.1.9) can be given as

$$\bar{x}(t, \varepsilon) = Ae^{t/\varepsilon} + Be^{-t/\varepsilon},$$

and the boundary conditions of (8.1.5) [and even (8.1.2)] always lead to uniquely determined values for A and B, for all small ε. Hence, the Dirichlet problem always has precisely one solution in this, the nonoscillatory, case.

In the remaining part of this section we prove that the result of Example 8.1.2 is typical in the general nonoscillatory case $c \geq 0$. That is, we show that various boundary-value problems for the equation (8.1.9) always have unique solutions if $c \geq 0$.

The positive parameter ε plays no role in this discussion, and so we may as well put $\varepsilon = 1$ and consider the differential equation

$$\frac{d^2 x}{dt^2} - c(t)x = f(t) \quad \text{for } t \in [t_1, t_2], \qquad (8.1.14)$$

where we have dropped the overbars from x and f. We shall later consider more general boundary conditions of the type (8.1.2), but we begin here with boundary conditions of Dirichlet type, given as

$$x(t_1) = \alpha_1 \quad \text{and} \quad x(t_2) = \alpha_2. \qquad (8.1.15)$$

This Dirichlet problem (8.1.14)–(8.1.15) is to be considered subject to the nonoscillatory condition

$$c(t) \geq 0 \quad \text{for } t \in [t_1, t_2].$$ (8.1.16)

We use an approach of Cochran (1968) that permits a direct *construction* of the solution of (8.1.14)–(8.1.15) in such a way as to give both existence and uniqueness. (Uniqueness follows in this case also directly from the maximum principle. For example, Theorem 0.3.6 can be applied to the difference of two solutions.)

The Dirichlet problem (8.1.14)–(8.1.15) is equivalent to the integral equation

$$x(t) = G_s(t, t_1)\alpha_1 - G_s(t, t_2)\alpha_2 - \int_{t_1}^{t_2} G(t, s)[f(s) + c(s)x(s)]\, ds,$$

(8.1.17)

where $G = G(t, s)$ is the Green function for the Dirichlet problem for the auxiliary operator d^2/dt^2 on $[t_1, t_2]$, given as

$$G(t, s) = \begin{cases} \dfrac{(t_2 - t)(s - t_1)}{t_2 - t_1} & \text{for } s \leq t, \\[2ex] \dfrac{(t_2 - t)(s - t_1)}{t_2 - t_1} + (t - s) & \text{for } s \geq t, \end{cases}$$ (8.1.18)

where the subscript denotes partial differentiation in (8.1.17), $G_s = \partial G(t, s)/\partial s$. The Green function $G(t, s)$ can be characterized here by the conditions $G_{ss}(t, s) = 0$ for $s \neq t$, $G(t, s) = 0$ at $s = t_1$ and at $s = t_2$, and $G_s(t, t-) - G_s(t, t+) = 1$, with $G(t, s)$ continuous everywhere including at $s = t$. See Exercises 8.1.2 and 8.1.3 for an indication of the derivation of such results as (8.1.17).

The integral equation (8.1.17) can be rewritten in the form

$$x(t) = F(t) + Q(x)(t - t_1) + \int_{t_1}^{t} V(t, s)x(s)\, ds, \quad (8.1.19)$$

with

$$F(t) := \frac{(t_2 - t)\alpha_1 + (t - t_1)\alpha_2}{t_2 - t_1} - \int_{t_1}^{t_2} G(t, s)f(s)\, ds,$$

(8.1.20)

$$Q(x) := -\int_{t_1}^{t_2} \frac{t_2 - s}{t_2 - t_1} c(s)x(s)\, ds, \quad (8.1.21)$$

and

$$V(t, s) := (t - s)c(s). \quad (8.1.22)$$

Note that the quantity $Q = Q(x)$ defined by (8.1.21) is a mapping of continuous functions x into the reals. Also, we see directly that the function $F(t)$ defined by (8.1.20) satisfies the specified boundary conditions, $F(t_j) = \alpha_j$ $(j = 1, 2)$.

The integral equation (8.1.17) is a nondegenerate Fredholm equation that is generally difficult to solve, whereas (8.1.19) gives a decomposition of the integral operator there into a sum of a *degenerate* Fredholm part and a *Volterra* part, each of which is easy to invert.

The Volterra part of (8.1.19) can be inverted to give [see (0.2.1) and (0.2.4)]

$$x(t) = F(t) + \int_{t_1}^{t} V^*(t, s) F(s) \, ds$$

$$+ Q(x)\left[t - t_1 + \int_{t_1}^{t} V^*(t, s)(s - t_1) \, ds \right], \qquad (8.1.23)$$

and now t can be put equal to t_2 in (8.1.23), and the boundary condition at t_2 can be used to find that *the Dirichlet problem* (8.1.14)–(8.1.15) *has a solution if and only if*

$$Q(x)\left[t_2 - t_1 + \int_{t_1}^{t_2} V^*(t_2, s)(s - t_1) \, ds \right]$$

$$= - \int_{t_1}^{t_2} V^*(t_2, s) F(s) \, ds. \qquad (8.1.24)$$

We can easily see that this last equation always has a unique solution for $Q(x)$. Indeed, the results (8.1.16) and (8.1.22) imply that the Volterra kernel here is nonnegative, $V(t, s) \geq 0$ for $t_1 \leq s \leq t \leq t_2$, and this implies, with the definition of the resolvent kernel V^*, the result [see (0.2.5)–(0.2.6)]

$$V^*(t, s) \geq 0 \quad \text{for } t_1 \leq s \leq t \leq t_2.$$

Hence,

$$t_2 - t_1 + \int_{t_1}^{t_2} V^*(t_2, s)(s - t_1) \, ds \geq t_2 - t_1 > 0,$$

and so the coefficient of $Q(x)$ is nonzero in (8.1.24). Hence, (8.1.24) can be solved for $Q(x)$ to give

$$Q(x) = - \frac{\displaystyle\int_{t_1}^{t_2} V^*(t_2, s) F(s) \, ds}{t_2 - t_1 + \displaystyle\int_{t_1}^{t_2} V^*(t_2, s)(s - t_1) \, ds}, \qquad (8.1.25)$$

and we have proved the following theorem.

Theorem 8.1.1: *The Dirichlet problem* (8.1.14)–(8.1.15) *has one and only one solution if the nonoscillatory condition* (8.1.16) *holds, and this solution can be given explicitly by* (8.1.23) *and* (8.1.25).

The foregoing proof of Theorem 8.1.1 is equally effective for more general boundary conditions other than Dirichlet conditions. For example, consider these boundary conditions of the type (8.1.2),

$$\xi_j x(t_j) + \eta_j x'(t_j) = \alpha_j \quad \text{for } j = 1, 2, \tag{8.1.26}$$

for specified constants ξ_j, η_j, and α_j $(j = 1, 2)$ subject to the normalization [see (8.1.3)]

$$(\xi_j)^2 + (\eta_j)^2 = 1 \quad \text{for } j = 1, 2. \tag{8.1.27}$$

We again use the Green function for the operator d^2/dt^2, where in the case of the boundary conditions (8.1.26) this Green function $G = G(t, s)$ is characterized by the conditions $G_{ss}(t, s) = 0$ for $s \neq t$, $\xi_j G(t, s) + \eta_j G_s(t, s) = 0$ at $s = t_j$ for $j = 1, 2$, and $G_s(t, t-) - G_s(t, t+) = 1$, with $G(t, s)$ continuous. This Green function can be given in the form

$$G(t, s) = \begin{cases} A(t)s + B(t) & \text{for } s < t, \\ A(t)s + B(t) + t - s & \text{for } s > t, \end{cases} \tag{8.1.28}$$

where the specified boundary conditions imply that the functions A and B must satisfy the linear system

$$\begin{bmatrix} \eta_1 + \xi_1 t_1 & \xi_1 \\ \eta_2 + \xi_2 t_2 & \xi_2 \end{bmatrix} \begin{bmatrix} A(t) \\ B(t) \end{bmatrix} = \begin{bmatrix} 0 \\ \eta_2 + \xi_2(t_2 - t) \end{bmatrix}. \tag{8.1.29}$$

We make the assumption

$$\xi_2 \eta_1 - \xi_1 \eta_2 - \xi_1 \xi_2(t_2 - t_1) \neq 0, \tag{8.1.30}$$

so that the system (8.1.29) is nonsingular, yielding a unique determination of $A(t)$ and $B(t)$.

The condition (8.1.30) eliminates the Neumann case $\xi_1 = \xi_2 = 0$ in (8.1.26), and, indeed, the Neumann problem is underdetermined for the operator d^2/dt^2, because any constant can be added to a solution. If the coefficient c is uniformly positive in (8.1.14), then (8.1.30) can be relaxed, as indicated in Exercises 8.1.3 and 8.1.4. In this latter case the Neumann problem has a unique solution for (8.1.14); see Exercise 8.1.4 along with (8.2.55) and (8.3.42).

The boundary-value problem (8.1.14) and (8.1.26) subject to the condition (8.1.30) is equivalent to the integral equation [compare with (8.1.19);

see Exercise 8.1.3 for an indication of the derivation here]

$$x(t) = F(t) + Q(x)[\xi_1(t - t_1) - \eta_1] + \int_{t_1}^{t} V(t, s)x(s)\, ds,$$

$$(8.1.31)$$

with

$$F(t): = -[\eta_1 G(t, t_1) - \xi_1 G_s(t, t_1)]\alpha_1$$
$$+ [\eta_2 G(t, t_2) - \xi_2 G_s(t, t_2)]\alpha_2 - \int_{t_1}^{t_2} G(t, s)f(s)\, ds,$$

$$(8.1.32)$$

$$Q(x): = -\int_{t_1}^{t_2}\left[\frac{\eta_2 + \xi_2(t_2 - s)}{\xi_1\eta_2 - \xi_2\eta_1 + \xi_1\xi_2(t_2 - t_1)}\right]c(s)x(s)\, ds,$$

$$(8.1.33)$$

and

$$V(t, s): = (t - s)c(s). \qquad (8.1.34)$$

We again obtain a unique solution for x for the given boundary-value problem in the nonoscillatory case $c(t) \geq 0$, and we obtain an explicit formula for this solution. See Exercise 8.1.2 for the details in the particular example

$$\xi_1 = 1, \qquad \eta_1 = 0,$$
$$\xi_2 = 0, \qquad \eta_2 = 1.$$

In the following two sections of this chapter, equation (8.1.1) is considered in certain nonoscillatory situations for which (8.1.16) holds, with [see (8.1.8)]

$$a(t, \varepsilon)^2 + 2\varepsilon\left[\frac{da(t, \varepsilon)}{dt} - 2b(t, \varepsilon)\right] \geq 0 \quad \text{for } t \in [t_1, t_2] \quad (8.1.35)$$

for all small $\varepsilon > 0$. Specifically, we shall consider the case

$$a(t, \varepsilon)^2 \geq \kappa^2 > 0, \qquad (8.1.36)$$

and also the case

$$a(t, \varepsilon) \equiv 0 \quad \text{with } b(t, \varepsilon) \leq -\kappa < 0, \qquad (8.1.37)$$

for a fixed positive constant κ, where, in either case, (8.1.36) or (8.1.37) is required to hold uniformly for all $t_1 \leq t \leq t_2$ and for all small $\varepsilon \geq 0$. The data are required to be smooth, and we see directly that either (8.1.36) or (8.1.37) implies the validity of the nonoscillatory condition (8.1.35).

We emphasize the Dirichlet boundary conditions (8.1.5) in the next two sections, although other conditions of the type (8.1.2) are also considered, and the same techniques suffice. More general boundary conditions are also included in Chapters 9 and 10.

The singularly perturbed eigenvalue problem is not considered here; see Moser (1955), Harris (1961), Boyce and Handelman (1961), Handelman and Keller (1962), Miranker (1963), Handelman, Keller, and O'Malley (1968), and O'Malley (1974*a*). Similarly, singularly perturbed boundary-value problems for partial differential equations are not considered her\, see Levinson (1950), Vishik and Lyusternik (1957), Tang (1972), Fife (1973*a*, 1973*b*), Eckhaus (1973, 1979), Tang and Fife (1975), van Harten (1978*a*), and Eckhaus and de Jager (1982).

Exercises

Exercise 8.1.1: Prove that any nontrivial solution of (8.1.9) can have at most one zero on $[t_1, t_2]$, and hence is nonoscillatory, if the coefficient function c is nonnegative-valued, $c(t, \varepsilon) \geq 0$ for $t \in [t_1, t_2]$. *Hint*: Theorem 0.3.5 implies the inequality

$$|\bar{x}(t, \varepsilon)| \leq \max\{|\bar{x}(t_3, \varepsilon)|, |\bar{x}(t_4, \varepsilon)|\} \quad \text{for } t \in [t_3, t_4] \subset [t_1, t_2],$$

so that the existence of two distinct zeros at, say, t_3 and t_4, implies $\bar{x} \equiv 0$ on $[t_3, t_4]$, from which we conclude also that $\bar{x} \equiv 0$ everywhere on $[t_1, t_2]$.

Exercise 8.1.2: **(a)** Derive the Green function for d^2/dt^2 on $[t_1, t_2]$ for the boundary conditions [see (8.1.26)–(8.1.30)]

$$x(t_1) = \alpha_1 \quad \text{and} \quad \frac{dx(t_2)}{dt} = \alpha_2.$$

(b) Let $G(t, s)$ be the Green function from part **(a)**, and derive the representation

$$x(t) = G_s(t, t_1)\alpha_1 + G(t, t_2)\alpha_2$$
$$- \int_{t_1}^{t_2} G(t, s)f(s)\, ds - \int_{t_1}^{t_2} G(t, s)c(s)x(s)\, ds \quad (8.1.38)$$

for any solution of the problem

$$\frac{d^2x}{dt^2} - c(t)x = f(t) \quad \text{for } t \in [t_1, t_2],$$

$$x(t_1) = \alpha_1 \quad \text{and} \quad \frac{dx(t_2)}{dt} = \alpha_2. \tag{8.1.39}$$

Hint: Begin with the relation (note that $G_{ss} = 0$)

$$\int_{t_1}^{t_2} G(t,s)[f(s) + c(s)x(s)] \, ds$$

$$= \int_{t_1}^{t_2} G(t,s) \frac{d^2x(s)}{ds^2} \, ds$$

$$= \int_{t_1}^{t} \left[G(t,s) \frac{d^2x(s)}{ds^2} - \frac{\partial^2 G(t,s)}{\partial s^2} x(s) \right] ds$$

$$+ \int_{t}^{t_2} \left[G(t,s) \frac{d^2x(s)}{ds^2} - \frac{\partial^2 G(t,s)}{\partial s^2} x(s) \right] ds$$

for any solution of the differential equation of (8.1.39), and then evaluate the integrals on the right side here by parts, using the appropriate boundary conditions for x and G.

(c) Show that the integral equation (8.1.38) can be rewritten as

$$x(t) = F(t) + Q(x)(t - t_1) + \int_{t_1}^{t} V(t,s)x(s) \, ds, \quad (8.1.40)$$

with

$$F(t): = \alpha_1 + (t - t_1)\alpha_2 - \int_{t_1}^{t_2} G(t,s)f(s) \, ds,$$

$$Q(x): = -\int_{t_1}^{t_2} c(s)x(s) \, ds,$$

and

$$V(t,s): = (t - s)c(s).$$

(d) For any solution x of (8.1.39), derive the results

$$x(t) = F(t) + \int_{t_1}^{t} V^*(t,s)F(s) \, ds$$

$$+ Q(x) \left[t - t_1 + \int_{t_1}^{t} V^*(t,s)(s - t_1) \, ds \right] \quad (8.1.41)$$

and

$$x'(t) = F'(t) + \int_{t_1}^{t} \frac{\partial V^*(t,s)}{\partial t} F(s) \, ds$$

$$+ Q(x) \left[1 + \int_{t_1}^{t} \frac{\partial V^*(t,s)}{\partial t} (s - t_1) \, ds \right]. \quad (8.1.42)$$

Hint: Note that $V^*(t,s) = 0$ at $s = t$.

(e) Show that the boundary-value problem (8.1.39) has a solution if and only if

$$Q(x)\left[1 + \int_{t_1}^{t_2} \frac{\partial V^*(t_2, s)}{\partial t}(s - t_1)\, ds\right] = -\int_{t_1}^{t_2} \frac{\partial V^*(t_2, s)}{\partial t} F(s)\, ds,$$

(8.1.43)

and show that this last equation always has a unique solution for $Q(x)$ in the nonoscillatory case $c \geq 0$. *Hint*: For the last result, we use the inequality $\partial V^*(t, s)/\partial t \geq 0$ (for $t \geq s$), which follows directly from the definition of V^* in the case $c \geq 0$. It follows that the problem (8.1.39) has precisely one solution in the case $c \geq 0$, and this solution is given by (8.1.41), with $Q(x)$ determined from (8.1.43).

Exercise 8.1.3: Let κ be a fixed positive constant, and let $\mathcal{L}$ denote the differential operator

$$\mathcal{L} := \frac{d^2}{dt^2} - \kappa^2.$$

(8.1.44)

(a) Derive the Green function $G = G(t, s)$ for $\mathcal{L}$ on $[t_1, t_2]$ for the boundary conditions

$$\xi_j x(t_j) + \eta_j x'(t_j) = \alpha_j \quad \text{for } j = 1, 2,$$

(8.1.45)

where ξ_j, η_j, and α_j are given constants subject to the normalization

$$(\xi_j)^2 + (\eta_j)^2 = 1 \quad \text{for } j = 1, 2,$$

(8.1.46)

and subject also to the condition

$$(\eta_1 \kappa + \xi_1)(\eta_2 \kappa - \xi_2) \neq (\eta_2 \kappa + \xi_2)(\eta_1 \kappa - \xi_1)e^{2\kappa(t_2 - t_1)}.$$

(8.1.47)

In particular, show that G can be given in the form

$$G(t, s) = \begin{cases} A(t)e^{\kappa(s-t)} + B(t)e^{-\kappa(s-t)} & \text{for } s < t, \\ A(t)e^{\kappa(s-t)} + B(t)e^{-\kappa(s-t)} + \dfrac{\sinh \kappa(t - s)}{\kappa} & \text{for } s > t, \end{cases}$$

(8.1.48)

for suitable functions A and B that are determined by the boundary conditions. *Hint*: $G(t, s)$ must be continuous and satisfy the conditions $G_{ss}(t, s) - \kappa^2 G(t, s) = 0$ for $s \neq t$, $\xi_j G(t, s) + \eta_j G_s(t, s) = 0$ for $s = t_j$ $(j = 1, 2)$, and $G_s(t, t-) - G_s(t, t+) = 1$.

(b) Show that the solution of $\mathcal{L}x = f(t)$ on $[t_1, t_2]$ subject to the boundary conditions (8.1.45)–(8.1.47) can be given as

$$x(t) = \left[\xi_1 G_s(t, t_1) - \eta_1 G(t, t_1)\right]\alpha_1 - \left[\xi_2 G_s(t, t_2) - \eta_2 G(t, t_2)\right]\alpha_2$$
$$- \int_{t_1}^{t_2} G(t, s) f(s)\, ds. \tag{8.1.49}$$

Hint: The terms involving the boundary values here are obtained, for example, as

$$G_s(t, t_1) x(t_1) - G(t, t_1) x'(t_1)$$
$$= \left(\xi_1^2 + \eta_1^2\right)\left[G_s(t, t_1) x(t_1) - G(t, t_1) x'(t_1)\right]$$
$$= \left[\xi_1 G_s(t, t_1) - \eta_1 G(t, t_1)\right]\alpha_1$$

if $G(t, s)$ satisfies the appropriate boundary relation at $s = t_1$. [The condition (8.1.47) will hold automatically for all small $\varepsilon > 0$ in the singularly perturbed problems considered later. Note that (8.1.47) includes the Neumann problem $\xi_j = 0$, $\eta_j = 1$ $(j = 1, 2)$ if $\kappa > 0$. Note also that this condition (8.1.47) reduces to the previous condition (8.1.30) in the limit $\kappa \to 0$.]

Exercise 8.1.4: Let the piecewise continuous function $c = c(t)$ be positive-valued, with

$$c(t) \geq \kappa^2 > 0 \quad \text{for } t \in [t_1, t_2] \tag{8.1.50}$$

for some positive constant $\kappa > 0$, and let $G = G(t, s)$ be the Green function for the operator $\mathcal{L}$ of (8.1.44) for the Neumann problem, with $\xi_j = 0$ and $\eta_j = 1$ $(j = 1, 2)$ in (8.1.45).

(a) Show that any solution x of the Neumann problem

$$\frac{d^2 x}{dt^2} - c(t)x = f(t) \quad \text{for } t \in [t_1, t_2],$$

$$\frac{dx}{dt} = \alpha_j \quad \text{at } t = t_j, \qquad \text{for } j = 1, 2, \tag{8.1.51}$$

must satisfy the integral equation

$$x(t) = F(t) + P(t)Q(x) + \int_{t_1}^{t} V(t, s) x(s)\, ds, \tag{8.1.52}$$

where

$$V(t, s) := \left[c(s) - \kappa^2\right]\frac{\sinh \kappa(t - s)}{\kappa}, \tag{8.1.53}$$

$$F(t) := \frac{-\alpha_1 \cdot \cosh \kappa(t_2 - t) + \alpha_2 \cdot \cosh \kappa(t - t_1)}{\kappa \sinh \kappa(t_2 - t_1)}$$
$$- \int_{t_1}^{t_2} G(t, s) f(s)\, ds, \tag{8.1.54}$$

and

$$P(t) := -\frac{\cosh \kappa (t - t_1)}{\kappa \sinh \kappa (t_2 - t_1)}, \tag{8.1.55}$$

with $Q(x) := \int_{t_1}^{t_2} [\cosh \kappa (t_2 - s)][c(s) - \kappa^2] x(s) \, ds$. *Hint:* Write the differential equation of (8.1.51) in terms of the auxiliary operator $\mathscr{L}$ of (8.1.44) as

$$\mathscr{L}x = f(t) + [c(t) - \kappa^2] x(t),$$

and apply the result (8.1.49), with f replaced there by $f + (c - \kappa^2) x$.

(b) Use the method of Cochran to prove that the Neumann problem (8.1.50)–(8.1.51) has one and only one solution, and obtain a representation for this solution in terms of the data. *Hint:* The Volterra kernel $V(t, s)$ of (8.1.53) is nonnegative along with its t-derivatives, and then we find also that the resolvent kernel $V^*(t, s)$ is nonnegative along with $\partial V^*(t, s)/\partial t$, for $s \le t$. The function P of (8.1.55) is nonpositive, and then we find that the equation

$$Q(x) \left[1 - \int_{t_1}^{t_2} \frac{\partial V^*(t_2, s)}{\partial t} P(s) \, ds \right] = \int_{t_1}^{t_2} \frac{\partial V^*(t_2, s)}{\partial t} F(s) \, ds \tag{8.1.56}$$

has a unique solution for the quantity $Q(x)$, from which the stated result can be shown to follow.

8.2 A problem with boundary layers at both endpoints

In this section we consider the previous differential equation (8.1.1) in the special case (8.1.37), so that the equation becomes

$$\varepsilon^2 \frac{d^2 x}{dt^2} + b(t, \varepsilon) x = f(t, \varepsilon) \quad \text{for } t \in [t_1, t_2], \tag{8.2.1}$$

where the coefficient function b is uniformly negative,

$$b(t, \varepsilon) \le -\kappa < 0 \quad \text{for } t \in [t_1, t_2], \text{ all small } \varepsilon \ge 0, \tag{8.2.2}$$

for a fixed positive constant κ, and where the parameter ε in (8.1.1) has been replaced here with ε^2 for notational convenience (as discussed further in Section 8.4). The given functions b and f in (8.2.1) are assumed to have asymptotic power-series expansions of the form

$$\begin{bmatrix} b(t, \varepsilon) \\ f(t, \varepsilon) \end{bmatrix} \sim \sum_{k=0}^{\infty} \begin{bmatrix} b_k(t) \\ f_k(t) \end{bmatrix} \varepsilon^k \tag{8.2.3}$$

for suitable smooth functions $b_k = b_k(t)$ and $f_k = f_k(t)$. Condition (8.2.2) can be replaced with the equivalent condition

$$b_0(t) < 0 \quad \text{for } t \in [t_1, t_2]. \tag{8.2.4}$$

Vasil'eva (1972) has shown that condition (8.2.2) can be relaxed inside boundary layers, where b can be permitted to take on positive values. Such boundary-layer "impulses" are important for certain related nonlinear problems, such as those discussed in Sections 10.2 and 10.3. As usual, expansions such as (8.2.3) need only hold to some specified finite order, for a partial sum with suitable remainder on the right side of (8.2.3), but this is not of primary interest here.

The techniques used here are effective for boundary conditions of the type (8.1.2) and more general boundary conditions as well, but for the most part, in this section we consider boundary conditions of Dirichlet type,

$$x(t_1, \varepsilon) = \alpha(\varepsilon) \quad \text{and} \quad x(t_2, \varepsilon) = \beta(\varepsilon), \tag{8.2.5}$$

where the boundary values $\alpha(\varepsilon), \beta(\varepsilon)$ are assumed to have asymptotic expansions of the form

$$\begin{bmatrix} \alpha(\varepsilon) \\ \beta(\varepsilon) \end{bmatrix} \sim \sum_{k=0}^{\infty} \begin{bmatrix} \alpha_k \\ \beta_k \end{bmatrix} \varepsilon^k \tag{8.2.6}$$

for suitable constants α_k and β_k.

Example 8.2.1 (Carrier and Pearson 1968, p. 193; Carrier 1970, pp. 176–7): The problem

$$\varepsilon^2 \frac{d^2x}{dt^2} - (2 - t^2)x = -1 \quad \text{for } -1 \le t \le 1,$$

$$x = 0 \text{ at the endpoints } t = -1, \qquad t = +1, \tag{8.2.7}$$

arises as a mathematical model in a certain thermodynamic-chemical process. The coefficient b of (8.2.1) is given here as

$$b(t, \varepsilon) \equiv b_0(t) := -(2 - t^2) \quad \text{for } -1 \le t \le 1,$$

with all other $b_k = 0$ in (8.2.3) for $k \ge 1$. The condition (8.2.2) is satisfied with $\kappa = 1$.

Theorem 8.1.1 guarantees that the Dirichlet problem for (8.2.1) has precisely one solution. The intention here is to obtain uniformly valid and easily interpretable approximations to the solution for small values of the positive parameter ε.

The reduced equation obtained by putting $\varepsilon = 0$ in the differential equation (8.2.1) is simply

$$b_0(t) X = f_0(t),$$

with unique solution

$$X = \frac{f_0(t)}{b_0(t)} \quad \text{for } t \in [t_1, t_2]. \tag{8.2.8}$$

In general, this "reduced solution" (8.2.8) satisfies neither of the specified boundary conditions of (8.2.5), and so the exact solution x of the original boundary-value problem must be expected to exhibit boundary-layer behavior at both endpoints.

The problem (8.2.1) and (8.2.5) can be easily solved in the special case in which $b(t, \varepsilon)$ reduces to a given *constant* function (independent of t), and in this special case we find that the solution $x(t, \varepsilon)$ depends in an essential way on the three quantities t, $(t - t_1)/\varepsilon$, and $(t_2 - t)/\varepsilon$ (see Exercise 8.2.1). The variable $(t - t_1)/\varepsilon$ is important in describing the behavior of $x(t, \varepsilon)$ near $t = t_1$, whereas $(t_2 - t)/\varepsilon$ is important near $t = t_2$.

In the general case in which $b(t, \varepsilon)$ is not constant, we shall use a three-variable approach in terms of the variables t, σ and τ, where the new variables σ and τ are to be given in the forms

$$\sigma := \frac{h(t)}{\varepsilon} \quad \text{and} \quad \tau := \frac{k(t)}{\varepsilon} \tag{8.2.9}$$

for suitable monotonic functions h and k that are to vanish respectively at $t = t_1$ and at $t = t_2$,

$$h(t_1) = 0 \quad \text{and} \quad k(t_2) = 0. \tag{8.2.10}$$

The variable σ will be important near the left boundary, and τ will be important near the right boundary. As mentioned earlier in Sections 5.2 and 5.4, there is a certain flexibility in the choice of such boundary-layer variables. For example, we could at the outset simply take

$$h(t) := t - t_1 \quad \text{and} \quad k(t) := t_2 - t,$$

but slightly different choices may yield quantitatively more accurate approximations (inside boundary layers) for such linear problems as considered here.

Hence, an expansion is sought for the solution x in the form

$$x(t, \varepsilon) \sim \sum_{k=0}^{\infty} x_k(t, \sigma, \tau) \varepsilon^k \tag{8.2.11}$$

for suitable functions x_k depending on the three variables t, σ, and τ, where the functions x_k are evaluated in (8.2.11) at $\sigma = h(t)/\varepsilon$ and $\tau = k(t)/\varepsilon$ in accordance with (8.2.9).

The expansions (8.2.3) and (8.2.11) are inserted into the differential equation (8.2.1), and we use the chain rule of differentiation in the form

$$\frac{d}{dt} = \frac{\partial}{\partial t} + \frac{h'(t)}{\varepsilon}\frac{\partial}{\partial \sigma} + \frac{k'(t)}{\varepsilon}\frac{\partial}{\partial \tau},$$

along with the usual procedure of equating coefficients of like powers of ε, from which we find the relation

$$\left\{\left[h'(t)\frac{\partial}{\partial \sigma} + k'(t)\frac{\partial}{\partial \tau}\right]^2 + b_0(t)\right\}x_k(t,\sigma,\tau)$$

$$= f_k(t) - x_{k-2,tt} - 2\left[h'(t)\frac{\partial}{\partial \sigma} + k'(t)\frac{\partial}{\partial \tau}\right]x_{k-1,t}$$

$$-\left[h''(t)\frac{\partial}{\partial \sigma} + k''(t)\frac{\partial}{\partial \tau}\right]x_{k-1} - \sum_{n=1}^{k} b_n(t)x_{k-n}(t,\sigma,\tau)$$

$$\tag{8.2.12}_k$$

for $k = 0, 1, 2, \ldots$, where x_k is set equal to zero here for negative k, and where $x_{k,t}$ denotes the partial derivative $x_{k,t} = \partial x_k(t,\sigma,\tau)/\partial t$. The last term on the right side of (8.2.12) involving the summation is put equal to zero in the case $k = 0$.

The equations of (8.2.12) simplify somewhat with the choices

$$h'(t) = k'(t) = \sqrt{-b_0(t)},\tag{8.2.13}$$

and so we impose these conditions (8.2.13), which, with (8.2.10), give

$$h(t) = \int_{t_1}^{t}\sqrt{-b_0(s)}\,ds \quad\text{and}\quad k(t) = \int_{t_2}^{t}\sqrt{-b_0(s)}\,ds. \tag{8.2.14}$$

These last results can be used to rewrite (8.2.12) as

$$b_0(t)\left[\left(\frac{\partial}{\partial \sigma} + \frac{\partial}{\partial \tau}\right)^2 - 1\right]x_k$$

$$= -f_k(t) + x_{k-2,tt} + \sum_{n=1}^{k} b_n(t)x_{k-n}$$

$$+\left(\frac{\partial}{\partial \sigma} + \frac{\partial}{\partial \tau}\right)\left(2\sqrt{-b_0(t)}\,x_{k-1,t} + \frac{d\sqrt{-b_0(t)}}{dt}x_{k-1}\right)$$

$$\tag{8.2.15}_k$$

for $k = 0, 1, 2, \ldots$, with the same remarks as follow (8.2.12).

This equation (8.2.15)$_k$ can be regarded as a partial differential equation for $x_k(t,\sigma,\tau)$ as function of σ and τ, for each fixed t. Actually, the change of variables

$$p = \frac{\sigma + \tau}{2} \left.\right\} \quad \text{with} \begin{cases} \sigma = p + q \\ \tau = p - q, \end{cases} \qquad (8.2.16)$$
$$q = \frac{\sigma - \tau}{2}$$

reduces $(8.2.15)_k$ to an ordinary differential equation in terms of the independent variable p, for each fixed t and q. The resulting differential equation can be solved, up to certain "constants" of integration that still depend on t and $q = (\sigma - \tau)/2$. In going recursively from $(8.2.15)_{k-1}$ to $(8.2.15)_k$, certain spurious secular terms appear that can be eliminated by suitable choices of these "constants." The boundary conditions from $(8.2.5)$–$(8.2.6)$ are also imposed, and altogether the procedure leads to a determination of the functions x_k in the form [see Smith (1975a) for an indication of the details]

$$x_k(t, \sigma, \tau) = X_k(t) + e^{-\sigma}(^*X_k(t)) + e^{\tau}X_k^*(t) \qquad (8.2.17)_k$$

for certain functions X_k, $^*X_k$, and X_k^* depending on t alone, where these functions are determined by the respective conditions

$$b_0(t) X_k(t) = f_k(t) - \frac{d^2 X_{k-2}(t)}{dt^2} - \sum_{n=1}^{k} b_n(t) X_{k-n}(t), \qquad (8.2.18)_k$$

$$2[-b_0(t)]^{1/4} \frac{d}{dt} \left\{ [-b_0(t)]^{1/4} {}^*X_k(t) \right\} - b_1(t) {}^*X_k(t)$$
$$= \frac{d^2(^*X_{k-1}(t))}{dt^2} + \sum_{n=1}^{k} b_{n+1}(t) {}^*X_{k-n}(t), \qquad (8.2.19)_k$$

and

$$2[-b_0(t)]^{1/4} \frac{d}{dt} \left\{ [-b_0(t)]^{1/4} X_k^*(t) \right\} + b_1(t) X_k^*(t)$$
$$= -\frac{d^2 X_{k-1}^*(t)}{dt^2} - \sum_{n=1}^{k} b_{n+1}(t) X_{k-n}^*(t) \qquad (8.2.20)_k$$

for $k = 0, 1, 2, \ldots$, along with the boundary conditions

$$X_k(t_1) + {}^*X_k(t_1) = \alpha_k \qquad (8.2.21)_k$$

and

$$X_k(t_2) + X_k^*(t_2) = \beta_k \qquad (8.2.22)_k$$

for $k = 0, 1, 2, \ldots$. The quantities X_k, $^*X_k$, and X_k^* are understood to be identically zero for negative k in $(8.2.18)$–$(8.2.20)$.

The derivation of $(8.2.17)$–$(8.2.22)$ described in Smith (1975a) parallels that of O'Malley (1968a), Erdélyi (1968b), and Searl (1971) for a related problem; see also Yarmish (1975). An alternative approach is to solve the original boundary-value problem explicitly in the special case in which b is a given constant (independent of t), in which case we see that the resulting solution function is consistent with (8.2.9), (8.2.10), (8.2.11), and (8.2.17). Hence, in the case of a more general, nonconstant function b, we can simply use (8.2.17) as an *ansatz* for the form of x_k in (8.2.11), and we are led directly to the conditions (8.2.18)–(8.2.22) (see Exercise 8.2.2). The ansatz is justified, after the fact, by suitable error estimates, as given in Theorem 8.2.1.

The conditions (8.2.18)–(8.2.22) can be used to determine the functions X_k, $*X_k$, and X_k^* recursively for $k = 0, 1, \ldots$. For example, in the case $k = 0$, we find directly

$$X_0(t) = \frac{f_0(t)}{b_0(t)}, \tag{8.2.23}$$

$$*X_0(t) = \left[\frac{b_0(t_1)}{b_0(t)}\right]^{1/4}\left[\alpha_0 - \frac{f_0(t_1)}{b_0(t_1)}\right]\exp\left(\frac{1}{2}\int_{t_1}^t \frac{b_1}{\sqrt{-b_0}}\right), \tag{8.2.24}$$

and

$$X_0^*(t) = \left[\frac{b_0(t_2)}{b_0(t)}\right]^{1/4}\left[\beta_0 - \frac{f_0(t_2)}{b_0(t_2)}\right]\exp\left(\frac{1}{2}\int_t^{t_2} \frac{b_1}{\sqrt{-b_0}}\right). \tag{8.2.25}$$

Then, after X_j, $*X_j$, and X_j^* are determined for $j = 0, 1, \ldots, k - 1$, we find X_k directly from (8.2.18)$_k$, and the first-order differential equations (8.2.19)$_k$ and (8.2.20)$_k$ can be integrated to give $*X_k$ and X_k^* up to constants of integration that are fixed by the conditions (8.2.21)$_k$ and (8.2.22)$_k$. Note that the outer functions X_k are determined by (8.2.18) independent of the boundary-layer correction terms. Note also that all odd-indexed X_k vanish, with $X_{2k+1} \equiv 0$ for all k, if the data functions $b(t, \varepsilon)$ and $f(t, \varepsilon)$ have expansions as in (8.2.3) that *proceed only in even powers of* ε (with $b_{2k+1} \equiv f_{2k+1} \equiv 0$ for all k). This latter result does not hold for the boundary-layer corrections, which in general involve both even and odd powers of ε.

In order to check the validity of the foregoing procedure, we truncate the resulting expansion (8.2.11) and let $x^N(t, \varepsilon)$ denote the partial sum, given, with (8.2.9), (8.2.14), and (8.2.17), as

$$x^N(t, \varepsilon) := \sum_{k=0}^N X_k(t)\varepsilon^k + \sum_{k=0}^N \left[*X_k(t)\exp\left(-\frac{1}{\varepsilon}\int_{t_1}^t \sqrt{-b_0}\right)\right.$$
$$\left. + X_k^*(t)\exp\left(-\frac{1}{\varepsilon}\int_t^{t_2} \sqrt{-b_0}\right)\right]\varepsilon^k. \tag{8.2.26}$$

We shall now check that this x^N satisfies the original boundary-value problem approximately, with small residuals, from which we shall find that x^N does in fact provide a good approximation to the exact solution function x.

We find directly from (8.2.6), (8.2.21), (8.2.22), and (8.2.26) the boundary relations

$$x^N(t_1, \varepsilon) = \alpha(\varepsilon) - \phi_1(\varepsilon),$$
$$x^N(t_2, \varepsilon) = \beta(\varepsilon) - \phi_2(\varepsilon),$$
(8.2.27)

for suitable residuals ϕ_1 and ϕ_2 given as

$$\phi_1(\varepsilon) := \alpha(\varepsilon) - \sum_{k=0}^{N} \alpha_k \varepsilon^k - e^{-\kappa/\varepsilon} \sum_{k=0}^{N} X_k^*(t_1)\varepsilon^k,$$
$$\phi_2(\varepsilon) := \beta(\varepsilon) - \sum_{k=0}^{N} \beta_k \varepsilon^k - e^{-\kappa/\varepsilon} \sum_{k=0}^{N} {}^*X_k(t_2)\varepsilon^k,$$
(8.2.28)

where the positive constant κ is defined as

$$\kappa := \int_{t_1}^{t_2} \sqrt{-b_0(t)} \, dt.$$
(8.2.29)

Similarly, using (8.2.18)–(8.2.20) and (8.2.26), we find for x^N the differential equation

$$\varepsilon^2 \frac{d^2 x^N}{dt^2} + b(t, \varepsilon)x^N = f(t, \varepsilon) - \rho(t, \varepsilon),$$
(8.2.30)

where the residual $\rho(t, \varepsilon)$ can be given as

$$\rho(t, \varepsilon) = \rho_1(t, \varepsilon) + \rho_2(t, \varepsilon) + \rho_3(t, \varepsilon),$$
(8.2.31)

with

$$\rho_1(t, \varepsilon) = f(t, \varepsilon) - \sum_{k=0}^{N} f_k(t)\varepsilon^k - b(t, \varepsilon) \sum_{k=0}^{N} X_k(t)\varepsilon^k$$
$$+ \sum_{k=0}^{N} \left[\sum_{n=0}^{k} b_n(t) X_{k-n}(t) \right]\varepsilon^k - \left(\frac{d^2 X_{N-1}}{dt^2} + \varepsilon \frac{d^2 X_N}{dt^2} \right)\varepsilon^{N+1},$$
(8.2.32)

$$\rho_2(t, \varepsilon) = -\exp\left(-\frac{1}{\varepsilon} \int_{t_1}^{t} \sqrt{-b_0} \right)$$
$$\times \left\{ b(t, \varepsilon) \sum_{k=0}^{N} {}^*X_k(t)\varepsilon^k - \sum_{k=0}^{N+1} \left(\sum_{n=0}^{k} b_n {}^*X_{k-n} \right)\varepsilon^k \right.$$
$$\left. + \left[b_0(t) {}^*X_{N+1} + \varepsilon \frac{d^2 {}^*X_N}{dt^2} \right]\varepsilon^{N+1} \right\},$$
(8.2.33)

and

$$\rho_3(t, \varepsilon) = -\exp\left(-\frac{1}{\varepsilon}\int_t^{t_2}\sqrt{-b_0}\right)$$

$$\times\left\{b(t, \varepsilon)\sum_{k=0}^{N}X_k^*(t)\varepsilon^k - \sum_{k=0}^{N+1}\left(\sum_{n=0}^{k}b_nX_{k-n}^*\right)\varepsilon^k\right.$$

$$\left. + \left[b_0(t)X_{N+1}^* + \varepsilon\frac{d^2X_N^*}{dt^2}\right]\varepsilon^{N+1}\right\}. \qquad (8.2.34)$$

The remainder R_N, defined as

$$R_N(t, \varepsilon) := x(t, \varepsilon) - x^N(t, \varepsilon), \qquad (8.2.35)$$

is now found to satisfy the boundary-value problem

$$\varepsilon^2R_N'' + b(t, \varepsilon)R_N = \rho(t, \varepsilon) \quad \text{for } t_1 < t < t_2,$$

$$R_N(t_1, \varepsilon) = \phi_1(\varepsilon) \quad \text{and} \quad R_N(t_2, \varepsilon) = \phi_2(\varepsilon), \quad (8.2.36)$$

with the residuals ϕ_1, ϕ_2, and ρ given earlier. Hence, we are led to the following theorem.

Theorem 8.2.1: *Let the data b, f, α, β possess asymptotic expansions of the form (8.2.3) and (8.2.6), and assume that the coefficient functions b_k, f_k of (8.2.3) are of class $C^{N-k+2}[t_1, t_2]$ for $k = 0, 1, \ldots, N + 1$. Assume also that the nonoscillatory condition (8.2.4) holds. Let $x(t, \varepsilon)$ be the exact solution of the Dirichlet problem (8.2.1) and (8.2.5), and let $x^N(t, \varepsilon)$ be defined by (8.2.26) in terms of the functions $X_k(t)$, $^*X_k(t)$, and $X_k^*(t)$ constructed earlier, for $k = 0, 1, \ldots, N$. Then there are positive constants D_N and ε_1 such that*

$$\left|x(t, \varepsilon) - x^N(t, \varepsilon)\right| \leq D_N\varepsilon^{N+1} \quad \text{for } t \in [t_1, t_2], 0 < \varepsilon \leq \varepsilon_1.$$

$$(8.2.37)$$

Moreover, if the data b, f, α, β have expansions of the form (8.2.3) and (8.2.6) proceeding in even powers of ε, with $b_{2k+1} = f_{2k+1} = \alpha_{2k+1} = \beta_{2k+1} = 0$ for $k = 0, 1, \ldots, N$, then there are positive constants E_{2N} and ε_1 such that

$$\left|x(t, \varepsilon) - x^{2N}(t, \varepsilon)\right| \leq E_{2N}\varepsilon^{2N+2} \quad \text{for } t \in [t_1, t_2], 0 < \varepsilon \leq \varepsilon_1,$$

$$(8.2.38)$$

and in this latter case $X_{2k+1} \equiv 0$ for $k = 0, 1, \ldots, N$.

Proof: The boundary residuals ϕ_j of (8.2.28)–(8.2.29) are seen to satisfy

$$|\phi_1(\varepsilon)|, |\phi_2(\varepsilon)| \leq \text{const. } \varepsilon^{N+1} \qquad (8.2.39)$$

for all small $\varepsilon > 0$, and the residuals ρ_j of (8.2.31)–(8.2.34) are seen to satisfy

$$|\rho_2(t,\varepsilon)|, |\rho_3(t,\varepsilon)| \leq \text{const. } \varepsilon^{N+2},$$
$$|\rho_1(t,\varepsilon)| \leq \text{const. } \varepsilon^{N+1} \quad \text{for } t \in [t_1, t_2], \qquad (8.2.40)$$

again for all small $\varepsilon > 0$ (see Exercise 8.2.3). The stated estimate (8.2.37) then follows directly by an application of the maximum principle (Theorem 0.3.5) to the remainder function R_N of (8.2.35)–(8.2.36).

If the data b, f, α, β have asymptotic expansions (8.2.3) and (8.2.6) proceeding in even powers of ε, then the odd-indexed outer coefficients X_{2k+1} are seen to vanish, and the residuals corresponding to R_{2N} are seen to be order(ε^{2N+2}); so the result (8.2.38) follows again from the maximum principle. ∎

As a consequence of Theorem 8.2.1, we have the asymptotic relation

$$x(t,\varepsilon) \sim \sum_{k=0}^{\infty} \left[X_k(t) + e^{-p(t)/\varepsilon}(^*X_k(t)) + e^{-q(t)/\varepsilon} X_k^*(t) \right] \varepsilon^k,$$
$$(8.2.41)$$

with

$$p(t) := \int_{t_1}^{t} \sqrt{-b_0}, \qquad q(t) := \int_{t}^{t_2} \sqrt{-b_0}, \qquad (8.2.42)$$

and with the functions X_k, $^*X_k$, and X_k^* as constructed earlier. This result (8.2.41)–(8.2.42) coincides with the *Liouville/Green approximation* for the given Dirichlet problem, often called the JWKB approximation [for Jeffreys (1924*a*), Wentzel (1926), Kramers (1926), and Brillouin (1926)]; see Olver (1974) for a careful study of the Liouville/Green approximation, including both the nonoscillatory and the oscillatory cases. [See also Taylor (1978, 1982) and Smith (1984*b*).]

As noted earlier, we could at the outset simply take the scaled variables σ and τ in (8.2.9) to be given as

$$\sigma = \frac{t - t_1}{\varepsilon} \quad \text{and} \quad \tau = \frac{t_2 - t}{\varepsilon}, \qquad (8.2.43)$$

and we could seek an approximation in the slightly different (though asymptotically equivalent) form

$$x \sim X(t,\varepsilon) + {}^*X(\sigma,\varepsilon) + X^*(\tau,\varepsilon), \qquad (8.2.44)$$

where X, $*X$, and X^* have asymptotic expansions in powers of ε, and the boundary-layer correction functions $*X(\sigma, \varepsilon)$ and $X^*(\tau, \varepsilon)$ are required to satisfy suitable matching conditions as $\sigma, \tau \to \infty$. The procedure is analogous to that described in Section 8.4. For example, if the data $f(t)$, $b(t)$, α, and β are independent of ε, then we find, to lowest order,

$$x(t, \varepsilon) = \frac{f(t)}{b(t)} + \left[\alpha - \frac{f(t_1)}{b(t_1)}\right] \exp\left[-\frac{1}{\varepsilon}\sqrt{-b(t_1)}\,(t - t_1)\right]$$
$$+ \left[\beta - \frac{f(t_2)}{b(t_2)}\right] \exp\left[-\frac{1}{\varepsilon}\sqrt{-b(t_2)}\,(t_2 - t)\right] + \text{order}(\varepsilon^2).$$

$$(8.2.45)$$

By way of comparison, again if the data are independent of ε, then the Liouville/Green result (8.2.41)–(8.2.42) gives, to lowest order [see (8.2.38) with $N = 0$],

$$x(t, \varepsilon) = \frac{f(t)}{b(t)} + \left[\frac{b(t_1)}{b(t)}\right]^{1/4}\left[\alpha - \frac{f(t_1)}{b(t_1)}\right]\exp\left(-\frac{1}{\varepsilon}\int_{t_1}^{t}\sqrt{-b}\right)$$
$$+ \left[\frac{b(t_2)}{b(t)}\right]^{1/4}\left[\beta - \frac{f(t_2)}{b(t_2)}\right]\exp\left(-\frac{1}{\varepsilon}\int_{t}^{t_2}\sqrt{-b}\right) + \text{order}(\varepsilon^2).$$

$$(8.2.46)$$

In practice, the Liouville/Green result (8.2.46) is often quantitatively more accurate than (8.2.45) within the boundary layers.

Example 8.2.2: For the problem of Carrier and Pearson given by (8.2.7), we see that the lowest-order Liouville/Green result (8.2.46) becomes

$$x(t, \varepsilon) = \frac{1}{2 - t^2} - \frac{e^{-p(t)/\varepsilon} + e^{-q(t)/\varepsilon}}{(2 - t^2)^{1/4}} + R_0(t, \varepsilon), \qquad (8.2.47)$$

where p and q are given by (8.2.42) as

$$p(t) = \int_{-1}^{t} (2 - s^2)^{1/2}\, ds = \frac{t(2 - t^2)^{1/2}}{2} + \sin^{-1}\frac{t}{\sqrt{2}} + \frac{2 + \pi}{4},$$
$$q(t) = \int_{t}^{1} (2 - s^2)^{1/2}\, ds = -\frac{t(2 - t^2)^{1/2}}{2} - \sin^{-1}\frac{t}{\sqrt{2}} + \frac{2 + \pi}{4}.$$

$$(8.2.48)$$

By way of comparison, the result (8.2.45) yields in this case

$$x(t, \varepsilon) = \frac{1}{2 - t^2} - (e^{-(1+t)/\varepsilon} + e^{-(1-t)/\varepsilon}) + S_0(t, \varepsilon), \qquad (8.2.49)$$

where in each case the remainders R_0 and S_0 are defined respectively by (8.2.47) and (8.2.49).

If we insert (8.2.47) and (8.2.49) separately back into (8.2.7), we obtain certain specific Dirichlet problems for R_0 and S_0 analogous to (8.2.36). The maximum principle can then be used in turn to estimate R_0 and S_0, and we find in this way, for example,

$$|R_0(t, \varepsilon)| \le 14\varepsilon^2 \quad \text{and} \quad |S_0(t, \varepsilon)| \le 100\varepsilon^2. \tag{8.2.50}$$

The Liouville/Green approximation (8.2.47)–(8.2.48) is an order of magnitude better than (8.2.49) within the boundary layers. On the other hand, (8.2.49) is simpler. The two results agree asymptotically (as $\varepsilon \to 0+$) *away* from the boundary regions.

Examples involving more general boundary conditions are illustrated in Exercises 8.2.5 and 8.2.6, and also in Chapter 9. Error estimates for such problems can be obtained using various methods, including the Green function approach of Chapter 9. Nonlinear problems are considered in Chapter 10. Such linear and nonlinear problems have been studied by many authors, including Brish (1954), Vasil'eva and Tupčiev (1960), Fife (1973a, 1973b), Dorr, Parter, and Shampine (1973), Howes (1976b), van Harten (1978a, 1978b), Kelley (1979), and the authors mentioned earlier in the paragraph before (8.1.6).

Note that the equation (8.2.1) considered in this section can be written as the first-order system

$$\frac{d}{dt} \begin{bmatrix} x \\ \varepsilon^2 y \end{bmatrix} = \begin{bmatrix} 0 & 1 \\ -b(t, \varepsilon) & 0 \end{bmatrix} \begin{bmatrix} x \\ y \end{bmatrix} + \begin{bmatrix} 0 \\ f(t, \varepsilon) \end{bmatrix}, \tag{8.2.51}$$

with $y := dx/dt$, so that equation (8.2.1) can be considered to be a special case of the conditionally stable system (7.1.1) of Chapter 7, with $m = n = 1$ in (7.1.1), along with

$$A(t, \varepsilon) = 0, \qquad\qquad B(t, \varepsilon) = 1,$$
$$C(t, \varepsilon) = -b(t, \varepsilon), \qquad D(t, \varepsilon) = 0, \tag{8.2.52}$$

in the notation of (7.1.1). Hence, the 1×1 "matrix" $B_0(t)C_0(t) = -b_0(t)$ has eigenvalue λ given as $\lambda = -b_0(t)$, which is positive because of (8.2.4), so that condition (7.1.9) is satisfied. The quantity ω of (7.2.1) is given as

$$\omega = \omega(t) = \sqrt{-b_0(t)}, \tag{8.2.53}$$

and we see that the lowest-order approximation obtained in Chapter 7 for the fundamental solution matrix (7.5.9) agrees in this case with an analogous fundamental solution that can be constructed here in terms of two independent Liouville/Green solutions $*X$ and $X*$ of boundary-layer type for the homogeneous equation (8.2.1) with $f \equiv 0$.

Exercises

Exercise 8.2.1: Give the exact solution of the Dirichlet problem

$$\varepsilon^2 \frac{d^2x}{dt^2} - x = -t \quad \text{for } 0 \le t \le 1,$$

$$x = \alpha \quad \text{at } t = 0, \quad \text{and} \quad x = \beta \quad \text{at } t = 1,$$

and verify that the solution depends in an essential way on the quantities t, t/ε, and $(1 - t)/\varepsilon$.

Exercise 8.2.2: Show that (8.2.15) and (8.2.17) imply the Liouville/Green relations (8.2.18)–(8.2.20).

Exercise 8.2.3: Give details of the verifications of the results (8.2.37) and (8.2.38) in the case $N = 0$.

Exercise 8.2.4: Give details of a derivation of the estimates of (8.2.50) for the problem of Carrier and Pearson considered in Example 8.2.2.

Exercise 8.2.5: Use a multivariable approach to find a uniformly valid, easily interpretable approximation for the solution of the Robin problem

$$\varepsilon^2 \frac{d^2x}{dt^2} - (2 - t^2)x = -1 \quad \text{for } -1 \le t \le 1,$$

$$x(-1, \varepsilon) = \frac{dx(+1, \varepsilon)}{dt} = 0. \tag{8.2.54}$$

Note that Exercise 8.1.2 shows that this problem has one and only one solution. What can we say in this case regarding the strength of the boundary layer for $x(t, \varepsilon)$ near $t = +1$? Compare the graphs of the solutions to problems (8.2.7) and (8.2.54). (The actual justification for the validity of the approximate solution hinges on some suitable estimate for the difference between the exact and approximate solutions, using perhaps the Green function approach of Exercise 9.3.11; we need not do this here.)

Exercise 8.2.6: Find an approximate solution to the Neumann problem

$$0.0001 \frac{d^2x}{dt^2} - (2 - t^2)x = -1 \quad \text{for } -1 \le t \le 1,$$

$$\frac{dx}{dt} = 100 \quad \text{at the endpoints } t = \pm 1. \tag{8.2.55}$$

In particular, find approximations for the boundary values of the solution $x(-1)$ and $x(+1)$, and sketch a (rough) graph of $x = x(t)$. *Hint*: Consider the problem

$$\varepsilon^2 x'' + b(t)x = f(t) \quad \text{for } -1 \le t \le +1,$$

$$x'(t_j, \varepsilon) = \frac{\alpha_j}{\varepsilon} \quad \text{for } j = 1, 2 \ (t_1 = -1, t_2 = +1),$$

for given constants α_j ($j = 1, 2$) and for given smooth functions b and f, with $b(t) < 0$ for $-1 \le t \le 1$. This latter Neumann problem has a unique solution for each small $\varepsilon > 0$, as follows from Exercise 8.1.4.

8.3 A problem with a single boundary layer

In this section we consider the previous differential equation (8.1.1) in the special case (8.1.36), so that the equation becomes

$$\varepsilon \frac{d^2 x}{dt^2} + a(t, \varepsilon) \frac{dx}{dt} + b(t, \varepsilon)x = f(t, \varepsilon) \quad \text{for } t \in [t_1, t_2], \quad (8.3.1)$$

where the smooth coefficient function a is everywhere nonzero,

$$a(t, \varepsilon)^2 \ge \kappa^2 > 0 \quad \text{for } t \in [t_1, t_2], \text{ all small } \varepsilon \ge 0, \quad (8.3.2)$$

for some fixed positive constant $\kappa > 0$. The data in (8.3.1) are assumed to have asymptotic power-series expansions of the form

$$\begin{bmatrix} a(t, \varepsilon) \\ b(t, \varepsilon) \\ f(t, \varepsilon) \end{bmatrix} \sim \sum_{k=0}^{\infty} \begin{bmatrix} a_k(t) \\ b_k(t) \\ f_k(t) \end{bmatrix} \varepsilon^k \quad (8.3.3)$$

for suitable smooth functions $a_k = a_k(t)$, $b_k = b_k(t)$, and $f_k = f_k(t)$. Condition (8.3.2) can be replaced with the equivalent condition

$$a_0(t)^2 > 0 \quad \text{for } t_1 \le t \le t_2. \quad (8.3.4)$$

In this section, for notational brevity, and without loss, we shall fix the interval as $[t_1, t_2] = [0, 1]$, with

$$t_1 = 0 \quad \text{and} \quad t_2 = 1. \quad (8.3.5)$$

Because the function a and its coefficients a_k in (8.3.3) are smooth, we see that condition (8.3.2) [or (8.3.4)] implies that a (and a_0) is either *everywhere positive*, with

$$a(t, \varepsilon) \ge \kappa > 0 \quad [\text{and } a_0(t) > 0] \quad \text{for } 0 \le t \le 1, \quad (8.3.6)$$

or *everywhere negative*, with

$$a(t, \varepsilon) \le -\kappa < 0 \quad [\text{and } a_0(t) < 0] \quad \text{for } 0 \le t \le 1. \quad (8.3.7)$$

where the constant κ is positive in either case. As noted in Coddington and Levinson (1952), it makes no difference which condition (8.3.6) or (8.3.7) we use, because the change of variable $t_{\text{new}} = 1 - t_{\text{old}}$ transforms either condition into the other. For definiteness, condition (8.3.6) will be assumed here. This condition can be relaxed inside a boundary layer, where $a(t, \varepsilon)$ can be permitted to change sign. Such boundary-layer impulses are important for certain related nonlinear problems such as those discussed in Section 10.4.

We shall discuss more general boundary conditions later in this section and in later chapters, but initially here we consider the Dirichlet conditions

$$x(0, \varepsilon) = \alpha(\varepsilon) \quad \text{and} \quad x(1, \varepsilon) = \beta(\varepsilon) \tag{8.3.8}$$

for specified boundary values $\alpha(\varepsilon)$ and $\beta(\varepsilon)$ that are assumed to have expansions of the form

$$\begin{bmatrix} \alpha(\varepsilon) \\ \beta(\varepsilon) \end{bmatrix} \sim \sum_{k=0}^{\infty} \begin{bmatrix} \alpha_k \\ \beta_k \end{bmatrix} \varepsilon^k \tag{8.3.9}$$

for suitable constants α_k and β_k.

The Sturm transformation (8.1.7) transforms the present Dirichlet problem into a related Dirichlet problem for the differential equation (8.1.6), and assumption (8.3.2) guarantees that the coefficient c of (8.1.8) is positive for all small $\varepsilon \geq 0$ and for all t. It follows from Theorem 8.1.1, along with the Sturm transformation, that the present Dirichlet problem for (8.3.1) has one and only one solution, for each small value of ε. The intention here is to obtain a uniformly valid and easily interpretable approximation for this solution.

The reduced equation obtained by putting $\varepsilon = 0$ in (8.3.1) is [see (8.3.3)]

$$a_0(t)\frac{dX}{dt} + b_0(t)X = f_0(t), \tag{8.3.10}$$

and this first-order equation can, in general, satisfy at most only one of the boundary conditions of (8.3.8). Hence, we expect the exact solution of (8.3.1) and (8.3.8) to exhibit boundary-layer behavior at one endpoint, corresponding to the boundary condition that is lost as $\varepsilon \to 0+$.

For definiteness we are here assuming the normalization $\varepsilon > 0$, and then the location of the boundary layer at either the left endpoint or the right endpoint is determined by the highest-order term in the reduced equation (8.3.10). Indeed, we have already noted a related phenomenon in Chapter 5, where the boundary-layer correction was determined with the differential equation (5.2.9) or (5.4.12) containing the highest-order terms [see p. 14 of J. Cole (1968)]; this matter is discussed further in

Section 8.4 [see (8.4.29)]. Hence, in the present case the differential equation

$$\varepsilon \frac{d^2 X^*}{dt^2} + a(t, \varepsilon) \frac{dX^*}{dt} = 0 \qquad (8.3.11)$$

is expected to determine the dominant boundary-layer behavior of solutions of (8.3.1). Solutions of this latter equation (8.3.11) can be represented in terms of a linear combination of the functions

$$\int_0^t \exp\left(-\frac{1}{\varepsilon}\int_0^s a\right) ds \quad \text{and} \quad \int_t^1 \exp\left(\frac{1}{\varepsilon}\int_s^1 a\right) ds. \qquad (8.3.12)$$

Because rapid decay, not growth, is required for boundary-layer behavior, it follows that we should expect the boundary layer to occur at $t = 0$ if (8.3.6) holds with $a_0 > 0$, whereas we expect the boundary layer to occur at $t = 1$ if (8.3.7) holds with $a_0 < 0$. These expectations are corroborated in the constant-coefficient case (a and b constants, independent of t), which can be solved explicitly (see Exercise 8.3.1).

As stated earlier, we shall assume that (8.3.6) holds, and so we expect a boundary layer at $t = 0$. Hence, we are led to seek a two-variable expansion for x in the form

$$x(t, \varepsilon) \sim \sum_{k=0}^{\infty} x_k(t, s)\varepsilon^k \quad \text{with } s := \frac{h(t)}{\varepsilon}, \qquad (8.3.13)$$

where the function h is at our disposal subject to the conditions that it be strictly monotonic and vanish at $t = 0$,

$$h' \neq 0, \qquad h(0) = 0. \qquad (8.3.14)$$

Possible choices include $h(t) := t$ and $h(t) := \int_0^t a_0$. The fact that the second variable s is taken to be proportional to $1/\varepsilon$ in (8.3.13) is suggested by an examination of the constant-coefficient case, which can be explicitly solved. Alternatively, this result follows from the discussion to be given in Section 8.4 [see (8.4.28)].

The expansions (8.3.13) and (8.3.3) are inserted into the differential equation (8.3.1), and the usual procedure leads to the following equations for the coefficients x_k:

$$h'(t)\big[h'(t)x_{k+1,ss} + a_0(t)x_{k+1,s}\big]$$

$$= f_k(t) - \bigg[h''(t)x_{k,s} + 2h'(t)x_{k,ts} + x_{k-1,tt}$$

$$+ h'(t)\sum_{n=0}^{k} a_{k+1-n}x_{n,s} + \sum_{n=0}^{k}\left(a_{k-n}x_{n,t} + b_{k-n}x_k\right)\bigg]$$

$$\text{for } k = -1, 0, 1, \ldots, \qquad (8.3.15)_k$$

where the right side of $(8.3.15)_{-1}$ is understood to be zero in the case $k = -1$. The equations of (8.3.15) simplify somewhat with the choice

$$h'(t) = a_0(t), \qquad\qquad (8.3.16)$$

and we shall impose (8.3.16) here, so that (8.3.15) becomes

$$a_0(t)^2[x_{k+1,ss} + x_{k+1,s}]$$

$$= f_k(t) - \left\{ \sum_{n=0}^{k} \left[a_0(t)a_{k+1-n}x_{n,s} + a_{k-n}x_{n,t} + b_{k-n}x_n \right] \right.$$

$$\left. + \left[a_0'(t)x_{k,s} + 2a_0(t)x_{k,ts} + x_{k-1,tt} \right] \right\}$$

$$\text{for } k = -1, 0, \ldots, \qquad (8.3.17)_k$$

where again the right side is put equal to zero in the case $k = -1$. The equations of $(8.3.17)_k$ can be solved recursively with respect to s, for each fixed t.

In the case $k = -1$, we have the differential equation

$$x_{0,ss} + x_{0,s} = 0,$$

which can be solved to give

$$x_0(t, s) = A_0(t) + B_0(t)e^{-s}$$

for suitable "constants" of integration A_0, B_0 that can still depend on t. We now insert this result for x_0 back into the right side of $(8.3.17)_0$ and find for x_1 the differential equation

$$a_0(t)^2(x_{1,ss} + x_{1,s}) = f_0(t) - [a_0(t)A_0'(t) + b_0(t)A_0(t)]$$

$$+ e^{-s}[a_0'(t)B_0(t)$$

$$+ a_0(t)B_0'(t) + (a_0a_1 - b_0)B_0(t)].$$

$$(8.3.18)$$

Hence, the differential equation for x_1 can be written as

$$x_{1,ss} + x_{1,s} = k_1 + k_2 \cdot e^{-s}$$

for suitable quantities k_1 and k_2 independent of s, depending only on t. The forcing term $k_1 + k_2 \cdot e^{-s}$ is itself a solution of the homogeneous equation $x_{ss} + x_s = 0$. Such a forcing term produces spurious secular components in $x_1(t, s)$ that are relatively large for large s, and such terms should be eliminated or at least controlled if possible. In the present case it is natural to impose the conditions $k_1 \equiv k_2 \equiv 0$, or [see (8.3.18)]

$$a_0(t)A_0'(t) + b_0(t)A_0(t) = f_0(t),$$
$$\frac{d}{dt}[a_0(t)B_0(t)] = [b_0(t) - a_0(t)a_1(t)]B_0(t). \tag{8.3.19}$$

Then the previous differential equation for x_1 becomes homogeneous, with solution

$$x_1(t,s) = A_1(t) + B_1(t)e^{-s}$$

for suitable functions A_1 and B_1 depending only on t.

This procedure can be continued recursively, and we find, in general,

$$x_k(t,s) = A_k(t) + B_k(t)e^{-s} \tag{8.3.20}_k$$

for suitable functions A_k and B_k that are taken to satisfy the equations [compare with (8.3.19) in the case $k = 0$]

$$a_0(t)A_k'(t) + b_0(t)A_k(t)$$
$$= f_k(t) - A_{k-1}''(t) - \sum_{n=0}^{k-1}\left[a_{k-n}(t)A_k'(t) + b_{k-n}(t)A_n(t)\right] \tag{8.3.21}_k$$

and

$$\frac{d}{dt}[a_0(t)B_k(t)] + [a_0(t)a_1(t) - b_0(t)]B_k(t)$$
$$= B_{k-1}''(t) + \sum_{n=0}^{k-1}\left[(-a_0a_{k+1-n} + b_{k-n})B_n(t) + a_{k-n}B_n'(t)\right] \tag{8.3.22}_k$$

for $k = 0, 1, \ldots$, where the summations are put equal to zero on the right sides of $(8.3.21)_k$ and $(8.3.22)_k$ in the case $k = 0$, and where we also use here the convention $A_k \equiv B_k \equiv 0$ for negative indices $k < 0$.

We should note here that, as usual, there is some flexibility in the determination of the boundary-layer correction terms. The condition $(8.3.22)_k$ is natural here for B_k, but weaker conditions also suffice. For example, it would suffice to impose a modified version of $(8.3.22)_k$ obtained by adding any fixed bounded function to the right side of the present $(8.3.22)_k$. In this latter case we obtain a final approximation for x that is asymptotically equivalent to the approximation obtained here. There is no reason to modify (8.3.22) in the present case, but the point becomes relevant for various analogous *nonlinear* problems related to the present problem. By way of comparison, the condition $(8.3.21)_k$ is essential for the outer approximation.

From (8.3.13), (8.3.14), (8.3.16), and (8.3.20) we are led to seek the following asymptotic representation for x:

$$x(t, \varepsilon) \sim \sum_{k=0}^{\infty} A_k(t) \varepsilon^k + e^{-h(t)/\varepsilon} \sum_{k=0}^{\infty} B_k(t) \varepsilon^k, \qquad (8.3.23)$$

with

$$h(t) = \int_0^t a_0. \qquad (8.3.24)$$

This representation (8.3.23)–(8.3.24) can be inserted into the boundary conditions of (8.3.8)–(8.3.9), and we find the conditions

$$A_k(0) + B_k(0) = \alpha_k,$$
$$A_k(1) = \beta_k \quad \text{for } k = 0, 1, \ldots, \qquad (8.3.25)_k$$

where we have neglected the exponentially small term $e^{-h(1)/\varepsilon}$ (which is asymptotically zero) in obtaining the boundary condition of $(8.3.25)_k$ at $t = 1$.

The differential equations of $(8.3.21)$–$(8.3.22)$ can be solved recursively along with the boundary conditions of (8.3.25) to determine the functions A_k and B_k for $k = 0, 1, \ldots$. For example, we find that A_0 is the unique solution of the reduced differential equation (8.3.10) satisfying the original boundary condition at $t = 1$ (to first order, with $\varepsilon = 0$ in that boundary condition), and B_0 is determined then by the problem

$$\frac{d}{dt}[a_0(t)B_0(t)] = [b_0(t) - a_0(t)a_1(t)]B_0(t),$$
$$B_0(0) = \alpha_0 - A_0(0),$$

where A_0 is already determined. Then, after $A_n(t)$ and $B_n(t)$ are determined for $n = 0, 1, \ldots, k - 1$, we determine $A_k(t)$ as the solution of the differential equation $(8.3.21)_k$ subject to the second boundary condition of $(8.3.25)_k$ at $t = 1$, and $B_k(t)$ is then determined as the solution of $(8.3.22)_k$ subject to the first boundary condition of $(8.3.25)_k$ at $t = 0$.

In order to check the validity of the foregoing procedure, we truncate the resulting expansion (8.3.23) and define the function $x^N = x^N(t, \varepsilon)$ as

$$x^N(t, \varepsilon) := \sum_{k=0}^{N} A_k(t) \varepsilon^k + e^{-h(t)/\varepsilon} \sum_{k=0}^{N} B_k(t) \varepsilon^k, \qquad (8.3.26)$$

with $h(t)$ given by (8.3.24). We find directly from (8.3.24)–(8.3.26) the boundary relations

$$x^N(0, \varepsilon) = \alpha(\varepsilon) - \sigma_0(\varepsilon),$$
$$x^N(1, \varepsilon) = \beta(\varepsilon) - \sigma_1(\varepsilon),$$

(8.3.27)

for residuals $\sigma_0(\varepsilon)$ and $\sigma_1(\varepsilon)$ defined as

$$\sigma_0(\varepsilon) := \alpha(\varepsilon) - \sum_{k=0}^{N} \alpha_k \varepsilon^k,$$

$$\sigma_1(\varepsilon) := \beta(\varepsilon) - \sum_{k=0}^{N} \beta_k \varepsilon^k - e^{-h(1)/\varepsilon} \sum_{k=0}^{N} B_k(1) \varepsilon^k.$$

(8.3.28)

Note that we are suppressing the dependence of $\sigma_0(\varepsilon)$ and $\sigma_1(\varepsilon)$ on N.

Similarly, using (8.3.21), (8.3.22), (8.3.24), and (8.3.26), we find for x^N the differential equation

$$\varepsilon \frac{d^2 x^N}{dt^2} + a(t, \varepsilon) \frac{dx^N}{dt} + b(t, \varepsilon) x^N = f(t, \varepsilon) - \rho_1(t, \varepsilon) - \rho_2(t, \varepsilon),$$

(8.3.29)

where the residuals $\rho_1(t, \varepsilon)$ and $\rho_2(t, \varepsilon)$ are defined as

$$\rho_1(t, \varepsilon) := f(t, \varepsilon) - A_N''(t) \varepsilon^{N+1}$$
$$- \sum_{k=0}^{N} \left[A_{k-1}''(t) + a(t, \varepsilon) A_k'(t) + b(t, \varepsilon) A_k(t) \right] \varepsilon^k$$

(8.3.30)

and

$$\rho_2(t, \varepsilon) := - e^{-h(t)/\varepsilon}$$

$$\times \left\{ a_0 \frac{a_0 - a}{\varepsilon} B_0(t) + a_0(a - a_0) B_{N+1}(t) \varepsilon^N + B_N''(t) \varepsilon^{N+1} \right.$$

$$+ \sum_{k=0}^{N} \left[B_{k-1}''(t) + (a - 2a_0) B_k'(t) + (b - a_0') B_k(t) \right.$$

$$\left. + a_0(a_0 - a) B_{k+1}(t) \right] \varepsilon^k \Bigg\},$$

(8.3.31)

where we also suppress the dependence of ρ_1 and ρ_2 on N.

The remainder $R_N = R_N(t, \varepsilon)$, defined as

$$R_N(t, \varepsilon) := x(t, \varepsilon) - x^N(t, \varepsilon),$$

(8.3.32)

is now found to satisfy the boundary conditions

$$R_N(0, \varepsilon) = \sigma_0(\varepsilon), \qquad R_N(1, \varepsilon) = \sigma_1(\varepsilon), \qquad (8.3.33)$$

along with the differential equation

$$\varepsilon R_N'' + a(t, \varepsilon) R_N' + b(t, \varepsilon) R_N = \rho_1(t, \varepsilon) + \rho_2(t, \varepsilon), \quad (8.3.34)$$

with the residuals $\sigma_0(\varepsilon)$, $\sigma_1(\varepsilon)$, $\rho_1(t, \varepsilon)$, and $\rho_2(t, \varepsilon)$ defined by (8.3.28), (8.3.30), and (8.3.31).

We find directly from the definitions of the residuals, along with the foregoing construction of the functions A_k and B_k, the estimates (see Exercise 8.3.2)

$$|\sigma_0(\varepsilon)|, |\sigma_1(\varepsilon)| \le \text{const. } \varepsilon^{N+1}, \qquad (8.3.35)$$

$$|\rho_1(t, \varepsilon)| \le \text{const. } \varepsilon^{N+1}, \qquad (8.3.36)$$

and

$$|\rho_2(t, \varepsilon)| \le \text{const. } \varepsilon^{N+1} e^{-h(t)/\varepsilon}, \qquad (8.3.37)$$

where the last two estimates are valid for $0 \le t \le 1$, and all estimates are valid for all small positive values of ε (see Exercise 8.3.2). The estimate (8.3.37) for ρ_2 is stronger than generally required. For example, the slightly different choice $s = t/\varepsilon$ for the boundary-layer variable in (8.3.13) leads to the weaker estimate given later by (8.3.57), with ε^{N+1} replaced by ε^N on the right side of (8.3.37). In particular, this latter choice $s = t/\varepsilon$ is customarily used in the study of analogous nonlinear boundary-value problems. The resulting weaker estimate (8.3.57) is sufficient for the purpose at hand.

It can be shown to follow from (8.3.33)–(8.3.34), along with the estimates (8.3.35)–(8.3.37), that the remainder $R_N(t, \varepsilon)$ is uniformly of order ε^{N+1}, so that

$$|x(t, \varepsilon) - x^N(t, \varepsilon)| \le \text{const. } \varepsilon^{N+1} \qquad (8.3.38)$$

uniformly for $0 \le t \le 1$ and uniformly for all small $\varepsilon > 0$. This assertion regarding R_N can be proved by any one of several different approaches. For example, we can use the method of Cochran, already used in Section 8.1, to prove the existence of a unique solution for nonoscillatory boundary-value problems [see Cochran (1968) or Smith (1975a)], or we can give a proof using comparison techniques and differential inequalities [see Weinstein and Smith (1975)]. As a third alternative, we can use a direct Green function approach. We shall present one version of this latter approach in Chapter 9 so as to complete the proof of (8.3.38). We also obtain estimates analogous to (8.3.38) for the derivatives, such as the following estimate on the first derivative:

$$\left| \frac{dx(t,\varepsilon)}{dt} - \frac{dx^N(t,\varepsilon)}{dt} \right| \leq \text{const. } \varepsilon^N, \qquad (8.3.39)$$

again uniformly for $0 \leq t \leq 1$ and uniformly for all small $\varepsilon > 0$.

Example 8.3.1: The boundary-value problem

$$\varepsilon x'' + x' + x = 0 \quad \text{for } 0 \leq t \leq 1,$$
$$x(0,\varepsilon) = 0 \quad \text{and} \quad x(1,\varepsilon) = 1, \qquad (8.3.40)$$

is of the foregoing type, with $\alpha(\varepsilon) = \alpha_0 = 0$, $\beta(\varepsilon) = \beta_0 = 1$, $a(t,\varepsilon) = a_0(t) = 1$, $b(t,\varepsilon) = b_0(t) = 1$, and $f(t,\varepsilon) = 0$. From $(8.3.21)_0$, $(8.3.22)_0$, and $(8.3.25)_0$ we find in this case

$$A_0(t) = e^{1-t} \quad \text{and} \quad B_0(t) = -e^{1+t},$$

with $h(t) = 1$. Hence, (8.3.26) yields for x^0 the result

$$x^0(t,\varepsilon) = e^{1-t} - e^{1+t}e^{-t/\varepsilon}. \qquad (8.3.41)$$

Of course, we can also easily give the exact solution in this case. The graph of the solution is indicated in Figure 8.1 for small positive ε.

The present approach can be easily used to handle problems with types of boundary conditions other than the Dirichlet conditions considered earlier. As an illustrative example, we consider the Neumann problem

$$\varepsilon x'' + a(t,\varepsilon)x' + b(t,\varepsilon)x = f(t,\varepsilon) \quad \text{for } 0 \leq t \leq 1,$$
$$x'(0,\varepsilon) = \alpha(\varepsilon) \quad \text{and} \quad x'(1,\varepsilon) = \beta(\varepsilon), \qquad (8.3.42)$$

where the data are again assumed to satisfy (8.3.3), (8.3.6), and (8.3.9).

Because of the positivity assumption (8.3.6), the solution will again exhibit boundary-layer behavior at the left endpoint $t = 0$, and the leading term A_0 in the outer approximation will again satisfy a terminal-value problem for the reduced equation subject to the specified boundary condition at $t = 1$; see (8.3.46). In this case we must also include the following additional assumption on the data at $t = 1$:

$$b_0(1) \neq 0. \qquad (8.3.43)$$

This assumption guarantees that the terminal-value problem (8.3.46) has a solution, and moreover it guarantees similarly that the original Neumann problem has one and only one solution for all small $\varepsilon > 0$. [The simple example with $b \equiv 0$ shows that the Neumann problem (8.3.42) may have no solution or infinitely many solutions in the case $\varepsilon > 0$ and $a_0 > 0$ if (8.3.43) is violated.]

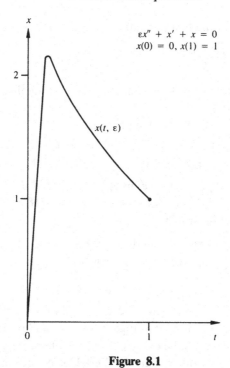

Figure 8.1

Subject to these assumptions, the problem (8.3.42) has a unique solution $x = x(t, \varepsilon)$ for all sufficiently small $\varepsilon > 0$, and we shall obtain a uniformly valid approximation to this solution for $0 \leq t \leq 1$. In particular, it is of interest to obtain information on the resulting boundary values for $x(0, \varepsilon)$ and $x(1, \varepsilon)$.

Guided by the situation for similar problems with constant coefficients, we again seek a representation in the form (8.3.23)–(8.3.24). We again find the differential equations (8.3.21)–(8.3.22) for determination of the coefficients $A_k(t)$ and $B_k(t)$, along with the boundary conditions

$$B_k(0) = \frac{1}{a_0(0)} \left[-\alpha_{k-1} + A'_{k-1}(0) + B'_{k-1}(0) \right],$$

$$A'_k(1) = \beta_k \quad \text{for } k = 0, 1, \ldots, \tag{8.3.44}_k$$

where the right side of the first equation here is put equal to zero in the case $k = 0$, with $B_0(0) = 0$.

This last boundary condition $B_0(0) = 0$, along with $(8.3.22)_0$, implies

$$B_0(t) \equiv 0 \quad \text{for } 0 \leq t \leq 1. \tag{8.3.45}$$

From $(8.3.21)_0$ and the second equation of $(8.3.44)_0$, we now have for $A_0(t)$ the problem

$$a_0(t) A_0'(t) + b_0(t) A_0(t) = f_0(t) \quad \text{for } 0 < t < 1,$$
$$A_0'(1) = \beta_0. \tag{8.3.46}$$

This terminal-value Neumann problem (8.3.46), subject to the assumption (8.3.43), serves to determine $A_0(t)$ uniquely for $0 \leq t \leq 1$, and then $B_1(t)$ is determined by the conditions [see $(8.3.22)_1$, $(8.3.44)_1$, and (8.3.45)]

$$\frac{d}{dt} [a_0(t) B_1(t)] + [a_0(t) a_1(t) - b_0(t)] B_1(t) = 0, \tag{8.3.47a}$$

with

$$a_0(0) B_1(0) = -\alpha_0 + A_0'(0). \tag{8.3.47b}$$

The procedure can be continued recursively in the usual manner to determine as many coefficients A_k and B_k as desired.

The Green function approach of Chapter 9 can be used to prove that the resulting truncated partial sum from (8.3.23) actually provides a suitable uniform approximation to the exact solution. In particular,

$$x(t, \varepsilon) = A_0(t) + O(\varepsilon) \quad \text{as } \varepsilon \to 0+,$$

uniformly for $0 \leq t \leq 1$, where $A_0(t)$ is the solution of the reduced Neumann terminal-value problem (8.3.46). In particular, the resulting boundary values $A_0(0)$ and $A_0(1)$ give useful approximations for the exact boundary values $x(0, \varepsilon)$ and $x(1, \varepsilon)$.

Example 8.3.2: The problem

$$0.001x'' + 2(1 + t)x' + 4(1 + t)^2 x = 4(1 + t)^2 \quad \text{for } 0 \leq t \leq 1,$$
$$x'(0) = 1 \quad \text{and} \quad x'(1) = -1, \tag{8.3.48}$$

is of the type (8.3.42), with $\varepsilon = 0.001$, $a = a_0(t) = 2(1 + t)$, $b = b_0(t) = 4(1 + t)^2$, $f = f_0(t) = 4(1 + t)^2$, $\alpha = \alpha_0 = 1$, and $\beta = \beta_0 = -1$. From (8.3.46) we find

$$A_0(t) = 1 + \frac{\exp[4 - (1 + t)^2]}{4} \quad \text{for } 0 \leq t \leq 1,$$

and then (8.3.47) yields

$$B_1(t) = -(1 + \tfrac{1}{2}e^3) \frac{\exp[-1 + (1 + t)^2]}{2(1 + t)} \quad \text{for } 0 \leq t \leq 1.$$

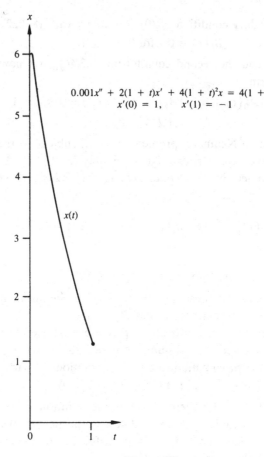

Figure 8.2

In particular, we have

$$x(0) \doteq A_0(0) = 1 + \frac{e^3}{4} \doteq 6.02 \quad \text{and} \quad x(1) \doteq A_0(1) = 1.25.$$

The graph of $x(t)$ is indicated in Figure 8.2. The first derivative $x'(t)$ exhibits boundary-layer behavior near $t = 0$, but $x(t)$ itself is uniformly approximated by the smooth function $A_0(t)$.

Example 8.3.3: The Robin problem

$$0.001x'' + 2(1 + t)x' + 4(1 + t)^2 x = 4(1 + t)^2 \quad \text{for } 0 \le t \le 1,$$

$$x(0) = 0 \quad \text{and} \quad x'(1) = -1, \tag{8.3.49}$$

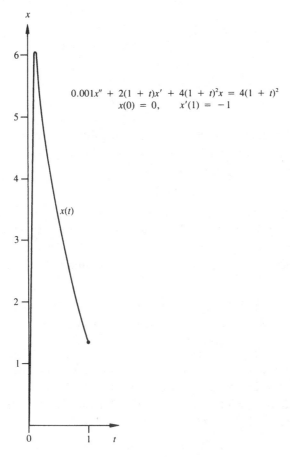

$$0.001x'' + 2(1 + t)x' + 4(1 + t)^2x = 4(1 + t)^2$$
$$x(0) = 0, \qquad x'(1) = -1$$

Figure 8.3

is only slightly different from the previous Neumann problem (8.3.48). However, in this case the solution $x(t)$ has a stronger boundary layer at $t = 0$, as indicated in Figure 8.3. The details of the approximation in this case are left to the reader; see Exercise 8.3.4.

As mentioned earlier after (8.3.14), and analogous to the situation in Section 8.2, there is some flexibility in the choice of the boundary-layer scaled variable s in (8.3.13). We could at the outset take s to be given as

$$s = t/\varepsilon. \qquad (8.3.50)$$

In this case, we simply seek an approximation in the form [compare with (6.2.8)]

$$x(t, \varepsilon) \sim X(t, \varepsilon) + X^*(s, \varepsilon) \quad \text{with } s := t/\varepsilon, \qquad (8.3.51)$$

and with

$$\begin{bmatrix} X(t, \varepsilon) \\ X^*(s, \varepsilon) \end{bmatrix} \sim \sum_{k=0}^{\infty} \begin{bmatrix} X_k(t) \\ X_k^*(s) \end{bmatrix} \varepsilon^k, \tag{8.3.52}$$

subject to the matching condition

$$\lim_{s \to \infty} X_k^*(s) = 0 \quad \text{for } k = 0, 1, \dots . \tag{8.3.53}$$

The outer approximation $X(t, \varepsilon)$ is required to satisfy the original differential equation, and we find that the outer coefficients $X_k(t)$ agree with the previous functions A_k:

$$X_k(t) = A_k(t) \quad \text{for } 0 \le t \le 1 \text{ and } k = 0, 1, \dots . \tag{8.3.54}$$

However, the boundary-layer correction terms $X_k^*(t/\varepsilon)$ differ from the previous corrections $e^{-h(t)/\varepsilon} B_k(t)$, although the resulting approximations are asymptotically equivalent. For example, for the Dirichlet problem subject to the boundary conditions (8.3.8), we find from (8.3.51)–(8.3.53), the result

$$X_0^*(s) = \left[\alpha_0 - A_0(0) \right] e^{-a_0(0)s} \quad \text{for } s \ge 0, \tag{8.3.55}$$

so that the lowest-order approximation becomes

$$x^0(t, \varepsilon) = A_0(t) + \left[\alpha_0 - A_0(0) \right] e^{-a_0(0)t/\varepsilon}, \tag{8.3.56}$$

which differs slightly from the corresponding previous result (8.3.26) in the case $N = 0$. In the present case the analogue of the previous boundary-layer residual $\rho_2(t, \varepsilon)$ of (8.3.31) satisfies an estimate of the type [compare with (8.3.37)]

$$|\rho_2(t, \varepsilon)| \le \text{const.} \ \varepsilon^N e^{-a_0(0)t/\varepsilon}. \tag{8.3.57}$$

This latter estimate, along with related estimates similar to the previous results (8.3.35) and (8.3.36), is sufficient in the present case to imply a result of the type (8.3.38) for the truncated expansion x^N obtained from (8.3.51)–(8.3.53).

Various linear and nonlinear problems related to those of this section have been studied by many authors, including Tschen (1935), von Mises (1950), Coddington and Levinson (1952), Brish (1954), Wasow (1956), Vasil'eva (1959), Cochran (1962), Willett (1966), Macki (1967), Erdélyi (1968a, 1968b), O'Malley (1968a, 1969a), Hoppensteadt (1971b), Searl (1971), Dorr, Parter, and Shampine (1973), Eckhaus (1973), Chang (1974, 1975, 1976), Habets (1974), Yarmish (1975), Howes (1976a, 1978), van Harten (1978a, 1978b), and the authors previously mentioned in the paragraph preceding (8.1.6). Certain additional problems related to those of this section will be considered in Chapters 9 and 10.

Note finally that equation (8.3.1) subject to the positivity condition (8.3.6), as considered in this section (for small $\varepsilon > 0$), can be viewed as a special case of the first-order system of Chapter 6, where the initial-value problem was considered for $t \geq 0$; see Examples 6.1.1 and 6.2.1. The alternative condition (8.3.7) corresponds to the terminal-value problem for the system of Chapter 6, for $t \leq 0$.

Exercises

Exercise 8.3.1: Give the exact solution for the Dirichlet problem

$$\varepsilon x'' + ax' + bx = f(t) \quad \text{for } 0 \leq t \leq 1,$$
$$x = \alpha \quad \text{at } t = 0, \quad \text{and} \quad x = \beta \quad \text{at } t = 1, \tag{8.3.58}$$

where a, b, α, and β are fixed given constants with $a \neq 0$. Use the exact solution directly to determine the location of the boundary layer for small $\varepsilon > 0$, and compute the limit

$$\lim_{\varepsilon \to 0+} x(t, \varepsilon)$$

for each fixed t ($0 \leq t \leq 1$).

Exercise 8.3.2: Verify directly the estimates of (8.3.35)–(8.3.37) in the two cases $N = 0$ and $N = 1$, arguing directly from the definitions of the residuals (8.3.28), (8.3.30), and (8.3.31).

Exercise 8.3.3: Construct the lowest-order approximation $x^0(t, \varepsilon)$ of (8.3.26) for the problem

$$\varepsilon x'' - x' + x = 0 \quad \text{for } 0 \leq t \leq 1,$$
$$x = 0 \quad \text{at } t = 0, \quad \text{and} \quad x = 1 \quad \text{at } t = 1. \tag{8.3.59}$$

Sketch the graph of the solution, and compare with the result of Example 8.3.1.

Exercise 8.3.4: Find an easily interpretable, uniformly valid approximation $x^0(t, \varepsilon)$ for the exact solution $x(t, \varepsilon)$ of the problem

$$\varepsilon x'' + 2(1 + t)x' + 4(1 + t)^2 x = 4(1 + t)^2 \quad \text{for } 0 \leq t \leq 1,$$
$$x = 0 \quad \text{at } t = 0, \quad \text{and} \quad x' = -1 \quad \text{at } t = 1.$$

(The error $x - x^0$ can be estimated using the Green function approach of Chapter 9, but we need not do this here.)

Exercise 8.3.5: Find the leading approximation

$$x^0 = A_0(t) + B_0(t)e^{-h(t)/\varepsilon}$$

from (8.3.26) for the problem ($\varepsilon > 0$)

$$\varepsilon x'' + (1 + t)x' - 2x = (1 + t)^3 \quad \text{for } 0 \le t \le 1,$$
$$x = 5 \quad \text{at } t = 0, \quad \text{and} \quad x = 12 \quad \text{at } t = 1. \tag{8.3.60}$$

Exercise 8.3.6: Let $M(\varepsilon)$ denote the average or mean value of the solution $x = x(t, \varepsilon)$ of the Dirichlet problem (8.3.1), (8.3.3), (8.3.5), (8.3.6), (8.3.8), and (8.3.9), given as

$$M(\varepsilon): = \int_0^1 x(t, \varepsilon)\, dt. \tag{8.3.61}$$

(a) Show that $M(\varepsilon)$ has an asymptotic expansion of the form

$$M(\varepsilon) \sim \sum_{k=0}^{\infty} M_k \varepsilon^k, \tag{8.3.62}$$

where the constants M_k can be given as

$$M_k = \int_0^1 X_k(t)\, dt + \int_0^{\infty} X_{k-1}^*(s)\, ds,$$

with X_k and X_{k-1}^* determined suitably as in (8.3.51)–(8.3.53).

(b) Compute the leading term $M_0 = \int_0^1 X_0$ in the expansion (8.3.62) for the problem (8.3.60).

8.4 Higher-order problems

The problems discussed in Sections 8.2 and 8.3 can be generalized in several ways. We can consider more general boundary conditions, and we can also consider analogous problems involving vector systems of differential equations, such as the second-order vector systems considered in Chang (1976) and Kelley (1979), or related problems involving first-order systems such as the systems of Chapters 6 and 7. We shall consider certain of these generalizations later in Chapter 9 and Chapter 10. In this section we shall discuss a generalization in terms of the scalar linear differential equation involving two given differential operators $\mathscr{L}_1$ and $\mathscr{L}_2$,

$$\varepsilon \mathscr{L}_1 x + \mathscr{L}_2 x = f(t, \varepsilon) \quad \text{for } t \in [t_1, t_2], \tag{8.4.1}$$

where this equation is to be solved for $x = x(t, \varepsilon)$ subject to suitable boundary conditions that will be given later, and where the order of the operator $\mathscr{L}_1$ exceeds the order of $\mathscr{L}_2$, so that the reduced equation

$$\mathscr{L}_2 x = f \tag{8.4.2}$$

is of lower order than (8.4.1). For simplicity here we assume that the operators $\mathscr{L}_1$ and $\mathscr{L}_2$ are linear. Such problems were studied by Wasow (1941, 1944) and O'Malley (1969b, 1969c); see also Wasow (1965, 1976),

O'Malley (1974*a*), and Tschen (1935). Of course, the scalar equation (8.4.1) can also be rewritten in terms of a suitable first-order system, but in this section we consider (8.4.1) explicitly as a single, higher-order scalar equation.

If the operators $\mathscr{L}_1$ and $\mathscr{L}_2$ are given as

$$\mathscr{L}_1 := \frac{d^2}{dt^2} \quad \text{and} \quad \mathscr{L}_2 := a(t,\varepsilon)\frac{d}{dt} + b(t,\varepsilon), \qquad (8.4.3)$$

then (8.4.1) reduces to the equation

$$\varepsilon x'' + a(t,\varepsilon)x' + b(t,\varepsilon)x = f(t,\varepsilon),$$

as considered in Section 8.3 subject to the condition [see (8.3.2)]

$$a(t,\varepsilon)^2 \geq \kappa^2 > 0 \quad \text{for } t \in [t_1, t_2] \qquad (8.4.4)$$

for a fixed positive constant κ. Similarly, if $\mathscr{L}_1$ and $\mathscr{L}_2$ are given as

$$\mathscr{L}_1 := \frac{d^2}{dt^2} \quad \text{and} \quad \mathscr{L}_2 := b(t,\varepsilon), \qquad (8.4.5)$$

then (8.4.1) reduces to the equation

$$\varepsilon x'' + b(t,\varepsilon)x = f(t,\varepsilon),$$

as considered in Section 8.2 subject to the condition

$$b(t,\varepsilon) \leq -\kappa < 0 \quad \text{for } t \in [t_1, t_2]. \qquad (8.4.6)$$

In this latter case, in Section 8.2 the small parameter ε was replaced with ε^2, and we shall see that this is consistent with the approach discussed here; see (8.4.18) and (8.4.30).

Example 8.4.1: If $\mathscr{L}_1$ and $\mathscr{L}_2$ are defined as

$$\mathscr{L}_1 x := \frac{d^2}{dt^2}\left[a(t)\frac{d^2 x}{dt^2}\right] \quad \text{and} \quad \mathscr{L}_2 x := -\frac{d^2 x}{dt^2} \qquad (8.4.7)$$

for any suitable function $x = x(t)$, and where $a = a(t)$ is a suitable given function, then (8.4.1) reduces to the fourth-order equation

$$\varepsilon\frac{d^2}{dt^2}\left[a(t)\frac{d^2 x}{dt^2}\right] - \frac{d^2 x}{dt^2} = f(t,\varepsilon). \qquad (8.4.8)$$

This differential equation is the equation for the deflection of an elastic beam with small, variable flexural rigidity (e.g., variable thickness) under tension subject to a specified load f, according to the linearized Euler/Bernoulli beam theory. The differential equation is to be solved subject to suitable boundary conditions that may be of a variety of

different types, as discussed, for example, in J. Cole (1968, p. 70) and
von Kármán and Biot (1940, pp. 259–322). The equation can be easily
solved in the case of constant rigidity, $a(t) = $ constant, and in this way
we are led to expect, in the general case, a three-variable expansion of the
form

$$
x(t, \varepsilon) \sim \exp\left[-\frac{(t - t_1)}{\varepsilon^{1/2}}\right] \sum_{k=0}^{\infty} {}^*X_k(t)(\varepsilon^{1/2})^k
$$

$$
+ \exp\left[-\frac{(t_2 - t)}{\varepsilon^{1/2}}\right] \sum_{k=0}^{\infty} X_k^*(t)(\varepsilon^{1/2})^k + \sum_{k=0}^{\infty} X_k(t)(\varepsilon^{1/2})^k
$$

$$(8.4.9)$$

for suitable functions X_k, ${}^*X_k$, and X_k^*. As usual, the boundary-layer
correction terms here can be packaged alternatively in several different
forms. For example, we can also write

$$
x(t, \varepsilon) \sim \sum_{k=0}^{\infty} X_k(t)(\varepsilon^{1/2})^k + \sum_{k=0}^{\infty} {}^*X_k(\sigma)(\varepsilon^{1/2})^k + \sum_{k=0}^{\infty} X_k^*(\tau)(\varepsilon^{1/2})^k,
$$

$$(8.4.10)$$

with

$$
\sigma := \frac{t - t_1}{\varepsilon^{1/2}} \quad \text{and} \quad \tau := \frac{t_2 - t}{\varepsilon^{1/2}}, \tag{8.4.11}
$$

where the functions ${}^*X_k$ and X_k^* are not the same in (8.4.9) and (8.4.10).
Further detailed results depend on the nature of the specified boundary
conditions.

Example 8.4.2: If $\mathscr{L}_1$ and $\mathscr{L}_2$ are defined as

$$
\mathscr{L}_1 := \frac{d^4}{dt^4} \quad \text{and} \quad \mathscr{L}_2 := -\frac{d}{dt}, \tag{8.4.12}
$$

then (8.4.1) becomes the fourth-order equation

$$
\varepsilon \frac{d^4 x}{dt^4} - \frac{dx}{dt} = f(t). \tag{8.4.13}
$$

For definiteness we take the interval to be the unit interval, $0 \le t \le 1$,
and we take the particular homogeneous boundary conditions

$$
x = \frac{dx}{dt} = 0 \quad \text{at both } t = 0 \text{ and } t = 1. \tag{8.4.14}
$$

This problem (8.4.13)–(8.4.14) is a simplified one-dimensional model of a
two-dimensional problem studied by Munk (1950) and Munk and

Carrier (1950) involving the wind-driven ocean circulation. The actual two-dimensional partial differential equation studied by Munk and Carrier is

$$\varepsilon\Delta^2 x - \frac{\partial x}{\partial t_1} = f(t_1, t_2), \qquad (8.4.15)$$

where the independent variable $t = (t_1, t_2)$ ranges over a given domain in the plane, with the positive t_1 axis extending toward the east, and where Δ denotes the two-dimensional Laplacian, $\Delta := (\partial^2/\partial t_1^2) + (\partial^2/\partial t_2^2)$. The dependent variable x is a stream function for mass transport and is to be determined by the differential equation (8.4.15) subject to certain homogeneous boundary conditions related to (8.4.14). The forcing term f in (8.4.15) is the vertical component of the curl of the wind stress, and the parameter ε is the cube of the ratio of the wind wave-number to the Coriolis friction wave-number and is typically small, of order 10^{-4} to 10^{-5}. The stated problem for (8.4.15) predicts a strong boundary layer at the western (but not the eastern) boundary, in agreement with the existence of such currents as the Gulf Stream in the western Atlantic and the Kuroshio Current in the western Pacific.

Each of these examples [excluding the partial differential equation (8.4.15)] involves a differential equation of the type (8.4.1) for suitable operators $\mathscr{L}_1$ and $\mathscr{L}_2$ of the forms

$$\mathscr{L}_1 = \sum_{k=0}^{m} a_k(t, \varepsilon) \frac{d^k}{dt^k} \qquad (8.4.16)$$

and

$$\mathscr{L}_2 = \sum_{k=0}^{n} b_k(t, \varepsilon) \frac{d^k}{dt^k} \qquad (8.4.17)$$

for suitable regular functions a_k, b_k subject to various conditions such as (8.4.4) and (8.4.6). Here the order m of $\mathscr{L}_1$ is assumed to exceed the order n of $\mathscr{L}_2$:

$$m := \text{order } \mathscr{L}_1 > n := \text{order } \mathscr{L}_2. \qquad (8.4.18)$$

Also, singular points are excluded here, both for the full differential equation (8.4.1) and also for the reduced equation (8.4.2), so that the respective highest-order coefficient functions a_m and b_n are assumed to be everywhere nonzero. There is no loss, then, in assuming

$$a_m(t, \varepsilon) \equiv 1, \qquad (8.4.19)$$

and as already indicated, we assume also

$$b_n(t, \varepsilon)^2 \geq \kappa^2 > 0 \quad \text{for all } t \text{ and all small } \varepsilon \geq 0, \qquad (8.4.20)$$

for a fixed positive constant κ.

The reduced differential equation (8.4.2) is of lower order than the full equation (8.4.1), and so we expect the exact solution of the boundary-value problem for (8.4.1) to exhibit boundary-layer behavior at one or both endpoints, for small $\varepsilon > 0$. Based on our experience in the special examples of earlier sections, we expect that a suitable boundary-layer variable for the left (right) endpoint $t = t_1$ ($t = t_2$) can be given as

$$\sigma := \frac{t - t_1}{\varepsilon^\nu} \quad \text{for the left endpoint,}$$

$$\tau := \frac{t_2 - t}{\varepsilon^\nu} \quad \text{for the right endpoint,}$$

(8.4.21)

for some suitable positive exponent ν. The solution x is then represented as

$$x(t, \varepsilon) \sim X(t, \varepsilon) + {}^*X(\sigma, \varepsilon) + X^*(\tau, \varepsilon) \qquad (8.4.22)$$

for a suitable outer solution $X = X(t, \varepsilon)$, a suitable left boundary-layer correction ${}^*X = {}^*X(\sigma, \varepsilon)$, and a suitable right boundary-layer correction $X^* = X^*(\tau, \varepsilon)$.

The boundary-layer correction terms are required to satisfy the matching conditions

$$\lim_{\sigma \to \infty} {}^*X(\sigma, \varepsilon) = \lim_{\tau \to \infty} X^*(\tau, \varepsilon) = 0. \qquad (8.4.23)$$

Indeed, the boundary-layer corrections

$${}^*X\left(\frac{t - t_1}{\varepsilon^\nu}, \varepsilon\right) \quad \text{and} \quad X^*\left(\frac{t_2 - t}{\varepsilon^\nu}, \varepsilon\right)$$

are expected to be negligible (asymptotic to zero, as $\varepsilon \to 0+$) away from the endpoints, and the outer solution is required to satisfy the full equation as

$$\varepsilon \mathscr{L}_1 X + \mathscr{L}_2 X = f(t, \varepsilon) \quad \text{for } t_1 < t < t_2. \qquad (8.4.24)$$

Subtracting (8.4.24) from (8.4.1), we are led with (8.4.22) to the requirement

$$\left(\varepsilon \mathscr{L}_1 + \mathscr{L}_2\right) {}^*X + \left(\varepsilon \mathscr{L}_1 + \mathscr{L}_2\right) X^* \sim 0.$$

Then, because each of ${}^*X$ and X^* is expected to be negligible where the other is not, we can achieve this last condition by imposing the separate requirements that ${}^*X$ and X^* must each satisfy the homogeneous equation

$$\left(\varepsilon \mathscr{L}_1 + \mathscr{L}_2\right) {}^*X = 0 \qquad (8.4.25)$$

and

$$\left(\varepsilon \mathscr{L}_1 + \mathscr{L}_2\right) X^* = 0. \qquad (8.4.26)$$

The formulas (8.4.16)–(8.4.21) can be used to rewrite each of the equations (8.4.25) and (8.4.26) in terms of the respective boundary-layer variables σ and τ. For example, (8.4.25) can be rewritten in terms of the appropriate boundary-layer variable σ as (e.g., $d/dt = \varepsilon^{-\nu} d/d\sigma$, etc.)

$$\left[\frac{d^m}{d\sigma^m} + \varepsilon^{-1+(m-n)\nu} b_n(t_1 + \varepsilon^\nu \sigma, \varepsilon) \frac{d^n}{d\sigma^n} \right] *X(\sigma, \varepsilon)$$

$$+ \left(\sum_{k=0}^{m-1} \varepsilon^{(m-k)\nu} a_k \frac{d^k}{d\sigma^k} + \sum_{k=0}^{n-1} \varepsilon^{-1+(m-k)\nu} b_k \frac{d^k}{d\sigma^k} \right) *X(\sigma, \varepsilon) = 0.$$

$$(8.4.27)$$

This equation has the possibility of a nontrivial boundary-layer solution $*X$ if the parameter ν is taken to satisfy

$$-1 + (m - n)\nu = 0,$$

or

$$\nu = \frac{1}{m - n}. \qquad (8.4.28)$$

Equation (8.4.26) for X^* leads to the same choice (8.4.28) (see Exercise 8.4.1), and so this choice is imposed on ν for the boundary-layer variables σ and τ in (8.4.21).

Note that this choice (8.4.28) yields both

$$(m - k)\nu = \frac{m - k}{m - n} > 0$$

and

$$-1 + (m - k)\nu = \frac{n - k}{m - n} > 0$$

for the respective values $k = 0, 1, \ldots, m - 1$ and $k = 0, 1, \ldots, n - 1$. Hence, the dominant parts $*X(\sigma, 0)$ and $X^*(\tau, 0)$ (with $\varepsilon = 0$) of the boundary-layer correction terms $*X(\sigma, \varepsilon)$ and $X^*(\tau, \varepsilon)$ are found to satisfy the differential equations

$$\left[\frac{d^m}{d\sigma^m} + b_n(t_1, 0) \frac{d^n}{d\sigma^n} \right] *X(\sigma, 0) = 0 \quad \text{for } \sigma > 0,$$

$$\left[\frac{d^m}{d\tau^m} + (-1)^{m-n} b_n(t_2, 0) \frac{d^n}{d\tau^n} \right] X^*(\tau, 0) = 0 \quad \text{for } \tau > 0,$$

$$(8.4.29)$$

so that the highest-order terms in the operators $\mathcal{L}_1$ and $\mathcal{L}_2$ determine the boundary-layer structure. Equations such as those of (8.4.27) and (8.4.29) are called *boundary-layer correction equations*.

The choice (8.4.28) indicates that it is convenient, for bookkeeping purposes, to replace the small parameter ε in (8.4.1) with ε^{m-n}. Hence, we replace (8.4.1) with the equation

$$\varepsilon^{m-n}\mathscr{L}_1 x + \mathscr{L}_2 x = f(t, \varepsilon) \quad \text{for } t \in [t_1, t_2], \qquad (8.4.30)$$

where the differential operators $\mathscr{L}_1$ and $\mathscr{L}_2$ are given by (8.4.16) and (8.4.17), with orders m and n satisfying (8.4.18), and where $\varepsilon = \varepsilon_{\text{new}}$ is given as

$$\varepsilon_{\text{new}} = (\varepsilon_{\text{old}})^{1/(m-n)}.$$

In terms of $\varepsilon = \varepsilon_{\text{new}}$, the boundary-layer variables σ and τ of (8.4.21) are now given as

$$\sigma = \frac{t - t_1}{\varepsilon} \quad \text{and} \quad \tau = \frac{t_2 - t}{\varepsilon}, \qquad (8.4.31)$$

and we see that exactly the same equations of (8.4.29) characterize $^*X(\sigma, 0)$ and $X^*(\tau, 0)$, where now σ and τ are given by (8.4.31).

Equation (8.4.30) is, for each fixed positive value of ε, a differential equation of mth order. Hence, a typical two-point boundary-value problem of classical type for (8.4.30) should include a total of m specified boundary conditions involving the values of certain derivatives of x at the endpoints (where the value of x itself is included as the zeroth-order derivative). These boundary conditions may be of a variety of different types, including conditions that mix the boundary values at the two different endpoints, and various nonlinear boundary conditions as well. Following Wasow (1941, 1944) and O'Malley (1969a, 1969c), we shall, for simplicity, consider only the linear, separated boundary conditions given as

$$\frac{d^{k_j}x}{dt^{k_j}} = \alpha_j \quad \text{for } j = \begin{cases} 1, 2, \ldots, s & \text{at } t = t_1, \\ s+1, s+2, \ldots, m & \text{at } t = t_2, \end{cases}$$

$$(8.4.32)$$

where the positive integer s denotes the number of boundary conditions specified at the left end point $t = t_1$, and then $m - s$ denotes the number of boundary conditions specified at the right endpoint $t = t_2$. The quantities α_j are given constants in (8.4.32), and the integers k_j give the respective orders of differentiation. For convenience, and without loss, we assume that the given boundary conditions are arranged at each endpoint in *decreasing orders of differentiation*, with

$$m > k_1 > k_2 > \cdots > k_s \geq 0,$$

$$m > k_{s+1} > k_{s+2} > \cdots > k_m \geq 0. \qquad (8.4.33)$$

Hence, we consider the boundary-value problem consisting of the differential equation (8.4.30) along with the boundary conditions (8.4.32)–(8.4.33), and we assume that this problem has a unique solution, at least for all small enough values of the positive parameter ε. [We can prove the existence of a unique solution provided that the data satisfy any one of various suitable conditions analogous to the previous conditions (8.4.4) and (8.4.6).] We now seek an asymptotic representation for the solution x in the form (8.4.22) with $\varepsilon = \varepsilon_{\text{new}}$, where the functions X, $*X$, and X^* are sought in the forms

$$\begin{bmatrix} X(t,\varepsilon) \\ *X(\sigma,\varepsilon) \\ X^*(\tau,\varepsilon) \end{bmatrix} \sim \sum_{k=0}^{\infty} \begin{bmatrix} X_k(t) \\ *X_k(\sigma) \\ X_k^*(\tau) \end{bmatrix} \varepsilon^k \qquad (8.4.34)$$

for suitable functions X_k, $*X_k$, and X_k^*, where the boundary-layer coefficient functions $*X_k$ and X_k^* are required to satisfy [see (8.4.23)]

$$\lim_{\sigma \to \infty} *X_k(\sigma) = \lim_{\tau \to \infty} X_k^*(\tau) = 0. \qquad (8.4.35)$$

Of course, as mentioned earlier, such a solution as indicated here will exist for the given boundary-value problem only subject to certain restrictions or conditions, analogous to the earlier conditions (8.3.43), (8.4.4), and (8.4.6).

Before considering this boundary-value problem for the general equation (8.4.30), it will be useful first to consider again the special cases given earlier with the operators $\mathscr{L}_1$ and $\mathscr{L}_2$, as in (8.4.3) and (8.4.5).

Consider first the case (8.4.3), with $\mathscr{L}_1$ and $\mathscr{L}_2$ given as

$$\mathscr{L}_1 = \frac{d^2}{dt^2} \quad \text{and} \quad \mathscr{L}_2 = a(t,\varepsilon)\frac{d}{dt} + b(t,\varepsilon), \qquad (8.4.36)$$

so that $m = 2$ and $n = 1$, and the coefficient $b_n \equiv b_1$ of (8.4.17) is given by the coefficient $a(t,\varepsilon)$ of (8.4.36):

$$b_n(t,\varepsilon) := a(t,\varepsilon).$$

Hence, in this case the boundary-layer correction equations of (8.4.29) become

$$\left[\frac{d^2}{d\sigma^2} + a(t_1,0)\frac{d}{d\sigma}\right]*X_0(\sigma) = 0 \quad \text{for } \sigma > 0,$$

$$\left[\frac{d^2}{d\tau^2} - a(t_2,0)\frac{d}{d\tau}\right]X_0^*(\tau) = 0 \quad \text{for } \tau > 0,$$

$$(8.4.37)$$

with $*X_0(\sigma) := *X(\sigma,0)$ and $X_0^*(\tau) := X^*(\tau,0)$ as in (8.4.34). The most general solutions of these differential equations of (8.4.37) can be given as

$$^*X_0(\sigma) = c_1 + c_2 \cdot e^{-a(t_1,0)\sigma},$$
$$X_0^*(\tau) = c_3 + c_4 \cdot e^{+a(t_2,0)\tau}, \qquad (8.4.38)$$

for suitable integration constants c_j ($j = 1, 2, 3, 4$).

It follows then from (8.4.38), along with the matching conditions of (8.4.35), that a suitable boundary-layer function exists either for $^*X_0$ or for X_0^* if and only if there holds, respectively,

$$a(t_1,0) > 0 \Leftrightarrow \begin{cases} ^*X_0 \text{ exists with } \lim_{\sigma \to \infty} {}^*X_0(\sigma) = 0, \\ \text{and with } c_1 = 0 \text{ in } (8.4.38), \end{cases} \qquad (8.4.39)$$

or

$$a(t_2,0) < 0 \Leftrightarrow \begin{cases} X_0^* \text{ exists with } \lim_{\tau \to \infty} X_0^*(\tau) = 0, \\ \text{and with } c_3 = 0 \text{ in } (8.4.38). \end{cases} \qquad (8.4.40)$$

Because we are here excluding turning points, with $a(t, \varepsilon)$ assumed to be everywhere of one sign [see (8.4.4) or (8.4.20)], it follows that we obtain a boundary layer at the left (right) endpoint if and only if $a > 0$ ($a < 0$), and in this case the resulting boundary-layer function $^*X_0(\sigma)$ [$X_0^*(\tau)$] involves only a single free parameter, denoted as c_2 (c_4) in (8.4.38). Moreover, precisely one (not both) of the conditions $a > 0$ and $a < 0$ occurs, so that precisely one boundary-layer function exists, either $^*X$ or X^*. In this case the outer solution X generally drops the specified boundary condition at the endpoint where the boundary layer occurs, so that X drops the boundary condition at the left (right) endpoint if $a > 0$ ($a < 0$). The single boundary-layer function, with its single free parameter, can handle precisely one boundary condition and can thus provide a single boundary-layer correction.

Consider now the special case (8.4.5), with $\mathcal{L}_1$ and $\mathcal{L}_2$ given as

$$\mathcal{L}_1 = \frac{d^2}{dt^2} \quad \text{and} \quad \mathcal{L}_2 = b(t, \varepsilon), \qquad (8.4.41)$$

so that $m = 2$ and $n = 0$, and the coefficient $b_n \equiv b_0$ of (8.4.17) is given by the coefficient $b(t, \varepsilon)$ of (8.4.41):

$$b_n(t, \varepsilon) := b(t, \varepsilon).$$

Hence, in this case the boundary-layer correction equations of (8.4.29) become

$$
\left[\frac{d^2}{d\sigma^2} + b(t_1,0)\right] {}^*X_0(\sigma) = 0,
$$

$$
\left[\frac{d^2}{d\tau^2} + b(t_2,0)\right] X_0^*(\tau) = 0,
$$

(8.4.42)

with general solutions

$$
{}^*X_0(\sigma) = c_1 \cdot \exp\left[-\sqrt{-b(t_1,0)}\,\sigma\right] + c_2 \cdot \exp\left[+\sqrt{-b(t_1,0)}\,\sigma\right],
$$

$$
X_0^*(\tau) = c_3 \cdot \exp\left[-\sqrt{-b(t_2,0)}\,\tau\right] + c_4 \cdot \exp\left[+\sqrt{-b(t_2,0)}\,\tau\right],
$$

(8.4.43)

for suitable constants c_j ($j = 1, 2, 3, 4$). These solutions are oscillatory if the function b is positive-valued, and in this case the matching conditions of (8.4.35) cannot be satisfied. Moreover, a typical two-point boundary-value problem for (8.4.30) fails to have a unique solution in this case. On the other hand, if b is negative-valued, as in (8.4.6), then the solutions given by (8.4.43) provide suitable boundary-layer functions subject to (8.4.35) with

$$
c_2 = c_4 = 0.
$$

In this latter case, both boundary-layer functions exist, and each boundary-layer function involves a single remaining free parameter [either c_1 or c_3 in (8.4.43)], so that each boundary layer can handle one boundary condition at the appropriate endpoint. A boundary-layer correction is available at each endpoint, and the outer solution X generally drops a single boundary condition at each endpoint. In this latter case both boundary conditions are cancelled for the two-point boundary-value problem.

Example 8.4.3: Consider again the one-dimensional model of the problem of Munk and Carrier discussed briefly in Example 8.4.2. We replace the previous parameter ε of (8.4.13) by $\varepsilon^3 = \varepsilon^{m-n}$, as in (8.4.30), so that the differential equation is written now as

$$
\varepsilon^3\frac{d^4x}{dt^4} - \frac{dx}{dt} = f(t) \quad \text{for } 0 < t < 1,
$$

(8.4.44)

subject also to the boundary conditions

$$
x = \frac{dx}{dt} = 0 \quad \text{at both } t = 0 \text{ and } t = 1.
$$

(8.4.45)

The operators $\mathscr{L}_1$ and $\mathscr{L}_2$ are given by (8.4.12), and we have $m = 4$ and $n = 1$. The boundary-layer correction equations of (8.4.29) become

$$\left[\frac{d^4}{d\sigma^4} - \frac{d}{d\sigma}\right] *X_0(\sigma) = 0 \quad \text{for } \sigma > 0,$$

$$\left[\frac{d^4}{d\tau^4} + \frac{d}{d\tau}\right] X_0^*(\tau) = 0 \quad \text{for } \tau > 0,$$

(8.4.46)

where the equation here for $*X_0$ has two linearly independent solutions vanishing as $\sigma \to \infty$, and the equation for X_0^* has only one independent solution vanishing as $\tau \to \infty$. The resulting general solutions for $*X_0$ and X_0^* satisfying the matching conditions of (8.4.35) can be given as

$$*X_0(\sigma) = c_1 \cdot e^{\lambda_1 \sigma} + c_2 \cdot e^{\lambda_2 \sigma}, \qquad X_0^*(\tau) = c_3 \cdot e^{\mu_1 \tau}, \quad (8.4.47)$$

where λ_1 and λ_2 are the appropriate solutions, with negative real parts, of the characteristic equation

$$\lambda^4 - \lambda = 0,$$

and μ_1 is the solution, with negative real part, of the characteristic equation

$$\mu^4 + \mu = 0.$$

Hence,

$$\lambda_1 = \frac{-1 + i\sqrt{3}}{2}, \qquad \lambda_2 = \frac{-1 - i\sqrt{3}}{2}, \qquad \mu_1 = -1. \quad (8.4.48)$$

It follows in this case that the boundary layer at the left endpoint can handle two (cancelled) boundary conditions, and the boundary layer at the right endpoint can handle only one boundary condition. The outer solution X drops two boundary conditions at $t = 0$ and drops one condition at $t = 1$, starting with the highest-order boundary conditions at each endpoint. In particular, the leading term $X_0(t)$ in the outer expansion from (8.4.34) is determined by the reduced, first-order differential equation

$$-\frac{dX_0}{dt} = f(t) \quad \text{for } 0 < t < 1, \quad (8.4.49)$$

subject to the remaining (uncancelled) boundary condition at $t = 1$:

$$X_0 = 0 \quad \text{at } t = 1. \quad (8.4.50)$$

Hence

$$X_0(t) = \int_t^1 f.$$

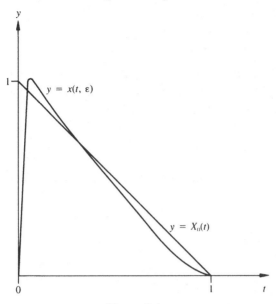

Figure 8.4

The graph of X_0 is indicated in Figure 8.4 along with the graph of the exact solution function $x = x(t, \varepsilon)$, for small $\varepsilon > 0$ and for the special choice $f(t) \equiv 1$ for all t.

For small ε, we find that the exact solution $x(t, \varepsilon)$ exhibits a strong boundary-layer effect involving both x and x' near the left endpoint, whereas the boundary-layer effect near the right endpoint is much weaker, involving only x'. That is, the strong boundary-layer effect occurs at the "western" boundary, as discussed in Example 8.4.2.

In each of the foregoing special cases given by (8.4.36), (8.4.41), and (8.4.44), we find that the boundary-layer correction equations of (8.4.29) yield the *cancellation rule* that determines the number of boundary conditions to be dropped or cancelled at each endpoint by the outer solution. In general, at each endpoint we seek to cancel a certain number of boundary conditions, with this number determined for each endpoint by the number of (stable) characteristic values having negative real part for the appropriate equation of (8.4.29) corresponding to the given endpoint. Alternatively, we need only consider the following *boundary-layer characteristic polynomial equation* for the differential equation (8.4.30):

$$\lambda^{m-n} + b_n(t, \varepsilon) = 0, \qquad (8.4.51)$$

where $\lambda = \lambda(t, \varepsilon)$ is a corresponding characteristic value obtained from

(8.4.51) as

$$\lambda = \left[-b_n(t, \varepsilon)\right]^{1/(m-n)}. \tag{8.4.52}$$

The boundary-layer characteristic equation (8.4.51) has $m - n$ roots, given by (8.4.52). These roots are distributed in the complex plane like the $m - n$ distinct $(m - n)$th roots of either $+1$ or -1. Let N and Π denote the numbers of roots given by (8.4.52) having, respectively, negative and positive real parts:

$$\left.\begin{matrix} N \\ \Pi \end{matrix}\right\} := \text{number of roots } \lambda \text{ of } (8.4.51)-(8.4.52)$$

$$\text{having}\left\{\begin{matrix} \text{negative} \\ \text{positive} \end{matrix}\right. \text{real parts.} \tag{8.4.53}$$

It is possible that there are no roots having zero real part, in which case

$$N + \Pi = m - n. \tag{8.4.54}$$

On the other hand, if there are some roots with zero real part, then there are precisely two such roots, in which case

$$N + \Pi = m - n - 2. \tag{8.4.55}$$

Hence, precisely these two possibilities (8.4.54) and (8.4.55), and no others, exist. Note that roots of (8.4.51) having positive real parts correspond to characteristic values having negative real parts for the second equation of (8.4.29). The coefficient function b_n appearing in (8.4.51)–(8.4.52) is assumed here to be of one sign [see (8.4.20)], uniformly for all $t \in [t_1, t_2]$ and for all small $\varepsilon > 0$, so that the actual values for N and Π in (8.4.53) are given by suitable fixed integers that are independent of t and ε.

In both cases (8.4.54) and (8.4.55), the cancellation rule for the two-point boundary-value problem (8.4.30) and (8.4.32) asks that we cancel N (Π) boundary conditions at the left (right) endpoint, proceeding always with those boundary conditions involving the highest-order derivatives. In the case (8.4.55), we also cancel successively from the remaining boundary conditions those two with the successive highest orders of differentiation, requiring that they both belong to the same endpoint and that their selection must be unambiguous, as will be illustrated by example later.

In this way we seek to obtain a suitable *reduced problem* for the leading outer function X_0, consisting of the reduced nth-order differential equation (8.4.2) with $\varepsilon = 0$, along with a suitable collection of n

boundary conditions obtained presumably as a subset of the m conditions of (8.4.32). Such a reduced problem may or may not exist. Moreover, if it exists, it may have no solution, a unique solution, or many solutions. In general, we must introduce suitable stability conditions such as (8.4.4) or (8.4.6) [see also (8.3.43)] in order to ensure that a suitable reduced problem exists that characterizes the limiting "full" solution away from the endpoints. A complete discussion of the (somewhat complicated) cancellation rule for such problems can be found in Wasow (1944, 1965, 1976), O'Malley (1968b, 1969b, 1969c, 1974a), and O'Malley and Keller (1968) and is omitted here. We shall be content to illustrate certain aspects of the cancellation rule by example.

Example 8.4.4: The differential equation

$$\varepsilon^2 \frac{d^4x}{dt^4} + \frac{d^2x}{dt^2} = 2 \quad \text{for } 0 \le t \le 1 \tag{8.4.56}$$

has $m = 4$ and $n = 2$, and the boundary-layer characteristic polynomial equation (8.4.51) is

$$\lambda^2 + 1 = 0, \tag{8.4.57}$$

with purely imaginary roots $\lambda_1 = \sqrt{-1} = i$ and $\lambda_2 = -i$. Hence, we have [see (8.4.53)]

$$N = \Pi = 0, \tag{8.4.58}$$

with $N + \Pi = m - n - 2$, so that we have the case (8.4.55). In the present case, then, the cancellation rule merely asks that we cancel two boundary conditions with the highest orders of differentiation, requiring that they belong to the same endpoint and that their selection be unambiguous.

For example, if we consider equation (8.4.56) subject to the boundary conditions

$$x'' = x' = x = 0 \quad \text{at } t = 0,$$
$$x = 0 \quad \text{at } t = 1, \tag{8.4.59}$$

then, according to the cancellation rule, we cancel the two boundary conditions $x'' = x' = 0$ at $t = 0$, so that the leading term $X_0(t)$ in the outer solution is characterized by the reduced problem

$$\frac{d^2X_0}{dt^2} = 2 \quad \text{for } 0 < t < 1,$$
$$X_0(0) = 0 \quad \text{and} \quad X_0(1) = 0. \tag{8.4.60}$$

Hence, we find

$$X_0(t) = t^2 - t. \tag{8.4.61}$$

The exact solution $x = x(t, \varepsilon)$ of (8.4.56) subject to the boundary conditions of (8.4.59) is

$$x(t, \varepsilon) = \varepsilon \frac{1 - 2\varepsilon^2\left(1 - \cos\dfrac{1}{\varepsilon}\right)}{1 - \varepsilon \sin\dfrac{1}{\varepsilon}} \sin\frac{t}{\varepsilon} - 2\varepsilon^2\left(1 - \cos\frac{t}{\varepsilon}\right)$$

$$+ t^2 + \frac{-1 + 2\varepsilon^2\left(1 - \cos\dfrac{1}{\varepsilon}\right)}{1 - \varepsilon \sin\dfrac{1}{\varepsilon}} t, \tag{8.4.62}$$

and we find the result

$$\lim_{\varepsilon \to 0+} x(t, \varepsilon) = X_0(t).$$

In this case the exact solution oscillates rapidly about $X_0(t)$ as $\varepsilon \to 0$, and the matching conditions of (8.4.35) do not apply.

Suppose now that we consider again the same differential equation (8.4.56), but subject now to the boundary conditions

$$\begin{aligned} x' = x = 0 \quad &\text{at } t = 0, \\ x' = x = 0 \quad &\text{at } t = 1. \end{aligned} \tag{8.4.63}$$

The cancellation rule again asks that we cancel two boundary conditions of highest orders at a single endpoint, with the requirement that their selection be unambiguous. But in the present case this is not possible, because there is no natural way to distinguish between the two boundary conditions of (8.4.63) at $t = 0$ and the two conditions at $t = 1$. Hence, in this case the cancellation rule is said to be *undefined*, which simply means that there is *no well-defined reduced problem* for X_0. And, in fact, the exact solution $x(t, \varepsilon)$ of (8.4.56) subject to (8.4.63) fails to exist if $\sin(1/\varepsilon) = 0$, but if $\sin(1/\varepsilon) \neq 0$,

$$x(t, \varepsilon) = \varepsilon \sin\frac{t}{\varepsilon} + \frac{\varepsilon\left(1 + \cos\dfrac{1}{\varepsilon}\right)}{\sin\dfrac{1}{\varepsilon}} \cos\frac{t}{\varepsilon}$$

$$- \frac{\varepsilon\left(1 + \cos\dfrac{1}{\varepsilon}\right)}{\sin\dfrac{1}{\varepsilon}} + t^2 - t, \quad \text{if } \sin\frac{1}{\varepsilon} \neq 0. \tag{8.4.64}$$

In this case the limit of $x(t, \varepsilon)$ as $\varepsilon \to 0$ *fails to exist*.

Example 8.4.5: Consider the boundary-value problem

$$\varepsilon^3 \frac{d^3x}{dt^3} - x = 1 \quad \text{for } 0 < t < 1,$$

(8.4.65)

$$x = 0 \quad \text{at } t = 0, \quad \text{and} \quad x' = x = 0 \quad \text{at } t = 1.$$

The third-order differential equation here is an integrated version of a special case of the previous fourth-order equation (8.4.44). The boundary-layer characteristic equation (8.4.51) becomes

$$\lambda^3 - 1 = 0,$$

with roots

$$\lambda_1 = \frac{-1 + i\sqrt{3}}{2}, \qquad \lambda_2 = \frac{-1 - i\sqrt{3}}{2}, \qquad \lambda_3 = +1,$$

so that there are two roots with negative real part and one root with positive real part. Hence, the cancellation rule asks that we cancel two boundary conditions at $t = 0$ and one boundary condition at $t = 1$. This is clearly not possible for the problem (8.4.65), because there is only one specified boundary condition at $t = 0$. Hence, the cancellation rule is undefined for the problem (8.4.65): there is no well-defined reduced problem for X_0. We find that the exact solution $x = x(t, \varepsilon)$ becomes exponentially unbounded as $\varepsilon \to 0+$, for each fixed $0 < t < 1$.

Example 8.4.6: The problem

$$\varepsilon x'' - x' = 0 \quad \text{for } 0 < t < 1,$$
$$x' = 1 \quad \text{at } t = 0, \quad \text{and} \quad x = 0 \quad \text{at } t = 1,$$

(8.4.66)

has the boundary-layer characteristic equation $\lambda - 1 = 0$, with root $\lambda_1 = 1$. Hence, we cancel one boundary condition at $t = 1$, so that the reduced problem is

$$-\frac{dX_0}{dt} = 0 \quad \text{for } 0 < t < 1,$$
$$\frac{dX_0}{dt} = 1 \quad \text{at} \quad t = 0.$$

(8.4.67)

The reduced problem is well defined, but inconsistent. *It has no solution.* The exact "full" solution is

$$x = x(t, \varepsilon) = -\varepsilon e^{1/\varepsilon}\left(1 - e^{-(1-t)/\varepsilon}\right),$$

and we see that $x(t, \varepsilon)$ becomes exponentially unbounded as $\varepsilon \to 0+$, for each fixed $0 \le t < 1$.

O'Malley (1974a) gives additional examples along with a general discussion of the problem (8.4.30)–(8.4.33). O'Malley also discusses related problems involving several small parameters.

These boundary-layer methods have been used in developing algorithms for numerical solution of boundary-value problems for singularly perturbed equations, including both higher-order scalar equations and various first-order systems, for both linear and suitable nonlinear problems, as discussed in Chapter 10.

Exercises

Exercise 8.4.1: Give the details of the derivation of the second equation of (8.4.29), for $X^*(\tau, 0)$.

Exercise 8.4.2: Consider the problem [compare with (8.4.44)–(8.4.45)]

$$\varepsilon^3 \frac{d^4 x}{dt^4} - (1 + t)^2 \frac{dx}{dt} = 1 \quad \text{for } 0 < t < 1,$$

$$\frac{dx}{dt} = x = 0 \quad \text{at both } t = 0 \text{ and } \quad t = 1. \tag{8.4.68}$$

Find the leading terms in a suitable approximation for the exact solution, obtained from (8.4.22) and (8.4.34). In particular, find

$$x^0(t, \varepsilon) := X_0(t) + {}^*X_0\left(\frac{t}{\varepsilon}\right) + X_0^*\left(\frac{1 - t}{\varepsilon}\right).$$

This function x^0 actually provides a uniform approximation to the exact solution function x, but you need not prove this here.

Exercise 8.4.3: Find a uniform approximation to the solution x of the problem

$$\varepsilon^3 \frac{d^4 x}{dt^4} - (1 + t)^2 \frac{dx}{dt} = 1 \quad \text{for } 0 < t < 1,$$

$$x'' = x' = x = 0 \quad \text{at } t = 0, \quad \text{and} \quad x' = 0 \quad \text{at } t = 1. \tag{8.4.69}$$

In particular, find the limiting value of $x(1, \varepsilon)$ at $t = 1$ as $\varepsilon \to 0+$.

8.5 Examples on interior transition points

It was seen in Section 8.1 that the second-order equation

$$\varepsilon \frac{d^2 x}{dt^2} + a(t) \frac{dx}{dt} + b(t) x = 0 \quad \text{for } t \in [t_1, t_2] \tag{8.5.1}$$

is transformed into

$$\varepsilon^2 \frac{d^2 \bar{x}}{dt^2} - c(t, \varepsilon) \bar{x} = 0 \quad \text{for } t \in [t_1, t_2] \tag{8.5.2}$$

by the Sturm transformation (8.1.7), with the coefficient c given by (8.1.8) as

$$c(t, \varepsilon) = \tfrac{1}{4}\left\{ a(t)^2 + 2\varepsilon[a'(t) - 2b(t)] \right\}. \tag{8.5.3}$$

As noted in Section 8.1, equation (8.5.2) is oscillatory if the coefficient c is everywhere uniformly negative, and the equation is nonoscillatory if c is everywhere nonnegative.

Isolated zeros of c are known as *transition points* or *turning points* (both for the given homogeneous differential equation and also for the corresponding inhomogeneous equation); see Wasow (1965, 1984). Solutions of the given differential equation may undergo rapid variations of boundary-layer type across a transition point. Such "boundary layers" located in the interior away from the boundary are known as *interior layers*. The study of equations with transition points is of interest partly because of applications in various fields such as quantum mechanics, elasticity theory, and fluid mechanics. For example, in the flow of a viscous fluid between two coaxial rotating disks, the angular velocity of the fluid satisfies a certain nonlinear equation related to (8.5.1), with ε representing the kinematic viscosity and with the transition point corresponding to a change in sign of the angular velocity [see Watts (1971) and Lakin (1972) and additional references listed there].

In this section we list several examples from Wasow (1941) and O'Malley (1970*b*, 1974*a*) [see also Dorr (1970)] involving a single interior transition point t_0, and in fact we consider only equation (8.5.1) on the interval $[-1, +1]$, with the transition point at the origin, for the special case in which the coefficient function a is given either as

$$a(t) = +2t \quad \text{for } -1 \leq t \leq 1 \tag{8.5.4}$$

or as

$$a(t) = -2t \quad \text{for } -1 \leq t \leq 1, \tag{8.5.5}$$

along with various special choices for the coefficient function b. Note that the function a of (8.5.4) or (8.5.5) changes sign with a simple zero at $t_0 = 0$. These special cases (8.5.4) and (8.5.5) can in fact be used to study equation (8.5.1) in the slightly more general case of an arbitrary smooth function a possessing an isolated interior simple zero; see O'Malley (1970*b*, 1974*a*) for a discussion, with further references, on the reduction of a more general equation to a suitable special case through the use of suitable transformations.

In the following we shall consider boundary conditions of Dirichlet type, as

$$x = \alpha \quad \text{at } t = -1, \quad \text{and} \quad x = \beta \quad \text{at } t = +1. \tag{8.5.6}$$

Note that the solution of (8.5.1) and (8.5.6) can be given explicitly as

$$x(t,\varepsilon) = \frac{\alpha \int_t^1 \exp\left(-\frac{1}{\varepsilon}\int_0^s a\right) ds + \beta \int_{-1}^t \exp\left(-\frac{1}{\varepsilon}\int_0^s a\right) ds}{\int_{-1}^1 \exp\left(-\frac{1}{\varepsilon}\int_0^s a\right) ds} \qquad (8.5.7)$$

in the case $b \equiv 0$ in (8.5.1).

Example 8.5.1: The equation

$$\varepsilon x'' + 2tx' = 0 \qquad (8.5.8)$$

corresponds to the case (8.5.4), along with $b \equiv 0$. The solution (8.5.7) of the Dirichlet problem becomes

$$x(t,\varepsilon) = \frac{\alpha \int_t^1 \exp\left(-\frac{s^2}{\varepsilon}\right) ds + \beta \int_{-1}^t \exp\left(-\frac{s^2}{\varepsilon}\right) ds}{\int_{-1}^1 \exp\left(-\frac{s^2}{\varepsilon}\right) ds}, \qquad (8.5.9)$$

and we easily obtain by direct calculation the result (see Exercise 8.5.1)

$$X(t) := \lim_{\substack{\varepsilon \to 0+ \\ \text{fixed } t}} x(t,\varepsilon) = \begin{cases} \alpha & \text{for } -1 \le t < 0, \\ \dfrac{\alpha + \beta}{2} & \text{for } \quad t = 0, \\ \beta & \text{for } \quad 0 < t \le 1, \end{cases} \qquad (8.5.10)$$

as indicated in Figure 8.5. There are no boundary layers at the endpoints, but there is an internal layer at the transition point $t = 0$. The limiting solution X of (8.5.10) satisfies the reduced equation $2tX'(t) = 0$ away from the transition point.

Example 8.5.2: The equation

$$\varepsilon x'' - 2tx' = 0 \qquad (8.5.11)$$

corresponds to the case (8.5.5), along with $b \equiv 0$. The solution (8.5.7) now becomes

$$x(t,\varepsilon) = \frac{\alpha \int_t^1 \exp\left(\frac{s^2}{\varepsilon}\right) ds + \beta \int_{-1}^t \exp\left(\frac{s^2}{\varepsilon}\right) ds}{\int_{-1}^1 \exp\left(\frac{s^2}{\varepsilon}\right) ds}, \qquad (8.5.12)$$

and in this case direct calculation gives (see Exercise 8.5.2)

$$X(t) := \lim_{\substack{\varepsilon \to 0+ \\ \text{fixed } t}} x(t,\varepsilon) = \begin{cases} \alpha & \text{for } \quad t = -1, \\ \dfrac{\alpha + \beta}{2} & \text{for } -1 < t < 1, \\ \beta & \text{for } \quad t = 1, \end{cases} \qquad (8.5.13)$$

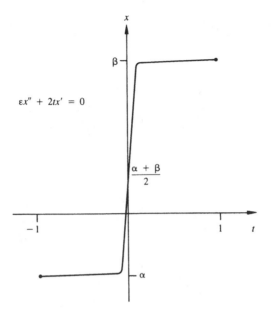

Figure 8.5

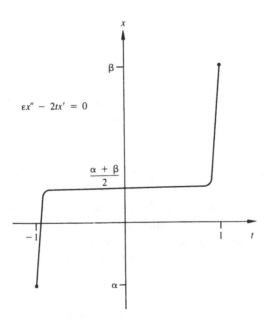

Figure 8.6

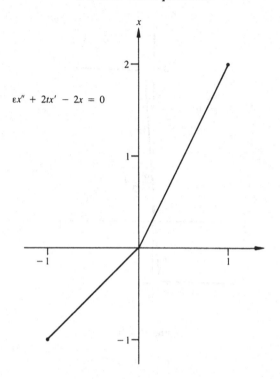

$$\varepsilon x'' + 2tx' - 2x = 0$$

Figure 8.7

as indicated in Figure 8.6. Hence, for small $\varepsilon > 0$, we have $x(t, \varepsilon) \sim$ $(\alpha + \beta)/2$ everywhere except near the endpoints, and boundary layers occur at each endpoint. There is no interior layer at the transition point. The limiting solution again satisfies the reduced equation away from the layers.

The coefficient b of (8.5.1) vanishes in the equations (8.5.8) and (8.5.11) of the two previous examples. We now list further examples with $b \neq 0$ so as to illustrate a diversity of possible behaviors. We first consider several examples with a given as $a(t) = 2t$, as in (8.5.4), and then later we take a as in (8.5.5). In most cases we list only the results, without details. Proofs are given in O'Malley (1970b, 1974a).

Example 8.5.3: The problem

$$\varepsilon x'' + 2tx' - 2x = 0 \quad \text{for } -1 < t < 1, \qquad (8.5.14)$$

$$x(-1) = -1 \quad \text{and} \quad x(1) = 2, \qquad (8.5.15)$$

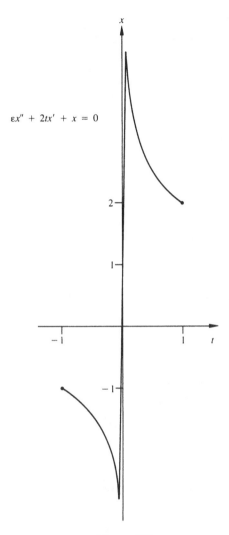

$$\varepsilon x'' + 2tx' + x = 0$$

Figure 8.8

has a solution $x(t) = x(t, \varepsilon)$ that satisfies

$$X(t) := \lim_{\substack{\varepsilon \to 0+ \\ \text{fixed } t}} x(t, \varepsilon) = \begin{cases} t & \text{for } -1 \le t < 0, \\ 2t & \text{for } 0 < t \le 1. \end{cases} \quad (8.5.16)$$

The limiting solution X of (8.5.16) satisfies the reduced equation $2tX' - 2X = 0$, except at the transition point. The exact solution x is uniformly continuous, but its first derivative experiences an interior layer at the transition point $t = 0$, as indicated in Figure 8.7.

Example 8.5.4: The problem

$$\varepsilon x'' + 2tx' + x = 0 \quad \text{for } -1 < t < 1,$$
$$x(-1) = -1 \quad \text{and} \quad x(1) = 2,$$

(8.5.17)

has a solution $x(t) = (x(t, \varepsilon)$ that satisfies

$$X(t) := \lim_{\substack{\varepsilon \to 0+ \\ \text{fixed } t}} x(t, \varepsilon) = \begin{cases} -(-t)^{-1/2} & \text{for } -1 \le t < 0, \\ 2t^{-1/2} & \text{for } 0 < t \le 1. \end{cases}$$

(8.5.18)

The limiting solution satisfies the reduced equation away from the transition point, and the exact solution x has a severe interior layer there, as indicated in Figure 8.8.

Example 8.5.5: The problem

$$\varepsilon x'' + 2tx' + 2x = 0 \quad \text{for } -1 < t < 1,$$
$$x(-1) = \alpha \quad \text{and} \quad x(1) = \beta,$$

(8.5.19)

can be solved explicitly to find [note that $2tx' + 2x \equiv 2(tx)'$]

$$x(t, \varepsilon) = \frac{\alpha \int_t^1 \exp\left(\dfrac{s^2 - t^2}{\varepsilon}\right) ds + \beta \int_{-1}^t \exp\left(\dfrac{s^2 - t^2}{\varepsilon}\right) ds}{\int_{-1}^1 \exp\left(\dfrac{s^2 - 1}{\varepsilon}\right) ds}.$$

(8.5.20)

A direct calculation then gives the result

$$X(t) := \lim_{\substack{\varepsilon \to 0+ \\ \text{fixed } t}} x(t, \varepsilon) = \begin{cases} \alpha & \text{for} \quad t = -1, \\ \infty & \text{for} \quad -1 < t < 1, \\ \beta & \text{for} \quad t = 1, \end{cases}$$

(8.5.21)

if $\alpha + \beta \ne 0$. For example, we find at $t = 0$ the result $x(0, \varepsilon) = [(\alpha + \beta)/2]e^{1/\varepsilon}$, so that $x(0, \varepsilon)$ becomes exponentially unbounded as $\varepsilon \to 0+$ if $\alpha \ne -\beta$. Note that there are boundary layers at the end-points, and in this case the limiting solution (8.5.21) does not satisfy the reduced differential equation. The graph of the exact solution is indicated in Figure 8.9 for the boundary values $\alpha = -1$ and $\beta = +2$.

We now list a few examples with the coefficient $a = -2t$, as in Example 8.5.2, but with $b \ne 0$.

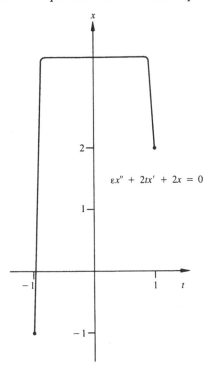

Figure 8.9

Example 8.5.6: The problem

$$\varepsilon x'' - 2tx' + x = 0 \quad \text{for } -1 < t < 1,$$
$$x(-1) = \alpha \quad \text{and} \quad x(1) = \beta, \qquad (8.5.22)$$

has a solution $x(t) = x(t, \varepsilon)$ that satisfies

$$X(t) := \lim_{\substack{\varepsilon \to 0+ \\ \text{fixed } t}} x(t, \varepsilon) = \begin{cases} \alpha & \text{for} \quad t = -1, \\ 0 & \text{for} \quad -1 < t < 1, \\ \beta & \text{for} \quad t = 1, \end{cases} \qquad (8.5.23)$$

as indicated in Figure 8.10. There are boundary layers at the endpoints. Away from the endpoints, the solution decays toward zero, and the limiting solution (8.5.23) satisfies the reduced equation there.

In the previously mentioned application in fluid mechanics for the problem of the flow of a viscous fluid between coaxial rotating disks, there is experimental and theoretical evidence [see Stewartson (1953)] that the result of Example 8.5.6 is in qualitative agreement with the corresponding result for the angular velocity in the case of two counter rotating disks (rotating in opposite directions), where the main body of the fluid is almost at rest. See Parter (1982) for a survey on this problem.

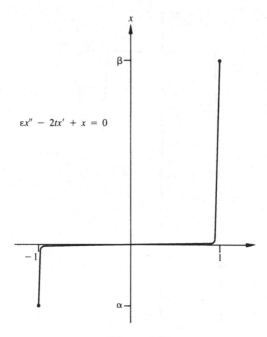

Figure 8.10

Example 8.5.7: The problem

$$\varepsilon x'' - 2tx' + 2x = 0 \quad \text{for } -1 < t < 1,$$
$$x(-1) = \alpha \quad \text{and} \quad x(1) = \beta, \tag{8.5.24}$$

has a solution $x(t) = x(t, \varepsilon)$ that satisfies

$$X(t) := \lim_{\substack{\varepsilon \to 0+ \\ \text{fixed } t}} x(t, \varepsilon) = \begin{cases} \alpha & \text{for } t = -1, \\ \dfrac{\beta - \alpha}{2} t & \text{for } -1 < t < 1, \\ \beta & \text{for } t = 1, \end{cases} \tag{8.5.25}$$

as indicated in Figure 8.11. The reduced differential equation is the same here as in Example 8.5.3, where there were no boundary layers at the endpoints in Example 8.5.3. However, in the present case there are boundary layers at the endpoints. The limiting solution (8.5.25) satisfies the reduced equation away from the endpoints.

The foregoing examples are all special cases of the problem

$$\varepsilon x'' + tA(t, \varepsilon)x' + B(t, \varepsilon)x = 0 \quad \text{for } -1 < t < 1,$$
$$x(-1) = \alpha \quad \text{and} \quad x(1) = \beta, \tag{8.5.26}$$

for given smooth functions A and B, with A everywhere nonzero on $[-1, 1]$. The general situation regarding (8.5.26) depends on the sign of

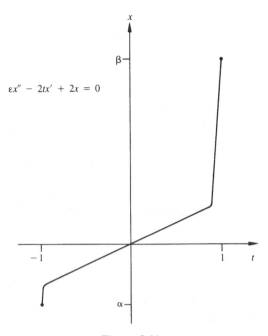

$$\varepsilon x'' - 2tx' + 2x = 0$$

Figure 8.11

$A(t, \varepsilon)$. When A is everywhere negative, it is shown in O'Malley (1970b) that the solution generally decays toward zero on compact subsets of $[-1, 1]$ as $\varepsilon \to 0+$, as in Example 8.5.6. However, Ackerberg and O'Malley (1970) point out that there are certain special cases in which the solution fails to decay toward zero, as in Examples 8.5.2 and 8.5.7. Ackerberg and O'Malley show that the condition

$$-\frac{B(0,0)}{A(0,0)} = \text{integer } k \in \{0, 1, 2, \ldots\} \qquad (8.5.27)$$

in the case $A < 0$ is a necessary condition for this special behavior known today as *Ackerberg/O'Malley resonance*. Note that the quantity $-B(0,0)/A(0,0)$ has the values 0 and 1 in the respective cases of Examples 8.5.2 and 8.5.7.

When A is everywhere positive, it is shown in O'Malley (1970b) that the solution generally is bounded as $\varepsilon \to 0+$, except near the transition point $t = 0$, where the solution may become *algebraically* unbounded, like $1/\varepsilon^\kappa$ for some positive constant κ. Such typical behavior occurs in Examples 8.5.1, 8.5.3, and 8.5.4. There are exceptional cases, however, in which the solution exhibits a certain exponential growth, a special case of which is illustrated in Example 8.5.5. Ackerberg and O'Malley show in this latter case that the condition

$$+\frac{B(0,0)}{A(0,0)} = \text{integer } k \in \{1,2,\dots\} \qquad (8.5.28)$$

in the case $A > 0$ is a necessary condition for such special behavior, which is again called Ackerberg/O'Malley resonance. Note that the quantity $B(0,0)/A(0,0)$ has the value $+1$ in Example 8.5.5.

There is a large literature on the study on (8.5.26) with A nonzero, of which we mention only Pearson (1968*a*), Ackerberg and O'Malley (1970), O'Malley (1970*b*), Watts (1971), Zauderer (1972), Lakin (1972), Cook and Eckhaus (1973), Kreiss and Parter (1974), Matkowsky (1975, 1976), Olver (1978, 1980), de Groen (1980), and Kopell (1980); see also Wasow (1984).

Exercises

Exercise 8.5.1: Give a direct verification of the limiting results of (8.5.10). *Hint:* In the case of negative t, with $-1 \le t \le -\delta < 0$ for a fixed $\delta > 0$, write (8.5.9) in the form

$$x(t,\varepsilon) = \alpha + R(t,\varepsilon), \qquad (8.5.29)$$

with

$$R(t,\varepsilon) = (\beta - \alpha)\frac{\displaystyle\int_{-1}^{t}\exp\!\left(-\frac{s^2}{\varepsilon}\right)ds}{\displaystyle\int_{-1}^{1}\exp\!\left(-\frac{s^2}{\varepsilon}\right)ds}. \qquad (8.5.30)$$

Show the results

$$\int_{-1}^{t}\exp\!\left(-\frac{s^2}{\varepsilon}\right)ds = \text{order}\!\left[\exp\!\left(-\frac{\delta^2}{\varepsilon}\right)\right] \quad \text{for } -1 \le t \le -\delta < 0, \ \varepsilon > 0,$$
$$(8.5.31)$$

and

$$\int_{-1}^{1}\exp\!\left(-\frac{s^2}{\varepsilon}\right)ds = \text{order}(\sqrt{\varepsilon}) \quad \text{for } \varepsilon > 0, \qquad (8.5.32)$$

and then use these latter results with (8.5.29) to obtain the desired limit in (8.5.10). The case of positive t is similar. For $t = 0$, an easy calculation from (8.5.9) gives $x(0,\varepsilon) = (\alpha + \beta)/2$.

Exercise 8.5.2: Give a direct verification of the limiting results of (8.5.13).

Hint: Write (8.5.12) in the form

$$x(t, \varepsilon) = \frac{\alpha + \beta}{2} + R(t, \varepsilon), \tag{8.5.33}$$

with

$$R(t, \varepsilon) = (\beta - \alpha) \frac{\int_0^t \exp\left(\frac{s^2}{\varepsilon}\right) ds}{\int_{-1}^1 \exp\left(\frac{s^2}{\varepsilon}\right) ds}. \tag{8.5.34}$$

Use direct estimations to show the result

$$\lim_{\varepsilon \to 0+} R(t, \varepsilon) = 0 \quad \text{for } |t| \le 1 - \delta, \tag{8.5.35}$$

for any fixed, small $\delta > 0$. The desired result (8.5.13) follows directly for $-1 < t < 1$, and the results are clear for $t = -1$ and $t = 1$.

Exercise 8.5.3: Discuss the boundary-value problem

$$\begin{aligned} \varepsilon x'' + 3t^2 x' = 0 \quad & \text{for } -1 < t < 1, \\ x = \alpha \quad \text{at } t = -1, \quad \text{and} \quad & x = \beta \quad \text{at } t = +1, \end{aligned} \tag{8.5.36}$$

for small positive ε. *Hint*: The solution can be given in closed form up to a quadrature. Note that the differential equation here is of the form of (8.5.1), with

$$a(t) := 3t^2. \tag{8.5.37}$$

Hence, $a(t)$ does not have a simple zero at $t = 0$, but rather in this case it has a double zero there.

9

Linear first-order systems

In this chapter we consider boundary-value problems for a linear first-order system that includes many of the linear problems arising in practice. In particular, the system studied here includes the previous equations of Chapter 8. This system is studied subject to an assumption that excludes turning points. A multivariable approach is used, as in Chapter 8, to obtain uniformly valid approximations for solutions to appropriate boundary-value problems. We use a Green function approach for the proof of the validity of the resulting approximations. For this latter purpose we follow the elegant development of Chang (1972) and Harris (1973) that employs a Riccati transformation so as to transform the original problem into a suitable canonical form. The resulting Green function representation can also be used to provide appropriate cancellation rules similar to those discussed in Section 8.4. An application in optimal-control theory is discussed in Section 9.4.

9.1 Introduction

Consider the system [cf. (6.1.16)]

$$\frac{dx}{dt} = A(t, \varepsilon)x + B(t, \varepsilon)y + f(t, \varepsilon),$$

$$\varepsilon\frac{dy}{dt} = C(t, \varepsilon)x + D(t, \varepsilon)y + g(t, \varepsilon) \quad \text{for } 0 < t < 1,$$

(9.1.1)

for solution functions $x = x(t, \varepsilon)$ and $y = y(t, \varepsilon)$ that are respectively m-dimensional and n-dimensional real vector-valued functions, and where the given data functions A, B, C, D, f, and g are real matrix-valued and vector-valued functions with appropriate compatible orders. As usual,

330

these data functions are assumed to be smooth and to have asymptotic power-series expansions in terms of the small positive parameter ε, with coefficients that are smooth functions of t. The components of x and y are the slow and fast variables, respectively.

The system (9.1.1) is to be considered subject to boundary conditions of the type

$$L(\varepsilon)\begin{bmatrix} x(0,\varepsilon) \\ y(0,\varepsilon) \end{bmatrix} + R(\varepsilon)\begin{bmatrix} x(1,\varepsilon) \\ y(1,\varepsilon) \end{bmatrix} = \begin{bmatrix} \alpha(\varepsilon) \\ \beta(\varepsilon) \end{bmatrix} \qquad (9.1.2)$$

for given $(m + n) \times (m + n)$ matrices $L = L(\varepsilon)$ and $R = R(\varepsilon)$ that have asymptotic expansions in ε, and for given vectors $\alpha(\varepsilon)$ and $\beta(\varepsilon)$ of respective dimensions m and n.

It is convenient here, as in Section 6.1, to introduce fundamental solution matrices for the systems [see (6.1.32)–(6.1.37)]

$$\frac{dx}{dt} = A(t,\varepsilon)x \qquad (9.1.3)$$

and

$$\varepsilon\frac{dy}{dt} = D(t,\varepsilon)y. \qquad (9.1.4)$$

Such fundamental solutions were denoted as X and Y in Section 6.1, but we modify the notation here and denote such fundamental solution matrices now as $\xi = \xi(t,\varepsilon)$ and $\eta = \eta(t,\varepsilon)$, respectively, so that, for example,

$$\varepsilon\frac{d\eta}{dt} = D(t,\varepsilon)\eta \quad \text{for } 0 \le t \le 1, \qquad (9.1.5)$$

where the $n \times n$ matrix $\eta(t,\varepsilon)$ is nonsingular on the region

$$0 \le t \le 1, \qquad 0 < \varepsilon \le \varepsilon_0, \qquad (9.1.6)$$

for some fixed positive ε_0.

We assume throughout this chapter (except where explicitly stated otherwise, as in Example 9.1.4 and Exercises 9.3.7 and 9.3.10) that the matrix-valued function D is uniformly continuous on the region of (9.1.6) and that all eigenvalues $\lambda = \lambda(t,\varepsilon)$ of D satisfy

$$|\operatorname{Re}\lambda(t,\varepsilon)| \ge 2\kappa_1 > 0 \qquad (9.1.7)$$

uniformly on the region of (9.1.6). This condition (9.1.7) states that each eigenvalue satisfies precisely one of the two inequalities

$$\operatorname{Re}\lambda \le -2\kappa_1 \quad \text{or} \quad \operatorname{Re}\lambda \ge +2\kappa_1, \qquad (9.1.8)$$

where the smoothness of D guarantees that no eigenvalue $\lambda(t,\varepsilon)$ can "switch" from one of these inequalities to the other as t and ε range over the region of (9.1.6). In particular, turning points are excluded.

According to a result of Chang and Coppel (1969) [see also Ferguson (1975)], this assumption (9.1.7) on D implies that the system (9.1.4) has an exponential dichotomy. That is, there is a fixed projection $P: \mathbb{R}^n \to \mathbb{R}^n$ and a suitable fundamental solution η for (9.1.5) such that the inequalities

$$\left| \eta(t, \varepsilon) P \eta^{-1}(s, \varepsilon) \right|, \left| \eta(s, \varepsilon)(I_n - P)\eta^{-1}(t, \varepsilon) \right| \le \kappa_0 \cdot e^{-\kappa_1 \cdot (t-s)/\varepsilon}$$

$$\text{for } 0 \le s \le t \le 1, 0 < \varepsilon \le \varepsilon_0, \quad (9.1.9)$$

hold for fixed positive constants κ_0 and κ_1. For definiteness we use here the earlier matrix norm of Section 6.1:

$$|E| := \max_{i,j} |e_{ij}| \quad \text{for any matrix } E = (e_{ij}).$$

A proof of this result of Chang and Coppel is indicated in Exercise 9.2.7 for differentiable D, using a Riccati transformation.

The projection P can be taken to be given as

$$P = \begin{bmatrix} I_k & 0 \\ 0 & 0 \end{bmatrix},$$

where the rank k of the projection is the sum of the algebraic multiplicities of all eigenvalues of D having negative real parts, whereas the rank $n - k$ of the projection $I_n - P$ is the sum of the algebraic multiplicities of all eigenvalues having positive real parts. Note that if, in a special case, *all* eigenvalues have negative real parts, then $P = I_n$, and in such a case the stated result of Chang and Coppel [that (9.1.7) implies (9.1.9)] reduces to the previous result of Flatto and Levinson (1955) discussed in Section 6.1 [see (6.1.33), (6.1.35), (6.1.40), and (6.1.42)].

The condition (9.1.9) has appeared prominently in studies of conditionally stable singularly perturbed initial-value problems, as in Chang (1969b) and Chang and Coppel (1969), and we shall see later that this condition (9.1.9) is also important in the development of Chang (1972) and Harris (1973) for the present boundary-value problem. Related questions of stability and dichotomy are discussed in Coppel (1965, 1978).

Example 9.1.1: Various boundary-value problems were considered in Section 8.3 for the second-order scalar equation [see (8.3.1), where the previous forcing term f is relabelled now as g]

$$\varepsilon \frac{d^2 x}{dt^2} + a(t, \varepsilon)\frac{dx}{dt} + b(t, \varepsilon)x = g(t, \varepsilon), \quad (9.1.10)$$

subject to the condition [see (8.3.2)]

$$|a(t, \varepsilon)| \geq \kappa > 0, \qquad (9.1.11)$$

for a fixed positive constant κ, uniformly for all relevant t and ε.

We put $y: = dx/dt$ and rewrite the second-order equation (9.1.10) as the first-order system

$$\dot{x} = y,$$
$$\varepsilon \dot{y} = -bx - ay + g, \qquad (9.1.12)$$

which is a system of the form (9.1.1), with

$$m = n = 1, \qquad A = f = 0, \qquad B = 1,$$
$$C = -b(t, \varepsilon), \qquad D = -a(t, \varepsilon). \qquad (9.1.13)$$

In this case the system (9.1.5) reduces to the scalar equation

$$\varepsilon \dot{\eta} = -a(t, \varepsilon)\eta, \qquad (9.1.14)$$

with a particular solution

$$\eta(t, \varepsilon) = \exp\left[-\frac{1}{\varepsilon} \int_0^t a(\tau, \varepsilon)\, d\tau \right]. \qquad (9.1.15)$$

There are precisely two possibilities to consider according to assumption (9.1.11): The given function a must be either everywhere negative or everywhere positive. We can take the projection P of (9.1.9) to be given as follows:

$$P = \begin{cases} 1 & \text{if } a \geq \kappa > 0, \\ 0 & \text{if } a \leq -\kappa < 0. \end{cases} \qquad (9.1.16)$$

We see directly that suitable inequalities are satisfied as in (9.1.9).

Example 9.1.2: Boundary-value problems were considered in Section 8.2 for a scalar equation of the form [see (8.2.1)]

$$\varepsilon^2 \frac{d^2 x}{dt^2} + b(t, \varepsilon)x = g(t, \varepsilon), \qquad (9.1.17)$$

subject to the condition [see (8.2.2)]

$$b(t, \varepsilon) \leq -\kappa < 0 \qquad (9.1.18)$$

for a fixed positive constant κ, uniformly for all relevant t and ε.

We put $y_1: = x$ and $y_2: = \varepsilon \dot{y}_1$, and then the second-order equation (9.1.17) can be rewritten as the first-order system

$$\varepsilon \dot{y}_1 = y_2,$$
$$\varepsilon \dot{y}_2 = -b(t, \varepsilon)y_1 + g(t, \varepsilon), \qquad (9.1.19)$$

which is a system of the form (9.1.1), with

$$m = 0, \qquad x = A = f = 0, \qquad B = (0,0),$$

$$n = 2, \qquad y = \begin{bmatrix} y_1 \\ y_2 \end{bmatrix}, \qquad C = \begin{bmatrix} 0 \\ 0 \end{bmatrix}, \qquad D = \begin{bmatrix} 0 & 1 \\ -b & 0 \end{bmatrix}. \qquad (9.1.20)$$

In this case, condition (9.1.18) implies that the matrix D of (9.1.20) has two distinct eigenvalues λ_1 and λ_2 that can be given as

$$\lambda_1 = -\sqrt{-b(t,\varepsilon)}, \qquad \lambda_2 = +\sqrt{-b(t,\varepsilon)}. \qquad (9.1.21)$$

The general case can be handled directly here using a Riccati transformation as in Section 9.2, but for the moment we consider here the special case in which b is constant in (9.1.17) and (9.1.20)–(9.1.21), say

$$b(t,\varepsilon) = -\kappa_1^2 = \text{constant}, \qquad (9.1.22)$$

for a fixed positive constant κ_1 independent of t and ε. Then the system (9.1.5) becomes

$$\varepsilon \dot{\eta} = \begin{bmatrix} 0 & 1 \\ \kappa_1^2 & 0 \end{bmatrix} \eta, \qquad (9.1.23)$$

and we easily find a suitable particular fundamental solution matrix, such as

$$\eta = \eta(t,\varepsilon) = \begin{bmatrix} e^{-\kappa_1 t/\varepsilon} & e^{\kappa_1 t/\varepsilon} \\ -\kappa_1 e^{-\kappa_1 t/\varepsilon} & \kappa_1 e^{\kappa_1 t/\varepsilon} \end{bmatrix}. \qquad (9.1.24)$$

The first and second column vectors of (9.1.24) are eigensolutions of (9.1.23) corresponding respectively to the negative eigenvalue $\lambda_1 = -\kappa_1$ and the positive eigenvalue $\lambda_2 = +\kappa_1$. We can take the projection P of (9.1.9) to be given as

$$P = \begin{bmatrix} 1 & 0 \\ 0 & 0 \end{bmatrix}, \qquad (9.1.25)$$

and then we see directly that suitable inequalities are satisfied as in (9.1.9) (see Exercise 9.1.1).

Example 9.1.3: As a final example of the system (9.1.1), consider the case in which D is a given diagonal matrix,

$$D = D(t) = \text{diag}\{\lambda_1(t), \lambda_2(t), \ldots, \lambda_n(t)\}, \qquad (9.1.26)$$

with ν eigenvalues $\lambda_1, \ldots, \lambda_\nu$ having negative real parts, and $n - \nu$

eigenvalues $\lambda_{\nu+1}, \ldots, \lambda_n$ having positive real parts,

$$\begin{aligned}
\operatorname{Re}\lambda_j(t) &\le -2\kappa_1 \quad \text{for } j = 1, \ldots, \nu, \\
\operatorname{Re}\lambda_j(t) &\ge +2\kappa_1 \quad \text{for } j = \nu + 1, \ldots, n,
\end{aligned} \tag{9.1.27}$$

uniformly for $0 \le t \le 1$, for a fixed positive constant κ_1, and for a given integer ν ($0 \le \nu \le n$). The matrix D can be written in block form as

$$D = \begin{bmatrix} D_1 & 0 \\ 0 & D_2 \end{bmatrix}, \tag{9.1.28}$$

where the $\nu \times \nu$ matrix D_1 and the $(n - \nu) \times (n - \nu)$ matrix D_2 are given as

$$\begin{aligned}
D_1(t) &= \operatorname{diag}\{\lambda_1(t), \ldots, \lambda_\nu(t)\}, \\
D_2(t) &= \operatorname{diag}\{\lambda_{\nu+1}(t), \ldots, \lambda_n(t)\}.
\end{aligned} \tag{9.1.29}$$

A fundamental solution for (9.1.5) can be given as

$$\eta(t, \varepsilon) = \begin{bmatrix} \exp\left(\dfrac{1}{\varepsilon}\int_0^t D_1\right) & 0 \\ 0 & \exp\left(\dfrac{1}{\varepsilon}\int_0^t D_2\right) \end{bmatrix}, \tag{9.1.30}$$

with

$$\eta^{-1}(t, \varepsilon) = \begin{bmatrix} \exp\left(-\dfrac{1}{\varepsilon}\int_0^t D_1\right) & 0 \\ 0 & \exp\left(-\dfrac{1}{\varepsilon}\int_0^t D_2\right) \end{bmatrix}.$$

The projection P can be taken as

$$P = \begin{bmatrix} I_\nu & 0 \\ 0 & \begin{matrix} 0 \\ (n-\nu)\times(n-\nu) \end{matrix} \end{bmatrix}, \tag{9.1.31}$$

and we again find suitable inequalities as in (9.1.9) (see Exercise 9.1.2).

Note that this last example includes the conditionally stable system (7.2.13), for which the initial-value problem was considered in Chapter 7.

According to the *Fredholm alternative* (see Exercise 0.1.2), the linear boundary-value problem (9.1.1)–(9.1.2) may have precisely one solution, or it is possible that either existence or uniqueness may fail, so that the

problem may have no solution or infinitely many solutions, depending on
the given data. We wish to study the situation in which the problem has a
unique solution for every sufficiently small $\varepsilon > 0$, and we wish to study
the resulting solution as $\varepsilon \to 0+$. We use perturbation techniques to
obtain uniformly valid approximations to the exact solution functions
$x = x(t, \varepsilon)$ and $y = y(t, \varepsilon)$, from which we find in particular the result

$$\lim_{\substack{\varepsilon \to 0+ \\ \text{fixed } 0 < t < 1}} \begin{bmatrix} x(t, \varepsilon) \\ y(t, \varepsilon) \end{bmatrix} = \begin{bmatrix} X_0(t) \\ Y_0(t) \end{bmatrix}, \tag{9.1.32}$$

where the limiting functions X_0, Y_0 satisfy the following reduced system
obtained by putting $\varepsilon = 0$ in (9.1.1):

$$\frac{dX_0}{dt} = A_0(t) X_0 + B_0(t) Y_0 + f_0(t),$$
$$0 = C_0(t) X_0 + D_0(t) Y_0 + g_0(t) \quad \text{for } 0 < t < 1. \tag{9.1.33}$$

These limiting outer solution functions X_0 and Y_0 also satisfy a certain
reduced boundary condition obtained effectively by cancelling a certain
part of the original boundary condition (9.1.2). The situation can be more
complicated if the exponential dichotomy (9.1.9) fails to hold, as illustrated later in Example. 9.1.4. Related nonlinear problems subject to
an exponential dichotomy (hyperbolic splitting) are studied in O'Malley
(1980).

A Green function approach can be used effectively to study the
boundary-value problem (9.1.1)–(9.1.2) as $\varepsilon \to 0+$. Such an approach
yields both the appropriate reduced boundary condition and suitable
error estimates for approximate solutions as provided by perturbation
techniques. The multivariable technique can be used for the present
boundary-value problem to provide suitable approximate solutions when
a unique solution exists. We shall not emphasize such multivariable
calculations in this chapter, because such calculations have been illustrated amply in previous chapters in the contexts of numerous different problems. Rather, beginning in Section 9.2, we follow the elegant
approach of Chang (1972) and Harris (1973) for study of the Green
function for (9.1.1)–(9.1.2) as $\varepsilon \to 0+$, leading to the (cancellation rule
for the) reduced boundary condition, and also yielding suitable error
estimates, as discussed earlier. The approach employs a Riccati transformation to diagonalize the system (9.1.1), thereby transforming the original problem to a convenient canonical form. The problem (9.1.1)–(9.1.2)
is also studied in Ferguson (1975). A related nonlinear problem subject to
a certain special type of boundary condition was studied also in Levin

(1956, 1957) in connection with the conditionally stable initial-value problem as in Chapter 7; see also Tupčiev (1962), Vasil'eva and Butuzov (1973), O'Malley (1984), and the references cited there.

Example 9.1.4: The boundary-value problem

$$\frac{dx_1}{dt} = -3y + 1, \qquad \frac{dx_2}{dt} = y - 1,$$

$$\varepsilon \frac{dy}{dt} = -x_1 + x_2 + 1 \quad \text{for } 0 < t < 1,$$

with

$$x_1(0) = x_2(0), \qquad x_1(1) = 1, \quad \text{and} \quad x_2(1) = \tfrac{1}{2},$$

is a special case of a linearized version of a problem in the physical theory of semiconducting devices to be studied in Section 10.5. The corresponding reduced system of differential equations here is

$$\frac{dX_1}{dt} = -3Y + 1, \qquad \frac{dX_2}{dt} = Y - 1,$$

$$0 = -X_1 + X_2 + 1 \quad \text{for } 0 < t < 1,$$

with general solution

$$X_1(t) = -\tfrac{1}{2}t + 1 + k, \qquad X_2(t) = -\tfrac{1}{2}t + k \qquad Y(t) = \tfrac{1}{2},$$

where k is a constant of integration that is to be determined by a suitable boundary condition. We see that the given boundary condition for x at $t = 0$ cannot be satisfied by the outer solution functions X_1 and X_2, although it would be possible to choose the integration constant k so as to satisfy one of the two specified boundary conditions at $t = 1$. The exact solution functions x_1, x_2, and y can be computed, and we find, on letting $\varepsilon \to 0+$ in the exact solution functions, that the constant k appearing in the functions X_1 and X_2 is given as $k = \tfrac{7}{8}$. Hence, all of the original boundary conditions are cancelled for the outer solution, because $X_1(1) = \tfrac{11}{8}$ and $X_2(1) = \tfrac{3}{8}$. Note that the exponential dichotomy (9.1.9) does not hold in this example.

A general study of certain linear problems of the form (9.1.1)–(9.1.2) for which the $n \times n$ matrix $D_0 = D(t,0)$ is singular (and of constant rank for $0 \le t \le 1$), so that the exponential dichotomy (9.1.9) fails to hold, is given in O'Malley (1979), where we see that such problems may lead to thinner boundary layers inside other layers. Certain examples with D_0 singular are considered also by Vasil'eva and Butuzov (1980).

Exercises

Exercise 9.1.1: Show that suitable estimates of the type (9.1.9) are satisfied by the fundamental solution (9.1.24) in Example 9.1.2.

Exercise 9.1.2: Show that suitable estimates of the type (9.1.9) are satisfied by the fundamental solution (9.1.30) in Example 9.1.3.

Exercise 9.1.3: Let $\xi = \xi(t, \varepsilon)$ be a fundamental solution for (9.1.3), so that the nonsingular $m \times m$ matrix-valued ξ satisfies

$$\frac{d\xi}{dt} = A(t, \varepsilon)\xi \quad \text{for } 0 \leq t \leq 1, 0 \leq \varepsilon \leq \varepsilon_0. \quad (9.1.34)$$

Show that ξ satisfies the estimate [compare with (9.1.9)]

$$\left|\xi(t, \varepsilon)\xi^{-1}(s, \varepsilon)\right| \leq |I_m| \cdot e^{\|A\| \cdot |t-s|} \quad (9.1.35)$$

for all $0 \leq \varepsilon \leq \varepsilon_0$ and for all $0 \leq s, t \leq 1$, where

$$\|A\| := \max_{0 \leq t \leq 1} |A(t, \varepsilon)|.$$

Hint: Use the suggestion given earlier for (6.1.45), applied here to the matrix function $\xi(t, \varepsilon)\xi^{-1}(s, \varepsilon)$, considered as a function of t. Note that the norm of the identity matrix appearing on the right side of (9.1.35) equals 1 if we use the matrix norm suggested in the paragraph of (9.1.9).

9.2 The Riccati transformation

In this section we follow Chang (1972) and Harris (1973) in simplifying the linear system

$$\frac{dx}{dt} = A(t, \varepsilon)x + B(t, \varepsilon)y + f(t, \varepsilon),$$

$$\varepsilon \frac{dy}{dt} = C(t, \varepsilon)x + D(t, \varepsilon)y + g(t, \varepsilon) \quad \text{for } 0 < t < 1, \quad (9.2.1)$$

using a linear transformation

$$\begin{bmatrix} x \\ y \end{bmatrix} = H(t, \varepsilon)\begin{bmatrix} u \\ v \end{bmatrix} \quad (9.2.2)$$

for a nonsingular $(m + n) \times (m + n)$ matrix $H = H(t, \varepsilon)$ suitably chosen so as to reduce the system (9.2.1) to the diagonal form

$$\frac{du}{dt} = (A - BT)u + (I_m + \varepsilon ST)f + Sg,$$

$$\varepsilon\frac{dv}{dt} = (D + \varepsilon TB)v + \varepsilon Tf + g,$$

(9.2.3)

where $S = S(t, \varepsilon)$ and $T = T(t, \varepsilon)$ are the matrix-valued functions appearing in the following definition of the transformation matrix H:

$$H(t, \varepsilon) := \begin{bmatrix} I_m & 0 \\ -T & I_n \end{bmatrix}\begin{bmatrix} I_m & -\varepsilon S \\ 0 & I_n \end{bmatrix}$$

$$= \begin{bmatrix} I_m & \overset{m \times n}{-\varepsilon S(t, \varepsilon)} \\ \overset{n \times m}{-T(t, \varepsilon)} & I_n + \varepsilon T(t, \varepsilon)S(t, \varepsilon) \end{bmatrix}.$$

(9.2.4)

We see directly that the transformation (9.2.2) and (9.2.4) does in fact transform (9.2.1) into (9.2.3) provided that the functions S and T are taken to satisfy the nonlinear equations

$$\varepsilon\frac{dS}{dt} = \varepsilon(A - BT)S - S(D + \varepsilon TB) - B \quad \text{for } 0 < t < 1 \quad (9.2.5)$$

and

$$\varepsilon\frac{dT}{dt} = DT - \varepsilon TA + \varepsilon TBT - C \qquad \text{for } 0 < t < 1. \quad (9.2.6)$$

Note that the transformation matrix H of (9.2.4) is always nonsingular, with inverse

$$H^{-1}(t, \varepsilon) = \begin{bmatrix} I_m & \varepsilon S \\ 0 & I_n \end{bmatrix}\begin{bmatrix} I_m & 0 \\ T & I_n \end{bmatrix} = \begin{bmatrix} I_m + \varepsilon ST & \varepsilon S \\ T & I_n \end{bmatrix}. \quad (9.2.7)$$

In fact, H is unimodular. As an unrelated aside, we note that the result (9.2.4) and (9.2.7) with $\varepsilon = 1$ can be conveniently used in practice to produce *explicitly* invertible (and unimodular) pairs H, H^{-1} of "large" dimension, and this is sometimes useful in various different contexts.

The matrix-valued functions S and T must be obtained as solutions of the nonlinear system (9.2.5)–(9.2.6). Specifically, the function T is first obtained as a solution of the matrix Riccati equation (9.2.6), and the resulting solution T is inserted back into (9.2.5), which then becomes a linear equation for S. We shall see later that the exponential dichotomy (9.1.9) implies that this system (9.2.5)–(9.2.6) can be solved using the Banach/Picard fixed-point theorem so as to obtain well-behaved solution functions S, T that are bounded as $\varepsilon \to 0+$, uniformly for $0 \le t \le 1$. Matrix Riccati equations such as (9.2.6) appear in many different contexts [see Reid (1972)], including optimal-control theory (see Section 9.4).

In the study of (9.2.1), Chang (1969b) introduced a transformation of the type (9.2.2) and (9.2.4), but involving only T, with $S \equiv 0$ and with T taken to satisfy the Riccati equation (9.2.6). There is a long tradition, going back to Riccati (1724), on the use of the *scalar* Riccati equation to "reduce the order" of a second-order linear scalar differential equation; see Davis (1962) and Smith (1984a). This classical Riccati transformation corresponds to (9.2.2) and (9.2.4) with $S \equiv 0$; see Exercise 9.2.2. The foregoing general transformation (9.2.2) and (9.2.4)–(9.2.6) appears in Chang (1972, 1975, 1976) and Harris (1973), and an earlier related transformation appears in Sibuya (1958, 1963b, 1966); see also Harris (1960, 1962a) and O'Malley (1969c). A special case of the transformation (9.2.2)–(9.2.6) with $m = n$ appears in Wilde and Kokotović (1972, 1973) in the context of a problem in optimal-control theory; see Section 9.4.

The general transformation (9.2.2)–(9.2.6) provides a generalization to the vector system (9.2.1) of the technique of Riccati for the reduction of order for the linear second-order equation. Note that the parameter ε plays no role in the transformation. We can put $\varepsilon = 1$ in the foregoing formulas, and we see that the Riccati transformation provides a general method for the reduction by diagonalization of any given linear first-order system (9.2.1) (with $\varepsilon = 1$), provided that we can handle the nonlinear Riccati equation (9.2.6) (see Exercise 9.2.6). The general transformation (9.2.2)–(9.2.6) should perhaps bear the names of Sibuya, Chang, Harris, Riccati, and others, but for brevity we refer to it simply as the Riccati transformation. This transformation is also used in Habets (1974, 1983), Mattheij (1982), O'Malley and Anderson (1982), O'Malley (1983b), Weiss (1984), Mattheij and O'Malley (1984), and Smith (1984a, 1984b, 1984c), where additional references are given, including references on the use of the *algebraic* Riccati transformation in control theory for time-invariant systems, as in Kokotović and Haddad (1975), Kokotović (1975, 1984), and Saksena, O'Reilly, and Kokotović (1984).

We turn now to the study of (9.2.6) as $\varepsilon \to 0+$, subject to the assumptions that the data functions and their first derivatives with respect to t are bounded and continuous on the region of (9.1.6) and that

D satisfies the eigenvalue condition (9.1.7) so that D is invertible and the exponential dichotomy (9.1.9) holds. For this purpose we follow Harris (1973) and use the identity, obtained by integration by parts,

$$W(t) = \eta(t)\left[P\eta^{-1}(0)W(0)\xi(0) + (I_n - P)\eta^{-1}(1)W(1)\xi(1)\right]\xi^{-1}(t)$$
$$+ \int_0^t \eta(t)P\eta^{-1}(s)\left(\dot{W} - \frac{1}{\varepsilon}DW + WA\right)(s)\xi(s)\xi^{-1}(t)\,ds$$
$$- \int_t^1 \eta(t)(I_n - P)\eta^{-1}(s)$$
$$\times \left(\dot{W} - \frac{1}{\varepsilon}DW + WA\right)(s)\xi(s)\xi^{-1}(t)\,ds \qquad (9.2.8)$$

for any $n \times m$ matrix-valued function $W = W(t)$ of class C^1, where $\xi(t) = \xi(t, \varepsilon)$ is a fundamental solution for (9.1.3), $\eta(t) = \eta(t, \varepsilon)$ is a fundamental solution for (9.1.4), and P is any fixed constant matrix taken here to coincide with the projection P of (9.1.9). See Exercise 9.2.3 for an indication of a derivation of (9.2.8).

We now apply (9.2.8) to an arbitrary solution $W = T$ of the Riccati equation (9.2.6) and find

$$T(t) = \eta(t)\left[P\eta^{-1}(0)T(0)\xi(0) + (I_n - P)\eta^{-1}(1)T(1)\xi(1)\right]\xi^{-1}(t)$$
$$+ \int_0^t \eta(t)P\eta^{-1}(s)\left[T(s)B(s,\varepsilon)T(s) - \frac{1}{\varepsilon}C(s,\varepsilon)\right]\xi(s)\xi^{-1}(t)\,ds$$
$$- \int_t^1 \eta(t)(I_n - P)\eta^{-1}(s)$$
$$\times \left[T(s)B(s,\varepsilon)T(s) - \frac{1}{\varepsilon}C(s,\varepsilon)\right]\xi(s)\xi^{-1}(t)\,ds \qquad (9.2.9)$$

for any solution $T = T(t) = T(t, \varepsilon)$ of (9.2.6). Conversely, a direct calculation shows that any solution of (9.2.9) satisfies the Riccati equation (9.2.6), and so the two equations are equivalent. Note that this equation (9.2.9) is related to the system of integral equations used in the study of the conditionally stable problem of Chapter 7, where such integral equations are constructed so as to exclude unwanted solution components, following an approach that has long been used in the study of asymptotic behavior of solutions of differential equations.

If the Riccati equation (9.2.6) in fact has a solution $T = T(t, \varepsilon)$ that is well-behaved for $0 \le t \le 1$ as $\varepsilon \to 0+$, then we would expect to be able to pass to the limit $\varepsilon \to 0+$ in (9.2.6) and find [see (9.2.17)]

$$\lim_{\varepsilon \to 0+} T(t, \varepsilon) \overset{?}{=} D^{-1}(t, 0)C(t, 0). \qquad (9.2.10)$$

We seek such a "well-behaved" solution T, and so we are led to replace the boundary values $T(0)$ and $T(1)$ in the right side of (9.2.9) with appropriate prescribed values consistent with (9.2.10). Specifically, we impose the following boundary conditions on T:

$$P\eta^{-1}(0)T(0): = P\eta^{-1}(0)D^{-1}(0,\varepsilon)C(0,\varepsilon),$$
$$(I_n - P)\eta^{-1}(1)T(1): = (I_n - P)\eta^{-1}(1)D^{-1}(1,\varepsilon)C(1,\varepsilon). \qquad (9.2.11)$$

Chang (1969b) replaces (9.2.11) with the corresponding homogeneous boundary conditions given by replacing the right sides of (9.2.11) with zero matrices. The resulting solution need not satisfy (9.2.10) near the endpoints $t = 0$ and/or $t = 1$. We can specify any arbitrary bounded (as $\varepsilon \to 0+$) boundary values on the right sides of (9.2.11), and obtain a corresponding bounded solution for T, as is done later. The special choice (9.2.11) of Harris (1973) has the merit of yielding a solution for which (9.2.10) is valid for $0 \le t \le 1$ and for which dT/dt is also bounded; see (9.2.17) and (9.2.18).

From (9.2.9) and (9.2.11) we are led to consider the integral equation

$$T(t) = T_0(t,\varepsilon) + \int_0^t \eta(t)P\eta^{-1}(s)T(s)B(s,\varepsilon)T(s)\xi(s)\xi^{-1}(t)\,ds$$
$$- \int_t^1 \eta(t)(I_n - P)\eta^{-1}(s)T(s)B(s,\varepsilon)T(s)\xi(s)\xi^{-1}(t)\,ds,$$

$$(9.2.12)$$

where the given function $T_0 = T_0(t,\varepsilon)$ is defined in terms of the data as

$$T_0(t,\varepsilon): = \eta(t)P\eta^{-1}(0)D^{-1}(0,\varepsilon)C(0,\varepsilon)\xi(0)\xi^{-1}(t)$$
$$- \frac{1}{\varepsilon}\int_0^t \eta(t)P\eta^{-1}(s)C(s,\varepsilon)\xi(s)\xi^{-1}(t)\,ds$$
$$+ \eta(t)(I_n - P)\eta^{-1}(1)D^{-1}(1,\varepsilon)C(1,\varepsilon)\xi(1)\xi^{-1}(t)$$
$$+ \frac{1}{\varepsilon}\int_t^1 \eta(t)(I_n - P)\eta^{-1}(s)C(s,\varepsilon)\xi(s)\xi^{-1}(t)\,ds,$$

$$(9.2.13)$$

and where here as elsewhere we often suppress the dependence of such functions as ξ and η on ε. Note that T_0 is at least of class C^1 with respect to t and satisfies a linearized version of (9.2.6) with $B \equiv 0$.

We assume that the data functions, including D and C and also D^{-1}, are bounded on the region

$$0 \leq t \leq 1, \qquad 0 < \varepsilon \leq \varepsilon_0, \qquad (9.2.14)$$

for some fixed $\varepsilon_0 > 0$. [The boundedness of D^{-1} actually follows from the boundedness of D and the assumption (9.1.7).] We also have the estimates of (9.1.9) and (9.1.35), and then (9.2.13) leads directly to a uniform bound on T_0 of the form (see Exercise 9.2.4)

$$|T_0(t, \varepsilon)| \leq \hat{\kappa} \quad \text{as } \varepsilon \to 0+, \text{ uniformly for } 0 \leq t \leq 1, \quad (9.2.15)$$

for a fixed positive constant $\hat{\kappa}$. Using these results, we find the following lemma.

Lemma 9.2.1 (Chang 1972, 1975, 1976; Harris 1973): *Let the data functions A, B, C, and D of (9.2.1) be bounded, along with their first derivatives with respect to t, uniformly on the region (9.2.14), with D^{-1} also bounded on (9.2.14), and assume that the exponential dichotomy (9.1.9) holds. Then, for small ε ($\varepsilon \to 0+$), the matrix Riccati integral equation (9.2.12)–(9.2.13) has a unique solution satisfying the bound*

$$|T(t, \varepsilon)| \leq 2\hat{\kappa} \quad \text{as } \varepsilon \to 0+, \text{ uniformly for } 0 \leq t \leq 1, \quad (9.2.16)$$

where $\hat{\kappa}$ is the constant of (9.2.15). This solution function is of class C^1 with respect to t and provides a solution for the Riccati differential equation (9.2.6). Moreover,

$$|T(t, \varepsilon) - D^{-1}(t, \varepsilon)C(t, \varepsilon)| \leq \text{const. } \varepsilon \quad \text{as } \varepsilon \to 0+, \quad (9.2.17)$$

uniformly for $0 \leq t \leq 1$, so that the conjectured result (9.2.10) is valid. Also, the solution satisfies the boundary conditions of (9.2.11) for each $\varepsilon \to 0+$, and dT/dt is also bounded as

$$\left| \frac{dT(t, \varepsilon)}{dt} \right| \leq \text{constant}, \quad \text{as } \varepsilon \to 0+, \quad (9.2.18)$$

uniformly for $0 \leq t \leq 1$.

Proof: We consider the right side of (9.2.12) as an operator, defined for continuous matrix-valued functions T, with the given term T_0 specified by (9.2.13). We prove then directly with the Banach/Picard fixed-point theorem that the integral equation (9.2.12)–(9.2.13) has a unique solution satisfying (9.2.16); see Exercise 9.2.5. The boundary conditions of (9.2.11) follow directly from (9.2.12)–(9.2.13).

We turn now to the proof of the result (9.2.17). To this end, we put $W = D^{-1}C$ in the identity (9.2.8) and find

$$D^{-1}(t, \varepsilon)C(t, \varepsilon) = T_0(t, \varepsilon) + \int_0^t \eta(t)P\eta^{-1}(s)$$

$$\times \left[\frac{d}{ds}(D^{-1}C) + D^{-1}CA \right](s)\xi(s)\xi^{-1}(t)\, ds$$

$$- \int_t^1 \eta(t)(I_n - P)\eta^{-1}(s)$$

$$\times \left[\frac{d}{ds}(D^{-1}C) + D^{-1}CA \right](s)\xi(s)\xi^{-1}(t)\, ds, \quad (9.2.19)$$

with T_0 as in (9.2.13). We subtract (9.2.19) from the previous result (9.2.12) for $T(t) = T(t, \varepsilon)$ and find the difference

$$T(t, \varepsilon) - D^{-1}(t, \varepsilon)C(t, \varepsilon)$$

$$= + \int_0^t \eta(t)P\eta^{-1}(s)$$

$$\times \left[TBT + \frac{d}{ds}(D^{-1}C) + D^{-1}CA \right](s)\xi(s)\xi^{-1}(t)\, ds$$

$$- \int_t^1 \eta(t)(I_n - P)\eta^{-1}(s)$$

$$\times \left[TBT + \frac{d}{ds}(D^{-1}C) + D^{-1}CA \right](s)\xi(s)\xi^{-1}(t)\, ds.$$

$$(9.2.20)$$

The bound (9.2.16), along with the assumed regularity of the data, yields a uniform bound on the terms in square brackets in the integrals here,

$$\left| T(t, \varepsilon)B(t, \varepsilon)T(t, \varepsilon) + \frac{d}{dt}\left[D^{-1}(t, \varepsilon)C(t, \varepsilon) \right] + D^{-1}(t, \varepsilon)C(t, \varepsilon)A(t, \varepsilon) \right|$$

$$\leq \text{constant}, \quad \text{as } \varepsilon \to 0+, \quad (9.2.21)$$

uniformly for $0 \leq t \leq 1$. Note that the required bound here on the derivative dD^{-1}/dt follows from the boundedness of the functions dD/dt and D^{-1} along with the relation $dD^{-1}/dt = -D^{-1}(dD/dt)D^{-1}$, which follows from the result $DD^{-1} = I_n$, on differentiation. The stated result (9.2.17) now follows directly from (9.2.20), along with the bounds (9.2.21), (9.1.9), and (9.2.41).

Finally, the result (9.2.18) follows easily from these previous results, along with (9.2.6). ∎

In addition to the exponential dichotomy (9.1.9), the foregoing proof of the existence of a uniformly bounded solution T, as in (9.2.16), requires only that the data functions be continuous and uniformly bounded. The assumed differentiability of D and C is used in (9.2.19) in the proof of (9.2.17) and (9.2.18). A representation for $D^{-1}C$ that is slightly different from (9.2.19) and that does not require differentiability for D and C is used in Chang (1972, 1975, 1976) and Harris (1973) to obtain a slightly modified version of (9.2.17) for a wider class of problems with data functions that need not be differentiable.

The solution T of the Riccati equation (9.2.6) given by Lemma 9.2.1 can be inserted back into (9.2.5), which then provides a linear system for $S = S(t, \varepsilon)$ that can be solved in just the same manner as (9.2.6). For this purpose it is convenient to introduce fundamental solution matrices $\hat{\xi}(t) = \hat{\xi}(t, \varepsilon)$ and $\hat{\eta}(t) = \hat{\eta}(t, \varepsilon)$, respectively, for the following systems:

$$\frac{d\hat{\xi}}{dt} = [A(t, \varepsilon) - B(t, \varepsilon)T(t, \varepsilon)]\hat{\xi} \tag{9.2.22}$$

and

$$\varepsilon \frac{d\hat{\eta}}{dt} = [D(t, \varepsilon) + \varepsilon T(t, \varepsilon)B(t, \varepsilon)]\hat{\eta}. \tag{9.2.23}$$

We find directly the result of the following lemma.

Lemma 9.2.2: *Let the same assumptions hold as in Lemma* 9.2.1. *Then the differential equation* (9.2.5), *with* T *given by Lemma* 9.2.1, *has a solution* $S = S(t, \varepsilon)$ *that is uniformly bounded, along with* dS/dt, *given in terms of the data as*

$$S(t, \varepsilon) = -\hat{\xi}(t)\hat{\xi}^{-1}(0)B(0, \varepsilon)D^{-1}(0, \varepsilon)\hat{\eta}(0)(I_n - P)\hat{\eta}^{-1}(t)$$

$$- \frac{1}{\varepsilon}\int_0^t \hat{\xi}(t)\hat{\xi}^{-1}(s)B(s, \varepsilon)\hat{\eta}(s)(I_n - P)\hat{\eta}^{-1}(t)\,ds$$

$$- \hat{\xi}(t)\hat{\xi}^{-1}(1)B(1, \varepsilon)D^{-1}(1, \varepsilon)\hat{\eta}(1)P\hat{\eta}^{-1}(t)$$

$$+ \frac{1}{\varepsilon}\int_t^1 \hat{\xi}(t)\hat{\xi}^{-1}(s)B(s, \varepsilon)\hat{\eta}(s)P\hat{\eta}^{-1}(t)\,ds. \tag{9.2.24}$$

This solution satisfies the boundary conditions

$$S(0, \varepsilon)\hat{\eta}(0, \varepsilon)(I_n - P) = -B(0, \varepsilon)D^{-1}(0, \varepsilon)\hat{\eta}(0, \varepsilon)(I_n - P),$$

$$S(1, \varepsilon)\hat{\eta}(1, \varepsilon)P = -B(1, \varepsilon)D^{-1}(1, \varepsilon)\hat{\eta}(1, \varepsilon)P, \tag{9.2.25}$$

along with the results

$$S(t, \varepsilon) = -B(t, \varepsilon)D^{-1}(t, \varepsilon) + \text{order}(\varepsilon),$$

$$\frac{dS(t, \varepsilon)}{dt} = \text{order}(1), \quad \text{as } \varepsilon \to 0+, \tag{9.2.26}$$

uniformly for $0 \le t \le 1$.

Proof: The proof is similar to that of Lemma 9.2.1, but with the previous identity (9.2.8) replaced here with the identity

$$W(t) = \hat{\xi}(t)\left[\hat{\xi}^{-1}(0)W(0)\hat{\eta}(0)(I_n - P) + \hat{\xi}^{-1}(1)W(1)\hat{\eta}(1)P\right]\hat{\eta}^{-1}(t)$$

$$+ \int_0^t \hat{\xi}(t)\hat{\xi}^{-1}(s)\left[\dot{W} + \frac{1}{\varepsilon}W(D + \varepsilon TB) - (A - BT)W\right](s)$$

$$\times \hat{\eta}(s)(I_n - P)\hat{\eta}^{-1}(t)\,ds$$

$$- \int_t^1 \hat{\xi}(t)\hat{\xi}^{-1}(s)\left[\dot{W} + \frac{1}{\varepsilon}W(D + \varepsilon TB) - (A - BT)W\right](s)$$

$$\times \hat{\eta}(s)P\hat{\eta}^{-1}(t)\,ds \tag{9.2.27}$$

for any $m \times n$ matrix-valued function $W = W(t)$ of class C^1.

The representation (9.2.24) follows directly with (9.2.5) and (9.2.25) by taking $W = S$ in (9.2.27). Moreover, we see directly that (9.2.24) does in fact satisfy (9.2.5) and (9.2.25). Finally, we use (9.2.27) again, with $W = -BD^{-1}$, in obtaining (9.2.26). For this latter purpose we use the fact that the fundamental solution $\hat{\xi}$ for (9.2.22) satisfies a bound of the type (9.1.35), but with $\|A\|$ replaced on the right side of (9.1.35) with $\|A - BT\|$. Similarly, from Theorem 2 of Coppel (1967), we find that the fundamental solution $\hat{\eta}$ for (9.2.23) satisfies the previous exponential dichotomy (9.1.9) with the same projection P, but with the previous constant κ_1 of (9.1.9) replaced now by any fixed positive $\hat{\kappa}_1 < \kappa_1$. Details are omitted. ∎

This completes the proof of the existence of a suitable Riccati transformation (9.2.2) and (9.2.4) that reduces the linear system (9.2.1) to the diagonal form (9.2.3). It follows from (9.2.17) and (9.2.26) that the transformation matrix H and its inverse satisfy

$$H(t, \varepsilon) = \left[\begin{array}{c:c} I_m & \varepsilon B_0(t) D_0^{-1}(t) + O(\varepsilon^2) \\ \hdashline - D_0^{-1}(t) C_0(t) + O(\varepsilon) & \begin{array}{c} I_n - \varepsilon D_0^{-1}(t) C_0(t) B_0(t) \\ \times D_0^{-1}(t) + O(\varepsilon^2) \end{array} \end{array} \right]$$

(9.2.28)

and

$$H^{-1}(t, \varepsilon) = \left[\begin{array}{c:c} \begin{array}{c} I_m - \varepsilon B_0(t) D_0^{-2}(t) \\ \times C_0(t) + O(\varepsilon^2) \end{array} & - \varepsilon B_0(t) D_0^{-1}(t) + O(\varepsilon^2) \\ \hdashline D_0^{-1}(t) C_0(t) + O(\varepsilon) & I_n \end{array} \right]$$

(9.2.29)

uniformly for $0 \le t \le 1$ as $\varepsilon \to 0+$, with $B_0(t) := B(t, 0)$, $C_0(t) := C(t, 0)$, and $D_0(t) := D(t, 0)$. The related expected results for the derivatives dH/dt and dH^{-1}/dt are also valid, but these results are omitted here.

Exercises

Exercise 9.2.1: Prove that the matrix $H(t, \varepsilon)$ of (9.2.4) is unimodular, with $\det H = 1$.

Exercise 9.2.2: The homogeneous second-order scalar equation

$$\varepsilon \frac{d^2 x}{dt^2} + p(t, \varepsilon) \frac{dx}{dt} + q(t, \varepsilon) x = 0 \qquad (9.2.30)$$

is equivalent to the system

$$\frac{dx}{dt} = y,$$

$$\varepsilon \frac{dy}{dt} = -q(t, \varepsilon) x - p(t, \varepsilon) y, \qquad (9.2.31)$$

which is of the form of (9.2.1), with $A := 0$, $B := 1$, $C := -q$, and $D := -p$. Show that the transformation (9.2.2) and (9.2.4), with $S := 0$, permits the general solution of (9.2.30) to be given in the form

$$x = \text{const.} \exp\left(-\int T \right), \qquad (9.2.32)$$

where T denotes the general solution of the appropriate scalar Riccati equation. (This is the classical Riccati transformation. The parameter ε plays no role here and can be put equal to 1 if desired.)

Exercise 9.2.3: Give a derivation of the identity (9.2.8). *Hint:* Compute the integrals

$$\int_0^t \frac{\partial}{\partial s} \left[P\eta^{-1}(s)W(s)\xi(s)\xi^{-1}(t) \right] ds$$

and

$$\int_1^t \frac{\partial}{\partial s} \left[(I_n - P)\eta^{-1}(s)W(s)\xi(s)\xi^{-1}(t) \right] ds,$$

and then add the results and solve for $W(t)$. Note that $d\eta^{-1}(s)/ds = -(1/\varepsilon)\eta^{-1}(s)D(s,\varepsilon)$.

Exercise 9.2.4: Show that the function $T_0 = T_0(t,\varepsilon)$ of (9.2.13) is bounded on the region (9.2.14); that is, derive (9.2.15) for some suitable constant $\hat{\kappa}$.

Exercise 9.2.5: Use the Banach/Picard fixed-point theorem to prove that equation (9.2.12)–(9.2.13) has a unique solution satisfying (9.2.16), as $\varepsilon \to 0+$. *Hint:* Use the Banach space consisting of all continuous $n \times m$ matrix-valued functions $T = T(t) = [T_{ij}(t)]$ on the interval $0 \le t \le 1$, with norm

$$\|T\| := \max_{i,j} \left[\max_{[0,1]} |T_{ij}(t)| \right].$$

Show that the right side of (9.2.12) maps the ball $\|T\| \le 2\hat{\kappa}$ into itself and is contracting on this ball, for all small enough $\varepsilon > 0$.

Exercise 9.2.6 (Sibuya/Chang/Harris/Riccati transformation): **(i)** Show that the system

$$\frac{d}{dt}\begin{bmatrix} x \\ y \end{bmatrix} = \Gamma \begin{bmatrix} x \\ y \end{bmatrix}, \qquad \Gamma := \begin{bmatrix} A & B \\ C & D \end{bmatrix}, \tag{9.2.33}$$

is transformed into

$$\frac{d}{dt}\begin{bmatrix} u \\ v \end{bmatrix} = \Delta \begin{bmatrix} u \\ v \end{bmatrix}, \qquad \Delta = \left(H^{-1}\Gamma + \frac{dH^{-1}}{dt} \right) H, \tag{9.2.34}$$

by the nonsingular transformation

$$\begin{bmatrix} x \\ y \end{bmatrix} = H \begin{bmatrix} u \\ v \end{bmatrix}. \tag{9.2.35}$$

(ii) Compute the matrix

$$\Delta = \begin{bmatrix} \Delta_{11} & \Delta_{12} \\ \Delta_{21} & \Delta_{22} \end{bmatrix}$$

of (9.2.34) in the case

$$H := \begin{bmatrix} I_m & -S \\ -T & I_n + TS \end{bmatrix}, \tag{9.2.36}$$

and show that Δ is block diagonal with $\Delta_{12} = 0$ and $\Delta_{21} = 0$ if and only if T and S satisfy the system

$$\frac{dS}{dt} = (A - BT)S - S(D + TB) - B,$$
$$\frac{dT}{dt} = (DT - TA) + TBT - C. \tag{9.2.37}$$

Show that Δ is given as

$$\Delta = \begin{bmatrix} A - BT & 0 \\ 0 & D + TB \end{bmatrix} \tag{9.2.38}$$

if (9.2.37) holds.

Exercise 9.2.7: Let the given $n \times n$ matrix $D = D(t, \varepsilon)$ and its first derivative be uniformly bounded and continuous on the region of (9.2.14), and let D have the block structure

$$D(t, \varepsilon) = \begin{bmatrix} \overset{n_1 \times n_1}{D_{11}(t, \varepsilon)} & \varepsilon D_{12}(t, \varepsilon) \\ \varepsilon D_{21}(t, \varepsilon) & \overset{n_2 \times n_2}{D_{22}(t, \varepsilon)} \end{bmatrix}, \tag{9.2.39}$$

with $n = n_1 + n_2$, where the diagonal blocks D_{11} and D_{22} have eigenvalues with negative and positive real parts, respectively. That is, all eigenvalues of D_{11} (D_{22}) have negative (positive) real parts, uniformly on the region of (9.2.14). Prove that the system [see (9.1.4)]

$$\varepsilon \frac{dy}{dt} = D(t, \varepsilon) y, \qquad y = \begin{bmatrix} \overset{n_1 \times 1}{y_1} \\ \overset{n_2 \times 1}{y_2} \end{bmatrix}, \tag{9.2.40}$$

has an exponential dichotomy of the type (9.1.9), with

$$P = \begin{bmatrix} I_{n_1} & 0 \\ 0 & 0 \end{bmatrix}, \tag{9.2.41}$$

and with fundamental solution $\eta = \eta(t, \varepsilon)$ of the form

$$\eta(t, \varepsilon) = \begin{bmatrix} W_1(t)W_1^{-1}(0) & -S(t)W_2(t)W_2^{-1}(1) \\ -T(t)W_1(t)W_1^{-1}(0) & \left[I_{n_2} + T(t)S(t)\right]W_2(t)W_2^{-1}(1) \end{bmatrix},$$

$$(9.2.42)$$

with inverse

$$\eta^{-1}(t, \varepsilon)$$

$$= \begin{bmatrix} W_1(0)W_1^{-1}(t) & \vdots & W_1(0)W_1^{-1}(t)\left[I_{n_1} + S(t)T(t)\right] \\ \times\left[I_{n_1} + S(t)T(t)\right] & \vdots & \times S(t)\left[I_{n_2} + T(t)S(t)\right]^{-1} \\ \cdots\cdots\cdots\cdots\cdots\cdots & \vdots & \cdots\cdots\cdots\cdots\cdots\cdots\cdots \\ W_2(1)W_2^{-1}(t)T(t) & \vdots & W_2(1)W_2^{-1}(t) \end{bmatrix},$$

$$(9.2.43)$$

where

$$\overset{n_1 \times n_1}{W_1} \quad \text{and} \quad \overset{n_2 \times n_2}{W_2}$$

are fundamental solutions of the respective systems

$$\varepsilon\frac{dW_1}{dt} = (D_{11} - \varepsilon D_{12}T)W_1 \quad \text{and} \quad \varepsilon\frac{dW_2}{dt} = (D_{22} + \varepsilon TD_{12})W_2,$$

$$(9.2.44)$$

and where T and S are determined by the respective conditions

$$\varepsilon\frac{dT}{dt} = D_{22}T - TD_{11} + \varepsilon TD_{12}T - \varepsilon D_{21} \quad \text{for } 0 < t < 1,$$

$$(9.2.45)$$

$$T(1) = \overset{n_2 \times n_1}{0} \quad \text{at } t = 1,$$

and

$$\varepsilon\frac{dS}{dt} = (D_{11} - \varepsilon D_{12}T)S - S(D_{22} + \varepsilon TD_{12}) - \varepsilon D_{12} \quad \text{for } 0 < t < 1,$$

$$S(0) = \overset{n_1 \times n_2}{0} \quad \text{at } t = 0. \qquad (9.2.46)$$

Hint: Use a suitable Riccati transformation to diagonalize (9.2.40). The problems (9.2.45) and (9.2.46) can be converted into suitable integral equations and handled by the techniques used in the proofs of Lemma 9.2.1 and Lemma 9.2.2. We find in this case the results

$$T(t) \equiv T(t, \varepsilon) = O(\varepsilon),$$

$$S(t) \equiv S(t, \varepsilon) = O(\varepsilon) \quad \text{as } \varepsilon \to 0+, \qquad (9.2.47)$$

uniformly for $0 \leq t \leq 1$. A well-known result of Flatto and Levinson

(1955) (see Exercise 6.1.5) guarantees that W_1 and W_2 satisfy estimates of the type

$$\left| W_1(t) W_1^{-1}(s) \right|, \left| W_2(s) W_2^{-1}(t) \right| \leq \text{const.} \, e^{-\kappa_1(t-s)/\varepsilon}$$

as $\varepsilon \to 0+$, uniformly for $0 \leq s \leq t \leq 1$. (9.2.48)

The desired result (9.1.9) follows now by direct calculation from (9.2.41)–(9.2.43) using (9.2.48). [A more general D can be block-diagonalized as on p. 279 of Chang and Coppel (1969), and then the present approach leads to the result (9.1.7)–(9.19) for D of class C^1.]

Exercise 9.2.8: Let $p = p(t, \varepsilon)$ be a given real-valued function for $t \geq t_0$, $0 < \varepsilon \leq \varepsilon_0$, satisfying the positivity condition

$$p(t, \varepsilon) \geq \kappa > 0 \quad \text{for } t \geq t_0, 0 < \varepsilon \leq \varepsilon_0, \qquad (9.2.49)$$

for a fixed constant $\kappa > 0$. Assume also that the derivative $p' = (d/dt)p(t, \varepsilon)$ is small in the sense that p' can be decomposed into the sum of a part $q_1(t, \varepsilon)$ that is globally integrable and a part $q_2(t, \varepsilon)$ that is uniformly small, as

$$p'(t, \varepsilon) = q_1(t, \varepsilon) + q_2(t, \varepsilon), \qquad (9.2.50)$$

with

$$\int_{t_0}^{\infty} \left| q_1(t, \varepsilon) \right| dt \leq \kappa_1(\varepsilon) \quad \text{and} \quad \left| q_2(t, \varepsilon) \right| \leq \kappa_2(\varepsilon) \quad \text{for } t \geq t_0 \quad (9.2.51)$$

for given functions $\kappa_1 = \kappa_1(\varepsilon)$ and $\kappa_2 = \kappa_2(\varepsilon)$ satisfying $\kappa_1(\varepsilon) = O(1)$ and $\kappa_2(\varepsilon) = o(1)$, as $\varepsilon \to 0+$. Prove that the second-order scalar equation

$$\frac{d^2 x}{dt^2} = p(t, \varepsilon)^2 x \quad \text{for } t \geq t_0, \text{ small enough } \varepsilon > 0, \qquad (9.2.52)$$

has linearly independent solutions $x_1(t, \varepsilon) = x_1(t)$ and $x_2(t, \varepsilon) = x_2(t)$ satisfying

$$x_1(t) = \left[1 + b_1(t, \varepsilon) \right] \exp\left(-\int_{t_0}^{t} p \right),$$

$$x_1'(t) = -p(t, \varepsilon)\left[1 + c_1(t, \varepsilon) \right] \exp\left(-\int_{t_0}^{t} p \right),$$

$$x_2(t) = \left[1 + b_2(t, \varepsilon) \right] \exp\left(+\int_{t_0}^{t} p \right), \qquad (9.2.53)$$

$$x_2'(t) = +p(t, \varepsilon)\left[1 + c_2(t, \varepsilon) \right] \exp\left(+\int_{t_0}^{t} p \right),$$

for suitable $C^1[t_0, \infty)$-functions b_j, c_j ($j = 1, 2$) satisfying

$$|b_j(t, \varepsilon)|, |c_j(t, \varepsilon)| \leq \text{constant} \left[\int_t^\infty |q_1(s, \varepsilon)| \, ds + \kappa_2(\varepsilon) \right]$$
$$\to 0 \quad \text{as } t \to \infty, \varepsilon \to 0+. \tag{9.2.54}$$

Hint: For the equivalent system

$$\frac{d}{dt} \begin{bmatrix} x \\ y \end{bmatrix} = \begin{bmatrix} 0 & 1 \\ p^2 & 0 \end{bmatrix} \begin{bmatrix} x \\ y \end{bmatrix}, \tag{9.2.55}$$

introduce the eigenvector transformation (see Section 7.2)

$$\begin{bmatrix} \hat{x} \\ \hat{y} \end{bmatrix} = \begin{bmatrix} 1 & -1/p \\ 1 & 1/p \end{bmatrix} \begin{bmatrix} x \\ y \end{bmatrix}, \tag{9.2.56}$$

and find the system

$$\frac{d}{dt} \begin{bmatrix} \hat{x} \\ \hat{y} \end{bmatrix} = \left[\begin{bmatrix} -p & 0 \\ 0 & p \end{bmatrix} + \frac{p'}{2p} \begin{bmatrix} -1 & 1 \\ 1 & -1 \end{bmatrix} \right] \begin{bmatrix} \hat{x} \\ \hat{y} \end{bmatrix}. \tag{9.2.57}$$

Use a suitable Riccati transformation

$$\begin{bmatrix} \hat{x} \\ \hat{y} \end{bmatrix} = H(t) \begin{bmatrix} u \\ v \end{bmatrix}$$

as in (9.2.35)–(9.2.37) to reduce (9.2.57) to the equivalent (uncoupled) diagonal system

$$\frac{du}{dt} = -\left[p + \frac{p'}{2p}(T + 1) \right] u, \quad \frac{dv}{dt} = \left[p + \frac{p'}{2p}(T - 1) \right] v, \tag{9.2.58}$$

provided that $T = T(t) = T(t, \varepsilon)$ satisfies the Riccati equation

$$\frac{dT}{dt} = 2pT + \frac{p'}{2p}(T^2 - 1). \tag{9.2.59}$$

Solve this latter equation, subject to the homogeneous condition $T = 0$ at $t = \infty$, by considering the equivalent integral equation

$$T(t) = T_0(t) - \int_t^\infty \left[\exp\left(-2 \int_t^s p \right) \right] \frac{p'(s, \varepsilon)}{2p(s, \varepsilon)} T(s)^2 \, ds,$$
$$T_0(t) := \int_t^\infty \left[\exp\left(-2 \int_t^s p \right) \right] \frac{p'(s, \varepsilon)}{2p(s, \varepsilon)} \, ds. \tag{9.2.60}$$

The Banach/Picard fixed-point theorem can be used to solve (9.2.60) for $t \geq t_1, 0 < \varepsilon \leq \varepsilon_1$, for large enough $t_1 \geq t_0$ and small enough $\varepsilon_1 > 0$, for a solution T satisfying

$$|T(t, \varepsilon)| \le \frac{1}{\kappa} \int_t^\infty |q_1(s, \varepsilon)| \, ds + \frac{\kappa_2(\varepsilon)}{2\kappa^2} \quad \text{for } t \ge t_1, 0 < \varepsilon \le \varepsilon_1.$$

$$(9.2.61)$$

The analogous equation of (9.2.37) for S can be easily solved subject to the initial condition $S(t_1) = 0$. The uncoupled system (9.2.58) is then easily solved, and the resulting solutions lead directly to corresponding solutions x_1, x_2 for (9.2.52) of the form

$$x_1(t) = \frac{1 - T(t)}{2} \exp\left\{ - \int_{t_1}^t \left[p + \frac{p'}{2p}(T + 1) \right] \right\}, \quad (9.2.62)$$

with a related formula for $x_2(t)$. These solutions x_1, x_2 can be multiplied by appropriate constants and then relabelled as x_1, x_2 to find suitable results of the form

$$x_1(t) = [1 + b_1(t)] \exp\left(- \int_{t_1}^t p \right), \quad (9.2.63)$$

along with related results for x_1', x_2, and x_2'. A standard continuation result permits us to replace t_1 with t_0 as the lower limit of integration in these latter results, which (up to a further constant multiple) yields (9.2.53). The method of proof here yields additional related results; see Smith (1984b). See Bellman (1953), Hartman and Wintner (1955), Olver (1961, 1974), Coppel (1965), Harris and Lutz (1974, 1977), and Taylor (1978, 1982) for further related results.

9.3 The Green function

In this section we construct a Green function for the previous boundary-value problem consisting of the system [see (9.1.1)–(9.1.2)]

$$\frac{d}{dt}\begin{bmatrix} x \\ y \end{bmatrix} = \begin{bmatrix} A(t, \varepsilon) & B(t, \varepsilon) \\ \dfrac{1}{\varepsilon} C(t, \varepsilon) & \dfrac{1}{\varepsilon} D(t, \varepsilon) \end{bmatrix} \begin{bmatrix} x \\ y \end{bmatrix} + \begin{bmatrix} f(t, \varepsilon) \\ \dfrac{1}{\varepsilon} g(t, \varepsilon) \end{bmatrix} \quad (9.3.1)$$

for $0 < t < 1$, along with the boundary conditions

$$L(\varepsilon)\begin{bmatrix} x(0, \varepsilon) \\ y(0, \varepsilon) \end{bmatrix} + R(\varepsilon)\begin{bmatrix} x(1, \varepsilon) \\ y(1, \varepsilon) \end{bmatrix} = \begin{bmatrix} \alpha(\varepsilon) \\ \beta(\varepsilon) \end{bmatrix}, \quad (9.3.2)$$

where the notation is as in Section 9.1.

According to the result of Exercise 0.1.2, the problem (9.3.1)–(9.3.2) has a unique solution that can be represented as

$$
\begin{bmatrix} x(t,\varepsilon) \\ y(t,\varepsilon) \end{bmatrix} = Z(t,\varepsilon)M^{-1}\begin{bmatrix} \alpha(\varepsilon) \\ \beta(\varepsilon) \end{bmatrix} + \int_0^1 G(t,s,\varepsilon)\begin{bmatrix} f(s,\varepsilon) \\ \frac{1}{\varepsilon}g(s,\varepsilon) \end{bmatrix} ds,
$$

(9.3.3)

provided that the matrix

$$
M = M(\varepsilon) := L(\varepsilon)Z(0,\varepsilon) + R(\varepsilon)Z(1,\varepsilon) \tag{9.3.4}
$$

is nonsingular, where $Z = Z(t,\varepsilon)$ is a fundamental solution matrix for the homogeneous system

$$
\frac{dx}{dt} = A(t,\varepsilon)x + B(t,\varepsilon)y,
$$

$$
\varepsilon\frac{dy}{dt} = C(t,\varepsilon)x + D(t,\varepsilon)y \quad \text{for } 0 < t < 1,
$$

(9.3.5)

and where $G = G(t,s) = G(t,s,\varepsilon)$ is the Green function defined as

$$
G(t,s) := \begin{cases} Z(t,\varepsilon)M^{-1}(\varepsilon)L(\varepsilon)Z(0,\varepsilon)Z(s,\varepsilon)^{-1} & \text{for } s < t, \\ -Z(t,\varepsilon)M^{-1}(\varepsilon)R(\varepsilon)Z(1,\varepsilon)Z(s,\varepsilon)^{-1} & \text{for } s > t. \end{cases}
$$

(9.3.6)

We use the Riccati transformation of Section 9.2 to construct a suitable fundamental solution $Z(t,\varepsilon)$ for (9.3.5) in a convenient form, subject to the assumptions of Section 9.1. In particular, we assume that the given matrix $D = D(t,\varepsilon)$ is such that the associated system (9.1.5) has a fundamental solution that satisfies the exponential dichotomy (9.1.9) for some fixed projection P.

We solve the homogeneous system (9.3.5) for a fundamental solution Z by solving the corresponding homogeneous *diagonalized* system (9.2.3), given as

$$
\frac{du}{dt} = (A - BT)u,
$$

$$
\varepsilon\frac{dv}{dt} = (D + \varepsilon TB)v,
$$

(9.3.7)

where $T = T(t,\varepsilon)$ is the matrix-valued function appearing in the Riccati transformation (9.2.2) and (9.2.4), as given in Lemma 9.2.1. This system (9.3.7) coincides with the combined equations of (9.2.22) and (9.2.23), which have respective fundamental solutions $U = U(t,\varepsilon)$ and $V = V(t,\varepsilon)$

that satisfy the inequalities (see the remarks in the proof of Lemma 9.2.2, where these fundamental solutions were labelled as $\hat{\xi} = U$ and $\hat{\eta} = V$)

$$\left| U(t, \varepsilon) U^{-1}(s, \varepsilon) \right| \leq |I_m| \cdot e^{\|A - BT\| \cdot |t-s|}$$

$$\text{for all } 0 \leq \varepsilon \leq \varepsilon_0 \text{ and for all } 0 \leq s, t \leq 1, \qquad (9.3.8)$$

where

$$\| A - BT \| := \max_{0 \leq t \leq 1} \left| A(t, \varepsilon) - B(t, \varepsilon) T(t, \varepsilon) \right|,$$

and

$$\left| V(t, \varepsilon) P V^{-1}(s, \varepsilon) \right|, \left| V(s, \varepsilon)(I_n - P) V^{-1}(t, \varepsilon) \right|$$

$$\leq \kappa_0 \cdot e^{-\kappa_1 \cdot (t-s)/\varepsilon} \quad \text{for } 0 \leq s \leq t \leq 1, 0 < \varepsilon \leq \varepsilon_0, \qquad (9.3.9)$$

for fixed positive constants κ_0 and κ_1 that are different from, but related to, the earlier constants of (9.1.9).

Following Chang (1972) and Harris (1973), we use now the following particular fundamental solution for (9.3.7):

$$\begin{bmatrix} U(t, \varepsilon) & \overset{m \times n}{0} \\ \overset{n \times m}{0} & V(t, \varepsilon) P V^{-1}(0, \varepsilon) + V(t, \varepsilon)(I_n - P) V^{-1}(1, \varepsilon) \end{bmatrix}$$

$$= \text{fundamental solution for (9.3.7)}, \qquad (9.3.10)$$

where U and V are the respective fundamental solutions of the separate equations of (9.3.7) that satisfy (9.3.8) and (9.3.9). The transformation (9.2.2) and (9.2.4), along with (9.3.10), yields now the following fundamental solution Z for (9.3.5):

$$Z(t, \varepsilon) = H(t, \varepsilon)$$

$$\times \begin{bmatrix} U(t, \varepsilon) & 0 \\ 0 & V(t, \varepsilon) P V^{-1}(0, \varepsilon) + V(t, \varepsilon)(I_n - P) V^{-1}(1, \varepsilon) \end{bmatrix},$$

$$(9.3.11)$$

where $H = H(t, \varepsilon)$ is given by (9.2.4).

Example 9.3.1: Consider the problem (9.3.1)–(9.3.2) subject to the assumptions of Section 9.1, and assume here that all eigenvalues of $D_0 = D_0(t) := D(t, 0)$ have negative real parts, uniformly for $0 \leq t \leq 1$, so that the projection P can be taken to coincide with the identity,

$$P = I_n. \qquad (9.3.12)$$

Moreover, assume that the dimensions of x and y coincide, with

$$m = n, \qquad (9.3.13)$$

and assume that the matrix $B_0 = B_0(t)$ is nonsingular at $t = 0$,

$$B_0(t): = B(t, 0) \quad \text{is nonsingular at } t = 0, \qquad (9.3.14)$$

so that the matrix $S = S(t, \varepsilon)$ of the Riccati transformation is also nonsingular at $t = 0$ for all sufficiently small $\varepsilon \geq 0$ [see (9.2.26)]. Finally, consider the Dirichlet problem with [see (9.3.2)]

$$L(\varepsilon): = \begin{bmatrix} I_n & 0 \\ 0 & 0 \end{bmatrix}, \qquad R(\varepsilon): = \begin{bmatrix} 0 & 0 \\ I_n & 0 \end{bmatrix}, \qquad (9.3.15)$$

so that the boundary conditions of (9.3.2) reduce to

$$x(0, \varepsilon) = \alpha(\varepsilon), \qquad x(1, \varepsilon) = \beta(\varepsilon). \qquad (9.3.16)$$

In this case the fundamental solution Z of (9.3.11) becomes

$$Z(t, \varepsilon) = H(t, \varepsilon) \begin{bmatrix} U(t, \varepsilon) & 0 \\ 0 & V(t, \varepsilon) V^{-1}(0, \varepsilon) \end{bmatrix}, \qquad (9.3.17)$$

and then the matrix $M(\varepsilon)$ of (9.3.4) is found, with (9.3.15), (9.3.17), and (9.2.4), to be

$$M(\varepsilon) = \begin{bmatrix} U(0, \varepsilon) & -\varepsilon S(0, \varepsilon) \\ U(1, \varepsilon) & -\varepsilon S(1, \varepsilon) V(1, \varepsilon) V^{-1}(0, \varepsilon) \end{bmatrix}. \qquad (9.3.18)$$

Each block submatrix on the right side of (9.3.18) is nonsingular (for $\varepsilon > 0$), and we find also that $M(\varepsilon)$ is itself nonsingular, with (see Exercises 9.3.1 and 9.3.2)

$$M^{-1}(\varepsilon)$$
$$= \sum_{k=0}^{\infty} \begin{bmatrix} -U(1)^{-1} \mu^{k+1} U(1) U^{-1}(0) & U^{-1}(1) \mu^k \\ -\dfrac{1}{\varepsilon} S^{-1}(0) U(0) U^{-1}(1) \mu^k U(1) U^{-1}(0) & \dfrac{1}{\varepsilon} S^{-1}(0) U(0) U^{-1}(1) \mu^k \end{bmatrix}, \qquad (9.3.19)$$

where μ is the matrix-valued quantity defined as

$$\mu = \mu(\varepsilon): = S(1) V(1) V^{-1}(0) S^{-1}(0) U(0) U^{-1}(1), \qquad (9.3.20)$$

and where we are here suppressing the dependence on ε, as $U(1) \equiv U(1, \varepsilon)$,

$S(0) \equiv S(0, \varepsilon)$, and so forth. The quantities U and S, and their inverses, are uniformly bounded as $\varepsilon \to 0+$, so that (9.3.20), along with (9.3.9) and (9.3.12), yields

$$\mu(\varepsilon) = O\big[V(1, \varepsilon)V^{-1}(0, \varepsilon)\big] = O(e^{-\kappa_1/\varepsilon}) \quad \text{as } \varepsilon \to 0+. \quad (9.3.21)$$

A direct calculation using the foregoing results yields (see Exercise 9.3.3)

$$Z(t, \varepsilon) M^{-1}(\varepsilon) = \begin{bmatrix} \gamma_{11}(t, \varepsilon) & \gamma_{12}(t, \varepsilon) \\ \gamma_{21}(t, \varepsilon) & \gamma_{22}(t, \varepsilon) \end{bmatrix} + \begin{bmatrix} O(\mu^2) & O(\mu) \\ O\left(\dfrac{\mu}{\varepsilon}\right) & O\left(\dfrac{\mu}{\varepsilon}\right) \end{bmatrix}$$

$$\text{as } \varepsilon \to 0+, \quad (9.3.22)$$

with

$$\gamma_{11}(t, \varepsilon) = -U(t)U_1^{-1}S_1V_1V_0^{-1}S_0^{-1} + S(t)V(t)V_0^{-1}S_0^{-1},$$

$$\gamma_{12}(t, \varepsilon) = U(t)U_1^{-1} - S(t)V(t)V_0^{-1}S_0^{-1}U_0U_1^{-1},$$

$$\gamma_{21}(t, \varepsilon) = T(t)U(t)U_1^{-1}S_1V_1V_0^{-1}S_0^{-1}$$
$$\qquad - \frac{1}{\varepsilon}\big[I + \varepsilon T(t)S(t)\big]V(t)V_0^{-1}S_0^{-1}, \qquad (9.3.23)$$

$$\gamma_{22}(t, \varepsilon) = -T(t)U(t)U_1^{-1} + \frac{1}{\varepsilon}\big[I + \varepsilon T(t)S(t)\big]V(t)V_0^{-1}S_0^{-1}U_0U_1^{-1},$$

where $U(t) \equiv U(t, \varepsilon)$, $S(t) \equiv S(t, \varepsilon)$, and so forth, and for brevity here $U_1 := U(1) = U(1, \varepsilon)$, $U_0 := U(0) = U(0, \varepsilon)$, and so forth. Similarly, a straightforward but tedious calculation yields for the Green function [see (9.3.6), (9.3.15), (9.3.17), (9.2.4), and (9.3.22)]

$$G(t, s, \varepsilon) = \begin{bmatrix} O(1) & O(\varepsilon) \\ O\left(\dfrac{1}{\varepsilon}\right) & O(1) \end{bmatrix} \quad \text{as } \varepsilon \to 0+, \quad (9.3.24)$$

uniformly for $0 \le t, s \le 1$. Slightly better estimates than those incorporated in (9.3.24) can be obtained from the foregoing calculation, but (9.3.24) suffices for our purposes.

The preceding results yield directly the cancellation rule for the present Dirichlet problem subject to (9.3.12). For example, if the boundary data $\alpha(\varepsilon)$, $\beta(\varepsilon)$ have asymptotic expansions in powers of ε, and if, for simplicity, the system (9.3.1) is homogeneous, with $f = 0$ and $g = 0$, then we see directly, with (9.3.3) and (9.3.21)–(9.3.23), that

$$\lim_{\substack{\varepsilon \to 0+ \\ \text{fixed } 0 < t < 1}} x(t, \varepsilon) = U(t, 0)U^{-1}(1, 0)\beta(0), \quad (9.3.25)$$

so that the boundary condition is cancelled at the left endpoint and retained at the right, as $\varepsilon \to 0+$.

The foregoing results also yield error estimates for approximate solutions to the given boundary-value problem. For example, the leading terms $x^0(t, \varepsilon)$ and $y^0(t, \varepsilon)$ in a two-variable approximation can be obtained in the form

$$x^0(t, \varepsilon) := X_0(t) + X_0^*\left(\frac{t}{\varepsilon}\right),$$

$$y^0(t, \varepsilon) := Y_0(t) + \frac{1}{\varepsilon}Y_0^*\left(\frac{t}{\varepsilon}\right),$$

(9.3.26)

for suitable outer solutions X_0, Y_0 and suitable boundary-layer correction terms X_0^*, Y_0^*, where these functions can be chosen so that the resulting differences

$$\hat{x} := x - x^0, \quad \hat{y} := y - y^0,$$

(9.3.27)

satisfy the system (see Exercise 9.3.4)

$$\frac{d}{dt}\begin{bmatrix} \hat{x} \\ \hat{y} \end{bmatrix} = \begin{bmatrix} A(t, \varepsilon) & B(t, \varepsilon) \\ \frac{1}{\varepsilon}C(t, \varepsilon) & \frac{1}{\varepsilon}D(t, \varepsilon) \end{bmatrix}\begin{bmatrix} \hat{x} \\ \hat{y} \end{bmatrix} + \begin{bmatrix} \rho_1(t, \varepsilon) \\ \frac{1}{\varepsilon}\rho_2(t, \varepsilon) \end{bmatrix}$$

(9.3.28)

for $0 < t < 1$, subject to the boundary conditions

$$\hat{x} = \sigma_1(\varepsilon) \quad \text{at } t = 0,$$

$$\hat{x} = \sigma_2(\varepsilon) \quad \text{at } t = 1,$$

(9.3.29)

for suitable residuals ρ_j and σ_j satisfying (see Exercise 9.3.5)

$$|\rho_j(t, \varepsilon)| \le C \cdot \left[\varepsilon + \left(1 + \frac{t}{\varepsilon}\right)e^{-\kappa_1 t/\varepsilon}\right],$$

$$|\sigma_j(\varepsilon)| \le C \cdot \varepsilon \quad \text{for } j = 1, 2, \quad \text{as } \varepsilon \to 0+,$$

(9.3.30)

for a fixed constant C, uniformly for $0 \le t \le 1$, where κ_1 is the constant appearing in (9.3.9) and (9.3.21).

It follows that the representation (9.3.3) can be applied to $\hat{x}, \hat{y}$, with f and g replaced by ρ_1 and ρ_2, and with α and β replaced by σ_1 and σ_2. The estimates (9.3.21)–(9.3.23), (9.3.24), (9.3.30), and (9.3.9) (with $P = I_n$) then yield directly the results $\hat{x} = O(\varepsilon)$ and $\hat{y} = O(1)$, or, with (9.3.27),

$$|x(t, \varepsilon) - x^0(t, \varepsilon)| \le \text{const. } \varepsilon,$$

$$|y(t, \varepsilon) - y^0(t, \varepsilon)| \le \text{constant}, \quad \text{as } \varepsilon \to 0+,$$

(9.3.31)

uniformly for $0 \le t \le 1$. Hence, for small $\varepsilon > 0$, the functions x^0, y^0 provide useful information on the exact solution functions, uniformly for all t.

Example 9.3.2: Consider the Dirichlet problem for the second-order vector system

$$\varepsilon x'' = C(t,\varepsilon)x + D(t,\varepsilon)x' + g(t,\varepsilon) \quad \text{for } 0 \le t \le 1, \quad (9.3.32)$$

for a real n-dimensional vector-valued solution function $x = x(t,\varepsilon)$, where C and D are given matrix-valued functions and g is a given vector-valued function. The data are assumed to be bounded and continuous along with their first derivatives on the region (9.2.14), and the data are also assumed to have asymptotic expansions in powers of ε. The eigenvalues of the matrix $D_0 = D(t,0)$ are assumed to have negative real parts, uniformly for $0 \le t \le 1$.

The boundary-value problem (9.3.16) and (9.3.32) can be written in the form

$$\frac{dx}{dt} = y,$$

$$\varepsilon \frac{dy}{dt} = Cx + Dy + g \quad \text{for } 0 < t < 1, \quad (9.3.33)$$

subject to the boundary conditions (9.3.16). This problem is of the type considered in Example 9.3.1, with

$$A \equiv 0, \qquad B \equiv I_n, \quad \text{and} \quad f \equiv 0,$$

and so the results of that example apply. It follows that this Dirichlet problem for (9.3.32) has a unique solution for all small $\varepsilon > 0$, and uniform approximations can be obtained for the solution function using a two-variable approach, as discussed earlier. We obtain in this way a proof of the earlier estimates (8.3.38) and (8.3.39) that were given without proof in Section 8.3 for the scalar case $n = 1$. The Dirichlet problem for a nonlinear generalization of (9.3.32) is studied in Chang (1974b, 1976).

More general linear boundary-value problems of the type (9.3.1)–(9.3.2) are considered in Chang (1972) and Harris (1973); see also Harris (1960, 1962a) and O'Malley (1969c) for related studies. Several such linear problems are discussed in the following examples for the two-dimensional system ($m = n = 1$)

$$\frac{dx}{dt} = a(t,\varepsilon)x + b(t,\varepsilon)y + f(t,\varepsilon),$$

$$\varepsilon \frac{dy}{dt} = c(t,\varepsilon)x + d(t,\varepsilon)y + g(t,\varepsilon), \quad \text{for } 0 \le t \le 1, 0 < \varepsilon \le \varepsilon_0, \quad (9.3.34)$$

subject to the condition

$$d_0(t) := d(t,0) < 0 \quad \text{for } 0 \le t \le 1, \quad (9.3.35)$$

for real-valued solution functions $x = x(t,\varepsilon)$ and $y = y(t,\varepsilon)$, and subject to the assumption that the data functions have asymptotic expansions in powers of ε, with coefficients that are smooth functions of t. The condition $d_0 < 0$ can be replaced with $d_0 > 0$, with the obvious modifications.

The fundamental solution (9.3.17) for (9.3.34)–(9.3.35) becomes

$Z(t, \varepsilon)$ = fundamental solution for (9.3.34)–(9.3.35)

$$= \begin{bmatrix} u(t, \varepsilon) & -\varepsilon S(t, \varepsilon) v(t, \varepsilon) \\ -T(t, \varepsilon) u(t, \varepsilon) & [1 + \varepsilon T(t, \varepsilon) S(t, \varepsilon)] v(t, \varepsilon) \end{bmatrix}, \quad (9.3.36)$$

with

$$u(t, \varepsilon): = \exp\left[\int_0^t (a - bT)\right], \qquad v(t, \varepsilon): = \exp\left[\frac{1}{\varepsilon}\int_0^t (d + \varepsilon bT)\right],$$

$$(9.3.37)$$

where S and T are the functions appearing in the Riccati transformation, as in Lemma 9.2.1 and Lemma 9.2.2, satisfying

$$T(t, \varepsilon) = \frac{c_0(t)}{d_0(t)} + O(\varepsilon), \qquad \frac{dT(t, \varepsilon)}{dt} = O(1),$$

$$(9.3.38)$$

$$S(t, \varepsilon) = -\frac{b_0(t)}{d_0(t)} + O(\varepsilon), \qquad \frac{dS(t, \varepsilon)}{dt} = O(1),$$

as $\varepsilon \to 0+$, uniformly for $0 \le t \le 1$.

Example 9.3.3: Consider a Neumann-type boundary-value problem for (9.3.34)–(9.3.35) with boundary conditions $y(0, \varepsilon) = \alpha(\varepsilon)$ and $y(1, \varepsilon) = \beta(\varepsilon)$, so that the boundary matrices L and R of (9.3.2) are given as

$$L = \begin{bmatrix} 0 & 1 \\ 0 & 0 \end{bmatrix}, \qquad R = \begin{bmatrix} 0 & 0 \\ 0 & 1 \end{bmatrix}. \qquad (9.3.39)$$

Also, assume that the coefficient function $c = c(t, \varepsilon)$ is nonzero at $t = 1$, with

$$c_0(1) = c(1, 0) \ne 0. \qquad (9.3.40)$$

In this case the matrix $M(\varepsilon)$ of (9.3.4) is seen to be given as

$$M(\varepsilon) = \begin{bmatrix} -T(0, \varepsilon) & 1 + \varepsilon T(0, \varepsilon) S(0, \varepsilon) \\ -T(1, \varepsilon) u(1, \varepsilon) & [1 + \varepsilon T(1, \varepsilon) S(1, \varepsilon)] v(1, \varepsilon) \end{bmatrix}, $$

$$(9.3.41)$$

so that we have [see (9.3.37)–(9.3.38) and (9.3.40)]

$$\det M(\varepsilon) = \frac{c_0(1)}{d_0(1)} u(1, \varepsilon) + O(\varepsilon) \ne 0 \qquad (9.3.42)$$

as $\varepsilon \to 0+$. Hence, the given Neumann problem has a unique solution for small $\varepsilon > 0$ when (9.3.40) holds, and this solution can be represented as in (9.3.3).

A direct calculation yields [see (9.3.36)–(9.3.38) and (9.3.41)]

$$
\lim_{\substack{\varepsilon \to 0+ \\ \text{fixed } 0 < t < 1}} Z(t, \varepsilon) M^{-1}(\varepsilon) \begin{bmatrix} \alpha(\varepsilon) \\ \beta(\varepsilon) \end{bmatrix}
$$

$$
= \begin{bmatrix} -1 \\ \dfrac{c_0(t)}{d_0(t)} \end{bmatrix} \dfrac{d_0(1)}{c_0(1)} \left\{ \exp\left(-\int_t^1 \frac{a_0 d_0 - b_0 c_0}{d_0} \right) \right\} \beta(0), \quad (9.3.43)
$$

so that we see that the boundary condition at $t = 1$ ($y = \beta$ at $t = 1$) is retained as $\varepsilon \to 0+$, whereas the boundary condition is generally lost at $t = 0$ as $\varepsilon \to 0+$.

A direct, but tedious, calculation shows in this case that the Green function (9.3.6) satisfies

$$
\int_0^1 |G_{ij}(t, s, \varepsilon)| \, ds \leq \begin{cases} \text{constant}, & \text{for } j = 1, \\ \text{const. } \varepsilon, & \text{for } j = 2, \end{cases} \quad (9.3.44)
$$

for $i = 1, 2$. As usual, these results can be used to provide error estimates for approximate solutions to the given boundary-value problem; see Exercise 9.3.8. Note that it is sometimes more convenient to give estimates of integral type for G, as in (9.3.44), whereas in other situations it is convenient to give pointwise estimates, as in (9.3.24).

Example 9.3.4: Consider the system (9.3.34)–(9.3.35) subject to the coupled boundary conditions

$$
\begin{aligned}
x(1, \varepsilon) + y(0, \varepsilon) &= \alpha(\varepsilon), \\
x(1, \varepsilon) + y(1, \varepsilon) &= \beta(\varepsilon),
\end{aligned} \quad (9.3.45)
$$

so that the boundary matrices L and R of (9.3.2) are

$$
L = \begin{bmatrix} 0 & 1 \\ 0 & 0 \end{bmatrix}, \qquad R = \begin{bmatrix} 1 & 0 \\ 1 & 1 \end{bmatrix}. \quad (9.3.46)
$$

Assume also that

$$
c_0(1) \neq d_0(1). \quad (9.3.47)
$$

In this case the matrix $M(\varepsilon)$ of (9.3.4) is

$$M(\varepsilon) = \begin{bmatrix} u(1,\varepsilon) - T(0,\varepsilon) & 1 + \varepsilon[T(0,\varepsilon)S(0,\varepsilon) - S(1,\varepsilon)v(1,\varepsilon)] \\ [1 - T(1,\varepsilon)]u(1,\varepsilon) & \{1 + \varepsilon S(1,\varepsilon)[T(1,\varepsilon) - 1]\}v(1,\varepsilon) \end{bmatrix},$$

(9.3.48)

so that we find

$$\det M(\varepsilon) = \left[\frac{c_0(1) - d_0(1)}{d_0(1)} + O(\varepsilon)\right]\exp\left(\int_0^1 \frac{a_0 d_0 - b_0 c_0}{d_0}\right) \neq 0$$

(9.3.49)

as $\varepsilon \to 0+$ [see (9.3.47)]. Hence, the given boundary-value problem has a unique solution for small $\varepsilon > 0$ when (9.3.47) holds. As usual, the solution can be represented by (9.3.3).

In this case a direct calculation yields

$$\lim_{\substack{\varepsilon \to 0+ \\ \text{fixed } 0 < t < 1}} Z(t,\varepsilon)M^{-1}(\varepsilon)\begin{bmatrix} \alpha(\varepsilon) \\ \beta(\varepsilon) \end{bmatrix}$$

$$= \begin{bmatrix} -1 \\ \dfrac{c_0(t)}{d_0(t)} \end{bmatrix} \frac{d_0(1)}{c_0(1) - d_0(1)} \left\{\exp\left(-\int_t^1 \frac{a_0 d_0 - b_0 c_0}{d_0}\right)\right\}\beta(0).$$

(9.3.50)

We see that the specified boundary condition

$$x(1,\varepsilon) + y(0,\varepsilon) = \alpha$$

is generally lost as $\varepsilon \to 0+$, whereas the other boundary condition

$$x(1,\varepsilon) + y(1,\varepsilon) = \beta$$

is retained at $t = 1$ as $\varepsilon \to 0+$.

We can easily obtain multivariable approximations for the solution functions in the form

$$\begin{bmatrix} x(t,\varepsilon) \\ y(t,\varepsilon) \end{bmatrix} \sim \begin{bmatrix} X(t,\varepsilon) \\ Y(t,\varepsilon) \end{bmatrix} + \begin{bmatrix} \varepsilon^* X(\tau,\varepsilon) \\ {}^* Y(\tau,\varepsilon) \end{bmatrix},$$

(9.3.51)

with $\tau := t/\varepsilon$, for suitable outer functions X, Y and suitable boundary-layer correction functions $^*X, ^*Y$ that have asymptotic expansions in powers of ε. These expansions can be truncated as usual to provide approximate solution functions, and the expected error estimates are readily obtained using suitable estimates on the Green function (see Exercise 9.3.9).

Example 9.3.5: Consider the system (9.3.34)–(9.3.35) subject to the coupled boundary conditions

$$x(1, \varepsilon) + y(0, \varepsilon) = \alpha(\varepsilon),$$
$$x(0, \varepsilon) + y(1, \varepsilon) = \beta(\varepsilon),$$

(9.3.52)

so that the boundary matrices L and R of (9.3.2) are

$$L = \begin{bmatrix} 0 & 1 \\ 1 & 0 \end{bmatrix}, \quad R = \begin{bmatrix} 1 & 0 \\ 0 & 1 \end{bmatrix}.$$

(9.3.53)

Assume in this case that we also have

$$c_0(1) \exp\left(\int_0^1 \frac{a_0 d_0 - b_0 c_0}{d_0} \right) \neq d_0(1).$$

(9.3.54)

In this case each of the two boundary conditions of (9.3.52) involves the left endpoint $t = 0$, and it may not be clear which of these two boundary conditions should be retained by the cancellation law for the reduced, limiting outer approximation. The following calculation shows that we drop the boundary condition involving the function with the stronger boundary-layer effect at $t = 0$, namely the fast-variable quantity $y(0, \varepsilon)$, rather than the slow-variable quantity $x(0, \varepsilon)$.

In this case the matrix $M(\varepsilon)$ of (9.3.4) is

$$M(\varepsilon) = \begin{bmatrix} u(1, \varepsilon) - T(0, \varepsilon) & 1 + \varepsilon T(0, \varepsilon) S(0, \varepsilon) - \varepsilon S(1, \varepsilon) v(1, \varepsilon) \\ 1 - T(1, \varepsilon) u(1, \varepsilon) & [1 + \varepsilon S(1, \varepsilon) T(1, \varepsilon)] v(1, \varepsilon) - \varepsilon S(0, \varepsilon) \end{bmatrix},$$

(9.3.55)

so that we find

$$\det M(\varepsilon) = -1 + \frac{c_0(1)}{d_0(1)} \exp\left(\int_0^1 \frac{a_0 d_0 - b_0 c_0}{d_0} \right) + O(\varepsilon) \neq 0 \quad (9.3.56)$$

as $\varepsilon \to 0+$ [see (9.3.54)]. Hence, the given boundary-value problem has a unique solution for small $\varepsilon > 0$ when (9.3.54) holds. In this case the coupling of the boundary conditions is such that the resulting sufficiency condition (9.3.54) involves the data over the entire interval $[0, 1]$, not just data values at the endpoints as in (9.3.40) and (9.3.47). Note also that these conditions such as (9.3.54) are sufficient, but not necessary, for the existence of unique solutions for the given boundary-value problems. For example, if (9.3.54) fails to hold, then the matrix $M(\varepsilon)$ may still have a nonzero determinant if the first nonzero term in an asymptotic expansion for $M(\varepsilon)$ is nonzero. In this latter situation the asymptotic structure of the resulting solution is different than that indicated by (9.3.58).

We assume here that (9.3.54) holds, and in this case a direct calculation yields

$$
\lim_{\substack{\varepsilon \to 0+ \\ \text{fixed } 0<t<1}} Z(t,\varepsilon)M^{-1}(\varepsilon)\begin{bmatrix} \alpha(\varepsilon) \\ \beta(\varepsilon) \end{bmatrix}
$$

$$
= \begin{bmatrix} -1 \\ \dfrac{c_0(t)}{d_0(t)} \end{bmatrix} \frac{d_0(1)\beta(0)\exp\left(\displaystyle\int_0^t \frac{a_0 d_0 - b_0 c_0}{d_0} \right)}{c_0(1)\exp\left(\displaystyle\int_0^1 \frac{a_0 d_0 - b_0 c_0}{d_0} \right) - d_0(1)}. \quad (9.3.57)
$$

We see that the specified boundary condition

$$y(0,\varepsilon) + x(1,\varepsilon) = \alpha$$

is generally lost as $\varepsilon \to 0+$, whereas the other boundary condition

$$x(0,\varepsilon) + y(1,\varepsilon) = \beta$$

is retained as $\varepsilon \to 0+$.

In this case a suitable multivariable expansion can be given for the solution functions in the form

$$
\begin{bmatrix} x(t,\varepsilon) \\ y(t,\varepsilon) \end{bmatrix} \sim \begin{bmatrix} X(t,\varepsilon) \\ Y(t,\varepsilon) \end{bmatrix} + \begin{bmatrix} {}^*X(\tau,\varepsilon) \\ \dfrac{1}{\varepsilon}{}^*Y(\tau,\varepsilon) \end{bmatrix}, \quad (9.3.58)
$$

with $\tau := t/\varepsilon$, for suitable outer functions X, Y and suitable boundary-layer correction functions ${}^*X, {}^*Y$ that have asymptotic expansions in powers of ε. These expansions can be truncated as usual to provide approximate solution functions, and the expected error estimates are readily obtained using suitable estimates on the Green function.

See O'Malley (1980, 1984), Flaherty and O'Malley (1980), and Mattheij and O'Malley (1984) for further discussion on the cancellation law for linear (and related quasi-linear) problems such as (9.3.1)–(9.3.2) subject to the exponential dichotomy (hyperbolic splitting) (9.1.9).

Exercises

Exercise 9.3.1: Let m_{ij} be $n \times n$ matrices for $i, j = 1, 2$, with m_{12} and m_{21} invertible, and assume that the matrix $I - \mu$ is also invertible, where

$$\mu := m_{22}m_{12}^{-1}m_{11}m_{21}^{-1}. \quad (9.3.59)$$

Show that the $2n \times 2n$ matrix M given in block form as

$$M := \begin{bmatrix} m_{11} & \varepsilon m_{12} \\ m_{21} & \varepsilon m_{22} \end{bmatrix} \quad (9.3.60)$$

is invertible for $\varepsilon \neq 0$, with inverse

$$
M^{-1} = \begin{bmatrix} -m_{21}^{-1}(I - \mu)^{-1}m_{22}m_{12}^{-1} & m_{21}^{-1}(I - \mu)^{-1} \\ \dfrac{1}{\varepsilon}m_{12}^{-1}m_{11}m_{21}^{-1}(I - \mu)^{-1}m_{21}m_{11}^{-1} & -\dfrac{1}{\varepsilon}m_{12}^{-1}m_{11}m_{21}^{-1}(I - \mu)^{-1} \end{bmatrix}.
$$

$$(9.3.61)$$

Exercise 9.3.2: Derive the result (9.3.19) for the matrix $M(\varepsilon)$ of (9.3.18). *Hint:* Apply the result of Exercise 9.3.1 with suitable m_{ij} and with $\mu = \mu(\varepsilon)$ given by (9.3.20). This matrix μ is small (in fact, asymptotic to zero) as $\varepsilon \to 0+$, and the inverse of $I - \mu$ can be given by the Neumann series

$$
(I - \mu)^{-1} = \sum_{k=0}^{\infty} \mu^k, \qquad (9.3.62)
$$

from which the stated result follows.

Exercise 9.3.3: Derive the result (9.3.22)–(9.3.23) from (9.3.17) and (9.3.19).

Exercise 9.3.4: Carry out the two-variable calculation indicated by (9.3.26) in Example 9.3.1. *Hint:* We can show that the outer functions X_0 and Y_0 are determined as

$$
\frac{dX_0}{dt} = \left(A_0 - B_0 D_0^{-1} C_0 \right) X_0 + f_0 - B_0 D_0^{-1} g_0 \quad \text{for } 0 < t < 1,
$$
$$
X_0(1) = \beta_0 := \beta(0) \qquad \text{at } t = 1, \qquad (9.3.63)
$$

with $Y_0 = -D_0^{-1}(C_0 X_0 + g_0)$, and then the boundary-layer correction functions can be given as

$$
X_0^*(\tau) = B_0(0)e^{D_0(0)\tau}B_0^{-1}(0)\left[\alpha_0 - X_0(0)\right],
$$
$$
Y_0^*(\tau) = e^{D_0(0)\tau}D_0(0)B_0^{-1}(0)\left[\alpha_0 - X_0(0)\right], \qquad (9.3.64)
$$

for $\tau \geq 0$, where $\tau = t/\varepsilon$ in (9.3.26). Note that $B_0(0)$ is assumed to be nonsingular, and D_0 is assumed to have stable eigenvalues with negative real parts, so that the boundary-layer correction functions satisfy the appropriate decay properties.

Exercise 9.3.5: With X_0, Y_0, X_0^*, and Y_0^* as in Exercise 9.3.4, and with $\hat{x}$, $\hat{y}$ defined by (9.3.27), show that the residuals in (9.3.28)–(9.3.29) satisfy estimates of the type (9.3.30).

Exercise 9.3.6: Carry out the details in the derivation of the inequalities of (9.3.31), using (9.3.3) and (9.3.21)–(9.3.30).

Exercise 9.3.7: Show that the two-dimensional system

$$\frac{dx}{dt} = a(t, \varepsilon)x + b(t, \varepsilon)y,$$
$$\frac{dy}{dt} = \varepsilon^{-2}c(t, \varepsilon)x + d(t, \varepsilon)y \quad \text{for } 0 \le t \le 1, 0 < \varepsilon \le \varepsilon_0,$$

(9.3.65)

for real-valued solution functions $x = x(t, \varepsilon)$ and $y = y(t, \varepsilon)$, subject to the assumptions

$$b_0(t) \equiv b(t, 0) > 0, \qquad c_0(t) \equiv c(t, 0) > 0 \quad \text{for } 0 \le t \le 1, \quad (9.3.66)$$

has a fundamental solution

$$Z(t, \varepsilon) \equiv \begin{bmatrix} z_{11}(t, \varepsilon) & z_{12}(t, \varepsilon) \\ z_{21}(t, \varepsilon) & z_{22}(t, \varepsilon) \end{bmatrix}$$

of the form

$$Z(t, \varepsilon) = \begin{bmatrix} (1 - T)u & (1 + TS - S)v \\ -\frac{1}{\varepsilon}\left(\frac{c}{b}\right)^{1/2}(1 + T)u & \frac{1}{\varepsilon}\left(\frac{c}{b}\right)^{1/2}(1 + TS + S)v \end{bmatrix}$$

(9.3.67)

for suitable functions $S = S(t, \varepsilon)$, $T = T(t, \varepsilon)$, $u = u(t, \varepsilon)$, and $v = v(t, \varepsilon)$, where S and T are of order ε,

$$|S(t, \varepsilon)|, |T(t, \varepsilon)| \le \text{const. } \varepsilon,$$
$$\left|\frac{dS(t, \varepsilon)}{dt}\right|, \left|\frac{dT(t, \varepsilon)}{dt}\right| \le \text{constant,}$$

(9.3.68)

uniformly on the region indicated in (9.3.65), and where u and v have representations as

$$u(t, \varepsilon) = \exp\left\{-\frac{1}{\varepsilon}\int_0^t [\sqrt{bc} + \varepsilon(-E_{11} + E_{12}T)]\right\},$$
$$v(t, \varepsilon) = \exp\left\{-\frac{1}{\varepsilon}\int_t^1 [\sqrt{bc} + \varepsilon(E_{22} + E_{12}T)]\right\},$$

(9.3.69)

for suitable functions $E_{ij} = E_{ij}(t, \varepsilon)$ that are continuous and uniformly bounded on the region indicated in (9.3.65). Assume that the real-valued data functions a, b, c, and d are of class C^1 on the region $\{(t, \varepsilon)|0 \leq t \leq 1, 0 \leq \varepsilon \leq \varepsilon_0\}$. *Hint*: As in Section 7.2, use the eigenvector transformation

$$\begin{bmatrix} \xi \\ \eta \end{bmatrix} = \mathscr{S}\begin{bmatrix} x \\ \varepsilon y \end{bmatrix},$$

with

$$\mathscr{S} := \tfrac{1}{2}\begin{bmatrix} 1 & -\sqrt{b/c} \\ 1 & +\sqrt{b/c} \end{bmatrix}, \tag{9.3.70}$$

and find the system

$$\varepsilon\frac{d}{dt}\begin{bmatrix} \xi \\ \eta \end{bmatrix} = D(t, \varepsilon)\begin{bmatrix} \xi \\ \eta \end{bmatrix}, \tag{9.3.71}$$

with

$$D = D(t, \varepsilon) = \begin{bmatrix} -\sqrt{bc} & 0 \\ 0 & +\sqrt{bc} \end{bmatrix} + \varepsilon E(t, \varepsilon) \tag{9.3.72}$$

for a suitable bounded 2×2 matrix-valued function $E = E(t, \varepsilon)$. A suitable Riccati transformation can be used to study the system (9.3.71) as in Exercise 9.2.7, and we obtain the stated result.

Exercise 9.3.8: Consider the Neumann problem of Example 9.3.3 for the system (9.3.34)–(9.3.35) subject to the condition (9.3.40), and consider expansions of the form (9.3.51) for the solution functions, for suitable outer solutions X, Y of the form

$$\begin{bmatrix} X(t, \varepsilon) \\ Y(t, \varepsilon) \end{bmatrix} \sim \sum_{k=0}^{\infty}\begin{bmatrix} X_k(t) \\ Y_k(t) \end{bmatrix}\varepsilon^k, \tag{9.3.73}$$

and suitable boundary-layer corrections $\varepsilon {}^*X, {}^*Y$, with $(\tau = t/\varepsilon)$

$$\begin{bmatrix} {}^*X(\tau, \varepsilon) \\ {}^*Y(\tau, \varepsilon) \end{bmatrix} \sim \sum_{k=0}^{\infty}\begin{bmatrix} {}^*X_k(\tau) \\ {}^*Y_k(\tau) \end{bmatrix}\varepsilon^k, \tag{9.3.74}$$

subject to the matching condition

$$\lim_{\tau\to\infty}\begin{bmatrix} {}^*X_k(\tau) \\ {}^*Y_k(\tau) \end{bmatrix} = \begin{bmatrix} 0 \\ 0 \end{bmatrix} \tag{9.3.75}$$

and the boundary conditions

$$Y_k(0) + {}^*Y_k(0) = \alpha_k, \qquad Y_k(1) = \beta_k. \tag{9.3.76}$$

As usual, the outer solution functions are required to satisfy the original system of differential equations on the open interval $0 < t < 1$.

(i) Show that the coefficient functions X_k, Y_k, $*X_k$, and $*Y_k$ of (9.3.73) and (9.3.74) can be determined recursively in a unique manner.

(ii) Let $x^0(t, \varepsilon) := X_0(t) + \varepsilon *X_0(t/\varepsilon)$ and $y^0(t, \varepsilon) := Y_0(t) + *Y_0(t/\varepsilon)$ be the leading terms in the expansions (9.3.51) and (9.3.73)–(9.3.74). Show that $\hat{x}(t, \varepsilon) := x(t, \varepsilon) - x^0(t, \varepsilon)$ and $\hat{y}(t, \varepsilon) := y(t, \varepsilon) - y^0(t, \varepsilon)$ satisfy the original problem with data f, g, α, and β replaced respectively by suitable residuals ρ_1, ρ_2, σ_1, and σ_2, all of which are of order ε.

(iii) Use the Green function representation (9.3.3) along with the estimates (9.3.44) and the estimates of (ii) to obtain the bounds $\hat{x}$, $\hat{y} = O(\varepsilon)$, from which we have

$$\left| x(t, \varepsilon) - X_0(t) \right|, \left| y(t, \varepsilon) - y^0(t, \varepsilon) \right| \le \text{const. } \varepsilon \qquad (9.3.77)$$

as $\varepsilon \to 0+$, uniformly for $0 \le t \le 1$. [We can similarly show that the multivariable expansion is asymptotically correct to any order, always provided that the data are sufficiently regular. The result (9.3.77) verifies the correctness of the approximations for the Neumann problem of Example 8.3.2.]

Exercise 9.3.9: Consider the problem of Example 9.3.4 for the system (9.3.34)–(9.3.35) subject to the coupled boundary conditions of (9.3.45) and subject to condition (9.3.47). Show that the Green function for this problem, given by (9.3.6), satisfies the estimates of (9.3.44).

Exercise 9.3.10: Consider the boundary-value problem consisting of the two-dimensional system

$$\frac{dx}{dt} = a(t, \varepsilon)x + b(t, \varepsilon)y + f(t, \varepsilon),$$

$$\varepsilon^2 \frac{dy}{dt} = c(t, \varepsilon)x + \varepsilon^2 d(t, \varepsilon)y + g(t, \varepsilon), \qquad (9.3.78)$$

$$\text{subject to } b_0(t) > 0, \ c_0(t) > 0,$$

for real-valued solution functions $x = x(t, \varepsilon)$ and $y = y(t, \varepsilon)$ for $0 \le t \le 1$, $\varepsilon \to 0+$, subject to the boundary conditions

$$x(0, \varepsilon) = \alpha(\varepsilon) \quad \text{and} \quad y(1, \varepsilon) = \beta(\varepsilon), \qquad (9.3.79)$$

so that the boundary matrices L and R of (9.3.2) are

$$L = \begin{bmatrix} 1 & 0 \\ 0 & 0 \end{bmatrix}, \quad R = \begin{bmatrix} 0 & 0 \\ 0 & 1 \end{bmatrix}. \qquad (9.3.80)$$

Assume that the data have asymptotic expansions in powers of ε, with coefficients (in the case of a, b, c, d, f, g) that are smooth functions of t. [The conditions $b_0, c_0 > 0$ can be replaced with $b_0, c_0 < 0$ in (9.3.78).] The result of Exercise 0.1.2 shows that the solution of this problem can be represented in terms of the Green function as

$$\begin{bmatrix} x(t, \varepsilon) \\ y(t, \varepsilon) \end{bmatrix} = Z(t, \varepsilon) M^{-1} \begin{bmatrix} \alpha(\varepsilon) \\ \beta(\varepsilon) \end{bmatrix} + \int_0^1 G(t, s, \varepsilon) \begin{bmatrix} f(s, \varepsilon) \\ \varepsilon^{-2} g(s, \varepsilon) \end{bmatrix} ds,$$

$$(9.3.81)$$

provided that the matrix $M(\varepsilon)$ of (9.3.4) is nonsingular, where G is given by (9.3.6) in terms of the fundamental solution Z given by (9.3.67). Show directly that $M(\varepsilon)$ is nonsingular for this problem as $\varepsilon \to 0+$, and show that the Green function $G = [G_{ij}]$ satisfies the estimates

$$G(t, s, \varepsilon) = \begin{bmatrix} O\left[\exp\left(-\dfrac{\kappa_1}{\varepsilon}|t - s|\right)\right] & O\left[\varepsilon \exp\left(-\dfrac{\kappa_1}{\varepsilon}|t - s|\right)\right] \\ O\left[\dfrac{1}{\varepsilon}\exp\left(-\dfrac{\kappa_1}{\varepsilon}|t - s|\right)\right] & O\left[\exp\left(-\dfrac{\kappa_1}{\varepsilon}|t - s|\right)\right] \end{bmatrix}$$

$$(9.3.82)$$

as $\varepsilon \to 0+$, uniformly for $0 \le t, s \le 1$, where κ_1 is any fixed positive constant satisfying

$$b_0(t)c_0(t) \equiv b(t,0)c(t,0) \ge 4\kappa_1^2 \qquad (9.3.83)$$

for $0 \le t \le 1$. Note that (9.3.82) implies the further results

$$\int_0^1 |G_{11}(t, s, \varepsilon)| \, ds \le \text{const. } \varepsilon, \qquad \int_0^1 |G_{12}(t, s, \varepsilon)| \, ds \le \text{const. } \varepsilon^2,$$

$$\int_0^1 |G_{21}(t, s, \varepsilon)| \, ds \le \text{constant}, \qquad \int_0^1 |G_{22}(t, s, \varepsilon)| \, ds \le \text{const. } \varepsilon,$$

$$(9.3.84)$$

as $\varepsilon \to 0+$, uniformly for $0 \le t \le 1$.

Exercise 9.3.11: For the boundary-value problem (9.3.78)–(9.3.79), find (easily interpretable) approximate solution functions $x^0(t, \varepsilon)$ and $y^0(t, \varepsilon)$ satisfying estimates of the form

$$|x(t, \varepsilon) - x^0(t, \varepsilon)| \le \text{const. } \varepsilon,$$

$$|y(t, \varepsilon) - y^0(t, \varepsilon)| \le \text{constant}, \quad \text{as } \varepsilon \to 0+,$$

$$(9.3.85)$$

uniformly for $0 \le t \le 1$. *Hint:* Use appropriate leading terms in a multivariable expansion of the type

$$x(t, \varepsilon) \sim X(t, \varepsilon) + {}^*X(\tau, \varepsilon) + \varepsilon X^*(\sigma, \varepsilon),$$

(9.3.86)

$$y(t, \varepsilon) \sim Y(t, \varepsilon) + \frac{1}{\varepsilon}{}^*Y(\tau, \varepsilon) + Y^*(\sigma, \varepsilon), \qquad \tau := \frac{t}{\varepsilon}, \, \sigma := \frac{1 - t}{\varepsilon},$$

for suitable outer solution functions X and Y, suitable left boundary-layer corrections ${}^*X$ and ${}^*Y$, and suitable right boundary-layer corrections X^* and Y^*. As usual, the outer solution functions are taken to satisfy the full system (9.3.78) on the open interval $0 < t < 1$, and the boundary-layer correction functions satisfy suitable matching conditions as $\tau \to \infty$ and $\sigma \to \infty$. We can apply the results of Exercise 9.3.10 to $\hat{x} := x - x^0$ and $\hat{y} := y - y^0$ so as to obtain the estimates (9.3.85) provided x^0 and y^0 are suitably chosen. Note that the results of this exercise can be used to prove the validity of the approximation in Exercise 8.2.5. Also, the procedure here can be used to study (9.3.78) subject to more general boundary conditions of the type (9.3.2).

9.4 The linear state regulator in optimal control

Solutions for many problems in optimal-control theory entail considerations of certain auxiliary boundary-value problems for systems of differential equations. Hence, optimal-control theory is an important source of boundary-value problems, and in this section we consider one of the simplest such control problems, the linear state regulator problem over a specified bounded time interval taken here as $0 \le t \le 1$. Such problems are most often discussed today within the framework of the Hamiltonian formulation of the calculus of variations [see Kalman (1963)], but we shall follow the classical variational approach here.

Let $z = z(t)$ and $u = u(t)$ be the state and control vectors, respectively, of respective dimensions M and N, with state equation

$$\frac{dz}{dt} = \Omega(t)z + \Pi(t)u \quad \text{for } 0 \le t \le 1,$$

(9.4.1)

and with given initial state z^0:

$$z(0) = z^0 \quad \text{at } t = 0.$$

(9.4.2)

The given coefficients

$$\overset{M \times M}{\Omega = \Omega(t)} \quad \text{and} \quad \overset{M \times N}{\Pi = \Pi(t)}$$

in (9.4.1) are continuous matrix-valued functions of appropriate compatible orders. (We could equally well work with piecewise-continuous data and controls.)

For any suitable given control function $u = u(t)$, the initial-value problem (9.4.1)–(9.4.2) has a unique solution that depends on u and is denoted as

$$z = z(t) = z(t, u) = \text{solution of } (9.4.1)–(9.4.2), \qquad (9.4.3)$$

where we write $z(t, u)$ when we wish to emphasize the dependence of the state function on the control function.

In the linear state regulator problem we wish to choose the control in (9.4.1) so as to minimize a given real-valued quadratic cost function $J = J(u)$ of the form

$$J(u): = \tfrac{1}{2}z^{\mathrm{T}}(1, u)\rho z(1, u)$$
$$+ \tfrac{1}{2} \int_0^1 \left[z^{\mathrm{T}}(t, u)\sigma(t)z(t, u) + u^{\mathrm{T}}(t)\tau(t)u(t) \right] dt, \quad (9.4.4)$$

where ρ, $\sigma = \sigma(t)$, and $\tau = \tau(t)$ are given square matrices of appropriate orders, with ρ and σ symmetric and positive semidefinite, and with τ symmetric and positive definite. The superscript T denotes the matrix transpose in (9.4.4), so that z^{T} is the row vector given as the transpose of the column vector z, and so forth.

The state function (9.4.3) can be represented by the variation-of-parameters formula

$$z(t) \equiv z(t, u) = \Gamma(t, 0)z^0 + \int_0^t \Gamma(t, s)\Pi(s)u(s)\, ds, \quad (9.4.5)$$

where

$$\overset{M \times M}{\Gamma} = \Gamma(t, s) \qquad (9.4.6)$$

denotes the resolvent (Green) function for the initial-value problem for the operator $d/dt - \Omega(t)$, with

$$\frac{\partial \Gamma(t, s)}{\partial t} = \Omega(t)\Gamma(t, s) \quad \text{for } t \neq s,$$
$$\Gamma(t, s) = I_M \quad \text{at } t = s. \qquad (9.4.7)$$

This Green function is given as

$$\Gamma(t, s) = Z(t)Z^{-1}(s), \qquad (9.4.8)$$

where $Z = Z(t)$ is any fundamental solution for $dz/dt = \Omega(t)z$.

If $u^0 = u^0(t)$ is an optimal-control function that minimizes J for, say, $u \in C^0[0,1]$, then the (first) variation of J must vanish at u^0 with

$$\delta J(u^0, \Delta u) \equiv 0 \quad \text{for every } \Delta u \in C^0[0,1]. \tag{9.4.9}$$

In the present case it is an easy matter to compute the variation of J from (9.4.4), and we find [see Smith (1974)]

$$\delta J(u, \Delta u) = \Delta z(1, \Delta u)^{\mathrm{T}} \rho z(1, u)$$
$$+ \int_0^1 \left[\Delta z(t, \Delta u)^{\mathrm{T}} \sigma(t) z(t, u) + \Delta u(t)^{\mathrm{T}} \tau(t) u(t) \right] dt$$
$$\tag{9.4.10}$$

for any functions u and Δu of class $C^0[0,1]$, where $\Delta z = \Delta z(t, \Delta u)$ is given here by the formula

$$\Delta z(t, \Delta u) = \int_0^t \Gamma(t,s) \Pi(s) \Delta u(s) \, ds \quad \text{for } 0 \le t \le 1, \tag{9.4.11}$$

for any $\Delta u \in C^0[0,1]$. We can insert (9.4.11) into (9.4.10) and then interchange an order of integration to find

$$\delta J(u, \Delta u) = \int_0^1 \Delta u(t)^{\mathrm{T}} \left[\Pi(t)^{\mathrm{T}} \Gamma(1,t)^{\mathrm{T}} \rho z(1,u) + \Pi(t)^{\mathrm{T}} \right.$$
$$\left. \times \int_t^1 \Gamma(s,t)^{\mathrm{T}} \sigma(s) z(s,u) \, ds + \tau(t) u(t) \right] dt. \tag{9.4.12}$$

This result (9.4.12) can be used in (9.4.9) with $u = u^0$, and we find directly for u^0 the following necessary condition, which is the Euler/Lagrange equation for the linear state regulator problem:

$$\Pi(t)^{\mathrm{T}} \left[\Gamma(1,t)^{\mathrm{T}} \rho z(1, u^0) + \int_t^1 \Gamma(s,t)^{\mathrm{T}} \sigma(s) z(s, u^0) \, ds \right]$$
$$+ \tau(t) u^0(t) = 0 \quad \text{for } 0 \le t \le 1, \tag{9.4.13}$$

where (9.4.5) gives

$$z(t, u^0) = \Gamma(t,0) z^0 + \int_0^t \Gamma(t,s) \Pi(s) u^0(s) \, ds. \tag{9.4.14}$$

We wish to solve this system of integral equations (9.4.13)–(9.4.14) for the optimal-control function u^0, and for this purpose it is convenient to introduce the costate (adjoint, conjugate) function $\zeta = \zeta(t)$ defined as

$$\zeta(t) := \Gamma(1,t)^{\mathrm{T}} \rho z(1, u^0) + \int_t^1 \Gamma(s,t)^{\mathrm{T}} \sigma(s) z(s, u^0) \, ds \tag{9.4.15}$$

for $0 \le t \le 1$. In terms of the costate function, (9.4.13) becomes

$$\tau(t) u^0(t) + \Pi(t)^{\mathrm{T}} \zeta(t) = 0 \quad \text{for } 0 \le t \le 1. \tag{9.4.16}$$

Because τ is assumed to be positive definite, and hence invertible, this last equation can be solved for u^0 to give

$$u^0(t) = -\tau^{-1}(t)\Pi(t)^{\mathsf{T}}\zeta(t), \tag{9.4.17}$$

so that knowledge of the costate function ζ yields also u^0.

We now seek ζ as a suitable solution of the differential equation

$$\frac{d\zeta}{dt} = -\Omega(t)^{\mathsf{T}}\zeta - \sigma(t)z, \tag{9.4.18}$$

where this last equation follows directly by differentiation of (9.4.15), using the known results $\partial\Gamma(s,t)^{\mathsf{T}}/\partial t = -\Omega(t)^{\mathsf{T}}\Gamma(s,t)^{\mathsf{T}}$ and $\Gamma(t,t) = I$. This costate or adjoint differential equation (9.4.18) can be combined with the original state equation (9.4.1), with $u = u^0$ replaced in (9.4.1) by the result of (9.4.17), and we then have the system of differential equations

$$\frac{d}{dt}\begin{bmatrix} z \\ \zeta \end{bmatrix} = \begin{bmatrix} \Omega(t) & -\Pi(t)\tau^{-1}(t)\Pi^{\mathsf{T}}(t) \\ -\sigma(t) & -\Omega^{\mathsf{T}}(t) \end{bmatrix}\begin{bmatrix} z \\ \zeta \end{bmatrix}, \tag{9.4.19}$$

which is to be solved for z and ζ subject to the boundary conditions

$$z(0) = z^0 \quad \text{and} \quad \zeta(1) = \rho z(1). \tag{9.4.20}$$

The boundary condition here at $t = 0$ is given by (9.4.2), and the condition at $t = 1$ is a variational natural boundary condition obtained by putting $t = 1$ in (9.4.15) and using $\Gamma(1,1) = I_m$. The system (9.4.19) is usually obtained as the canonical system within the Hamiltonian formulation of the calculus of variations.

The boundary conditions of (9.4.20) are of the form

$$L\begin{bmatrix} z(0) \\ \zeta(0) \end{bmatrix} + R\begin{bmatrix} z(1) \\ \zeta(1) \end{bmatrix} = \begin{bmatrix} z^0 \\ 0 \end{bmatrix}, \tag{9.4.21}$$

with

$$L = \begin{bmatrix} I_m & 0 \\ 0 & 0 \end{bmatrix}, \quad R = \begin{bmatrix} 0 & 0 \\ -\rho & I_M \end{bmatrix}. \tag{9.4.22}$$

We have the $2M$-dimensional boundary-value problem (9.4.19) and (9.4.21)–(9.4.22) for the optimal state and costate functions z and ζ, and then the optimal-control function u is given by (9.4.17). In the present case we can use a Riccati transformation to reduce the boundary-value problem for (9.4.19) to an associated nonlinear terminal-value problem, and it can be proved thereby that a unique solution exists (Kalman 1960); see Exercise 9.4.1. In practice, the calculations must be carried out numerically [see Stoer and Bulirsch (1980)], and the practical numerical difficulties become severe if the dimension $2M$ is large.

Following Kokotović and Sannuti (1968), Sannuti and Kokotović (1969), and O'Malley (1972a), consider now the special case of such a linear state regulator for which the M-dimensional state function z can be decomposed into lower-dimensional components x and y as

$$
\overset{M\times 1}{z} = \begin{bmatrix} \overset{m\times 1}{x} \\[2mm] \overset{n\times 1}{y} \end{bmatrix}, \qquad M = m + n, \tag{9.4.23}
$$

with the coefficients $\Omega = \Omega(t, \varepsilon)$ and $\Pi = \Pi(t, \varepsilon)$ in (9.4.1) depending on a small parameter ε, where these coefficient matrices can be given in block form as

$$
\Omega(t) = \begin{bmatrix} \overset{m\times m}{\alpha(t, \varepsilon)} & \overset{m\times n}{\beta(t, \varepsilon)} \\[3mm] \overset{n\times m}{\frac{1}{\varepsilon}\gamma(t, \varepsilon)} & \overset{n\times n}{\frac{1}{\varepsilon}\delta(t, \varepsilon)} \end{bmatrix}, \qquad \Pi(t, \varepsilon) = \begin{bmatrix} \overset{m\times N}{\varphi(t, \varepsilon)} \\[3mm] \overset{n\times N}{\frac{1}{\varepsilon}\psi(t, \varepsilon)} \end{bmatrix},
$$

$$
\tag{9.4.24}
$$

for given matrix-valued functions α, β, γ, δ, φ, and ψ that are assumed to have asymptotic expansions in terms of powers of ε. In this case the state equation (9.4.1) can be written in the singularly perturbed form

$$
\frac{dx}{dt} = \alpha(t, \varepsilon)x + \beta(t, \varepsilon)y + \varphi(t, \varepsilon)u,
$$

$$
\varepsilon\frac{dy}{dt} = \gamma(t, \varepsilon)x + \delta(t, \varepsilon)y + \psi(t, \varepsilon)u, \tag{9.4.25}
$$

where the state components x and y are respectively the slow and fast state variables. The n-dimensional fast component y represents terms undergoing rapid "fast dynamics." The idea of Kokotović and Sannuti is to use singular-perturbation theory so as to effectively replace consideration of the higher-dimensional system (9.4.19) (dimension $= 2M = 2m + 2n$) with considerations of appropriate systems of respective lower dimensions $2m$ and $2n$.

Along with the decomposition (9.4.23) for the state function, we introduce this corresponding decomposition for the costate function ζ, where we follow O'Malley (1972*a*, 1974*a*),

$$
\overset{M \times 1}{\zeta} = \begin{bmatrix} \overset{m \times 1}{\xi} \\ \overset{n \times 1}{\varepsilon \eta} \end{bmatrix},
\tag{9.4.26}
$$

and we assume that the symmetric matrices ρ and σ have decompositions of the form

$$
\rho = \begin{bmatrix} \rho_1 & \varepsilon \rho_2 \\ \varepsilon \rho_2^{\mathrm{T}} & \varepsilon \rho_3 \end{bmatrix}, \qquad
\sigma = \begin{bmatrix} \sigma_1 & \sigma_2 \\ \sigma_2^{\mathrm{T}} & \sigma_3 \end{bmatrix},
\tag{9.4.27}
$$

for given matrices ρ_j and σ_j ($j = 1, 2, 3$) that have asymptotic expansions in powers of ε.

Then the system (9.4.19) can be written in the form

$$
\frac{d}{dt} \overset{2m \times 1}{\begin{bmatrix} x \\ \xi \end{bmatrix}} = A(t, \varepsilon) \begin{bmatrix} x \\ \xi \end{bmatrix} + B(t, \varepsilon) \begin{bmatrix} y \\ \eta \end{bmatrix},
$$

$$
\varepsilon \frac{d}{dt} \overset{2n \times 1}{\begin{bmatrix} y \\ \eta \end{bmatrix}} = C(t, \varepsilon) \begin{bmatrix} x \\ \xi \end{bmatrix} + D(t, \varepsilon) \begin{bmatrix} y \\ \eta \end{bmatrix},
\tag{9.4.28}
$$

with coefficients A, B, C, and D defined as

$$
A(t, \varepsilon) := \begin{bmatrix} \alpha & -\varphi \tau^{-1} \varphi^{\mathrm{T}} \\ -\sigma_1 & -\alpha^{\mathrm{T}} \end{bmatrix}, \qquad
B(t, \varepsilon) := \begin{bmatrix} \beta & -\varphi \tau^{-1} \psi^{\mathrm{T}} \\ -\sigma_2 & -\gamma^{\mathrm{T}} \end{bmatrix},
$$

$$
C(t, \varepsilon) := \begin{bmatrix} \gamma & -\psi \tau^{-1} \varphi^{\mathrm{T}} \\ -\sigma_2^{\mathrm{T}} & -\beta^{\mathrm{T}} \end{bmatrix}, \qquad
D(t, \varepsilon) := \begin{bmatrix} \delta & -\psi \tau^{-1} \psi^{\mathrm{T}} \\ -\sigma_3 & -\delta^{\mathrm{T}} \end{bmatrix}.
$$

$$
\tag{9.4.29}
$$

Similarly, the boundary conditions (9.4.20) or (9.4.21)–(9.4.22) become

$$
x(0, \varepsilon) = x^0(\varepsilon), \qquad \xi(1, \varepsilon) = \rho_1 x(1, \varepsilon) + \varepsilon \rho_2 y(1, \varepsilon),
$$

$$
y(0, \varepsilon) = y^0(\varepsilon), \qquad \eta(1, \varepsilon) = \rho_2^{\mathrm{T}} x(1, \varepsilon) + \rho_3 y(1, \varepsilon),
\tag{9.4.30}
$$

or

$$\mathscr{L}\left[\begin{bmatrix} \begin{bmatrix} x(0,\varepsilon) \\ \xi(0,\varepsilon) \end{bmatrix} \\ \begin{bmatrix} y(0,\varepsilon) \\ \eta(0,\varepsilon) \end{bmatrix} \end{bmatrix}\right] + \mathscr{R}\left[\begin{bmatrix} \begin{bmatrix} x(1,\varepsilon) \\ \xi(1,\varepsilon) \end{bmatrix} \\ \begin{bmatrix} y(1,\varepsilon) \\ \eta(1,\varepsilon) \end{bmatrix} \end{bmatrix}\right] = \begin{bmatrix} \begin{bmatrix} x^0 \\ 0 \end{bmatrix} \\ \begin{bmatrix} y^0 \\ 0 \end{bmatrix} \end{bmatrix}, \qquad (9.4.31)$$

where $x^0 = x^0(\varepsilon)$ and $y^0 = y^0(\varepsilon)$ are the slow and fast components of the initial state $z^0 = z^0(\varepsilon)$, and where the $(2m + 2n) \times (2m + 2n)$ boundary matrices $\mathscr{L}$ and $\mathscr{R}$ are given as

$$\mathscr{L} = \begin{bmatrix} \begin{bmatrix} I_m & 0 \\ & {}_{m \times m} \\ 0 & 0 \end{bmatrix} & 0 \\ 0 & \begin{bmatrix} I_n & 0 \\ & {}_{n \times n} \\ 0 & 0 \end{bmatrix} \end{bmatrix},$$

$$(9.4.32)$$

$$\mathscr{R} = \begin{bmatrix} \begin{bmatrix} 0 & 0 \\ -\rho_1 & I_m \end{bmatrix} & \begin{bmatrix} 0 & 0 \\ -\varepsilon\rho_2 & 0 \end{bmatrix} \\ \begin{bmatrix} 0 & 0 \\ -\rho_2^{\mathrm{T}} & 0 \end{bmatrix} & \begin{bmatrix} 0 & 0 \\ -\rho_3 & I_n \end{bmatrix} \end{bmatrix}.$$

This singularly perturbed boundary-value problem (9.4.28)–(9.4.32) is of the type considered in the previous sections, where we used the key assumption that all eigenvalues of the matrix D should have uniformly nonzero real parts, as indicated by (9.1.7). We make that same assumption here, so that each eigenvalue $\lambda = \lambda(t, \varepsilon)$ of the $2n \times 2n$ matrix $D(t, \varepsilon)$ of (9.4.29) is assumed to satisfy

$$|\lambda(t, \varepsilon)| \geq 2\kappa_1 > 0 \qquad (9.4.33)$$

for a fixed positive constant κ_1, uniformly for

$$0 \leq t \leq 1, \quad 0 < \varepsilon \leq \varepsilon_0, \qquad (9.4.34)$$

for a fixed $\varepsilon_0 > 0$. We also assume that D is continuous, so that each eigenvalue satisfies precisely one of the two inequalities of (9.1.8), uniformly on (9.4.34).

In the present case the eigenvalues of D occur in pairs λ, $-\lambda$, because the characteristic polynomial $f(\lambda)$ given as

$$f(\lambda) := \det \begin{bmatrix} \delta - \lambda I_n & -\psi\tau^{-1}\psi^{\mathrm{T}} \\ -\sigma_3 & -(\delta^{\mathrm{T}} + \lambda I_n) \end{bmatrix} \tag{9.4.35}$$

can be shown by a direct determinant argument to be an *even* function of λ, $f(-\lambda) = f(\lambda)$; see Exercise 9.4.2. It follows that the eigenvalues occur in pairs λ_+ and $\lambda_- = -\lambda_+$, with

$$\mathrm{Re}\,\lambda_+ \geq +2\kappa_1 \quad \text{and} \quad \mathrm{Re}\,\lambda_- \leq -2\kappa_1, \tag{9.4.36}$$

uniformly on (9.4.34). Hence, the sum of the algebraic multiplicities of all eigenvalues of D having positive real parts is n, and the same is true for the eigenvalues having negative real parts. In this case, (9.1.5) is a $2n$-dimensional system that satisfies the exponential dichotomy (9.1.9), and the projection P occurring in (9.1.9) satisfies

$$P: \mathbb{R}^{2n} \to \mathbb{R}^{2n}, \quad \mathrm{rank}\ P = n. \tag{9.4.37}$$

It follows that we generally have boundary layers at both endpoints, with n boundary conditions generally lost (cancelled) at each endpoint as $\varepsilon \to 0+$. The cancellation rule in this case is given by O'Malley (1972a); we cancel or drop from (9.4.30) the following two n-dimensional boundary conditions:

$$y(0, \varepsilon) = y^0(\varepsilon), \qquad \eta(1, \varepsilon) = \rho_2^{\mathrm{T}}x(1, \varepsilon) + \rho_3 y(1, \varepsilon). \tag{9.4.38}$$

That is, for the outer solution we drop these two boundary conditions that involve primarily the fast variables y and η.

We refer the reader to O'Malley (1972a), where it is shown that the given problem has a unique solution for all small enough $\varepsilon \to 0+$, and the resulting solution functions can be represented asymptotically as

$$\begin{bmatrix} x(t, \varepsilon) \\ \xi(t, \varepsilon) \end{bmatrix} \sim \begin{bmatrix} X(t, \varepsilon) \\ \Xi(t, \varepsilon) \end{bmatrix} + \varepsilon \begin{bmatrix} {}^*X(\mu, \varepsilon) \\ {}^*\Xi(\mu, \varepsilon) \end{bmatrix} + \varepsilon \begin{bmatrix} X^*(\nu, \varepsilon) \\ \Xi^*(\nu, \varepsilon) \end{bmatrix},$$

$$\begin{bmatrix} y(t, \varepsilon) \\ \eta(t, \varepsilon) \end{bmatrix} \sim \begin{bmatrix} Y(t, \varepsilon) \\ \hat{\eta}(t, \varepsilon) \end{bmatrix} + \begin{bmatrix} {}^*Y(\mu, \varepsilon) \\ {}^*\hat{\eta}(\mu, \varepsilon) \end{bmatrix} + \begin{bmatrix} Y^*(\nu, \varepsilon) \\ \hat{\eta}^*(\nu, \varepsilon) \end{bmatrix}, \tag{9.4.39}$$

for suitable outer solution functions X, Ξ, Y, and $\hat{\eta}$, suitable left boundary-layer functions ${}^*X$, ${}^*\Xi$, ${}^*Y$, and ${}^*\hat{\eta}$, and suitable right boundary-layer functions X^*, Ξ^*, Y^*, and $\hat{\eta}^*$, all of which have asymptotic expansions in powers of ε, and where the boundary-layer variables are

$$\mu := \frac{t}{\varepsilon}, \qquad \nu := \frac{1-t}{\varepsilon}. \tag{9.4.40}$$

An indication of these results is given in Exercise 9.4.3 in the simplest case $m = n = 1$.

O'Malley (1974*a*) discusses various generalizations and modifications of the foregoing linear state regulator problem, and gives additional references. Certain problems in optimal control lead to systems related to (9.4.28) for which the matrix $D_0(t) = D(t,0)$ is singular; the theory of such problems is considered in O'Malley (1979). See O'Malley (1983*a*) for a discussion of results on the *nonlinear* state regulator problem. General reviews of applications of singular-perturbation theory in optimal-control theory are given in Saksena, O'Reilly, and Kokotović (1984) and Kokotović (1984).

Exercises

Exercise 9.4.1: Show that the solution of the two-point boundary-value problem (9.4.19)–(9.4.20) can be obtained by solving the initial-value problem

$$\frac{dz}{dt} = \left[\Omega(t) - \Pi(t)\tau^{-1}(t)\Pi^{\mathrm{T}}(t)P(t) \right] z \quad \text{for } 0 \le t \le 1,$$
$$z(0) = z^0 \quad \text{at } t = 0,$$
(9.4.41)

with the costate solution function given in terms of z as

$$\zeta(t) = P(t)z(t),$$
(9.4.42)

and with the optimal-control function given similarly by the *feedback rule*,

$$u^0(t) = -\tau^{-1}(t)\Pi^{\mathrm{T}}(t)P(t)z(t),$$
(9.4.43)

where the $M \times M$ matrix-valued function P is the solution of the Riccati terminal-value problem

$$\frac{dP}{dt} + \Omega^{\mathrm{T}}P + P\Omega + \sigma = P\Pi\tau^{-1}\Pi^{\mathrm{T}}P \quad \text{for } 0 \le t \le 1,$$
$$P(1) = \rho \quad \text{at } t = 1.$$
(9.4.44)

[Recall that the given matrices ρ and $\sigma(t)$ are symmetric and positive semidefinite, and τ is symmetric and positive definite.] *Hint:* Diagonalize the problem (9.4.19)–(9.4.20) with the Riccati transformation (see Exercise 9.2.6)

$$\begin{bmatrix} z \\ \zeta \end{bmatrix} = \begin{bmatrix} I_M & -S \\ P & I_M - PS \end{bmatrix}\begin{bmatrix} \hat{z} \\ \hat{\zeta} \end{bmatrix}, \quad \begin{bmatrix} \hat{z} \\ \hat{\zeta} \end{bmatrix} = \begin{bmatrix} I_M - SP & S \\ -P & I_M \end{bmatrix}\begin{bmatrix} z \\ \zeta \end{bmatrix},$$
(9.4.45)

where we have relabelled $T = -P$ in (9.2.36), and where the matrix function S satisfies the appropriate differential equation of (9.2.37). Assuming for the moment that $P(t)$ exists for $0 \le t \le 1$, show that $\hat{z}$ and $\hat{\zeta}$ satisfy the respective systems

$$\frac{d\hat{z}}{dt} = (\Omega - \Pi\tau^{-1}\Pi^{\mathrm{T}}P)\hat{z} \quad \text{and} \quad \frac{d\hat{\zeta}}{dt} = -(\Omega^{\mathrm{T}} - P\Pi\tau^{-1}\Pi^{\mathrm{T}})\hat{\zeta}.$$

(9.4.46)

Impose on P the terminal condition indicated in (9.4.44), and impose on S the homogeneous initial condition $S(0) = 0$, so that the transformed boundary conditions on $\hat{z}$ and $\hat{\zeta}$ become

$$\hat{z}(0) = z^0 \quad \text{and} \quad \hat{\zeta}(1) = 0.$$ (9.4.47)

Conclude that $\hat{\zeta} \equiv 0$, so that (9.4.45) yields $z \equiv \hat{z}$ and $\zeta \equiv P\hat{z}$. The results (9.4.41) and (9.4.42) follow directly, and then (9.4.17) leads directly to (9.4.43). Finally, to prove the global existence of P, replace (9.4.44) with the equivalent integral equation [see (9.2.9), with the projection there put equal to zero]

$$P(t) = P_0(t) - \int_t^1 \Gamma(s,t)^{\mathrm{T}} P(s)(\Pi\tau^{-1}\Pi^{\mathrm{T}})(s) P(s) \Gamma(s,t)\, ds,$$

(9.4.48)

with

$$P_0(t) := \Gamma(1,t)^{\mathrm{T}} \rho \Gamma(1,t) + \int_t^1 \Gamma(s,t)^{\mathrm{T}} \sigma(s) \Gamma(s,t)\, ds, \qquad (9.4.49)$$

where Γ is the resolvent function of (9.4.6) that satisfies $\partial\Gamma(s,t)/\partial t = -\Gamma(s,t)\Omega(t)$. Note that P_0 is symmetric, as is any solution P of (9.4.48). Also, P_0 is clearly positive semidefinite, $P_0 \ge 0$ for $t \le 1$, and it is shown in Kalman (1960) that any solution P of (9.4.48) is also semidefinite, $P \ge 0$ for $t \le 1$. It follows then directly from (9.4.48) that any solution P must satisfy the a priori bound

$$0 \le P(t) \le P_0(t) \quad \text{for } 0 \le t \le 1.$$ (9.4.50)

Because P_0 is bounded, we obtain from (9.4.50) a suitable bound of the type $|P(t)| \le$ constant, uniformly for $0 \le t \le 1$. The global existence of a unique solution function P then follows directly from known results in differential equation theory. Note that the Riccati equation of (9.4.44) appears here in the Riccati transformation of the system (9.4.19), whereas this same Riccati equation of (9.4.44) occurs in the solution of the Hamilton/Jacobi partial differential equation within the Hamiltonian formulation of the calculus of variations.

Exercise 9.4.2: Let δ, s_1, and s_2 be given $n \times n$ matrices with s_1 and s_2 symmetric, and let $f(\lambda)$ be the characteristic polynomial of the $2n \times 2n$ matrix

$$\begin{bmatrix} \delta & -s_1 \\ -s_2 & -\delta^{\mathrm{T}} \end{bmatrix},$$

with

$$f(\lambda) := \det \begin{bmatrix} \delta - \lambda I_n & -s_1 \\ -s_2 & -\delta^{\mathrm{T}} - \lambda I_n \end{bmatrix}. \tag{9.4.51}$$

Show that f is an even function: $f(-\lambda) = f(\lambda)$. *Hint:* Interchange suitable rows and columns to obtain

$$
\begin{aligned}
f(-\lambda) &= (-1)^n \det \begin{bmatrix} -s_1 & \delta + \lambda I \\ -\delta^{\mathrm{T}} + \lambda I & -s_2 \end{bmatrix} \\
&= (-1)^{2n} \det \begin{bmatrix} -\delta^{\mathrm{T}} + \lambda I & -s_2 \\ -s_1 & \delta + \lambda I \end{bmatrix} \\
&= \det \begin{bmatrix} -\delta + \lambda I & -s_1 \\ -s_2 & \delta^{\mathrm{T}} + \lambda I \end{bmatrix},
\end{aligned}
$$

where the last equality follows from the fact that the determinant is invariant under transposition of the matrix. Finally, multiply each of the first n columns by -1, and then multiply each of the last n rows by -1, to find

$$
\begin{aligned}
\det \begin{bmatrix} -\delta + \lambda I & -s_1 \\ -s_2 & \delta^{\mathrm{T}} + \lambda I \end{bmatrix} &= (-1)^n \det \begin{bmatrix} \delta - \lambda I & -s_1 \\ s_2 & \delta^{\mathrm{T}} + \lambda I \end{bmatrix} \\
&= (-1)^{2n} \det \begin{bmatrix} \delta - \lambda I & -s_1 \\ -s_2 & -\delta^{\mathrm{T}} - \lambda I \end{bmatrix}.
\end{aligned}
$$

The desired result $f(-\lambda) = f(\lambda)$ follows directly from these results.

Exercise 9.4.3: Consider the boundary-value problem (9.4.28)–(9.4.32) in the case $m = n = 1$, so that A, B, C, and D are 2×2 matrices in (9.4.29). Assume that the data are sufficiently smooth, and assume that the given real-valued function $\delta = \delta(t, \varepsilon)$ is uniformly nonzero. Without loss, take δ to be positive, with

$$\delta(t, \varepsilon) \geq 2\kappa_1 > 0 \tag{9.4.52}$$

for a fixed positive constant κ_1, uniformly on (9.4.34).

(i) Verify that the characteristic polynomial (9.4.35) is a function of λ^2, with precisely two roots given as

$$\lambda_+ = -\lambda_- = \left(\delta^2 + \sigma_3 \tau^{-1}\psi^2\right)^{1/2} > 0. \tag{9.4.53}$$

Note that the real-valued quantities σ_3 and $\psi\tau^{-1}\psi^T = \tau^{-1}\psi^2$ are non-negative, and the eigenvalues here satisfy (9.4.36).

(ii) Show that the two-dimensional system [see (9.1.5)]

$$\varepsilon\frac{d}{dt}\begin{bmatrix} y \\ \eta \end{bmatrix} = D(t, \varepsilon)\begin{bmatrix} y \\ \eta \end{bmatrix} \tag{9.4.54}$$

has a fundamental solution $V = V(t, \varepsilon)$ of the form

$$V(t, \varepsilon)$$
$$= \begin{bmatrix} [\tau^{-1}\psi^2 + O(\varepsilon)]\exp\left[\dfrac{1}{\varepsilon}\displaystyle\int_0^t (\lambda_- + \varepsilon E_1)\right] & [\delta + \lambda_+ + O(\varepsilon)]\exp\left[-\dfrac{1}{\varepsilon}\displaystyle\int_t^1 (\lambda_+ + \varepsilon E_2)\right] \\ [\delta - \lambda_- + O(\varepsilon)]\exp\left[\dfrac{1}{\varepsilon}\displaystyle\int_0^t (\lambda_- + \varepsilon E_1)\right] & [-\sigma_3 + O(\varepsilon)]\exp\left[-\dfrac{1}{\varepsilon}\displaystyle\int_t^1 (\lambda_+ + \varepsilon E_2)\right] \end{bmatrix}$$

$$\tag{9.4.55}$$

uniformly on (9.4.34), for suitable functions E_1 and E_2 that are uniformly bounded on (9.4.34). *Hint*: The method of Exercise 9.3.7 can be used based on an initial transformation of the type involving (9.3.70), followed by a suitable Riccati transformation.

(iii) Show that (9.4.54) satisfies the exponential dichotomy (9.1.9) with the fundamental solution V of (9.4.55) and with projection

$$P = \begin{bmatrix} 1 & 0 \\ 0 & 0 \end{bmatrix}.$$

(iv) Apply a suitable Riccati transformation so as to reduce the 4×4 system (9.4.28) to two suitable 2×2 systems as in (9.3.7). Show that the resulting equation corresponding to the second equation of (9.3.7) has a fundamental solution of the type (9.4.55). In this way we can obtain an explicit representation for a suitable fundamental solution of (9.4.28), but the details need not be pursued further here.

(v) Compute the leading terms in the multivariable expansions of (9.4.39)–(9.4.40). You need not give error estimates.

10

Nonlinear problems

In this chapter we shall consider several examples of nonlinear boundary-value problems for which solutions of boundary-layer type exist without interior layers, except in Section 10.6, where examples involving interior layers are given. Such problems exhibiting solutions of boundary-layer type arise in many areas, including chemical-reactor theory, optimal-control theory, fluid dynamics, and the physical theory of semiconducting devices. Assuming that an exact solution of interest exists for such a problem, a corresponding appropriate approximate solution is obtained here by a multivariable construction. The original problem is then recast as a linearization about the proposed approximate solution, and we use the Green function for the resulting linearized problem so as to prove for the original problem the existence of an exact solution that is well approximated by the given approximate solution, provided that certain conditions hold. A discussion is also given on the numerical solution of such problems, with emphasis on an algorithm of Maier.

10.1 Introduction

In Section 8.2, various boundary-value problems were considered for a linear equation of the form

$$\varepsilon^2 \frac{d^2x}{dt^2} = H(t, x, \varepsilon),$$ (10.1.1)

where H is a suitable function that is linear in x, of the form

$$H(t, x, \varepsilon) := f(t, \varepsilon) - b(t, \varepsilon)x$$ (10.1.2)

382

for given functions f and b, with b negative, so that

$$H_x(t,x,0) \equiv \frac{\partial H(t,x,0)}{\partial x} = -b(t,0) > 0 \qquad (10.1.3)$$

for all t.

Similarly, in Exercise 9.3.10 we considered a related boundary-value problem that (in the special case $b \equiv 1$) includes the equation of (10.1.1), with the function $H = H(t,x,\varepsilon)$ of (10.1.1)–(10.1.2) replaced by

$$H(t,x,\dot{x},\varepsilon) := \left[c + \varepsilon^2(\dot{a} - ad)\right]x + \varepsilon^2(a+d)\dot{x} + g(t,\varepsilon),$$

$$(10.1.4)$$

where the given functions a, c, and d depend on t and ε, with $c = c(t,\varepsilon)$ uniformly positive, so that $\partial H/\partial x$ is again positive,

$$H_x(t,x,\dot{x},0) = c(t,0) > 0 \qquad (10.1.5)$$

for all t.

In Section 8.3, and also in Chapter 9, various boundary-value problems were considered for a linear equation of the form

$$\varepsilon \frac{d^2x}{dt^2} = H(t,x,\dot{x},\varepsilon), \qquad (10.1.6)$$

where H is linear in x and $\dot{x}$, with

$$H(t,x,\dot{x},\varepsilon) := f(t,\varepsilon) - a(t,\varepsilon)\dot{x} - b(t,\varepsilon)x \qquad (10.1.7)$$

for given functions f, a, and b, with the function a uniformly nonzero, so that

$$H_{\dot{x}}(t,x,\dot{x},0) \equiv \frac{\partial H(t,x,\dot{x},0)}{\partial \dot{x}} = -a(t,0) \neq 0 \qquad (10.1.8)$$

for all t. In particular, $H_{\dot{x}}$ in (10.1.8) is either everywhere positive or everywhere negative.

The conditions (10.1.3), (10.1.5), and (10.1.8) are stability conditions that also eliminate the possibility of oscillatory solutions for the given (homogeneous) differential equations (see Section 8.1). In this chapter we consider nonlinear generalizations involving scalar equations of the type (10.1.1) and (10.1.6), subject to appropriate stability conditions related to (10.1.3) and (10.1.8).

First, consider a nonlinear scalar equation of the type (10.1.1),

$$\varepsilon^2 \frac{d^2x}{dt^2} = H(t,x,\varepsilon) \quad \text{for } 0 \leq t \leq 1, \qquad (10.1.9)$$

for a given smooth function H that satisfies the global condition

$$H_x(t, x, \varepsilon) \equiv \frac{\partial H(t, x, \varepsilon)}{\partial x} > 0 \qquad (10.1.10)$$

for $0 \leq t \leq 1$ and for all small $\varepsilon \to 0+$, uniformly for all x. For example, the function $H(t, x, \varepsilon) := a(t, \varepsilon)x^3 + b(t, \varepsilon)x + c(t, \varepsilon)$ satisfies (10.1.10) if the given functions a and b satisfy $a \geq 0$, $b > 0$. The assumption (10.1.10) implies, in particular, that the Dirichlet problem for (10.1.9) has at most one solution; see Exercise 10.1.1.

Although some type of positivity condition is required on H_x, the global condition (10.1.10) is unnecessarily restrictive. Hence, rather than (10.1.10), assume instead that the reduced (algebraic) equation obtained by putting $\varepsilon = 0$ in (10.1.9),

$$0 = H(t, x, 0), \qquad (10.1.11)$$

has a smooth solution $x = X(t)$ satisfying

$$H[t, X(t), 0] \equiv 0, \quad \text{with } H_x[t, X(t), 0] > 0, \qquad (10.1.12)$$

for $0 \leq t \leq 1$. We shall see in Section 10.2 that such a local condition as (10.1.12) provides a partial basis for study of boundary-value problems for (10.1.9).

Example 10.1.1: Take H to be given as

$$H(t, x, \varepsilon) \equiv H(x) := (x^2 - 1)(x^2 - 4) \qquad (10.1.13)$$

for $x \in \mathbb{R}$, with $H(x)$ independent of t and ε, so that (10.1.9) becomes

$$\varepsilon^2 \frac{d^2 x}{dt^2} = (x^2 - 1)(x^2 - 4) \quad \text{for } 0 \leq t \leq 1. \qquad (10.1.14)$$

The reduced equation (10.1.11) is

$$0 = (x^2 - 1)(x^2 - 4), \qquad (10.1.15)$$

with four solutions X_1, X_2, X_3, X_4 given as

$$X_1(t) \equiv 1, \qquad X_2(t) \equiv -1, \qquad \qquad \qquad (10.1.16)$$
$$X_3(t) \equiv 2, \qquad X_4(t) \equiv -2, \quad \text{for } 0 \leq t \leq 1.$$

The function H of (10.1.13) does not satisfy the global condition (10.1.10), because, for example, $H_x = -6$ at $x = 1$. However, the condition (10.1.12) is satisfied by $x = X_2(t)$ and also by $x = X_3(t)$, and in this case the Dirichlet problem for (10.1.14) may have two stable solutions

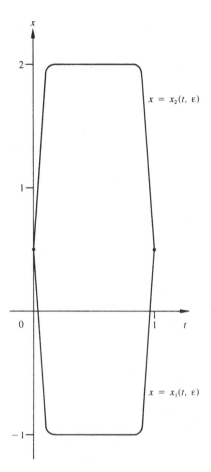

Figure 10.1

satisfying the same boundary conditions, as $\varepsilon \to 0$. For example, there are two solutions satisfying the Dirichlet conditions $x = 0.5$ at the endpoints, as illustrated in Figure 10.1.

Such nonlinear boundary-value problems for equations including (10.1.9) subject to (10.1.12) have been considered by many authors, including Brish (1954), Vasil'eva and Tupčiev (1960), Boglaev (1970), Vasil'eva and Butuzov (1973), Fife (1973a, 1973b), Yarmish (1975), Howes (1976a, 1976b, 1978), van Harten (1978a, 1978b), Chang and Howes (1984), and others. We consider such problems in Section 10.2. The Dirichlet problem for a related second-order vector system is studied in Kelley (1979) and in Howes and O'Malley (1980).

Consider now a quasi-linear scalar equation of the type

$$\varepsilon \frac{d^2x}{dt^2} = k(t, x, \varepsilon) \frac{dx}{dt} + h(t, x, \varepsilon) \qquad (10.1.17)$$

for given functions h and k, so that the equation is of the type (10.1.6), with H given as

$$H(t, x, \dot{x}, \varepsilon) := k(t, x, \varepsilon)\dot{x} + h(t, x, \varepsilon). \qquad (10.1.18)$$

Corresponding to (10.1.8) in the linear case, we have here the related condition

$$H_{\dot{x}}(t, x, \dot{x}, 0) = k(t, x, 0) \neq 0 \quad \text{for all } t, x. \qquad (10.1.19)$$

The Dirichlet problem for (10.1.17) has been studied extensively by many authors, including von Mises (1950), Coddington and Levinson (1952), Brish (1954), Wasow (1956), Cochran (1962), Willett (1966), Erdélyi (1968a, 1975), O'Malley (1968a, 1969a), Vasil'eva and Butuzov (1973), Chang (1974a, 1975, 1976), Habets (1974), Yarmish (1975), Howes (1976a, 1978), van Harten (1978a), Eckhaus (1979), and Chang and Howes (1984); see also Vasil'eva (1963) and O'Malley (1974a). With only a few exceptions, these works have generally made the assumption either that the "boundary-layer jump" is sufficiently small or that the function k satisfies a strong condition related to (10.1.19) implying that k is bounded uniformly away from zero on a suitable domain. These assumptions are unnecessarily restrictive, as discussed by Coddington and Levinson (1952), Howes (1978), and van Harten (1978a), where a weaker assumption [mentioned also in O'Malley (1974a)] suffices. These matters are considered more fully in Section 10.4. See also Chang (1974b) and Howes and O'Malley (1980), where related vector systems are considered.

Equation (10.1.17) also appears in problems with boundary conditions of Neumann or Robin type, as in the following example.

Example 10.1.2: Consider the equation (10.1.17) for $0 \le t \le 1$, with h and k given as

$$k(t, x, \varepsilon) := -1,$$
$$h(t, x, \varepsilon) := (x - 2)^2 \quad \text{for all } t, x, \varepsilon, \qquad (10.1.20)$$

subject to the boundary conditions

$$\frac{dx}{dt} - x = 1 \quad \text{at } t = 0,$$
$$\frac{dx}{dt} + \frac{x}{3} = 1 \quad \text{at } t = 1. \qquad (10.1.21)$$

In the present case the differential equation (10.1.17) becomes, with (10.1.20),

$$\varepsilon \frac{d^2x}{dt^2} + \frac{dx}{dt} = (x - 2)^2 \quad \text{for } 0 \le t \le 1. \qquad (10.1.22)$$

By analogy with the corresponding linear problems of Chapter 8, we expect any solution $x = x(t, \varepsilon)$ of (10.1.21)–(10.1.22) to exhibit a boundary layer at $t = 0$ and not at $t = 1$ as $\varepsilon \to 0+$ (because of the sign of the coefficient of the first derivative term in the differential equation). In particular, the leading term $X_0 = X_0(t)$ in an outer expansion is expected to satisfy the prescribed boundary condition at $t = 1$, with

$$X_0'(1) + \tfrac{1}{3}X_0(1) = 1 \quad \text{at } t = 1, \qquad (10.1.23)$$

and X_0 is also expected to satisfy the reduced, first-order differential equation obtained by putting $\varepsilon = 0$ in (10.1.22),

$$X_0'(t) = \left[X_0(t) - 2 \right]^2 \quad \text{for } 0 \le t \le 1. \qquad (10.1.24)$$

From these last two equations we find, at $t = 1$, the nonlinear equation

$$\left[X_0(1) - 2 \right]^2 + \tfrac{1}{3}X_0(1) = 1, \qquad (10.1.25)$$

and this latter quadratic equation has the two solutions

$$X_0(1) = \frac{11 - \sqrt{13}}{6} \quad \text{and} \quad X_0(1) = \frac{11 + \sqrt{13}}{6}. \qquad (10.1.26)$$

In fact, the given boundary-value problem (10.1.21)–(10.1.22) has two distinct solutions for small $\varepsilon \to 0+$, corresponding to the two choices of (10.1.26).

Nonlinear problems of the type (10.1.21)–(10.1.22), and similar problems for systems of equations, appear as mathematical models for certain chemical and biochemical flow reactors. The mathematical theory of such problems implies the existence of multiple stable solutions in certain cases, as in Example 10.1.2, and there is also experimental evidence for the existence of such multiple solutions; see Lapidus and Amundson (1977) and Aris (1975). Varma and Aris (1977) point out that "this has been one of those peculiar instances in engineering where extensive numerical and analytical work ... preceded the first published accounts of experimental observations." Mathematical studies of such problems are found in Parter (1972), Keller (1972), Chen (1972), O'Malley (1972b, 1974a), Chen and O'Malley (1974), and additional works listed in these references. These problems will be discussed briefly in Section 10.4.

Boundary-value problems can also be considered for more general equations of the form (10.1.6) that are not quasi-linear, so that $H(t, x, \dot{x}, \varepsilon)$ is not linear in $\dot{x}$. Wasow (1970), O'Malley (1974a), and Chang and Howes (1984) give references, along with discussions indicating a variety of different behaviors that can occur. For example, this problem introduced in Coddington and Levinson (1952),

$$\varepsilon x'' + x' + (x')^3 = 0 \quad \text{for } 0 < t < 1,$$
$$x(0) = 0 \quad \text{and} \quad x(1) = 0.5,$$

has no solution for sufficiently small ε, whereas the modified problem

$$\varepsilon x'' - x' + (x')^3 = 0 \quad \text{for } 0 < t < 1,$$
$$x(0) = 0 \quad \text{and} \quad x(1) = 0.5,$$

has a solution with an *angular transition point* at the interior point $t = 0.5$ as $\varepsilon \to 0$; see O'Malley (1970c, 1974a). An interesting nonlinear problem in elasticity theory is studied in Flaherty and O'Malley (1982). We are content here to confine our discussion to certain aspects of the equations (10.1.9) and (10.1.17), along with some related equations and systems.

Exercise

Exercise 10.1.1: Prove that the Dirichlet problem

$$\frac{d^2 x}{dt^2} = H(t, x) \quad \text{for } 0 \le t \le 1,$$
$$x(0) = \alpha \quad \text{at } t = 0, \quad \text{and} \quad x(1) = \beta \quad \text{at } t = 1, \tag{10.1.27}$$

has at most one solution if the given smooth function H satisfies the stability condition

$$H_x(t, x) \equiv \frac{\partial H(t, x)}{\partial x} \ge 0 \tag{10.1.28}$$

for all $0 \le t \le 1$, all $x \in \mathbf{R}$. *Hint*: Let x_1 and x_2 be any two solutions, and put

$$y = y(t) := x_2(t) - x_1(t) \quad \text{for } 0 \le t \le 1. \tag{10.1.29}$$

Show that y must be a solution of the problem

$$\frac{d^2 y}{dt^2} = A[t, y(t)] y \quad \text{for } 0 \le t \le 1,$$
$$y(0) = y(1) = 0, \tag{10.1.30}$$

where $A = A(t, y)$ is given as

$$A(t, y) := \int_0^1 H_x\big[t, x_1(t) + sy\big]\, ds. \tag{10.1.31}$$

For the fixed y of (10.1.29), use the maximum principle (see Theorem 0.3.5) to show that the associated linear problem

$$\frac{d^2z}{dt^2} = A[t, y(t)]z \quad \text{for } 0 \le t \le 1,$$
$$z(0) = z(1) = 0, \tag{10.1.32}$$

has only the trivial solution $z(t) \equiv 0$ for $0 \le t \le 1$. Conclude also that y must vanish, $y(t) \equiv 0$, from which the desired result follows.

10.2 A problem with boundary layers at both endpoints

Consider the scalar second-order equation (10.1.9), rewritten here for convenience,

$$\varepsilon^2 \frac{d^2x}{dt^2} = H(t, x, \varepsilon) \quad \text{for } 0 \le t \le 1, \tag{10.2.1}$$

for a given smooth function H. This equation can be written as a first-order system in several ways, such as

$$\frac{dx}{dt} = y,$$
$$\varepsilon^2 \frac{dy}{dt} = H(t, x, \varepsilon) \quad \text{for } 0 \le t \le 1, \tag{10.2.2}$$

or $\varepsilon\, dx/dt = z$, $\varepsilon\, dz/dt = H(t, x, \varepsilon)$. We can also consider more general semilinear systems, such as

$$\frac{dx}{dt} = F(t, x, \varepsilon) + G(t, x, \varepsilon)y,$$
$$\varepsilon^2 \frac{dy}{dt} = H(t, x, \varepsilon) + \varepsilon^2 K(t, x, \varepsilon)y \quad \text{for } 0 \le t \le 1, \tag{10.2.3}$$

for suitable functions F, G, H, K (see the linear problem in Exercise 9.3.10).

We shall be content here to consider only (10.2.1), subject to given fixed boundary conditions that we arbitrarily take here as

$$\frac{dx(0, \varepsilon)}{dt} = \alpha(\varepsilon) \quad \text{and} \quad x(1, \varepsilon) = \beta(\varepsilon). \tag{10.2.4}$$

Again, we can consider more general boundary conditions of the form

$$
L\begin{bmatrix} x(0,\varepsilon) \\ \dot{x}(0,\varepsilon) \end{bmatrix} + R\begin{bmatrix} x(1,\varepsilon) \\ \dot{x}(1,\varepsilon) \end{bmatrix} = \begin{bmatrix} \alpha(\varepsilon) \\ \beta(\varepsilon) \end{bmatrix}
\tag{10.2.5}
$$

for given boundary matrices L and R, which in the case (10.2.4) are given as

$$
L = \begin{bmatrix} 0 & 1 \\ 0 & 0 \end{bmatrix} \quad \text{and} \quad R = \begin{bmatrix} 0 & 0 \\ 1 & 0 \end{bmatrix} \quad \text{for (10.2.4).} \tag{10.2.6}
$$

Nonlinear boundary conditions of various types can also be considered; see van Harten (1978b). The specific kind of boundary conditions imposed plays an important role in determining the asymptotic structure of the boundary layers for the resulting solution functions. For example, the boundary condition of Dirichlet type at $t = 1$ in (10.2.4) leads to a stronger boundary layer at the right endpoint than at the left endpoint (see Exercise 9.3.11, where the stronger boundary layer occurs at the left). See the paragraph preceding (10.1.17) for references on (10.2.1).

We assume that the data quantities H, α, β of (10.2.1) and (10.2.4) have asymptotic power-series expansions in ε, as

$$
\begin{bmatrix} H(t,x,\varepsilon) \\ \alpha(\varepsilon) \\ \beta(\varepsilon) \end{bmatrix} \sim \sum_{k=0}^{\infty} \begin{bmatrix} H_k(t,x) \\ \alpha_k \\ \beta_k \end{bmatrix} \varepsilon^k,
\tag{10.2.7}
$$

for given coefficients H_k, α_k, and β_k. Moreover, H and H_k are assumed to be sufficiently smooth functions of their arguments.

The reduced equation obtained by putting $\varepsilon = 0$ in (10.2.1) is $H_0(t,x) = 0$, and we now take as given a fixed, smooth solution $x = X_0(t)$ of this reduced equation,

$$
H_0[t, X_0(t)] = 0 \quad \text{for } 0 \le t \le 1. \tag{10.2.8}
$$

Moreover, we assume that [see (10.1.12)]

$$
\left.\frac{\partial H_0(t,x)}{\partial x}\right|_{x = X_0(t)} \equiv H_{0,x}[t, X_0(t)] > 0 \quad \text{for } 0 \le t \le 1,
$$
$$
\tag{10.2.9}
$$

and we also require the additional condition (Fife 1973a)

$$
\int_0^x H_0[1, X_0(1) + z]\, dz \ge \tfrac{1}{2}\nu_1^2 x^2 \quad \text{for all } x \text{ between 0 and}
$$
$$
\beta_0 - X_0(1), \text{ and for a fixed}
$$
$$
\text{positive constant } \nu_1. \tag{10.2.10}
$$

Some such condition as (10.2.10) is required here at $t = 1$ because of the relatively strong boundary layer resulting at the right endpoint as a consequence of the Dirichlet-type boundary condition at $t = 1$. No such condition as (10.2.10) is required in the present case at the left endpoint, because the specified boundary condition at $t = 0$ is not of Dirichlet type. The conditions (10.2.9) and (10.2.10) shall imply the existence of a corresponding stable solution for the original boundary-value problem; see Howes (1978) and Chang and Howes (1984) for a general discussion of such stability conditions.

Example 10.2.1: Consider the problem (10.2.1) and (10.2.4) with H and β given as

$$\beta(\varepsilon) \equiv \beta_0 := 0.5,$$
$$H(t, x, \varepsilon) \equiv H_0(x) := (x^2 - 1)(x^2 - 4), \qquad (10.2.11)$$

for all x. The function

$$X_0 = X_0(t) := -1 \quad \text{for } 0 \le t \le 1 \qquad (10.2.12)$$

is easily seen to satisfy (10.2.8) and (10.2.9), and we also have in this case

$$\int_0^x H_0[X_0(1) + z] \, dz = 3x^2\left(1 + \frac{x}{9} - \frac{x^2}{3} + \frac{x^3}{15}\right), \qquad (10.2.13)$$

from which we see that (10.2.10) holds, with, for example, $\nu_1 = 1.96$.

Now consider instead the function

$$X_0 = X_0(t) := 2 \quad \text{for } 0 \le t \le 1. \qquad (10.2.14)$$

We again find directly that (10.2.8), (10.2.9), and (10.2.10) all hold in Example 10.2.1, for a suitable constant ν_1 in (10.2.10); see Exercise 10.2.1. Hence, in this case we have two distinct possibilities for the function X_0.

For given H_0 and X_0 satisfying (10.2.8)–(10.2.9), we see that the additional condition (10.2.10) amounts to a restriction on β_0, and in particular (10.2.10) places a limit on the size of the boundary-layer jump $\beta_0 - X_0(1)$. For example, the value $\beta_0 = 0$ is disqualified by (10.2.10) in conjunction with the X_0 of (10.2.14) in Example 10.2.1, although this same value $\beta_0 = 0$ is acceptable in conjunction with the X_0 of (10.2.12).

Given a suitable fixed function X_0 satisfying (10.2.8)–(10.2.10), we now seek a corresponding solution of the original problem (10.2.1) and (10.2.4) for small $\varepsilon \to 0+$, in the asymptotic form

$$x(t, \varepsilon) \sim X(t, \varepsilon) + \varepsilon(^*X(\tau, \varepsilon)) + X^*(\sigma, \varepsilon), \qquad (10.2.15)$$

with

$$\tau := \frac{t}{\varepsilon} \quad \text{and} \quad \sigma := \frac{1-t}{\varepsilon}, \tag{10.2.16}$$

and where the functions on the right side of (10.2.15) are to be determined in the form of asymptotic expansions as

$$\begin{bmatrix} X(t,\varepsilon) \\ {}^*X(\tau,\varepsilon) \\ X^*(\sigma,\varepsilon) \end{bmatrix} \sim \sum_{k=0}^{\infty} \begin{bmatrix} X_k(t) \\ {}^*X_k(\tau) \\ X_k^*(\sigma) \end{bmatrix} \varepsilon^k, \tag{10.2.17}$$

where the leading term $X_0(t)$ in the outer solution $X(t, \varepsilon)$ is taken to be the given, fixed function satisfying (10.2.8)–(10.2.10), and where the respective left and right boundary-layer correction terms ${}^*X_k(\tau)$ and $X_k^*(\sigma)$ are required to satisfy the usual matching conditions

$$\lim_{\tau \to \infty} {}^*X_k(\tau) = 0,$$
$$\lim_{\sigma \to \infty} X_k^*(\sigma) = 0 \quad \text{for } k = 0, 1, \ldots . \tag{10.2.18}$$

As in Exercise 9.3.11, the explicit factor of ε that multiplies ${}^*X(\tau, \varepsilon)$ in (10.2.15) reflects the relatively weaker boundary-layer effect here at $t = 0$ as compared with $t = 1$, because of the type of boundary conditions given in (10.2.4).

The boundary-layer correction terms are expected to be negligible away from the endpoints, and so the outer solution $X(t, \varepsilon)$ is required to satisfy the full differential equation (10.2.1):

$$\varepsilon^2 \frac{d^2 X}{dt^2} = H(t, X, \varepsilon) \quad \text{for } 0 < t < 1. \tag{10.2.19}$$

We insert into (10.2.19) the respective expansions from (10.2.7) and (10.2.17) for H and X, and we find directly that the higher-order terms $X_k(t)$ satisfy linear (algebraic) equations of the form

$$H_{0,x}[t, X_0(t)] X_k = P_{k-1}(t)$$
$$\text{for } 0 \le t \le 1, \ k = 1, 2, \ldots, \tag{10.2.20}_k$$

for suitable functions P_{k-1} that are known successively in terms of preceding coefficients X_j for $j = 0, 1, \ldots, k - 1$. For example, we have

$$P_0(t) = -H_1[t, X_0(t)],$$

$$P_1(t) = \frac{d^2 X_0(t)}{dt^2} - \left\{ H_2[t, X_0(t)] + H_{1,x}[t, X_0(t)] X_1(t) \right.$$
$$\left. + \tfrac{1}{2} H_{0,xx}[t, X_0(t)] X_1(t)^2 \right\}, \tag{10.2.21}$$

and so forth, with analogous formulas holding for higher-indexed P_k's.

The condition (10.2.9) guarantees that $X_k = X_k(t)$ is well determined by $(10.2.20)_k$ in terms of preceding coefficients, so that the coefficients in the expansion of (10.2.17) for the outer solution are uniquely determined recursively in terms of the fixed solution X_0.

In the special case that H has an asymptotic expansion in ε^2, we find directly that the outer solution X also has an expansion in ε^2, with

$$X_{2k+1} \equiv 0 \quad \text{for } 0, 1, 2, \ldots . \tag{10.2.22}$$

This latter result obtains in particular if H is independent of ε, as occurs in Example 10.2.1.

We turn now to determination of the boundary-layer correction terms $*X(t/\varepsilon, \varepsilon)$ and $X*[(1 - t)/\varepsilon, \varepsilon]$. Because each of these functions is negligible where the other is not, we require in (10.2.15) that each of the two functions $x = X + \varepsilon*X$ and $x = X + X*$ should separately satisfy the full equation (10.2.1), from which we find directly, with (10.2.16) and (10.2.19), the equations

$$\varepsilon \frac{d^2}{d\tau^2} *X(\tau, \varepsilon) = H\left[\varepsilon\tau, X(\varepsilon\tau, \varepsilon) + \varepsilon*X(\tau, \varepsilon), \varepsilon\right]$$

$$- H\left[\varepsilon\tau, X(\varepsilon\tau, \varepsilon), \varepsilon\right] \quad \text{for } \tau > 0 \tag{10.2.23}$$

and

$$\frac{d^2}{d\sigma^2} X*(\sigma, \varepsilon) = H\left[1 - \varepsilon\sigma, X(1 - \varepsilon\sigma, \varepsilon) + X*(\sigma, \varepsilon), \varepsilon\right]$$

$$- H\left[1 - \varepsilon\sigma, X(1 - \varepsilon\sigma, \varepsilon), \varepsilon\right] \quad \text{for } \sigma > 0. \tag{10.2.24}$$

Similarly, from (10.2.4) we obtain for $*X$ and $X*$ the following initial conditions:

$$\left.\frac{d}{d\tau} *X(\tau, \varepsilon)\right|_{\tau=0} = \alpha(\varepsilon) - X'(0, \varepsilon) \tag{10.2.25}$$

and

$$X*(0, \varepsilon) = \beta(\varepsilon) - X(1, \varepsilon). \tag{10.2.26}$$

For $*X$, we insert into (10.2.23) the respective expansions from (10.2.7) and (10.2.17) for H, X, and $*X$, and we find that the terms $*X_k(\tau)$ satisfy linear equations of the type

$$\frac{d^2}{d\tau^2} *X_k(\tau) = H_{0,x}\left[0, X_0(0)\right] *X_k(\tau) + *P_{k-1}(\tau) \tag{10.2.27}_k$$

for $k = 0, 1, \ldots$, for suitable functions $*P_{k-1} = *P_{k-1}(\tau)$ that are known successively in terms of preceding coefficients $*X_j$ for $j \le k - 1$. For example, we have

$$*P_{-1}(\tau) \equiv 0,$$

$$*P_0(\tau) = \left\{ H_{0,xx}[0, X_0(0)] \left[\tau X_0'(0) + X_1(0) + \tfrac{1}{2}*X_0(\tau) \right] \right.$$
$$\left. + \tau H_{0,tx}[0, X_0(0)] + H_{1,x}[0, X_0(0)] \right\} *X_0(\tau),$$
$$(10.2.28)$$

with similar formulas holding for higher-indexed $*P_k$'s.

The differential equation $(10.2.27)_k$ is to be solved for $*X_k$ subject to the appropriate matching condition at $\tau = \infty$, given by (10.2.18), along with the initial condition

$$\frac{d}{d\tau}*X_k(0) = \alpha_k - X_k'(0) \qquad (10.2.29)_k$$

for $k = 0, 1, \ldots,$ where these latter conditions follow directly from (10.2.25).

The unique solution for $*X_k$ is easily seen to be given as

$$*X_k(\tau) = \frac{X_k'(0) - \alpha_k}{\nu_0} e^{-\nu_0 \tau} - \int_0^\infty \frac{*P_{k-1}(s)}{2\nu_0} e^{-\nu_0(\tau+s)} \, ds$$
$$- \int_0^\tau \frac{*P_{k-1}(s)}{2\nu_0} e^{-\nu_0(\tau-s)} \, ds - \int_\tau^\infty \frac{*P_{k-1}(s)}{2\nu_0} e^{-\nu_0(s-\tau)} \, ds,$$
$$(10.2.30)_k$$

where the positive constant ν_0 is given as

$$\nu_0 := \sqrt{H_{0,x}[0, X_0(0)]} > 0 \qquad (10.2.31)$$

[see (10.2.9)]. Because $*P_{-1} \equiv 0$ [see (10.2.28)], we have, in particular, the result

$$*X_0(\tau) = \frac{X_0'(0) - \alpha_0}{\nu_0} e^{-\nu_0 \tau} \quad \text{for } \tau \geq 0, \qquad (10.2.32)$$

so that $*X_0(\tau)$ decays exponentially to zero as $\tau \to \infty$, with decay constant ν_0. Using $(10.2.30)_k$, along with the form of $*P_{k-1}$ as provided by the foregoing construction, we prove directly by induction that each $*X_k(\tau)$ decays exponentially, with

$$|*X_k(\tau)|, \left| \frac{d}{d\tau}*X_k(\tau) \right| \leq C_k e^{-\nu \tau} \quad \text{for } \tau \geq 0, \qquad (10.2.33)_k$$

for suitable constants C_k, and for any fixed positive constant ν less than ν_0,

$$0 < \nu < \nu_0. \qquad (10.2.34)$$

This completes the determination of the boundary-layer correction terms near the left endpoint $t = 0$.

Near $t = 1$, we use similar techniques to determine the terms $X_k^*(\sigma)$ in the expansion of (10.2.17) for $X^*(\sigma, \varepsilon)$. We insert into (10.2.24) the expansions from (10.2.7) and (10.2.17) for H, X, and X^*, and we find, in particular, that $X_0^*(\sigma)$ satisfies the nonlinear equation [we use (10.2.8) here at $t = 1$]

$$\frac{d^2}{d\sigma^2} X_0^* = H_0\big[1, X_0(1) + X_0^*\big] \quad \text{for } \sigma > 0, \tag{10.2.35}$$

and higher-indexed terms X_k^* satisfy linear equations of the form

$$\frac{d^2}{d\sigma^2} X_k^*(\sigma) = H_{0,x}\big[1, X_0(1) + X_0^*(\sigma)\big] X_k^*(\sigma) + P_{k-1}^*(\sigma) \tag{10.2.36}_k$$

for $k = 1, 2, \ldots$, for suitable functions $P_{k-1}^* = P_{k-1}^*(\sigma)$ that are known successively in terms of preceding coefficients X_j^* for $j \leq k - 1$. For example, we have

$$\begin{aligned}
P_0^*(\sigma) = &-\sigma\big\{ H_{0,t}\big[1, X_0(1) + X_0^*(\sigma)\big] - H_{0,t}\big[1, X_0(1)\big]\big\} \\
&+ \big\{ H_{0,x}\big[1, X_0(1) + X_0^*(\sigma)\big] - H_{0,x}\big[1, X_0(1)\big]\big\} \\
&\times \big[-\sigma X_0'(1) + X_1(1)\big] \\
&+ \big\{ H_1\big[1, X_0(1) + X_0^*(\sigma)\big] - H_1\big[1, X_0(1)\big]\big\}, \tag{10.2.37}
\end{aligned}$$

and similar, related expressions hold for higher-indexed P_k^*'s.

The differential equations of (10.2.35)–(10.2.36) are to be solved for the functions X_k^* subject to the appropriate matching conditions at $\sigma = \infty$, given by (10.2.18), along with the corresponding initial conditions of

$$X_k^*(0) = \beta_k - X_k(1) \tag{10.2.38}_k$$

for $k = 0, 1, \ldots$, where these latter conditions follow directly from (10.2.26).

Using the condition (10.2.10), it can be shown (Fife 1973a) that the given nonlinear problem for $X_0^*(\sigma)$ has a unique solution for $\sigma \geq 0$, satisfying

$$\begin{aligned}
&|X_0^*(\sigma)| \leq |\beta_0 - X_0(1)|e^{-\nu_1\sigma}, \\
&\left|\frac{dX_0^*(\sigma)}{d\sigma}\right| \leq \text{const.} \, |\beta_0 - X_0(1)|e^{-\nu_1\sigma}, \quad \text{for } \sigma \geq 0,
\end{aligned} \tag{10.2.39}$$

where ν_1 is the positive constant appearing in (10.2.10); see Exercise 10.2.2 for an indication of a proof of this result (10.2.39). The equation (10.2.70) in Exercise 10.2.2 illustrates the *necessity* of (at least a weakened version of) the earlier condition (10.2.10) for a solution of boundary-layer type as considered here.

The further coefficient functions X_k^* also exist and are given now by a formula analogous to $(10.2.30)_k$ (see Exercise 10.2.3), with which we find the estimates, analogous to (10.2.33) (see Exercise 10.2.4),

$$|X_k^*(\sigma)|, \left|\frac{dX_k^*(\sigma)}{d\sigma}\right| \leq C_k e^{-\nu\sigma} \quad \text{for } \sigma \geq 0, \qquad (10.2.40)_k$$

for suitable constants C_k [which are not the same here as in $(10.2.33)_k$] and for any fixed positive constant ν less than the constant ν_1 of (10.2.10),

$$0 < \nu < \nu_1. \qquad (10.2.41)$$

Hence, the coefficients in the expansion (10.2.15)–(10.2.17) can be determined recursively for the nonlinear boundary-value problem (10.2.1) and (10.2.4), provided that the assumed conditions of (10.2.8)–(10.2.10) hold and provided always that the given function H is sufficiently smooth. As usual, we wish to truncate the resulting asymptotic representation of (10.2.15)–(10.2.17) so as to obtain what is expected to be an approximate solution that can be used to prove that the original problem in fact has a solution that is well approximated by the truncated asymptotic solution. Hence, we let $x^N(t, \varepsilon)$ denote the Nth partial sum of the expansion (10.2.15)–(10.2.17), given as

$$x^N(t, \varepsilon) := \sum_{k=0}^{N} \left[X_k(t) + \varepsilon^* X_k\left(\frac{t}{\varepsilon}\right) + X_k^*\left(\frac{1-t}{\varepsilon}\right) \right] \varepsilon^k \qquad (10.2.42)$$

for $0 \leq t \leq 1$, where the coefficients X_k, $^*X_k$, and X_k^* are constructed as before.

The original problem (10.2.1) and (10.2.4) is now recast as a linearization about the given proposed approximation x^N of (10.2.42). Specifically, if x denotes a solution of (10.2.1) and (10.2.4), then the function $\hat{x}$, defined as

$$\hat{x}(t) \equiv \hat{x}(t, \varepsilon) := x(t, \varepsilon) - x^N(t, \varepsilon), \qquad (10.2.43)$$

is found to satisfy the boundary conditions

$$\frac{d\hat{x}}{dt} = \phi_1(\varepsilon) \quad \text{at } t = 0,$$

$$\hat{x} = \phi_2(\varepsilon) \quad \text{at } t = 1, \qquad (10.2.44)$$

along with the differential equation (see Sections 6.3 and 8.2, where similar calculations are given)

$$\varepsilon^2 \frac{d^2 \hat{x}}{dt^2} = H\big[t, x^N(t, \varepsilon) + \hat{x}, \varepsilon\big] - H\big[t, x^N(t, \varepsilon), \varepsilon\big] + \rho(t, \varepsilon)$$

$$= H_x\big[t, x^N(t, \varepsilon), \varepsilon\big]\hat{x} + E(t, \hat{x}, \varepsilon) + \rho(t, \varepsilon), \qquad (10.2.45)$$

where the quantities $\phi_1(\varepsilon)$, $\phi_2(\varepsilon)$, and $\rho(t, \varepsilon)$ are the residuals defined as

$$\phi_1(\varepsilon) := \alpha(\varepsilon) - \frac{dx^N(0, \varepsilon)}{dt}, \qquad \phi_2(\varepsilon) := \beta(\varepsilon) - x^N(1, \varepsilon),$$

$$(10.2.46)$$

and

$$\rho(t, \varepsilon) := H\big[t, x^N(t, \varepsilon), \varepsilon\big] - \varepsilon^2 \frac{d^2 x^N(t, \varepsilon)}{dt^2}, \qquad (10.2.47)$$

and where the quantity $E = E(t, \hat{x}, \varepsilon)$ is then defined by the right side of (10.2.45) and can be given by a Taylor/Cauchy formula as [see (4.1.39–(4.1.41)]

$$E(t, \hat{x}, \varepsilon) = \left\{ \int_0^1 (1 - s) H_{xx}\big[t, x^N(t, \varepsilon) + s\hat{x}, \varepsilon\big] \, ds \right\} \hat{x}^2$$

$$(10.2.48)$$

for any suitable numbers t, $\hat{x}$, and ε. The subscripts in (10.2.45) and (10.2.48) denote the appropriate partial derivative, as

$$H_x\big[t, x^N(t, \varepsilon), \varepsilon\big] = \frac{\partial H(t, x, \varepsilon)}{\partial x}\bigg|_{x = x^N(t, \varepsilon)},$$

with an analogous result for H_{xx}. Also, note that we generally suppress the obvious dependence of ϕ_1, ϕ_2, ρ, and E on the nonnegative integer N.

The definitions of the residuals (10.2.46) and (10.2.47), along with the construction of the coefficients X_k, $*X_k$, and X_k^*, yield directly estimates of the form

$$|\phi_1(\varepsilon)|, |\phi_2(\varepsilon)| \leq \text{const. } \varepsilon^{N+1}, \qquad (10.2.49)$$

and (Fife 1973a)

$$|\rho(t, \varepsilon)| \leq \text{const. } \varepsilon^{N+1} \quad \text{for } 0 \leq t \leq 1, \qquad (10.2.50)$$

as $\varepsilon \to 0+$, for fixed constants that depend on N. A proof of (10.2.49)–(10.2.50) is indicated in Exercise 10.2.5 for the case $N = 0$.

We could work directly with the second-order scalar problem (10.2.44)–(10.2.45), but we prefer instead to work with the equivalent first-order system

$$\frac{d}{dt}\begin{bmatrix}\hat{x}\\\hat{y}\end{bmatrix} = \frac{1}{\varepsilon}\begin{bmatrix}0 & 1\\H_x[t,x^N(t,\varepsilon),\varepsilon] & 0\end{bmatrix}\begin{bmatrix}\hat{x}\\\hat{y}\end{bmatrix} + \frac{1}{\varepsilon}\begin{bmatrix}0\\E(t,\hat{x},\varepsilon)+\rho(t,\varepsilon)\end{bmatrix}$$

(10.2.51)

for $0 \le t \le 1$, subject to the boundary conditions

$$L\begin{bmatrix}\hat{x}(0,\varepsilon)\\\hat{y}(0,\varepsilon)\end{bmatrix} + R\begin{bmatrix}\hat{x}(1,\varepsilon)\\\hat{y}(1,\varepsilon)\end{bmatrix} = \begin{bmatrix}\varepsilon\phi_1(\varepsilon)\\\phi_2(\varepsilon)\end{bmatrix},$$

(10.2.52)

with the given boundary matrices

$$L := \begin{bmatrix}0 & 1\\0 & 0\end{bmatrix} \quad \text{and} \quad R := \begin{bmatrix}0 & 0\\1 & 0\end{bmatrix}.$$

(10.2.53)

The nonlinearity of the problem is reflected in the term $E(t,\hat{x},\varepsilon)$ on the right side of (10.2.51).

The present nonlinear problem (10.2.51)–(10.2.53) is closely related to the previous linear problem (9.3.78)–(9.3.80) of Exercise 9.3.10, with slightly different boundary conditions and with slightly different "packaging" (y replaced by εy). An appropriate Green function was used in Exercise 9.3.10 to obtain error estimates for the linear problem there, and indeed a corresponding Green function representation can be used here to provide both the existence of an exact solution for the present nonlinear problem and useful error estimates on the difference $\hat{x}$ between the resulting exact solution x and the corresponding approximate solution x^N.

It is shown in Section 10.3 that the linearized homogeneous operator of (10.2.51) (with $E \equiv 0$, $\rho \equiv 0$) has a Green function $G = G(t,s,\varepsilon)$ corresponding to the boundary conditions of (10.2.52)–(10.2.53), and this Green function can be used to rewrite the present nonlinear boundary-value problem (10.2.51)–(10.2.53) as the equivalent nonlinear integral equation

$$\begin{bmatrix}\hat{x}(t)\\\hat{y}(t)\end{bmatrix} = \begin{bmatrix}\hat{x}_0(t,\varepsilon)\\\hat{y}_0(t,\varepsilon)\end{bmatrix} + \frac{1}{\varepsilon}\int_0^1 G(t,s,\varepsilon)\begin{bmatrix}0\\E[s,\hat{x}(s),\varepsilon]\end{bmatrix}ds,$$

(10.2.54)

with

$$\begin{bmatrix}\hat{x}_0(t,\varepsilon)\\\hat{y}_0(t,\varepsilon)\end{bmatrix} := Z(t,\varepsilon)M(\varepsilon)^{-1}\begin{bmatrix}\varepsilon\phi_1(\varepsilon)\\\phi_2(\varepsilon)\end{bmatrix} + \frac{1}{\varepsilon}\int_0^1 G(t,s,\varepsilon)\begin{bmatrix}0\\\rho(s,\varepsilon)\end{bmatrix}ds,$$

(10.2.55)

where we are suppressing the dependence of $\hat{x}$ and $\hat{y}$ on ε, and where $Z(t, \varepsilon)$ is a suitable fundamental solution for the homogeneous system (10.2.51) with $E = \rho = 0$. The matrix M is defined as [see (9.3.4)] $M(\varepsilon) := LZ(0, \varepsilon) + RZ(1, \varepsilon)$. [In a related problem, van Harten and Vader-Burger (1984) show that the exact Green function can be replaced by an *approximate* Green function.]

The required fundamental solution $Z = [Z_{ij}(t, \varepsilon)]_{i,\,j=1,2}$ and the Green function $G = [G_{ij}(t, s, \varepsilon)]_{i,\,j=1,2}$ are constructed in Section 10.3, where we have the estimates

$$|Z(t, \varepsilon) M(\varepsilon)^{-1}| \le \text{constant}, \quad \text{for } 0 \le t \le 1, \tag{10.2.56}$$

and

$$|G(t, s, \varepsilon)| \le \text{const. } \exp\left(-\frac{\nu_0 |t - s|}{\varepsilon}\right) \quad \text{for } 0 \le t, s \le 1, \tag{10.2.57}$$

uniformly as $\varepsilon \to 0+$, for a suitable fixed positive constant ν_0 obtained from the conditions (10.2.8)–(10.2.10).

It follows now directly from the estimates (10.2.49), (10.2.50), (10.2.56), and (10.2.57), along with a routine application of the Banach/Picard fixed-point theorem (see Exercise 10.2.6), that the integral equation (10.2.54)–(10.2.55) has a solution $\hat{x}(t) = \hat{x}(t, \varepsilon)$ and $\hat{y}(t) = \hat{y}(t, \varepsilon)$ satisfying estimates of the type

$$|\hat{x}(t, \varepsilon)|, |\hat{y}(t, \varepsilon)| \le \text{const. } \varepsilon^{N+1} \quad \text{as } \varepsilon \to 0+, \tag{10.2.58}$$

uniformly for $0 \le t \le 1$. Moreover, the solution functions $\hat{x}, \hat{y}$ are uniquely determined subject to (10.2.58); that is, there are no other solution functions for (10.2.54)–(10.2.55) satisfying (10.2.58).

In terms of the original boundary-value problem (10.2.1) and (10.2.4), we have the following result.

Theorem 10.2.1: *Let the data H, α, β of (10.2.1) and (10.2.4) possess asymptotic expansions as in (10.2.7), and assume that H and the coefficient functions $H_k = H_k(t, x)$ are sufficiently smooth. Let $X_0 = X_0(t)$ be a fixed, smooth solution of the reduced equation (10.2.8), and assume that the stability conditions (10.2.9) and (10.2.10) hold. Then there is a fixed number $\varepsilon_0 > 0$ such that the function $x^N = x^N(t, \varepsilon)$ of (10.2.42) is well defined by the foregoing multivariable construction on the region*

$$0 \le t \le 1, \quad 0 < \varepsilon \le \varepsilon_0. \tag{10.2.59}$$

Moreover, the problem (10.2.1) and (10.2.4) has an exact solution $x =$

x(t, ε) close to x^N on the region (10.2.59), *and*

$$|x(t, \varepsilon) - x^N(t, \varepsilon)| \leq \text{const. } \varepsilon^{N+1},$$

$$\left| \frac{dx(t, \varepsilon)}{dt} - \frac{dx^N(t, \varepsilon)}{dt} \right| \leq \text{const. } \varepsilon^N,$$

(10.2.60)

hold uniformly on (10.2.59). *The particular exact solution so constructed is unique subject to* (10.2.60).

Proof: Given the estimates of (10.2.56)–(10.2.57), the results of the theorem follow from the foregoing discussion, along with (10.2.43). The estimates (10.2.56)–(10.2.57) are derived in Section 10.3. ∎

The foregoing discussion borrows heavily from Fife (1973a), where the Dirichlet problem is considered for a semilinear elliptic partial differential equation that reduces to (10.2.1) in the case of a single independent variable. Fife obtains estimates of the type (10.2.60) using an analysis based in part on a variational study of the spectrum of the linearization (10.2.51). See also Berger and Fraenkel (1969–70), De Villiers (1973), Tang (1972), Tang and Fife (1975), Eckhaus (1973, 1979), and van Harten (1978a) for related studies of quasi-linear elliptic partial differential equations.

Example 10.2.2: We can apply the preceding results to the earlier boundary-value problem of Example 10.2.1, with X_0 given by (10.2.12). We can also consider this same boundary-value problem, but with X_0 given by (10.2.14). Because both of these functions X_0 satisfy the required conditions of (10.2.8)–(10.2.10), it follows that the given boundary-value problem has two distinct solutions, say $x = x_1(t, \varepsilon)$ and $x = x_2(t, \varepsilon)$ for small $\varepsilon > 0$. The graphs of these solution functions are indicated in Figure 10.2 for the case $\alpha_0 = 0$.

Note that the boundary values can play a crucial role regarding the solvability of a boundary-value problem. For example, the problem

$$\varepsilon^2 \frac{d^2x}{dt^2} = (x^2 - 1)(x^2 - 4) \quad \text{for } 0 < t < 1,$$

$$\frac{dx}{dt} = 0 \quad \text{at } t = 0, \quad \text{and} \quad x = \beta \quad \text{at } t = 1,$$

(10.2.61)

may have no solution of boundary-layer type, precisely one solution of

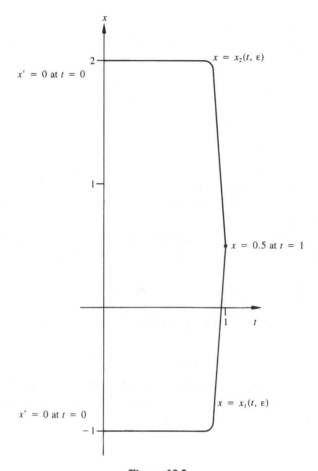

Figure 10.2

boundary-layer type, or precisely two such solutions, depending on the value of the specified number β, as $\varepsilon \to 0+$. Let β_1 and β_2 be the real roots of the respective cubics $q_1(\beta)$ and $q_2(\beta)$, where

$$q_1(\beta) := 3\beta^3 - 6\beta^2 - 16\beta + 38,$$
$$q_2(\beta) := 3\beta^3 + 12\beta^2 + 11\beta - 4,$$

(10.2.62)

with $\beta_1 \doteq -2.40569$ and $\beta_2 \doteq 0.275279$. Then, as $\varepsilon \to 0+$, the problem (10.2.61) has a solution $x_1(t, \varepsilon)$ corresponding to the outer solution $X_0(t) \equiv -1$ for $\beta > \beta_1$ but not for $\beta < \beta_1$, and similarly (10.2.61) has a solution $x_2(t, \varepsilon)$ corresponding to the outer solution $X_0(t) \equiv +2$ for $\beta > \beta_2$ but not for $\beta < \beta_2$, as indicated in Figure 10.3. The critical values

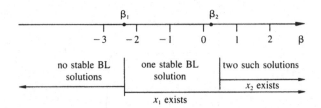

Figure 10.3

β_1 and β_2 are computed as the appropriate roots of the functions of (10.2.62), so that the stability condition (10.2.10) will hold for the appropriate values of β. See O'Malley (1976) and Howes (1978) for related results.

As mentioned earlier, the foregoing approach can be used to handle more general boundary conditions of the form (10.2.5).

Example 10.2.3: Consider the Dirichlet problem

$$\varepsilon^2 \frac{d^2x}{dt^2} = 1 + (2 - t^2)x - x^2 \quad \text{for } |t| < 1,$$

$$x = 0 \quad \text{at } t = -1 \text{ and at } t = +1. \tag{10.2.63}$$

This is a certain nonlinear version of the problem of Carrier and Pearson considered earlier in Examples 8.2.1 and 8.2.2. The differential equation of (10.2.63) is of the form (10.2.1), with

$$H(t, x, \varepsilon) \equiv H_0(t, x) := 1 + (2 - t^2)x - x^2 \tag{10.2.64}$$

for $t \in [-1, +1]$, $x \in \mathbb{R}$. The change of variable $t_{\text{new}} := (1 + t)/2$ could be used to transform the interval $[-1, +1]$ of (10.2.63) onto the previous interval $[0, 1]$, as in (10.2.1), but this is unimportant: We shall retain the interval $|t| \le 1$, as in (10.2.63).

For (10.2.63)–(10.2.64), the reduced equation (10.2.8) is a quadratic equation with the two solutions

$$X_0(t) = \frac{2 - t^2 + \left[(2 - t^2)^2 + 4\right]^{1/2}}{2},$$

$$X_0(t) = \frac{2 - t^2 - \left[(2 - t^2)^2 + 4\right]^{1/2}}{2}. \tag{10.2.65}$$

We find directly that the stability condition (10.2.9) is violated by the first choice in (10.2.65), but (10.2.9) holds for the second choice. Hence, we take as the leading term in an outer approximation

$$X_0(t) = \frac{2 - t^2 - \left[(2 - t^2)^2 + 4\right]^{1/2}}{2}, \tag{10.2.66}$$

which satisfies both (10.2.8) and (10.2.9).

Because at each endpoint the boundary condition is of Dirichlet type, we must impose a condition of the type (10.2.10) at each endpoint. We find in the present case, with (10.2.64) and (10.2.66),

$$\int_0^x H_0\left[\pm 1, X_0(\pm 1) + z\right] dz = \frac{\sqrt{5}\, x^2}{2}\left(1 - \frac{2x}{3\sqrt{5}}\right) \geq \frac{\nu_1^2 x^2}{2}$$

$$\tag{10.2.67}$$

for all x between 0 and $(\sqrt{5} - 1)/2$, with $\nu_1 = [(1 + 2\sqrt{5})/3]^{1/2} > 1.35$. Hence, the appropriate condition holds at each endpoint for the X_0 of (10.2.66).

We seek an expansion of the type

$$x(t, \varepsilon) \sim X(t, \varepsilon) + {}^*X(\tau, \varepsilon) + X^*(\sigma, \varepsilon), \tag{10.2.68}$$

with

$$\tau := \frac{1 + t}{\varepsilon} \quad \text{and} \quad \sigma := \frac{1 - t}{\varepsilon}, \tag{10.2.69}$$

and where the functions on the right side of (10.2.68) are to be determined in the asymptotic form (10.2.17)–(10.2.18) with X_0 given by (10.2.66). The coefficient functions X_k, ${}^*X_k$, and X_k^* can again be determined successively in terms of previous coefficients, and the results of Fife (1973a) show that the resulting truncated expansion from (10.2.68) can be used as a proposed approximate solution in a linearization so as to obtain a corresponding exact solution $x = x(t, \varepsilon)$ to the original Dirichlet problem. This result also follows by a Green function approach of the sort illustrated earlier for (10.2.1) and (10.2.4). We find in particular that the function X_0 of (10.2.66) provides an $O(\varepsilon^2)$ approximation to the corresponding exact solution $x(t, \varepsilon)$ away from the endpoints (see Example 8.2.2). Details are omitted.

Problems of the type considered in this section, including similar problems for systems of equations, appear as mathematical models in studies of chemical and biochemical kinetics without convection [see Aris (1969) and Lapidus and Amundson (1977) and the references listed there] and in studies of population (ecological) kinetics for competing species [see van Harten and Vader-Burger (1984) and the references listed there].

Exercise 10.2.1: Show that the function X_0 of (10.2.14) satisfies the conditions of (10.2.8)–(10.2.10), with data as in (10.2.11).

Exercise 10.2.2 (Fife 1973a): Let H_0 be a given function of class C^1 satisfying condition (10.2.10) for given numbers $X_0(1)$, β_0, and $\nu_1 > 0$, and assume that $X_0(1)$ satisfies (10.2.8) at $t = 1$. Show that the initial-value problem

$$\left(\frac{dX_0^*(\sigma)}{d\sigma} \right)^2 = 2 \int_0^{X_0^*(\sigma)} H_0\big[1, X_0(1) + x\big]\, dx \quad \text{for } \sigma \geq 0,$$

$$X_0^*(0) = \beta_0 - X_0(1) \quad \text{at } \sigma = 0, \tag{10.2.70}$$

has a solution X_0^* of class $C^2[0, \infty)$ satisfying

$$|X_0^*(\sigma)| \leq |\beta_0 - X_0(1)| e^{-\nu_1 \sigma}, \tag{10.2.71}$$

$$\left| \frac{dX_0^*(\sigma)}{d\sigma} \right| \leq \text{const.} \ |\beta_0 - X_0(1)| e^{-\nu_1 \sigma} \quad \text{for } \sigma \geq 0.$$

Moreover, if $\beta_0 \neq X_0(1)$ [so that $X_0^*(0) \neq 0$], then show that X_0^* and its first derivative are everywhere nonzero, with

$$|X_0^*(\sigma)| > 0 \quad \text{and} \quad \left| \frac{dX_0^*(\sigma)}{d\sigma} \right| > 0 \quad \text{for } \sigma \geq 0, \text{ if } X_0^*(0) \neq 0. \tag{10.2.72}$$

Note that the resulting solution X_0^* of (10.2.70)–(10.2.71) provides a solution for (10.2.35) and (10.2.38)$_0$ satisfying also the appropriate matching condition of (10.2.18) as $\sigma \to \infty$. *Hint:* If $\beta_0 = X_0(1)$, then we find $X_0^*(\sigma) \equiv 0$ for $\sigma \geq 0$, and the stated result is trivial in this case. Hence, consider the case $\beta_0 \neq X_0(1)$, and take an appropriate square root of the differential equation of (10.2.70) to find

$$\frac{dX_0^*(\sigma)}{d\sigma} = \begin{cases} -\left\{ 2 \int_0^{X_0^*(\sigma)} H_0\big[1, X_0(1) + x\big]\, dx \right\}^{1/2} & \text{if } \beta_0 > X_0(1), \\[3ex] +\left\{ 2 \int_0^{X_0^*(\sigma)} H_0\big[1, X_0(1) + x\big]\, dx \right\}^{1/2} & \text{if } \beta_0 < X_0(1). \end{cases} \tag{10.2.73}$$

Consider first the case in which

$$X_0^*(0) = \beta_0 - X_0(1) > 0, \tag{10.2.74}$$

so that the differential equation of (10.2.73) becomes

$$\frac{dX_0^*(\sigma)}{d\sigma} = -\left\{2\int_0^{X_0^*(\sigma)} H_0[1, X_0(1) + x]\, dx\right\}^{1/2} \quad \text{for } \sigma > 0,$$

$$(10.2.75)$$

subject to (10.2.74). The stated assumptions imply that a suitable uniqueness result holds, along with a (local) existence result, for the initial-value
problem (10.2.74)–(10.2.75), and we easily conclude that

$$X_0^*(\sigma) > 0 \qquad\qquad (10.2.76)$$

for all σ on a maximal interval of existence. From (10.2.10) and (10.2.75)
follows the differential inequality

$$\frac{dX_0^*(\sigma)}{d\sigma} \leq -\nu_1 X_0^*(\sigma), \qquad\qquad (10.2.77)$$

again for all σ on a maximal interval of existence. This last inequality can
be integrated to yield $X_0^*(\sigma) \leq X_0^*(0)e^{-\nu_1\sigma}$, which, with (10.2.74) and
(10.2.76), yields the first inequality of (10.2.71) for X_0^*. This inequality
implies that the solution exists and satisfies the given inequality for all
$\sigma \geq 0$. The derivation of the corresponding inequalities for $dX_0^*(\sigma)/d\sigma$
is left to the reader, along with the corresponding results for the case
$\beta_0 < X_0(1)$.

Exercise 10.2.3: Let H_0 be a given function of class C^2 satisfying the
condition (10.2.10) for given numbers $X_0(1)$, β_0, and $\nu_1 > 0$, and assume
that $X_0(1)$ satisfies (10.2.8) and (10.2.9) at $t = 1$. Let $X_0^* = X_0^*(\sigma)$ be the
solution of (10.2.70), as obtained in Exercise 10.2.2. Also, assume that

$$\beta_0 \neq X_0(1). \qquad\qquad (10.2.78)$$

The remaining case $\beta_0 = X_0(1)$ is simpler to handle, and this latter
simpler case is left as a further exercise that will not be considered here.

(a) Show that the linear homogeneous equation

$$\frac{d^2\xi}{d\sigma^2} = H_{0,x}[1, X_0(1) + X_0^*(\sigma)]\xi \quad \text{for } \sigma \geq 0 \qquad (10.2.79)$$

has linearly independent solutions ξ_1 and ξ_2 satisfying

$$\xi_1(\sigma) = [1 + b_1(\sigma)]e^{-\hat{\nu}\sigma}, \qquad \frac{d\xi_1(\sigma)}{d\sigma} = -\hat{\nu}[1 + c_1(\sigma)]e^{-\hat{\nu}\sigma},$$

$$\xi_2(\sigma) = [1 + b_2(\sigma)]e^{+\hat{\nu}\sigma}, \qquad \frac{d\xi_2(\sigma)}{d\sigma} = +\hat{\nu}[1 + c_2(\sigma)]e^{+\hat{\nu}\sigma},$$

$$\text{for } \sigma \geq 0, \quad \text{with } \hat{\nu} := \sqrt{H_{0,x}[1, X_0(1)]} \geq \nu_1,$$

$$(10.2.80)$$

for suitable C^2 functions b_j, c_j ($j = 1, 2$) satisfying

$$b_j(\sigma), c_j(\sigma) = O(e^{-\nu_1 \sigma}) \quad \text{as } \sigma \to \infty. \tag{10.2.81}$$

Also, show that ξ_1 can be taken to be everywhere positive, with

$$0 < \xi_1(\sigma) \le \text{const. } e^{-\nu_1 \sigma} \quad \text{for } \sigma \ge 0. \tag{10.2.82}$$

(b) Show that ξ_1 and ξ_2 satisfy the relation

$$\xi_1(\sigma) \frac{d\xi_2(\sigma)}{d\sigma} - \xi_2(\sigma) \frac{d\xi_1(\sigma)}{d\sigma} = 2\hat{\nu} \quad \text{for } \sigma \ge 0. \tag{10.2.83}$$

(c) Show that the problem $(10.2.36)_k$ and $(10.2.38)_k$ has a unique solution decaying at infinity, given as

$$X_k^*(\sigma) = \left[\beta_k - X_k(1)\right] \frac{\xi_1(\sigma)}{\xi_1(0)} + \frac{\xi_2(0)}{\xi_1(0)} \int_0^\infty \frac{\xi_1(\sigma)\xi_1(s)}{2\hat{\nu}} P_{k-1}^*(s)\, ds$$

$$- \int_0^\sigma \frac{\xi_1(\sigma)\xi_2(s)}{2\hat{\nu}} P_{k-1}^*(s)\, ds - \int_\sigma^\infty \frac{\xi_2(\sigma)\xi_1(s)}{2\hat{\nu}} P_{k-1}^*(s)\, ds \tag{10.2.84}$$

for $\sigma \ge 0$, provided that P_{k-1}^* is globally integrable. [Note the importance here of the result $\xi_1(0) \ne 0$, which is assured by (10.2.82).]

Hint: For (a), note first that the given assumptions, along with Taylor's theorem, yield

$$H_0[1, X_0(1) + z] = H_{0,x}[1, X_0(1)]z + O(z^2) \quad \text{as } z \to 0. \tag{10.2.85}$$

This result, along with (10.2.9) and (10.2.10), yields $\hat{\nu} \ge \nu_1$, as indicated in (10.2.80). Similarly, Taylor's theorem, along with (10.2.71), yields directly

$$H_{0,x}[1, X_0(1) + X_0^*(\sigma)] = \hat{\nu}^2 + O(e^{-\nu_1 \sigma}) \quad \text{as } \sigma \to \infty. \tag{10.2.86}$$

We also easily see that the derivative of $H_{0,x}[1, X_0(1) + X_0^*(\sigma)]$ with respect to σ is $O[\exp(-\nu_1\sigma)]$ as $\sigma \to \infty$, and so the technique employed in Exercise 9.2.8 yields results here of the form (10.2.80), but with the exponentials $e^{\pm\hat{\nu}\sigma}$ in (10.2.80) replaced by the corresponding exponentials

$$\exp\left[\pm \int_{\sigma_1}^\sigma \{H_{0,x}[1, X_0(1) + X_0^*(s)]\}^{1/2}\, ds\right] \tag{10.2.87}$$

for a sufficiently large, fixed σ_1. From (10.2.86) we find

$$\{H_{0,x}[1, X_0(1) + X_0^*(\sigma)]\}^{1/2} = \hat{\nu} + \mu(\sigma), \tag{10.2.88}$$

with suitable $\mu(\sigma) = O[\exp(-\nu_1\sigma)]$ as $\sigma \to \infty$, so that we have

$$\int_{\sigma_1}^{\sigma}\left\{H_{0,x}\left[1, X_0(1) + X_0^*(s)\right]\right\}^{1/2} ds = \hat{\nu}(\sigma - \sigma_1) + \int_{\sigma_1}^{\infty}\mu(s)\, ds$$

$$- \int_{\sigma}^{\infty}\mu(s)\, ds.$$

Hence, we find the result

$$\exp\left[\left[\pm\int_{\sigma_1}^{\sigma}\left\{H_{0,x}\left[1, X_0(1) + X_0^*(s)\right]\right\}^{1/2} ds\right]\right] = e^{\pm\hat{\nu}\sigma}e^{A}\left[1 + O(e^{-\nu_1\sigma})\right]$$

$$(10.2.89)$$

for $\sigma \geq \sigma_1$ and for a suitable fixed constant A. Hence, up to inessential multiplicative constants, we have the results of (10.2.80) for $\sigma \geq \sigma_1$, and hence also for $\sigma \geq 0$. To obtain (10.2.82), note first the fact (Fife 1973a) that the function $\xi := dX_0^*(\sigma)/d\sigma$ is a solution of (10.2.79), as follows directly on differentiation of (10.2.35). We see then that the previous solution ξ_1 of (10.2.80) must be a constant nontrivial multiple of $\xi = dX_0^*(\sigma)/d\sigma$. The multiplicative constant can be taken so as to achieve (10.2.82).

For (b), note that the left side of (10.2.83) is the wronskian of ξ_1 and ξ_2, which is constant, and hence equal to its limiting value as $\sigma \to \infty$, obtained from (10.2.80)–(10.2.81). Finally, the result (10.2.84) of (c) is obtained by variation of parameters.

Exercise 10.2.4: Derive the estimates of $(10.2.40)_k$–(10.2.41) in the case $k = 1$. *Hint:* Show that the function $P_0^* = P_0^*(\sigma)$ of (10.2.37) satisfies an estimate of the form

$$|P_0^*(\sigma)| \leq \text{const. } e^{-\nu\sigma} \quad \text{for } \sigma \geq 0, \qquad (10.2.90)$$

for any fixed positive number $\nu < \nu_1$. Now take $k = 1$ in (10.2.84), and use (10.2.90) to show that the solution function X_1^* satisfies the stated estimates of $(10.2.40)_1$. It can be shown that each successive function P_k^* satisfies an estimate of the same form as (10.2.90), and we then use (10.2.84) to obtain $(10.2.40)_k$ for each successive k; you need not do this here.

Exercise 10.2.5: Derive the estimates of (10.2.49)–(10.2.50) for the residuals ϕ_j and ρ in the case $N = 0$. *Hint:* For ρ, use (10.2.47), along with

(10.2.7), (10.2.16), (10.2.35), and (10.2.42), and find

$$\rho(t,\varepsilon) = H\big[t, X_0(t) + \varepsilon^* X_0(\tau) + X_0^*(\sigma), \varepsilon\big]$$

$$- \varepsilon^2 \frac{d^2 X_0(t)}{dt^2} - \varepsilon \frac{d^{2*} X_0(\tau)}{d\tau^2} - \frac{d^2 X_0^*(\sigma)}{d\sigma^2}$$

$$= H_0\big[t, X_0(t) + \varepsilon^* X_0(\tau) + X_0^*(\sigma)\big]$$

$$- H_0\big[1, X_0(1) + X_0^*(\sigma)\big] + O(\varepsilon)$$

$$= H_0\big[t, X_0(t) + X_0^*(\sigma)\big] - H_0\big[1, X_0(1) + X_0^*(\sigma)\big] + O(\varepsilon)$$

$$\text{as } \varepsilon \to 0+, \quad (10.2.91)$$

uniformly for $0 \le t \le 1$, where the last equality in (10.2.91) follows directly with Taylor's theorem. Take $f(t,r) := H_0[t, X_0(t) + r]$ in the identity

$$f(t,r) - f(1,r) = \int_1^t \int_0^r \frac{\partial^2}{\partial t_1 \, \partial r_1} f(t_1, r_1) \, dr_1 \, dt_1,$$

any smooth f with $f(t,0) \equiv 0 \left(\Rightarrow \dfrac{\partial f(t,0)}{\partial t} \equiv 0 \right),$ \qquad (10.2.92)

and find

$$H_0\big[t, X_0(t) + r\big] - H_0\big[1, X_0(1) + r\big]$$

$$= \int_1^t \int_0^r \big\{ H_{0,xt}\big[t_1, X_0(t_1) + r_1\big]$$

$$+ H_{0,xx}\big[t_1, X_0(t_1) + r_1\big] X_0'(t_1) \big\} \, dr_1 \, dt_1. \qquad (10.2.93)$$

Finally, take $r := X_0^*(\sigma)$ in (10.2.93), and conclude

$$\big| H_0\big[t, X_0(t) + X_0^*(\sigma)\big] - H_0\big[1, X_0(1) + X_0^*(\sigma)\big] \big|$$

$$\le \text{const.} \, |1 - t| \cdot |X_0^*(\sigma)| = \text{const.} \, \varepsilon |\sigma X_0^*(\sigma)| = O(\varepsilon)$$

$$(10.2.94)$$

as $\varepsilon \to 0+$, where we use $\sigma = (1 - t)/\varepsilon$. The desired result $\rho(t, \varepsilon) = O(\varepsilon)$ then follows directly from (10.2.91) and (10.2.94). The estimates of (10.2.49) for the boundary residuals are straightforward.

Exercise 10.2.6: Given the validity of the estimates of (10.2.56) and (10.2.57), show that the nonlinear integral equation (10.2.54)–(10.2.55) has a unique small solution $\left[\begin{smallmatrix}\hat{x}\\\hat{y}\end{smallmatrix}\right]$ satisfying suitable inequalities of the type (10.2.58). *Hint:* Use the Banach space consisting of all continuous vector-valued functions

$$\begin{bmatrix} \hat{x} \\ \hat{y} \end{bmatrix} = \begin{bmatrix} \hat{x}(t) \\ \hat{y}(t) \end{bmatrix}$$

on $0 \leq t \leq 1$, with norm

$$\left\| \begin{bmatrix} \hat{x} \\ \hat{y} \end{bmatrix} \right\| := \max \Big\{ \max_{0 \leq t \leq 1} |\hat{x}(t)|, \max_{0 \leq t \leq 1} |\hat{y}(t)| \Big\}.$$

Use the estimates (10.2.49), (10.2.50), (10.2.56), and (10.2.57) to show that the vector $\begin{bmatrix} \hat{x}_0 \\ \hat{y}_0 \end{bmatrix}$ of (10.2.55) satisfies

$$\left\| \begin{bmatrix} \hat{x}_0 \\ \hat{y}_0 \end{bmatrix} \right\| \leq \mu_0 \varepsilon^{N+1} \quad \text{as } \varepsilon \to 0+, \tag{10.2.95}$$

for a fixed constant $\mu_0 > 0$. Consider the right side of (10.2.54) as an operator defined for suitable vectors $\begin{bmatrix} \hat{x} \\ \hat{y} \end{bmatrix}$, with value

$$M \begin{bmatrix} \hat{x} \\ \hat{y} \end{bmatrix}(t) := \text{right side of (10.2.54)}, \tag{10.2.96}$$

and show that M maps the ball of radius $2\mu_0 \varepsilon^{N+1}$ centered at the origin into itself for all small enough $\varepsilon > 0$. Similarly, show that M is a contracting map on this same ball, again for all small enough $\varepsilon > 0$. The desired result follows from the Banach/Picard fixed-point theorem.

Exercise 10.2.7 (Fife 1973a): Let H_0 be a given function of class C^1 satisfying the condition (10.2.10) for given numbers $X_0(1)$, β_0, and $\nu_1 > 0$, and assume that $X_0(1)$ satisfies (10.2.8) at $t = 1$. Assume also that

$$\beta_0 \neq X_0(1). \tag{10.2.97}$$

Let δ be any fixed positive number satisfying

$$0 < \delta < \hat{\nu} := \sqrt{H_{0,x}[1, X_0(1)]} \quad (\hat{\nu} \geq \nu_1). \tag{10.2.98}$$

Show that the solution X_0^* of (10.2.70)–(10.2.71) satisfies

$$C_\delta^{-1} e^{-(\hat{\nu}+\delta)\sigma} \leq |X_0^*(\sigma)|, \left| \frac{dX_0^*(\sigma)}{d\sigma} \right| \leq C_\delta e^{-(\hat{\nu}-\delta)\sigma} \quad \text{for } \sigma \geq 0,$$

$$\tag{10.2.99}$$

for a suitable fixed positive constant C_δ. *Hint:* Take

$$\Phi(s) := H_0[1, X_0(1) + sx]$$

in the identity $\Phi(1) = \Phi(0) + \Phi'(0) + \int_0^1 \Phi''(s)(1 - s)\, ds$, and find

$$H_0[1, X_0(1) + x] = H_{0,x}[1, X_0(1)] x$$

$$+ \Big\{ \int_0^1 H_{0,xx}[1, X_0(1) + sx](1 - s)\, ds \Big\} x^2.$$

$$\tag{10.2.100}$$

The function $x = X_0^*(\sigma)$ satisfies (10.2.35), and now (10.2.100) can be used to rewrite the right side of (10.2.35), from which we see that $x = X_0^*(\sigma)$ is a solution of the linear equation

$$\frac{d^2x}{d\sigma^2} = [\hat{\nu}^2 + q(\sigma)]x \quad \text{for } \sigma \geq 0,$$

with $q(\sigma) := \left\{ \int_0^1 H_{0,xx}[1, X_0(1) + sX_0^*(\sigma)](1-s)\,ds \right\} X_0^*(\sigma),$

$$(10.2.101)$$

where $\hat{\nu}$ is defined in (10.2.98). Given any such δ as in (10.2.98), it follows directly from the smoothness of H, along with (10.2.71) and the definition of q in (10.2.101), that

$$0 < (\hat{\nu} - \delta)^2 \leq \hat{\nu}^2 + q(\sigma) \leq (\hat{\nu} + \delta)^2 \quad \text{for } \sigma \geq \sigma_1,$$

$$(10.2.102)$$

for a suitably large, fixed $\sigma_1 > 0$. We see directly also that the derivative of q satisfies $q'(\sigma) = O[\exp(-\nu_1\sigma)]$ for $\sigma \geq 0$, so that the technique of Exercise 9.2.8 can be applied to equation (10.2.101), giving solutions x_1 and x_2 in the form

$$x_1(\sigma) = b_1(\sigma)\exp\left[-\int_{\sigma_1}^{\sigma}(\hat{\nu}^2 + q)^{1/2}\right],$$

$$x_1'(\sigma) = -[\hat{\nu}^2 + q(\sigma)]^{1/2}c_1(\sigma)\exp\left[-\int_{\sigma_1}^{\sigma}(\hat{\nu}^2 + q)^{1/2}\right],$$

$$x_2(\sigma) = b_2(\sigma)\exp\left[+\int_{\sigma_1}^{\sigma}(\hat{\nu}^2 + q)^{1/2}\right],$$

$$x_2'(\sigma) = +[\hat{\nu}^2 + q(\sigma)]^{1/2}c_2(\sigma)\exp\left[+\int_{\sigma_1}^{\sigma}(\hat{\nu}^2 + q)^{1/2}\right],$$

$$(10.2.103)$$

for $\sigma \geq \sigma_1$ and for suitable smooth functions b_j, c_j ($j = 1, 2$) satisfying

$$b_j(\sigma), c_j(\sigma) = 1 + O[\exp(-\nu_1\sigma)] \quad \text{for } \sigma \geq \sigma_1. \quad (10.2.104)$$

In particular, the functions b_j and c_j are positive-valued for all large σ, say for all $\sigma \geq \sigma_2 \geq \sigma_1$, and we find then directly from (10.2.102) and (10.2.103) the results

$$0 < b_1(\sigma)e^{-(\hat{\nu}+\delta)(\sigma-\sigma_1)} \leq x_1(\sigma) \leq b_1(\sigma)e^{-(\hat{\nu}-\delta)(\sigma-\sigma_1)},$$

$$(10.2.105)$$

$$0 < c_1(\sigma)e^{-(\hat{\nu}+\delta)(\sigma-\sigma_1)} \leq \frac{-x_1'(\sigma)}{\sqrt{\hat{\nu}^2 + q}} \leq c_1(\sigma)e^{-(\hat{\nu}-\delta)(\sigma-\sigma_1)},$$

for $\sigma \geq \sigma_2$. We see that the particular solution $x = X_0^*(\sigma)$ of (10.2.101) is a constant multiple of $x_1(\sigma)$ for $\sigma \geq 0$, and the stated result (10.2.99) follows then with (10.2.105), for $\sigma \geq \sigma_2$. We easily extend the result for all $\sigma \geq 0$, using (10.2.72).

10.3 The Green function

This section, which is somewhat technical in nature, may be skimmed over lightly, or skipped, without causing later difficulties. The purpose of the section is to indicate certain direct techniques that can be used to obtain information on a suitable Green function for the linear system [see (10.2.51)]

$$\frac{d}{dt}\begin{bmatrix} x \\ y \end{bmatrix} = \frac{1}{\varepsilon}\begin{bmatrix} 0 & 1 \\ H_x[t, x^N(t,\varepsilon), \varepsilon] & 0 \end{bmatrix}\begin{bmatrix} x \\ y \end{bmatrix}, \qquad (10.3.1)$$

where the circumflexes have been dropped here from $\hat{x}$ and $\hat{y}$ in (10.2.51). The more tedious details are omitted at several points in the following discussion. The techniques illustrated here can be readily applied to other, related problems, but for definiteness we consider only this equation (10.3.1) subject to the boundary conditions of (10.3.5)–(10.3.6).

The given function $H = H(t, x, \varepsilon)$ is assumed to be sufficiently smooth and is assumed to have an asymptotic expansion as in (10.2.7),

$$H(t, x, \varepsilon) \sim H_0(t, x) + \sum_{k=1}^{\infty} H_k(t, x)\varepsilon^k, \qquad (10.3.2)$$

with coefficient functions H_k that are suitably smooth. The given func-. tion $x^N = x^N(t, \varepsilon)$ is the partial sum constructed in Section 10.2 [see (10.2.42)],

$$x^N(t, \varepsilon): = X_0(t) + \varepsilon^* X_0(\tau) + X_0^*(\sigma)$$

$$+ \sum_{k=1}^{N} \left[X_k(t) + \varepsilon^* X_k(\tau) + X_k^*(\sigma) \right]\varepsilon^k, \qquad (10.3.3)$$

with

$$\tau: = \frac{t}{\varepsilon} \quad \text{and} \quad \sigma: = \frac{1-t}{\varepsilon}. \qquad (10.3.4)$$

The given, smooth functions H_0 and X_0 are assumed to satisfy the previous conditions (10.2.8), (10.2.9), and (10.2.10) as in Section 10.2.

Finally, as in Section 10.2, the boundary conditions are taken as [see (10.2.52)–(10.2.53)]

$$L\begin{bmatrix} x(0) \\ y(0) \end{bmatrix} + R\begin{bmatrix} x(1) \\ y(1) \end{bmatrix} = \text{given vector}, \qquad (10.3.5)$$

with boundary matrices L and R given here as

$$L: = \begin{bmatrix} 0 & 1 \\ 0 & 0 \end{bmatrix} \quad \text{and} \quad R: = \begin{bmatrix} 0 & 0 \\ 1 & 0 \end{bmatrix}. \qquad (10.3.6)$$

Solution functions $x = x(t, \varepsilon)$ and $y = y(t, \varepsilon)$ for (10.3.1) and (10.3.5) will often be denoted simply as $x(t)$ and $y(t)$, where we often suppress the dependence on ε so as to lighten the notation.

We could replace y in (10.3.1) with εy, and in this way the system (10.3.1) would be recast as a special case of the linear system considered in Exercise 9.3.10, where a related Green function was obtained subject to certain conditions that for (10.3.1) would amount to the requirements

$$H_x\big[t, x^N(t, \varepsilon), \varepsilon\big] > 0,$$

$$\left| \frac{d}{dt} H_x\big[t, x^N(t, \varepsilon), \varepsilon\big] \right| \leq \text{constant}, \qquad (10.3.7)$$

where these conditions on $H_x \equiv \partial H / \partial x$ would be required to hold uniformly for all t, ε on a region of the form

$$0 \leq t \leq 1, \qquad 0 < \varepsilon \leq \varepsilon_0, \qquad (10.3.8)$$

for some fixed $\varepsilon_0 > 0$. However, for problems of the type considered here, one or both of the conditions of (10.3.7) typically fail within a (Dirichlet-type) boundary layer, where the function $H_x[t, x^N(t, \varepsilon), \varepsilon]$ undergoes a rapid transition, with $dH_x[t, x^N(t, \varepsilon), \varepsilon]/dt$ being of order $1/\varepsilon$ there. Special attention must be given to the boundary-layer "impulses" that may occur in these problems; indeed, inadequate attention to this point has marred more than one work. For example, further attention is required in Yarmish (1975) on the use of Theorem 4.1 in the proof of Theorem 4.2 there. See also the discussion on this point in Smith (1981). These difficulties do not arise with various types of non-Dirichlet boundary conditions. For example, there is no such difficulty in the foregoing problem at the left endpoint $t = 0$, where we have a Neumann boundary condition; see also van Harten (1978b), where semilinear boundary conditions of Neumann and mixed type are considered.

Example 10.3.1: Consider the boundary-value problem

$$\varepsilon^2 \frac{d^2x}{dt^2} = (x^2 - 1)(x^2 - 4) \quad \text{for } 0 < t < 1, \quad \text{with}$$

$$\frac{dx}{dt} = 0 \quad \text{at } t = 0, \quad \text{and} \quad x = 0.5 \quad \text{at } t = 1, \tag{10.3.9}$$

with $H \equiv H_0(x) := (x^2 - 1)(x^2 - 4)$. The conditions of (10.2.8)–(10.2.10) are seen to be satisfied with, for example, X_0 given as

$$X_0(t) = 2 \quad \text{for } 0 \le t \le 1. \tag{10.3.10}$$

The construction given in Section 10.2 yields the corresponding partial sum x^N of (10.3.3). In this case we have

$$x^N(1, \varepsilon) = 0.5 + O(\varepsilon) \quad \text{as } \varepsilon \to 0+, \quad \text{at } t = 1, \tag{10.3.11}$$

and we find directly the result

$$H_x\left[1, x^N(1, \varepsilon), \varepsilon\right] = -\tfrac{9}{2} + O(\varepsilon) \quad \text{at } t = 1, \quad \text{as } \varepsilon \to 0+. \tag{10.3.12}$$

Hence, the first condition of (10.3.7) fails near the right endpoint, and we see also that the other condition of (10.3.7) fails there. Indeed, equation (10.3.1) has a *turning point* here within the given boundary layer. Nevertheless, the conditions of (10.3.7) hold throughout most of the interval $0 \le t \le 1$, as indicated by the graph in Figure 10.4.

Even though such boundary-layer impulses can invalidate (10.3.7) within a boundary layer, a careful analysis shows that such boundary-layer impulses *cannot destroy the validity of the approximation*; that is, the original nonlinear boundary-value problem does in fact have an exact solution for which the partial sum x^N given by (10.3.3) provides a suitable approximation. This result follows from the analysis of Fife (1973a), which is based in part on a variational study of the eigenvalues of the Dirichlet problem for (10.3.1). We give here a different proof based on a direct construction of the appropriate Green function, with the aid of a device of Vasil'eva (1972). The appropriate Green function is constructed as in Exercise 0.1.2 in terms of a suitable fundamental solution $Z = Z(t, \varepsilon)$ for (10.3.1).

Following Vasil'eva, we decompose the interval $[0, 1]$ into subintervals as

$$[0, 1] = [0, t_1(\varepsilon)] \cup [t_1(\varepsilon), 1], \quad \text{with } t_1(\varepsilon) \equiv t_1 := 1 - \frac{\varepsilon}{\nu_1} \ln \frac{1}{\varepsilon}, \tag{10.3.13}$$

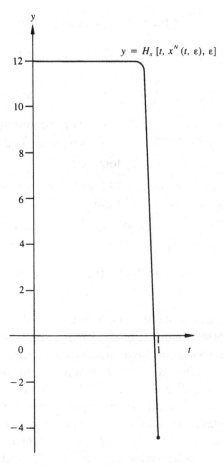

Figure 10.4

where ν_1 is the fixed positive constant appearing in the condition (10.2.10). The conditions of (10.3.7) shall hold on the larger subinterval $[0, t_1]$, and the boundary layer is confined to the thin boundary subinterval $[t_1, 1]$. A suitable fundamental solution Z is constructed on $[0, 1]$ by joining together (suitable combinations of) fundamental solutions $\overline{Z}$ and $\hat{Z}$, where $\overline{Z}$ is a fundamental solution on $[0, t_1]$, and $\hat{Z}$ is a fundamental solution on $[t_1, 1]$, with $t_1 = t_1(\varepsilon)$ as in (10.3.13). The approach here is reminiscent of a technique in the method of matched asymptotic expansions.

In terms of the appropriate boundary-layer variable σ of (10.3.4), we see that the t value $t_1(\varepsilon)$ of (10.3.13) corresponds to the σ value σ_1, given as

$$\sigma_1 \equiv \sigma_1(\varepsilon) = \frac{1}{\nu_1}\ln\frac{1}{\varepsilon}. \tag{10.3.14}$$

Note that $\sigma_1 \to +\infty$ as $\varepsilon \to 0+$.

We shall now see that the conditions of (10.3.7) hold on the t interval $[0, t_1]$.

Lemma 10.3.1: *The function $H_x[t, x^N(t, \varepsilon), \varepsilon]$ is uniformly positive for $0 \le t \le t_1$ (as $\varepsilon \to 0+$). That is, there is a fixed positive constant $\kappa_0 > 0$, such that*

$$H_x[t, x^N(t, \varepsilon), \varepsilon] \ge \kappa_0^2 > 0 \quad \text{as } \varepsilon \to 0+, \quad \text{uniformly for } 0 \le t \le t_1, \tag{10.3.15}$$

where t_1 is given as in (10.3.13). Moreover, the derivative of this function with respect to t is bounded,

$$\frac{d}{dt}H_x[t, x^N(t, \varepsilon), \varepsilon] = O(1) \quad \text{as } \varepsilon \to 0+, \quad \text{uniformly for } 0 \le t \le t_1. \tag{10.3.16}$$

Proof: The boundary-layer variable σ of (10.3.4) satisfies $\sigma \ge \sigma_1(\varepsilon) = (1/\nu_1)\ln(1/\varepsilon)$ for $t \le t_1$, so that the estimates of (10.2.39) imply

$$\left|X_0^*\left(\frac{1-t}{\varepsilon}\right)\right| \le |\beta_0 - X_0(1)|\varepsilon,$$

$$\left|\frac{d}{dt}X_0^*\left(\frac{1-t}{\varepsilon}\right)\right| \le \text{constant}, \quad \text{for } t \in [0, t_1]. \tag{10.3.17}$$

It follows with (10.3.3) and (10.3.17), along with the construction in Section 10.2, that the given function x^N is well approximated by the leading outer term X_0 on $[0, t_1]$, with

$$x^N(t, \varepsilon) = X_0(t) + O(\varepsilon),$$

$$\frac{dx^N(t, \varepsilon)}{dt} = O(1) \quad \text{as } \varepsilon \to 0+, \quad \text{uniformly for } t \in [0, t_1]. \tag{10.3.18}$$

The assumed smoothness of H_0, along with (10.3.3) and Taylor's theorem, yields

$$H_{0,x}[t, x^N(t, \varepsilon)] = H_{0,x}[t, X_0(t)] + O[x^N(t, \varepsilon) - X_0(t)],$$

from which, with (10.3.18),

$$H_{0,x}\big[t, x^N(t, \varepsilon)\big] = H_{0,x}\big[t, X_0(t)\big] + O(\varepsilon)$$

$$\text{as } \varepsilon \to 0+, \text{ uniformly for } t \in [0, t_1]. \quad (10.3.19)$$

The assumption (10.2.9) implies that the quantity $H_{0,x}[t, X_0(t)]$ is uniformly positive for $0 \le t \le 1$,

$$H_{0,x}\big[t, X_0(t)\big] \ge \text{constant} > 0 \quad \text{for } 0 \le t \le 1. \quad (10.3.20)$$

These last two results yield directly

$$H_{0,x}\big[t, x^N(t, \varepsilon)\big] \ge \text{constant} > 0$$

$$\text{as } \varepsilon \to 0+, \quad \text{uniformly for } 0 \le t \le t_1, \quad (10.3.21)$$

where the constant is generally not the same here as in (10.3.20). Finally, the assumed smoothness of H, along with (10.3.21), yields the stated result (10.3.15). The result (10.3.16) follows similarly using (10.3.18) and the smoothness of H. ∎

Using Lemma 10.3.1, we can construct a fundamental solution for the equation (10.3.1) in a convenient form on the interval $[0, t_1]$, as in the next lemma.

Lemma 10.3.2: *Let H and x^N be as before, with the conditions (10.2.8), (10.2.9), and (10.2.10) holding, and let t_1 be given as in (10.3.13). Then, on the interval $0 \le t \le t_1$, the system (10.3.1) has a fundamental solution $\bar{Z} = \bar{Z}(t, \varepsilon)$ of the form*

$$\bar{Z}(t, \varepsilon) = \big[\bar{Z}_{ij}(t, \varepsilon)\big]\big|_{i,j=1,2}$$

$$= \begin{bmatrix} [1 - \bar{T}(t, \varepsilon)]\bar{u}(t, \varepsilon) & \{1 + \bar{S}(t, \varepsilon)[\bar{T}(t, \varepsilon) - 1]\}\bar{v}(t, \varepsilon) \\ -\bar{p}(t, \varepsilon)[1 + \bar{T}(t, \varepsilon)]\bar{u}(t, \varepsilon) & \bar{p}(t, \varepsilon)\{1 + \bar{S}(t, \varepsilon)[\bar{T}(t, \varepsilon) + 1]\}\bar{v}(t, \varepsilon) \end{bmatrix},$$

$$(10.3.22)$$

where the positive-valued function $\bar{p}$ is defined as

$$\bar{p} = \bar{p}(t, \varepsilon) := \big\{ H_x\big[t, x^N(t, \varepsilon), \varepsilon\big] \big\}^{1/2} \quad \text{for } 0 \le t \le t_1,$$

$$(10.3.23)$$

and where the functions $\bar{u}$ and $\bar{v}$ are given as

$$\bar{u}(t, \varepsilon) := \exp\left\{ -\frac{1}{\varepsilon} \int_0^t \left[\bar{p} + \frac{\varepsilon \bar{p}'}{2\bar{p}}(\bar{T} + 1) \right] \right\},$$

$$\bar{v}(t, \varepsilon) := \exp\left\{ -\frac{1}{\varepsilon} \int_t^{t_1} \left[\bar{p} + \frac{\varepsilon \bar{p}'}{2\bar{p}}(\bar{T} - 1) \right] \right\},$$

$$(10.3.24)$$

for suitable fixed functions $\bar{S}$ and $\bar{T}$ satisfying estimates of the type

$$|\bar{S}(t, \varepsilon)|, |\bar{T}(t, \varepsilon)| \leq \text{const. } \varepsilon \quad \text{as } \varepsilon \to 0+, \qquad (10.3.25)$$

uniformly for $0 \leq t \leq t_1$.

Proof: It follows from (10.3.15)–(10.3.16) that the function $\bar{p}$ is well defined by (10.3.23) and satisfies

$$\bar{p}(t, \varepsilon) \geq \kappa_0 > 0, \qquad |\bar{p}'(t, \varepsilon)| \leq \text{constant}, \quad \text{as } \varepsilon \to 0+,$$

$$(10.3.26)$$

uniformly for $0 \leq t \leq t_1$. Hence, as in Exercise 9.3.7, the following transformation is one-to-one:

$$\begin{bmatrix} x \\ y \end{bmatrix} = \begin{bmatrix} 1 & 1 \\ -\bar{p}(t, \varepsilon) & \bar{p}(t, \varepsilon) \end{bmatrix} \begin{bmatrix} \xi \\ \eta \end{bmatrix} \qquad (10.3.27)$$

We now see that (10.3.27) transforms the system (10.3.1) into

$$\varepsilon \frac{d}{dt} \begin{bmatrix} \xi \\ \eta \end{bmatrix} = \left[\begin{bmatrix} -\bar{p} & 0 \\ 0 & \bar{p} \end{bmatrix} + \frac{\varepsilon \bar{p}'}{2\bar{p}} \begin{bmatrix} -1 & 1 \\ 1 & -1 \end{bmatrix} \right] \begin{bmatrix} \xi \\ \eta \end{bmatrix}. \qquad (10.3.28)$$

Finally, the Riccati transformation

$$\begin{bmatrix} \xi \\ \eta \end{bmatrix} = \begin{bmatrix} 1 & -\bar{S}(t, \varepsilon) \\ -\bar{T}(t, \varepsilon) & 1 + \bar{T}(t, \varepsilon)\bar{S}(t, \varepsilon) \end{bmatrix} \begin{bmatrix} \bar{u} \\ \bar{v} \end{bmatrix} \qquad (10.3.29)$$

diagonalizes (10.3.28), yielding the uncoupled system

$$\varepsilon \frac{d\bar{u}}{dt} = -\left[\bar{p} + \frac{\varepsilon \bar{p}'}{2\bar{p}}(\bar{T} + 1) \right] \bar{u},$$

$$\varepsilon \frac{d\bar{v}}{dt} = +\left[\bar{p} + \frac{\varepsilon \bar{p}'}{2\bar{p}}(\bar{T} - 1) \right] \bar{v}, \qquad (10.3.30)$$

provided that $\bar{T}$ and $\bar{S}$ satisfy the system

$$\varepsilon \frac{d\bar{T}}{dt} = 2\bar{p}\bar{T} + \frac{\varepsilon \bar{p}'}{2\bar{p}}(\bar{T}^2 - 1),$$

$$\varepsilon \frac{d\bar{S}}{dt} = -\left(2\bar{p} + \frac{\varepsilon \bar{p}'}{2\bar{p}}\bar{T} \right)\bar{S} - \frac{\varepsilon \bar{p}'}{2\bar{p}}. \qquad (10.3.31)$$

We solve (10.3.31) as usual (see Exercise 10.3.1) for $\bar{T}$ and $\bar{S}$ on $[0, t_1]$ subject to suitable homogeneous boundary conditions, and then (10.3.30) is solved for $\bar{u}$ and $\bar{v}$ subject to the conditions $\bar{u}(0, \varepsilon) = 1$ and $\bar{v}(t_1, \varepsilon) = 1$. We find directly the stated results of the lemma; details are omitted. (Note that we have $\bar{T} \equiv \bar{S} \equiv 0$ in the special case that $\bar{p}$ is independent of t with $\bar{p}' = 0$.) ∎

In the following we assume that

$$\beta_0 \neq X_0(1), \quad \text{with } X_0^*(0) \neq 0. \tag{10.3.32}$$

Indeed, if (10.3.32) fails to hold, so that $X_0^*(0) = 0$, then we have $X_0^*(\sigma) \equiv 0$ [see (10.2.71)], and in this latter case the required analysis is simpler: we can take $t_1 = 1$. We do not consider this simpler case here.

We turn now to a development of a fundamental solution $\hat{Z}$ for (10.3.1) in a convenient form on the boundary subinterval $[t_1, 1]$ in the case (10.3.32). In this case the definition of x^N, along with Taylor's theorem and the definition of $t_1 = t_1(\varepsilon)$, yields directly the following results:

$$x^N(t, \varepsilon) = X_0(1) + X_0^*(\sigma) + O\left(\varepsilon \ln \frac{1}{\varepsilon}\right),$$

$$\frac{d}{dt} x^N(t, \varepsilon) = -\frac{1}{\varepsilon} \frac{dX_0^*(\sigma)}{d\sigma}\bigg|_{\sigma = (1-t)/\varepsilon} + O(1) \quad \text{as } \varepsilon \to 0+,$$

uniformly for $t_1 = 1 - \dfrac{\varepsilon}{\nu_1} \ln \dfrac{1}{\varepsilon} \leq t \leq 1.$ \hfill (10.3.33)

With these results, we find the following lemma.

Lemma 10.3.3: *Let H and x^N be as before, with the conditions (10.2.8), (10.2.9), and (10.2.10) holding, and let t_1 be given as in (10.3.13). Then, on the interval $t_1 \leq t \leq 1$, the system (10.3.1) has a fundamental solution $\hat{Z} = \hat{Z}(t, \varepsilon)$ of the form*

$$\hat{Z}(t, \varepsilon) = \left[\hat{Z}_{ij}(t, \varepsilon)\right]\big|_{i,j=1,2}$$

$$= \begin{bmatrix} b_1(t, \varepsilon)\exp\left[-\hat{\nu}\left(\dfrac{1-t}{\varepsilon}\right)\right] & b_2(t, \varepsilon)\exp\left[+\hat{\nu}\left(\dfrac{1-t}{\varepsilon}\right)\right] \\[2mm] \hat{\nu}c_1(t, \varepsilon)\exp\left[-\hat{\nu}\left(\dfrac{1-t}{\varepsilon}\right)\right] & -\hat{\nu}c_2(t, \varepsilon)\exp\left[+\hat{\nu}\left(\dfrac{1-t}{\varepsilon}\right)\right] \end{bmatrix},$$

\hfill (10.3.34)

where the positive constant $\hat{\nu}$ satisfies

$$\hat{\nu} := \left\{H_{0,x}[1, X_0(1)]\right\}^{1/2} \geq \nu_1, \tag{10.3.35}$$

with ν_1 the constant appearing in the condition (10.2.10), and where $b_j(t, \varepsilon)$ and $c_j(t, \varepsilon)$ ($j = 1, 2$) are smooth functions that are uniformly bounded on the region

$$t_1 \leq t \leq 1, \quad 0 < \varepsilon \leq \varepsilon_1, \tag{10.3.36}$$

for a sufficiently small, fixed constant $\varepsilon_1 > 0$. Moreover, the functions b_j, c_j

satisfy

$$b_j(t, \varepsilon), c_j(t, \varepsilon) = 1 + o(1) \quad \text{as } t \to t_1, \, \varepsilon \to 0+. \tag{10.3.37}$$

In particular, for $t = t_1$,

$$b_j(t_1, \varepsilon), c_j(t_1, \varepsilon) = 1 + O\left(\varepsilon \ln \frac{1}{\varepsilon}\right) \quad \text{as } \varepsilon \to 0+. \tag{10.3.38}$$

Finally, at $t = 1$, *the function* b_1 *is uniformly positive, with*

$$b_1(1, \varepsilon) = \hat{Z}_{11}(1, \varepsilon) \geq \kappa_1 > 0 \quad \text{as } \varepsilon \to 0+, \tag{10.3.39}$$

for a fixed positive constant κ_1.

Proof: First note that the inequality for $\hat{\nu}$ of (10.3.35) has already been proved in Exercise 10.2.3. In order to derive (10.3.34), we rewrite the system (10.3.1) in terms of the boundary-layer variable σ of (10.3.4).

For t on the interval $t_1 \leq t \leq 1$, put

$$\begin{bmatrix} \hat{x}(\sigma, \varepsilon) \\ \hat{y}(\sigma, \varepsilon) \end{bmatrix} := \begin{bmatrix} x(t, \varepsilon) \\ y(t, \varepsilon) \end{bmatrix}\bigg|_{t = 1 - \varepsilon\sigma}, \tag{10.3.40}$$

and then the system (10.3.1) can be rewritten as

$$\frac{d}{d\sigma} \begin{bmatrix} \hat{x} \\ \hat{y} \end{bmatrix} = \begin{bmatrix} 0 & -1 \\ -Q(\sigma, \varepsilon) & 0 \end{bmatrix} \begin{bmatrix} \hat{x} \\ \hat{y} \end{bmatrix} \quad \text{for } 0 \leq \sigma \leq \sigma_1 = \frac{1}{\nu_1} \ln \frac{1}{\varepsilon}, \tag{10.3.41}$$

with Q defined as

$$Q(\sigma, \varepsilon) := H_x[t, x^N(t, \varepsilon), \varepsilon]|_{t = 1 - \varepsilon\sigma} \quad \text{for } 0 \leq \sigma \leq \sigma_1, 0 < \varepsilon \leq \varepsilon_1. \tag{10.3.42}$$

It follows from (10.3.33) and (10.3.42), along with the smoothness of H and Taylor's theorem, that Q satisfies

$$Q(\sigma, \varepsilon) = H_{0, x}[1, X_0(1) + X_0^*(\sigma)] + O\left(\varepsilon \ln \frac{1}{\varepsilon}\right)$$

$$= \hat{\nu}^2 + \left\{ \int_0^1 H_{0, xx}[1, X_0(1) + s X_0^*(\sigma)] \, ds \right\} X_0^*(\sigma) + O\left(\varepsilon \ln \frac{1}{\varepsilon}\right)$$

$$\text{as } \varepsilon \to 0+, \quad \text{uniformly for } 0 \leq \sigma \leq \sigma_1 = \frac{1}{\nu_1} \ln \frac{1}{\varepsilon}, \tag{10.3.43}$$

where $\hat{\nu}$ is given in (10.3.35) and where the second equality here follows by taking $\Phi(s) := H_{0, x}[1, X_0(1) + s X_0^*(\sigma)]$ in the identity $\Phi(1) = \Phi(0) + \int_0^1 \Phi'(s) \, ds$. It follows now from the exponential decay of $X_0^*(\sigma)$ [see (10.2.39)], along with (10.3.43) and the regularity of H, that $Q(\sigma, \varepsilon)$ is uniformly positive for large enough σ and small enough $\varepsilon > 0$. In particular, we see that there are fixed positive constants σ_0 and ε_0 such that

$$Q(\sigma,\varepsilon) \geq \left(\frac{\hat{p}}{2}\right)^2 \quad \text{for } \sigma \geq \sigma_0, 0 < \varepsilon \leq \varepsilon_0. \qquad (10.3.44)$$

This last result permits us to define the positive-valued quantity $\hat{p} = \hat{p}(\sigma,\varepsilon)$ as

$$\hat{p}(\sigma,\varepsilon) := \sqrt{Q(\sigma,\varepsilon)} \quad \text{for } \sigma \geq \sigma_0, 0 < \varepsilon \leq \varepsilon_0, \qquad (10.3.45)$$

where the previous result implies $\hat{p}(\sigma,\varepsilon) \geq \hat{v}/2$ for the stated range of values. Similarly, we show that the derivative of $\hat{p}$ is small, in the sense that (see Exercise 10.3.2)

$$\hat{p}'(\sigma,\varepsilon) \equiv \frac{d\hat{p}(\sigma,\varepsilon)}{d\sigma} = q_1(\sigma,\varepsilon) + q_2(\sigma,\varepsilon) \qquad (10.3.46)$$

for suitable functions q_1 and q_2 satisfying, respectively, $q_1(\sigma,\varepsilon) = O[\exp(-v_1\sigma)]$ and $q_2(\sigma,\varepsilon) = O(\varepsilon)$, both uniformly for $\sigma \geq \sigma_0, 0 < \varepsilon \leq \varepsilon_0$. It follows that we can study the system (10.3.41) using the techniques of Exercise 9.2.8 [see (9.2.55)], and in this way we find the stated results (10.3.34)–(10.3.38). Details are omitted; see Exercise 10.3.3.

It remains to consider the result (10.3.39). For this purpose we employ the comparison system [compare with (10.3.41)–(10.3.43)]

$$\frac{d}{d\sigma}\begin{bmatrix} \xi \\ \eta \end{bmatrix} = \begin{bmatrix} 0 & -1 \\ -Q_0(\sigma) & 0 \end{bmatrix}\begin{bmatrix} \xi \\ \eta \end{bmatrix} \quad \text{for } \sigma \geq 0, \qquad (10.3.47)$$

where Q_0 is defined as

$$Q_0(\sigma) := H_{0,x}\big[1, X_0(1) + X_0^*(\sigma)\big]. \qquad (10.3.48)$$

An exponentially decaying solution of (10.3.47)–(10.3.48) is given as (Fife 1973a)

$$\begin{bmatrix} \xi_1(\sigma) \\ \\ \eta_1(\sigma) \end{bmatrix} := \begin{bmatrix} -\dfrac{dX_0^*(\sigma)}{d\sigma} \\ \\ \dfrac{d^2X_0^*(\sigma)}{d\sigma} \end{bmatrix} \quad \text{for } \sigma \geq 0, \qquad (10.3.49)$$

as follows directly on differentiation of (10.2.35). The solution functions ξ_1, η_1 satisfy [see (10.2.35) and (10.2.39)]

$$|\xi_1(\sigma)|, |\eta_1(\sigma)| \leq \text{const. } e^{-v_1\sigma} \quad \text{for } \sigma \geq 0, \qquad (10.3.50)$$

and, moreover, ξ_1 is everywhere nonzero [see (10.2.72) and (10.3.32)], $|\xi_1(\sigma)| > 0$ for $\sigma \geq 0$. Without loss of generality we can assume that ξ_1 is positive. (Otherwise, replace ξ_1, η_1 with their negatives.) Hence, as in (10.2.99),

$$0 < C_\delta^{-1} e^{-(\hat{\nu}+\delta)\sigma} \leq \xi_1(\sigma) \leq C_\delta e^{-(\hat{\nu}-\delta)\sigma} \quad \text{for } \sigma \geq 0, \quad (10.3.51)$$

for any fixed $0 < \delta < \hat{\nu}$ [assuming always here that (10.3.32) holds].

The system (10.3.41) can be written as the second-order equation

$$\frac{d^2\hat{x}}{d\sigma^2} = Q(\sigma,\varepsilon)\hat{x} = Q_0(\sigma)\hat{x} + [Q(\sigma,\varepsilon) - Q_0(\sigma)]\hat{x}. \quad (10.3.52)$$

We obtain a suitable solution of (10.3.52) by variation of parameters in the form

$$\hat{x}(\sigma,\varepsilon) = w(\sigma,\varepsilon)\xi_1(\sigma) \quad \text{for } 0 \leq \sigma \leq \sigma_1 = \frac{1}{\nu_1}\ln\frac{1}{\varepsilon}, \quad (10.3.53)$$

for a suitable function w subject to the terminal conditions

$$w = 1, \qquad \frac{dw}{d\sigma} = 0 \quad \text{at } \sigma = \sigma_1. \quad (10.3.54)$$

We find for w the differential equation

$$\frac{d^2w}{d\sigma^2} + \left(\frac{2\xi_1'(\sigma)}{\xi_1(\sigma)}\right)\frac{dw}{d\sigma} = [Q(\sigma,\varepsilon) - Q_0(\sigma)]w, \quad (10.3.55)$$

which yields, with (10.3.54), on integration the equivalent integral equation

$$w(\sigma,\varepsilon) = 1 + \int_\sigma^{\sigma_1}\xi_1(t)^{-2}\int_t^{\sigma_1}\xi_1(s)^2[Q(s,\varepsilon) - Q_0(s)]w(s,\varepsilon)\,ds\,dt$$

$$(10.3.56)$$

for $0 \leq \sigma \leq \sigma_1$. The quantity $Q - Q_0$ is small, with [see (10.3.43) and (10.3.48)]

$$Q(\sigma,\varepsilon) - Q_0(\sigma) = O\left(\varepsilon\ln\frac{1}{\varepsilon}\right) \quad \text{as } \varepsilon \to 0+, \quad (10.3.57)$$

uniformly for $0 \leq \sigma \leq \sigma_1$. We then easily find, with (10.3.51), (10.3.57), and the Banach/Picard fixed-point theorem, that (10.3.56) has a unique solution satisfying (see Exercise 10.3.4)

$$\tfrac{1}{2} \leq w(\sigma,\varepsilon) \leq 2 \quad \text{for } 0 \leq \sigma \leq \sigma_1, 0 < \varepsilon \leq \varepsilon_1, \quad (10.3.58)$$

for a fixed $\varepsilon_1 > 0$. This last result, along with (10.3.40), (10.3.51), and (10.3.53), proves that (10.3.1) has a solution x_1, y_1 with

$$0 < \frac{1}{2}\xi_1\left(\frac{1-t}{\varepsilon}\right) \leq x_1(t,\varepsilon) \leq 2\xi_1\left(\frac{1-t}{\varepsilon}\right) \quad \text{as } \varepsilon \to 0+,$$

$$(10.3.59)$$

uniformly for $t_1 \leq t \leq 1$. This solution function x_1 must be a nontrivial constant multiple of $\hat{Z}_{11}(t, \varepsilon)$ in (10.3.34), and the desired result (10.3.39) follows then with (10.3.59). ∎

We now use the solution $\overline{Z}$ given by (10.3.22) on the subinterval $[0, t_1]$, along with the solution $\hat{Z}$ given by (10.3.34) on the subinterval $[t_1, 1]$, so as to form a suitable composite fundamental solution Z on $[0, 1]$. Specifically, we take Z as

$$Z(t, \varepsilon) := \begin{cases} \overline{Z}(t, \varepsilon) & \text{for } t \in [0, t_1], \\ \hat{Z}(t, \varepsilon)\hat{Z}(t_1, \varepsilon)^{-1}\overline{Z}(t_1, \varepsilon) & \text{for } t \in [t_1, 1], \end{cases}$$

$$(10.3.60)$$

where this Z is clearly continuous for $0 \leq t \leq 1$. Moreover, it follows directly from Lemma 10.3.2 and Lemma 10.3.3 that the matrix $\hat{Z}(t_1, \varepsilon)^{-1}\overline{Z}(t_1, \varepsilon)$ multiplying $\hat{Z}(t, \varepsilon)$ in (10.3.60) satisfies

$$\hat{Z}(t_1, \varepsilon)^{-1}\overline{Z}(t_1, \varepsilon)$$

$$= \begin{bmatrix} \varepsilon^{-\hat{p}/\nu_1}\bar{u}(t_1, \varepsilon)O\left(\varepsilon \ln\frac{1}{\varepsilon}\right) & \varepsilon^{-\hat{p}/\nu_1}\left[1 + O\left(\varepsilon \ln\frac{1}{\varepsilon}\right)\right] \\ \varepsilon^{\hat{p}/\nu_1}\bar{u}(t_1, \varepsilon)\left[1 + O\left(\varepsilon \ln\frac{1}{\varepsilon}\right)\right] & \varepsilon^{\hat{p}/\nu_1}O\left(\varepsilon \ln\frac{1}{\varepsilon}\right) \end{bmatrix}$$

$$(10.3.61)$$

as $\varepsilon \to 0+$. (Note that $e^{-\hat{p}\sigma_1} = \varepsilon^{\hat{p}/\nu_1}$.)

The required matrix $M = M(\varepsilon)$ appearing in the Green function representation (10.2.55) is defined as $M(\varepsilon) := LZ(0, \varepsilon) + RZ(1, \varepsilon)$, and in the present case it follows now directly from (10.3.6), (10.3.60), and the foregoing lemmas that

$$M(\varepsilon) = \begin{bmatrix} -\bar{p}(0, \varepsilon)[1 + O(\varepsilon)] & \bar{p}(0, \varepsilon)\bar{v}(0, \varepsilon)[1 + O(\varepsilon)] \\ \dfrac{b_1(1, \varepsilon)\bar{u}(t_1, \varepsilon)}{\varepsilon^{\hat{p}/\nu_1}}O\left(\varepsilon \ln\frac{1}{\varepsilon}\right) & \dfrac{b_1(1, \varepsilon)}{\varepsilon^{\hat{p}/\nu_1}}\left[1 + O\left(\varepsilon \ln\frac{1}{\varepsilon}\right)\right] \end{bmatrix}$$

$$(10.3.62)$$

as $\varepsilon \to 0+$. Hence, we find that M is invertible for all small enough $\varepsilon > 0$, with inverse satisfying [note that $\bar{u}(t_1, \varepsilon)\bar{v}(0, \varepsilon) \sim 0$]

$$M(\varepsilon)^{-1} = \begin{bmatrix} -\dfrac{1}{\bar{p}(0, \varepsilon)}\left[1 + O\left(\varepsilon \ln\frac{1}{\varepsilon}\right)\right] & \dfrac{\varepsilon^{\hat{p}/\nu_1}\bar{v}(0, \varepsilon)}{b_1(1, \varepsilon)}\left[1 + O\left(\varepsilon \ln\frac{1}{\varepsilon}\right)\right] \\ \dfrac{\bar{u}(t_1, \varepsilon)}{\bar{p}(0, \varepsilon)}O\left(\varepsilon \ln\frac{1}{\varepsilon}\right) & \dfrac{\varepsilon^{\hat{p}/\nu_1}}{b_1(1, \varepsilon)}\left[1 + O\left(\varepsilon \ln\frac{1}{\varepsilon}\right)\right] \end{bmatrix}$$

$$(10.3.63)$$

as $\varepsilon \to 0+$.

We find directly from (10.3.22), (10.3.34), (10.3.60), (10.3.61), and (10.3.63) the result

$$|Z(t, \varepsilon) M(\varepsilon)^{-1}| \leq \text{constant}, \quad \text{as } \varepsilon \to 0+, \qquad (10.3.64)$$

uniformly for $0 \leq t \leq 1$.

The Green function for the present problem is defined as [see (0.1.16)]

$$G(t, s, \varepsilon) = \begin{cases} Z(t, \varepsilon) M(\varepsilon)^{-1} L Z(0, \varepsilon) Z(s, \varepsilon)^{-1} & \text{for } s < t, \\ -Z(t, \varepsilon) M(\varepsilon)^{-1} R Z(1, \varepsilon) Z(s, \varepsilon)^{-1} & \text{for } s > t. \end{cases}$$

$$(10.3.65)$$

The foregoing results on Z and M, along with a lengthy, tedious calculation, yield the result

$$|G(t, s, \varepsilon)| \leq \text{const.} \exp\left(-\frac{\nu_0 |t - s|}{\varepsilon}\right) \quad \text{as } \varepsilon \to 0+, \qquad (10.3.66)$$

uniformly for $t, s \in [0, 1]$, for a suitable fixed positive constant ν_0.

Hence, the present approach leads to the previous estimates (10.2.56) and (10.2.57) used in Section 10.2 in the study of the original nonlinear boundary-value problem (10.2.1) and (10.2.4).

Exercises

Exercise 10.3.1: Show that the system (10.3.31) has solution functions $\bar{T}, \bar{S}$ on the interval $[0, t_1]$ satisfying the estimates of (10.3.25). *Hint:* Show that a solution $\bar{T}$ vanishing at $t = t_1$ can be obtained by solving the integral equation

$$\bar{T}(t) = \bar{T}_0(t, \varepsilon) - \int_t^{t_1} \exp\left(-\frac{2}{\varepsilon} \int_t^s \bar{p}\right) \frac{\bar{p}'(s, \varepsilon)}{2\bar{p}(s, \varepsilon)} \bar{T}(s)^2 \, ds,$$

$$(10.3.67a)$$

with

$$\bar{T}_0(t, \varepsilon) := \int_t^{t_1} \exp\left(-\frac{2}{\varepsilon} \int_t^s \bar{p}\right) \frac{\bar{p}'(s, \varepsilon)}{2\bar{p}(s, \varepsilon)} \, ds \quad \text{for } 0 \leq t \leq t_1,$$

$$(10.3.67b)$$

where the dependence of $\bar{T}(t, \varepsilon) = \bar{T}(t)$ on ε is suppressed to lighten the notation. Use the estimates of (10.3.26) to show that $\bar{T}_0$ is of order ε, uniformly on $0 \leq t \leq t_1$, and then use (10.3.26) again, along with the Banach/Picard fixed-point theorem, to study (10.3.67) for $\bar{T}$. Once $\bar{T}$ is in hand, solve the second (linear) equation of (10.3.31) for $\bar{S}$, with $\bar{S} = 0$ at $t = 0$.

Exercise 10.3.2: Show that the function $\hat{p}$ defined by (10.3.45) satisfies (10.3.46) for suitable functions q_1 and q_2 as described there. *Hint:* It suffices to verify the related result for the function Q of (10.3.42). That is, show that Q satisfies

$$Q'(\sigma, \varepsilon) \equiv \frac{dQ(\sigma, \varepsilon)}{d\sigma} = Q_1(\sigma, \varepsilon) + Q_2(\sigma, \varepsilon) \qquad (10.3.68)$$

for suitable functions Q_1 and Q_2 satisfying $Q_1(\sigma, \varepsilon) = O[\exp(-\nu_1\sigma)]$ and $Q_2(\sigma, \varepsilon) = O(\varepsilon)$. These latter results follow from the result [see (10.3.42)]

$$Q'(\sigma, \varepsilon) = -\varepsilon H_{xt}\left[t, x^N(t, \varepsilon), \varepsilon\right]\big|_{t=1-\varepsilon\sigma}$$

$$+ H_{xx}\left[t, x^N(t, \varepsilon), \varepsilon\right]\big|_{t=1-\varepsilon\sigma} \cdot \frac{dx^N}{d\sigma}, \qquad (10.3.69)$$

along with considerations related to those given for (10.3.42)–(10.3.44).

Exercise 10.3.3: Derive the fundamental solution (10.3.34) of Lemma 10.3.3 on $[t_1, 1]$. *Hint:* The successive transformations

$$\begin{bmatrix} \hat{x} \\ \hat{y} \end{bmatrix} = \tfrac{1}{2}\begin{bmatrix} 1 & 1 \\ \hat{p} & -\hat{p} \end{bmatrix}\begin{bmatrix} \hat{\xi} \\ \hat{\eta} \end{bmatrix} \quad [\hat{p} \text{ given by (10.3.45)}] \quad (10.3.70)$$

and

$$\begin{bmatrix} \hat{\xi} \\ \hat{\eta} \end{bmatrix} = \begin{bmatrix} 1 & -\hat{S} \\ -\hat{T} & 1+\hat{T}\hat{S} \end{bmatrix}\begin{bmatrix} \hat{u} \\ \hat{v} \end{bmatrix} \qquad (10.3.71)$$

transform the system (10.3.41) into a suitable diagonal form (for $\sigma \geq \sigma_0$) provided that $\hat{T}$ and $\hat{S}$ satisfy the Riccati system

$$\frac{d\hat{T}}{d\sigma} = 2\hat{p}\hat{T} + \frac{\hat{p}'}{2\hat{p}}(\hat{T}^2 - 1),$$

$$\qquad (10.3.72)$$

$$\frac{d\hat{S}}{d\sigma} = -\left(2\hat{p} + \frac{\hat{p}'\hat{T}}{2\hat{p}}\right)\hat{S} - \frac{\hat{p}'}{2\hat{p}} \quad \text{for } \sigma_0 \leq \sigma \leq \sigma_1 = \frac{1}{\nu_1}\ln\frac{1}{\varepsilon}.$$

Standard techniques can be used (as in Exercises 9.2.8 and 10.3.1) to solve this last system subject to the boundary conditions $\hat{T}(\sigma_1, \varepsilon) = 0$ and $\hat{S}(\sigma_0, \varepsilon) = 0$, and we find solutions

$$\begin{bmatrix} \hat{x}_1 \\ \hat{y}_1 \end{bmatrix} \quad \text{and} \quad \begin{bmatrix} \hat{x}_2 \\ \hat{y}_2 \end{bmatrix}$$

for (10.3.41) in the form

$$\hat{x}_1(\sigma, \varepsilon) = \hat{b}_1(\sigma, \varepsilon)\exp\left(-\int_{\sigma_0}^{\sigma}\hat{p}\right), \quad \hat{y}_1(\sigma, \varepsilon) = -\hat{p}(\sigma, \varepsilon)\hat{c}_1(\sigma, \varepsilon)\exp\left(-\int_{\sigma_0}^{\sigma}\hat{p}\right),$$

$$\hat{x}_2(\sigma, \varepsilon) = \hat{b}_2(\sigma, \varepsilon)\exp\left(+\int_{\sigma_0}^{\sigma}\hat{p}\right), \quad \hat{y}_2(\sigma, \varepsilon) = +\hat{p}(\sigma, \varepsilon)\hat{c}_2(\sigma, \varepsilon)\exp\left(+\int_{\sigma_0}^{\sigma}\hat{p}\right),$$

$$\qquad (10.3.73)$$

for suitable functions $\hat{b}_j, \hat{c}_j$ $(j = 1, 2)$ satisfying

$$\hat{b}_j(\sigma, \varepsilon), \hat{c}_j(\sigma, \varepsilon) = 1 + O\left(\varepsilon \ln \frac{1}{\varepsilon}\right) + O[\exp(-\nu_1 \sigma)]. \quad (10.3.74)$$

The function $\hat{p}$ satisfies [see (10.3.43)] $\hat{p}(\sigma, \varepsilon) = \hat{\nu} + O[\varepsilon \ln(1/\varepsilon)] + O[\exp(-\nu_1 \sigma)]$, and an argument of a type used in Exercise 10.2.3 permits (10.3.73) to be replaced with

$$\hat{x}_1(\sigma, \varepsilon) = \hat{b}_1(\sigma, \varepsilon) e^{-\hat{\nu}\sigma}, \qquad \hat{y}_1(\sigma, \varepsilon) = -\hat{\nu}\hat{c}_1(\sigma, \varepsilon) e^{-\hat{\nu}\sigma},$$
$$\hat{x}_2(\sigma, \varepsilon) = \hat{b}_2(\sigma, \varepsilon) e^{+\hat{\nu}\sigma}, \qquad \hat{y}_2(\sigma, \varepsilon) = +\hat{\nu}\hat{c}_2(\sigma, \varepsilon) e^{+\hat{\nu}\sigma}, \qquad (10.3.75)$$

for *different* functions $\hat{b}_j, \hat{c}_j$ that are not the same here as in (10.3.73), although (10.3.74) still holds. The solutions (10.3.75) can be continued for $0 \le \sigma \le \sigma_0$, and the solutions are uniformly bounded on this latter fixed initial interval. We find the result (10.3.75) on $0 \le \sigma \le \sigma_1$, from which (10.3.34) follows, with (10.3.40). The result (10.3.38) follows from the related result $\hat{b}_j(\sigma_1, \varepsilon), \hat{c}_j(\sigma_1, \varepsilon) = 1 + O[\varepsilon \ln(1/\varepsilon)]$, which follows from (10.3.74).

Exercise 10.3.4: Show that the equation (10.3.56) has a solution satisfying (10.3.58) if (10.3.57) holds. *Hint:* Write (10.3.56) in the form

$$w(\sigma) = 1 + \mathscr{L}w(\sigma), \qquad (10.3.76)$$

with

$$\mathscr{L}w(\sigma) := \int_\sigma^{\sigma_1} \xi_1(t)^{-2} \int_t^{\sigma_1} \xi_1(s)^2 [Q(s, \varepsilon) - Q_0(s)] w(s) \, ds \, dt \qquad (10.3.77)$$

for any continuous function w, where we are suppressing the dependence of $w(\sigma) \equiv w(\sigma, \varepsilon)$ on ε in (10.3.76). Use (10.3.57) and the estimates of (10.3.51), with $\delta := \frac{1}{8}\nu_1$, and find $|\mathscr{L}w(\sigma)| \le \text{const.} \sqrt{\varepsilon} \ln(1/\varepsilon) \cdot \|w\|$ as $\varepsilon \to 0+$, uniformly for $0 \le \sigma \le \sigma_1$, where

$$\|w\| := \max_{0 \le \sigma \le \sigma_1} |w(\sigma)|.$$

Use this last result, along with the Banach/Picard fixed-point theorem, to solve (10.3.76), and obtain the stated results.

Exercise 10.3.5: Show that, for small $\varepsilon \to 0+$, the boundary-value problem

$$\varepsilon^2 \frac{d^2 x}{dt^2} = x^2 - 1 \quad \text{for } 0 < t < 1,$$

$$x = \alpha \quad \text{at } t = 0, \quad \text{and} \quad \frac{dx}{dt} = \beta \quad \text{at } t = 1, \qquad (10.3.78)$$

has a solution $x_1 = x_1(t, \varepsilon)$ of boundary-layer type corresponding to the outer solution $X_0(t) \equiv 1$, provided that the boundary value α satisfies $\alpha > -2$.

10.4 Problems with a single boundary layer

Consider the scalar second-order equation (10.1.17), rewritten here for convenience,

$$\varepsilon \frac{d^2 x}{dt^2} = h(t, x, \varepsilon) + k(t, x, \varepsilon) \frac{dx}{dt} \quad \text{for } 0 \le t \le 1, \quad (10.4.1)$$

for given smooth functions h and k. This equation can be written as a first-order system as

$$\frac{dx}{dt} = y,$$
$$\varepsilon \frac{dy}{dt} = h(t, x, \varepsilon) + k(t, x, \varepsilon) y \quad \text{for } 0 \le t \le 1, \quad (10.4.2)$$

and, indeed, we can also consider more general quasi-linear systems such as

$$\frac{dx}{dt} = f(t, x, \varepsilon) + g(t, x, \varepsilon) y,$$
$$\varepsilon \frac{dy}{dt} = h(t, x, \varepsilon) + k(t, x, \varepsilon) y \quad \text{for } 0 \le t \le 1, \quad (10.4.3)$$

for suitable functions f, g, h, and k [see O'Malley (1980)]. The earlier linear system (9.3.34) is a special case of (10.4.3).

We are content here to consider only (10.4.1), subject to given fixed boundary conditions that for the moment are taken to be of Dirichlet type, as

$$x(0, \varepsilon) = \alpha(\varepsilon) \quad \text{and} \quad x(1, \varepsilon) = \beta(\varepsilon). \quad (10.4.4)$$

This problem (10.4.1) and (10.4.4) has an extensive literature, as indicated in Section 10.1 in the paragraph following (10.1.19). Again, we can consider boundary conditions other than (10.4.4), and indeed we shall do so later in this section.

We assume that the data quantities h, k, α, and β of (10.4.1) and (10.4.4) have asymptotic power-series expansions in ε, as

$$\begin{bmatrix} h(t, x, \varepsilon) \\ k(t, x, \varepsilon) \\ \alpha(\varepsilon) \\ \beta(\varepsilon) \end{bmatrix} \sim \sum_{j=0}^{\infty} \begin{bmatrix} h_j(t, x) \\ k_j(t, x) \\ \alpha_j \\ \beta_j \end{bmatrix} \varepsilon^j, \quad (10.4.5)$$

for given coefficients h_j, k_j, α_j, and β_j. Moreover, h and k, along with h_j and k_j, are assumed to be sufficiently smooth functions of their arguments. As usual, the expansion (10.4.5) need only hold to some specified finite order for a suitable partial sum on the right side of (10.4.5), but this is not of primary interest here.

The reduced equation obtained by putting $\varepsilon = 0$ in (10.4.1) is

$$k_0(t, X_0)\frac{dX_0}{dt} + h_0(t, X_0) = 0 \quad \text{for } 0 < t < 1, \qquad (10.4.6)$$

where h_0 and k_0 are the leading terms in the asymptotic expansions for h and k, and where X_0 shall be the leading term in a suitable outer expansion for x. We seek boundary-layer solutions here for the original problem, with a single boundary layer, and so we shall assume that this reduced, first-order nonlinear differential equation (10.4.6) has a fixed smooth solution X_0 for which [see (10.1.8) and (10.1.19)]

$$k_0[t, X_0(t)] \neq 0 \quad \text{for } 0 \leq t \leq 1. \qquad (10.4.7)$$

Then we seek an exact solution $x = x(t, \varepsilon)$ for (10.4.1) and (10.4.4) corresponding to such a fixed X_0. Because the functions k_0 and X_0 are assumed to be smooth, we see that (10.4.7) implies that $k_0[t, X_0(t)]$ is either everywhere positive, with

$$k_0[t, X_0(t)] > 0 \quad \text{for } 0 \leq t \leq 1, \qquad (10.4.8)$$

or everywhere negative, with

$$k_0[t, X_0(t)] < 0 \quad \text{for } 0 \leq t \leq 1. \qquad (10.4.9)$$

Example 10.4.1: The differential equation

$$\varepsilon\left(\frac{d^2x}{dt^2} + \frac{n-1}{t}\frac{dx}{dt}\right) + x\frac{dx}{dt} = 0 \qquad (10.4.10)$$

has appeared in the literature as a simplified scalar model for the steady-state Navier/Stokes equations [see Lagerstrom (1961), Bush (1971), and Smith (1975b)], where ε corresponds to the kinematic viscosity and n is a given fixed positive integer corresponding to the dimension of the physical space. The differential operator

$$\frac{d^2}{dt^2} + \frac{n-1}{t}\frac{d}{dt}$$

appearing in (10.4.10) is the spherically symmetric part of the n-dimensional Laplacian, so that the independent variable t of (10.4.10) corresponds to a spatial radial coordinate. Equation (10.4.10) can also be interpreted as a mathematical model for the equilibrium temperature distribution in a homogeneous medium with nonlinear heat loss represented by the term $x \cdot dx/dt$ [see Lagerstrom and Casten (1972)].

Equation (10.4.10) is of the type (10.4.1), with

$$h(t, x, \varepsilon) := 0 \quad \text{and} \quad k(t, x, \varepsilon) := -\left(x + \varepsilon \frac{n-1}{t}\right).$$

$$(10.4.11)$$

The reduced equation (10.4.6) is

$$X_0 \frac{dX_0}{dt} = 0 \quad \text{for } t_1 < t < t_2, \qquad (10.4.12)$$

with a one-parameter family of smooth solutions given as

$$X_0(t) \equiv \gamma \quad \text{for } t_1 \le t \le t_2 \qquad (10.4.13)$$

for any constant value of the parameter γ. The function $k_0 = k(t, x, 0)$ is obtained from (10.4.11) and is given as

$$k_0(t, x) = -x. \qquad (10.4.14)$$

Hence, in this case, (10.4.13) yields

$$k_0[t, X_0(t)] = -\gamma, \qquad (10.4.15)$$

so that the function X_0 of (10.4.13) satisfies the condition (10.4.7) for any given *nonzero* constant $\gamma \ne 0$, whereas the condition (10.4.7) fails to hold if $\gamma = 0$.

The reduced equation (10.4.6), subject to the condition (10.4.7), is a first-order differential equation for X_0. Hence, we expect to prescribe a single boundary condition for (10.4.6) chosen from the two given boundary conditions of (10.4.4). In view of the earlier results of Section 8.3 for the corresponding linear problem, we expect that X_0 should satisfy the reduced boundary condition at the left endpoint $[X_0(0) = \alpha_0]$ if (10.4.8) holds, and in this case the boundary layer occurs at the right endpoint. Similarly, we expect that X_0 should satisfy the reduced boundary condition at the right endpoint $[X_0(1) = \beta_0]$ if (10.4.9) holds, and then the boundary layer is at the left endpoint.

Following van Harten (1978a) and Howes and O'Malley (1980) [see also Coddington and Levinson (1952)], we shall assume that the data satisfy precisely one of the two conditions

$$X_0(0) = \alpha_0, \quad \text{with } k_0[t, X_0(t)] > 0 \quad \text{for } 0 \le t \le 1, \quad \text{and}$$

$$x \cdot \left\{ \int_0^1 k_0[1, X_0(1) + sx]\, ds \right\} x \ge \nu |x|^2 \qquad (10.4.16)$$

for all x between 0 and $\beta_0 - X_0(1)$,

or

$$X_0(1) = \beta_0, \quad \text{with } k_0[t, X_0(t)] < 0 \quad \text{for } 0 \le t \le 1, \quad \text{and}$$

$$x \cdot \left\{ \int_0^1 k_0[0, X_0(0) + sx] \, ds \right\} x \le -\nu |x|^2 \tag{10.4.17}$$

$$\text{for all } x \text{ between 0 and } \alpha_0 - X_0(0),$$

where in either case the quantity ν is a fixed positive constant. The stated inequality here involving the integral term and also involving the positive constant ν is required for a solution of boundary-layer type and is analogous to the earlier condition (10.2.10) of Fife (1973a) used in Section 10.2. These conditions are stability conditions, as discussed more generally in Howes (1978) and Chang and Howes (1984).

If the Dirichlet problem (10.4.1) and (10.4.4) is considered on an interval $[t_1, t_2]$ other than the unit interval $[0, 1]$, then the conditions (10.4.16) and (10.4.17) are modified in the obvious way for $t_1 \le t \le t_2$.

Example 10.4.2: Consider the Dirichlet problem for equation (10.4.10) of Example 10.4.1 on the interval $[1, 2]$,

$$\varepsilon \left(\frac{d^2 x}{dt^2} + \frac{n-1}{t} \frac{dx}{dt} \right) + x \frac{dx}{dt} = 0 \quad \text{for } 1 < t < 2,$$

$$x(1, \varepsilon) = \alpha(\varepsilon) \quad \text{and} \quad x(2, \varepsilon) = \beta(\varepsilon), \tag{10.4.18}$$

with α and β having asymptotic expansions as in (10.4.5). We see directly in this case that the outer solution X_0 of (10.4.13) satisfies (10.4.16) if and only if

$$\left. \begin{array}{c} X_0(t) \equiv \alpha_0 < 0, \\ \beta_0 < -\alpha_0 \end{array} \right\} \Leftrightarrow \left\{ \begin{array}{l} \text{Solution of boundary-layer} \\ \text{type exists with BL at right} \\ \text{endpoint.} \end{array} \right.$$

$$(10.4.19)$$

Similarly, we see that (10.4.17) holds in this case if and only if

$$\left. \begin{array}{c} X_0(t) \equiv \beta_0 > 0, \\ \alpha_0 > -\beta_0 \end{array} \right\} \Leftrightarrow \left\{ \begin{array}{l} \text{Solution of boundary-layer} \\ \text{type exists with BL at left} \\ \text{endpoint.} \end{array} \right.$$

$$(10.4.20)$$

Hence, just as in Section 10.2, we see here that the specific numerical values of the given bounday data can play a crucial role in determining the nature of possible solutions.

For the present problem (10.4.18), the situation is indicated in Figure 10.5 for typical values of α_0 and β_0 leading to a solution of boundary-layer

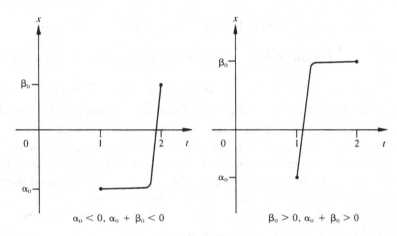

Figure 10.5

type. The boundary layer occurs at the right endpoint if the specified left boundary value α_0 is negative and if the average of the boundary values is also negative, $\frac{1}{2}(\alpha_0 + \beta_0) < 0$, whereas the boundary layer occurs at the left endpoint if the specified right boundary value β_0 is positive and if the average of the boundary values is also positive, $\frac{1}{2}(\alpha_0 + \beta_0) > 0$. Note that the coefficient $k[t, x(t, \varepsilon), \varepsilon]$ multiplying dx/dt in (10.4.10) may have a zero and change sign inside a boundary layer, but such a "boundary-layer turning point" causes no difficulty here. Finally, note that the problem (10.4.18) exhibits a certain structural discontinuity with respect to small changes in the boundary values for nonzero values of α_0 and β_0 satisfying $\alpha_0 + \beta_0 = 0$, $\beta_0 > 0$. Indeed, in this latter case a small random change in the boundary values may yield modified values $\bar{\alpha}_0$ and $\bar{\beta}_0$ satisfying $\bar{\alpha}_0 + \bar{\beta}_0 > 0$ or $\bar{\alpha}_0 + \bar{\beta}_0 < 0$, resulting in a solution of boundary-layer type with the corresponding layer at either the left or right endpoint.

   Following O'Malley (1969a, 1970a, 1974a), we now obtain an appropriate asymptotic expansion for a solution of the original problem (10.4.1) and (10.4.4) in terms of a fixed, given function $X_0 = X_0(t)$ satisfying the reduced equation (10.4.6), along with one of the conditions (10.4.16) or (10.4.17). For definiteness we consider the case (10.4.17), so that the resulting boundary layer occurs at the left endpoint. In this case we seek a representation for a solution in the form

$$x(t, \varepsilon) \sim X(t, \varepsilon) + {}^{*}\!X(\tau, \varepsilon), \qquad (10.4.21)$$

where the boundary-layer variable is [see (8.3.50) for the analogous linear case]

$$\tau := t/\varepsilon, \qquad (10.4.22)$$

and where the functions $X(t, \varepsilon)$ and $*X(\tau, \varepsilon)$ are to be determined in the form of asymptotic expansions as

$$\begin{bmatrix} X(t, \varepsilon) \\ *X(\tau, \varepsilon) \end{bmatrix} \sim \sum_{k=0}^{\infty} \begin{bmatrix} X_k(t) \\ *X_k(\tau) \end{bmatrix} \varepsilon^k, \qquad (10.4.23)$$

with the leading term $X_0(t)$ in the outer solution $X(t, \varepsilon)$ taken to be the given, fixed function satisfying (10.4.6) and (10.4.17), and where the boundary-layer coefficients $*X_k$ are required to satisfy the usual matching conditions

$$\lim_{\tau \to \infty} {}^*X_k(\tau) = 0 \quad \text{for } k = 0, 1, \dots . \qquad (10.4.24)$$

If X_0 were to satisfy (10.4.16) instead of (10.4.17), then the foregoing representation would be replaced with $x(t, \varepsilon) \sim X(t, \varepsilon) + X^*(\sigma, \varepsilon)$, with $\sigma := (1 - t)/\varepsilon$.

We indicate briefly the results for the case (10.4.17), omitting most of the details. The outer solution $X(t, \varepsilon)$ is required to satisfy the full differential equation (10.4.1),

$$\varepsilon \frac{d^2X}{dt^2} = h(t, X, \varepsilon) + k(t, X, \varepsilon) \frac{dX}{dt} \quad \text{for } 0 < t < 1, \quad (10.4.25)$$

along with the specified boundary condition from (10.4.4) at $t = 1$,

$$X(1, \varepsilon) = \beta(\varepsilon). \qquad (10.4.26)$$

We insert into (10.4.25) the respective expansions from (10.4.5) and (10.4.23) for h, k, and X, and we find with (10.4.6) that the higher-order outer coefficients $X_k(t)$ satisfy linear equations of the form [see O'Malley (1974a)]

$$k_0[t, X_0(t)] \frac{dX_k}{dt} + \{ k_{0,x}[t, X_0(t)] X_0'(t) + h_{0,x}[t, X_0(t)] \} X_k$$

$$= P_{k-1}(t) \quad \text{for } 0 < t < 1 \text{ and for } k = 1, 2, \dots, \qquad (10.4.27)_k$$

for suitable functions P_{k-1} that are determined successively in terms of the coefficients X_j for $j = 0, 1, \dots, k - 1$. For example, we have

$$P_0(t) = \frac{d^2X_0(t)}{dt^2} - \left\{ k_1[t, X_0(t)] \frac{dX_0(t)}{dt} + h_1[t, X_0(t)] \right\},$$

$$(10.4.28)$$

and analogous formulas hold for higher-indexed P_k's. Similarly, with

(10.4.26), we find the conditions

$$X_k(1) = \beta_k \quad \text{for } k = 1, 2, \ldots. \tag{10.4.29}_k$$

It follows from (10.4.17) that the coefficient of dX_k/dt is nonzero in (10.4.27)$_k$, and this, along with the assumed smoothness of the data, implies that the linear terminal-value problems (10.4.27)$_k$ and (10.4.29)$_k$ can be solved successively in terms of preceding coefficients.

Turning now to the boundary-layer correction $*X(\tau, \varepsilon)$, we require that $x = X + *X$ satisfy the full equation (10.4.1) along with the boundary condition from (10.4.4) at $t = 0$, from which we find the differential equation

$$\frac{d^2}{d\tau^2}*X(\tau, \varepsilon) = k\big[\varepsilon\tau, X(\varepsilon\tau, \varepsilon) + *X(\tau, \varepsilon), \varepsilon\big]\frac{d}{d\tau}*X(\tau, \varepsilon)$$

$$+ \varepsilon\Bigg[\big\{ h\big[\varepsilon\tau, X(\varepsilon\tau, \varepsilon) + *X(\tau, \varepsilon), \varepsilon\big] - h\big[\varepsilon\tau, X(\varepsilon\tau, \varepsilon), \varepsilon\big] \big\}$$

$$+ \big\{ k\big[\varepsilon\tau, X(\varepsilon\tau, \varepsilon) + *X(\tau, \varepsilon), \varepsilon\big]$$

$$- k\big[\varepsilon\tau, X(\varepsilon\tau, \varepsilon), \varepsilon\big] \big\}\frac{dX}{dt} \Bigg] \tag{10.4.30}$$

for $\tau > 0$, along with the boundary condition

$$X(0, \varepsilon) + *X(0, \varepsilon) = \alpha(\varepsilon) \quad \text{at } \tau = 0. \tag{10.4.31}$$

As usual, we insert into (10.4.30) the appropriate expansions, and we find in this case that $*X_0(\tau)$ satisfies the nonlinear equation

$$\frac{d^2}{d\tau^2}*X_0(\tau) = k_0\big[0, X_0(0) + *X_0(\tau)\big]\frac{d}{d\tau}*X_0(\tau) \quad \text{for } \tau > 0,$$

$$\tag{10.4.32}_0$$

and higher-indexed terms $*X_j$ satisfy linear equations of the form

$$\frac{d^2}{d\tau^2}*X_j = k_0\big[0, X_0(0) + *X_0(\tau)\big]\frac{d}{d\tau}*X_j$$

$$+ k_{0,x}\big[0, X_0(0) + *X_0(\tau)\big]\frac{d*X_0(\tau)}{d\tau}*X_j + *P_{j-1}(\tau)$$

$$\tag{10.4.32}_j$$

for $j = 1, 2, \ldots$, for suitable functions $*P_{j-1} = *P_{j-1}(\tau)$ that are known successively in terms of the coefficients $*X_i$ for $i \leq j - 1$. Similarly, with (10.4.31), we find the conditions

$$*X_j(0) = \alpha_j - X_j(0) \quad \text{for } j = 0, 1, \ldots. \tag{10.4.33}_j$$

Using condition (10.4.17) along with the appropriate matching condition from (10.4.24), it can be shown that the given nonlinear problem for

$^*X_0(\tau)$ has a unique solution satisfying [see O'Malley (1974a)]

$$|{}^*X_0(\tau)| \le |\alpha_0 - X_0(0)|e^{-\nu\tau},$$

$$\left|\frac{d}{d\tau}\,{}^*X_0(\tau)\right| \le \text{const. } |\alpha_0 - X_0(0)|e^{-\nu\tau}, \quad \text{for } \tau \ge 0,$$

$$(10.4.34)$$

where ν is the positive constant appearing in (10.4.17); see Exercise 10.4.1 for an indication of a proof of (10.4.34). Following O'Malley (1969a, 1974a), we can now solve the corresponding linear problems (10.4.24) and (10.4.32)$_j$–(10.4.33)$_j$ for $j \ge 1$ to obtain

$$^*X_j(\tau) = \left[\alpha_j - X_j(0)\right]\exp\left\{\int_0^\tau k_0[0, X_0(0) + {}^*X_0(s)]\,ds\right\}$$

$$- \int_0^\tau \exp\left\{\int_\sigma^\tau k_0[0, X_0(0) + {}^*X_0(s_1)]\,ds_1\right\}\int_\sigma^\infty {}^*P_{j-1}(s_2)\,ds_2\,d\sigma$$

$$(10.4.35)$$

for $\tau \ge 0$ and for $j = 1, 2, \ldots$, with which we find directly estimates of the type

$$|{}^*X_j(\tau)|, \left|\frac{d}{d\tau}\,{}^*X_j(\tau)\right| \le C_j e^{-\nu_1\tau} \quad \text{for } \tau \ge 0, \qquad (10.4.36)$$

and for $j = 1, 2, \ldots$, for suitable constants C_j and for any fixed positive constant ν_1 less than the constant ν of (10.4.17),

$$0 < \nu_1 < \nu. \qquad (10.4.37)$$

Hence, the coefficients in the expansion (10.4.21)–(10.4.24) can be determined recursively for the nonlinear boundary-value problem (10.4.1) and (10.4.4), provided that the given function X_0 satisfies the reduced differential equation (10.4.6) and provided that condition (10.4.17) holds, and provided also that the given functions h and k are sufficiently smooth. As usual, we wish to truncate the resulting asymptotic representation of (10.4.21)–(10.4.23) so as to obtain a proposed approximate solution that can be used to prove that the original problem has an exact solution close to the proposed approximate solution. Hence, we let $x^N(t, \varepsilon)$ denote the Nth partial sum of the expansion (10.4.21)–(10.4.23), given as

$$x^N(t, \varepsilon) := \sum_{j=0}^N \left[X_j(t) + {}^*X_j\!\left(\frac{t}{\varepsilon}\right)\right]\varepsilon^j \qquad (10.4.38)$$

for $0 \le t \le 1$, where the coefficients X_j and $^*X_j$ are constructed as before.

The original problem (10.4.1) and (10.4.4) is now recast as a linearization about x^N. If x denotes a solution of the original problem, then the function $\hat{x}$, defined as

$$\hat{x}(t) \equiv \hat{x}(t, \varepsilon) := x(t, \varepsilon) - x^N(t, \varepsilon), \qquad (10.4.39)$$

is found to satisfy boundary conditions of the form

$$\hat{x}(0, \varepsilon) = \phi_1(\varepsilon) \quad \text{and} \quad \hat{x}(1, \varepsilon) = \phi_2(\varepsilon), \qquad (10.4.40)$$

along with a differential equation of the form

$$\varepsilon \frac{d^2\hat{x}}{dt^2} = k\big[t, x^N(t, \varepsilon), \varepsilon\big] \frac{d\hat{x}}{dt}$$

$$+ \left\{ h_x\big[t, x^N(t, \varepsilon), \varepsilon\big] + k_x\big[t, x^N(t, \varepsilon), \varepsilon\big] \frac{dx^N}{dt} \right\} \hat{x}$$

$$+ \rho(t, \varepsilon) + E\left(t, \hat{x}, \frac{d\hat{x}}{dt}, \varepsilon\right) \quad \text{for } 0 < t < 1, \qquad (10.4.41)$$

for suitable known residuals ϕ_1, ϕ_2, and ρ that satisfy

$$|\phi_1(\varepsilon)|, |\phi_2(\varepsilon)|, \int_0^1 |\rho(t, \varepsilon)| \, dt \le C_N \varepsilon^{N+1} \quad \text{as } \varepsilon \to 0+, \qquad (10.4.42)$$

for a suitable constant C_N. A proof of (10.4.42) is indicated in Exercise 10.4.2 for the case $N = 0$. Note that we are suppressing the obvious dependence of the residuals on N so as to lighten the notation. Finally, the function E in (10.4.41) is a given function of four real variables and can be given as [compare with (10.2.48)]

$$E(t, x, z, \varepsilon) = \int_0^1 \left[\left\{ \frac{d}{ds} k\big[t, x^N(t, \varepsilon) + sx, \varepsilon\big] \right\} z \right.$$

$$+ (1 - s) \frac{d^2}{ds^2} \left\{ h\big[t, x^N(t, \varepsilon) + sx, \varepsilon\big] \right.$$

$$\left. \left. + \frac{dx^N}{dt} k\big[t, x^N(t, \varepsilon) + sx, \varepsilon\big] \right\} \right] ds$$

$$(10.4.43)$$

for any suitable numbers t, x, z, and ε, with E evaluated in (10.4.41) at $x = \hat{x}$ and $z = d\hat{x}/dt$. The derivatives with respect to s in (10.4.43) can be evaluated as

$$\frac{d}{ds} k\big[t, x^N(t, \varepsilon) + sx, \varepsilon\big] = k_x\big[t, x^N(t, \varepsilon) + sx, \varepsilon\big] x,$$

and so forth. We are suppressing the dependence of E on the nonnegative integer N.

The derivative $dx^N(t, \varepsilon)/dt$ is obtained from (10.4.38) and is seen to be of order $1/\varepsilon$, uniformly for $0 \le t \le 1$. Then (10.4.43) leads directly to the estimate

$$|E(t, x, z, \varepsilon)| \leq \text{const.} \left(\frac{1}{\varepsilon} \cdot |x|^2 + |x| \cdot |z| \right)$$

$$\leq \text{const.} \left(\frac{1}{\varepsilon} \cdot |x|^2 + \varepsilon |z|^2 \right) \qquad (10.4.44)$$

for a fixed constant, uniformly for $0 \leq t \leq 1$, uniformly for x on a fixed compact set, uniformly for all z, and uniformly as $\varepsilon \to 0+$.

Certain a priori estimates are derived and used in van Harten (1978*a*) to show that the problem (10.4.40)–(10.4.42) has a solution satisfying $\hat{x} = O(\varepsilon^{N+1})$ if (10.4.17) holds. We shall indicate here an alternative proof based on a direct construction of a suitable Green function, patterned after the approach used in Section 10.3 for related problems in the case $k \equiv 0$. For this purpose we prefer to work with the equivalent first-order system

$$\frac{d}{dt} \begin{bmatrix} \hat{x} \\ \hat{y} \end{bmatrix} = \frac{1}{\varepsilon} \begin{bmatrix} 0 & 1 \\ A(t, \varepsilon) & B(t, \varepsilon) \end{bmatrix} \begin{bmatrix} \hat{x} \\ \hat{y} \end{bmatrix} + \begin{bmatrix} 0 \\ \rho(t, \varepsilon) + E\left(t, \hat{x}, \frac{1}{\varepsilon}\hat{y}, \varepsilon\right) \end{bmatrix}$$

$$(10.4.45)$$

for $0 \leq t \leq 1$, subject to the boundary conditions

$$L \begin{bmatrix} \hat{x}(0, \varepsilon) \\ \hat{y}(0, \varepsilon) \end{bmatrix} + R \begin{bmatrix} \hat{x}(1, \varepsilon) \\ \hat{y}(1, \varepsilon) \end{bmatrix} = \begin{bmatrix} \phi_1(\varepsilon) \\ \phi_2(\varepsilon) \end{bmatrix}, \qquad (10.4.46)$$

where the coefficients A and B in (10.4.45) are defined as

$$A(t, \varepsilon) := \varepsilon \left\{ h_x[t, x^N(t, \varepsilon), \varepsilon] + k_x[t, x^N(t, \varepsilon), \varepsilon] \frac{dx^N(t, \varepsilon)}{dt} \right\},$$

$$B(t, \varepsilon) := k[t, x^N(t, \varepsilon), \varepsilon] \quad \text{for } 0 \leq t \leq 1, 0 < \varepsilon \leq \varepsilon_0, \quad (10.4.47)$$

for a suitably small, fixed $\varepsilon_0 > 0$, and where the given boundary matrices L and R in (10.4.46) are defined as

$$L := \begin{bmatrix} 1 & 0 \\ 0 & 0 \end{bmatrix} \quad \text{and} \quad R := \begin{bmatrix} 0 & 0 \\ 1 & 0 \end{bmatrix}. \qquad (10.4.48)$$

The methods of Section 10.3 can be used to show that (for small enough $\varepsilon > 0$) the linearized homogeneous operator of (10.4.45) (corresponding to $\rho \equiv 0$ and $E \equiv 0$) has a Green function $G = G(t, s, \varepsilon)$ corresponding to the boundary conditions of (10.4.46) and (10.4.48); see Smith (1984*c*). This Green function can then be used to rewrite the boundary-value problem (10.4.40)–(10.4.41) as the equivalent nonlinear integral equation

$$\begin{bmatrix} \hat{x}(t) \\ \hat{y}(t) \end{bmatrix} = \begin{bmatrix} \hat{x}_0(t,\varepsilon) \\ \hat{y}_0(t,\varepsilon) \end{bmatrix} + \int_0^1 G(t,s,\varepsilon) \begin{bmatrix} 0 \\ E\left[s, \hat{x}(s), \frac{1}{\varepsilon}\hat{y}(s), \varepsilon\right] \end{bmatrix} ds,$$

$$(10.4.49)$$

with

$$\begin{bmatrix} \hat{x}_0(t,\varepsilon) \\ \hat{y}_0(t,\varepsilon) \end{bmatrix} := Z(t,\varepsilon)M(\varepsilon)^{-1} \begin{bmatrix} \phi_1(\varepsilon) \\ \phi_2(\varepsilon) \end{bmatrix} + \int_0^1 G(t,s,\varepsilon) \begin{bmatrix} 0 \\ \rho(s,\varepsilon) \end{bmatrix} ds,$$

$$(10.4.50)$$

where the dependence of $\hat{x}$ and $\hat{y}$ on ε and N is suppressed in (10.4.49), and where $Z = Z(t,\varepsilon)$ is a suitable fundamental solution for the homogeneous system (10.4.45) with $E \equiv \rho \equiv 0$. The matrix $M = M(\varepsilon)$ is defined as usual as $M(\varepsilon) := LZ(0,\varepsilon) + RZ(1,\varepsilon)$.

The required fundamental solution Z and the Green function G are given by a direct construction in Smith (1984c), and we have the estimates

$$|Z(t,\varepsilon)M(\varepsilon)^{-1}| \leq \text{constant}, \quad \text{for } 0 \leq t \leq 1,$$
$$|G(t,s,\varepsilon)| \leq \text{constant}, \quad \text{for } 0 \leq t, s \leq 1,$$

$$(10.4.51)$$

as $\varepsilon \to 0+$.

It follows now directly from the estimates (10.4.42), (10.4.44), and (10.4.51), along with (10.4.43) and a routine application of the Banach/Picard fixed-point theorem, that for $N \geq 1$, the integral equation (10.4.49)–(10.4.50) has a solution $\hat{x}(t) = \hat{x}(t,\varepsilon)$ and $\hat{y}(t) = \hat{y}(t,\varepsilon)$ satisfying estimates of the type

$$|\hat{x}(t,\varepsilon)|, |\hat{y}(t,\varepsilon)| \leq \text{const. } \varepsilon^{N+1} \quad \text{as } \varepsilon \to 0+, \qquad (10.4.52)$$

uniformly for $0 \leq t \leq 1$. The solution functions are uniquely determined subject to (10.4.52); that is, there are no other solution functions for (10.4.49)–(10.4.50) satisfying (10.4.52).

In terms of the original boundary-value problem (10.4.1) and (10.4.4), we then have the following result.

Theorem 10.4.1: *Let the data* h, k, α, *and* β *of (10.4.1) and (10.4.4) possess asymptotic expansions as in (10.4.5), and assume that h, k and the coefficient functions h_j, k_j are sufficiently smooth. Let $X_0 = X_0(t)$ be a fixed, given smooth solution of the reduced equation (10.4.6), and assume that (10.4.17) holds. Then there is a fixed number $\varepsilon_0 > 0$ such that* *the function* $x^N = x^N(t, \varepsilon)$ *of* (10.4.38) *is well defined by the foregoing O'Malley construction on the region*

$$0 \leq t \leq 1, \qquad 0 < \varepsilon \leq \varepsilon_0. \tag{10.4.53}$$

Moreover, the problem (10.4.1) *and* (10.4.4) *has an exact solution* $x = x(t, \varepsilon)$ *close to* x^N *on the region* (10.4.53), *and*

$$|x(t, \varepsilon) - x^N(t, \varepsilon)| \leq \text{const. } \varepsilon^{N+1},$$

$$\left| \frac{dx(t, \varepsilon)}{dt} - \frac{dx^N(t, \varepsilon)}{dt} \right| \leq \text{const. } \varepsilon^N, \tag{10.4.54}$$

uniformly on (10.4.53). *The particular exact solution so constructed is unique subject to* (10.4.54).

Proof: The stated results follow directly from the foregoing discussion in the case $N \geq 1$, and then the case $N = 0$ follows directly also as usual, with $x = x^1 + O(\varepsilon^2)$ and $x^1 = x^0 + O(\varepsilon)$. ■

The theorem remains true, with the obvious modifications, if condition (10.4.17) is replaced by (10.4.16). Theorem 10.4.1 has a long history, as discussed earlier in the paragraph following (10.1.19). See Levinson (1950), Eckhaus (1973, 1979), and van Harten (1978a) for related results for elliptic partial differential equations.

Example 10.4.3: Consider the problem

$$\varepsilon \frac{d^2 x}{dt^2} = x \frac{dx}{dt} - x^3 \quad \text{for } 0 < t < 1,$$

$$x(0, \varepsilon) = \tfrac{1}{2} \quad \text{and} \quad x(1, \varepsilon) = -\tfrac{1}{2}. \tag{10.4.55}$$

Condition (10.4.16) is easily seen to be satisfied by the solution X_0 of the reduced differential equation given as

$$X_0(t) := \frac{1}{2 - t}, \tag{10.4.56}$$

with $\nu = \tfrac{1}{4}$ in (10.4.16). Hence, for all small enough $\varepsilon > 0$, the problem (10.4.55) has an exact solution $x_1 = x_1(t, \varepsilon)$ satisfying

$$\lim_{\substack{\varepsilon \to 0+ \\ \text{fixed } 0 < t < 1}} x_1(t, \varepsilon) = \frac{1}{2 - t}. \tag{10.4.57}$$

Similarly, condition (10.4.17) is seen to be satisfied by the solution X_0 of the reduced equation given as

$$X_0(t) := -\frac{1}{1 + t}, \tag{10.4.58}$$

again with $\nu = \tfrac{1}{4}$. Hence, for all small enough $\varepsilon > 0$, the problem

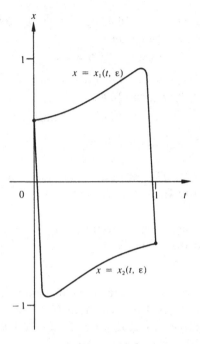

Figure 10.6

(10.4.55) also has an exact solution $x_2 = x_2(t, \varepsilon)$ satisfying

$$\lim_{\substack{\varepsilon \to 0+ \\ \text{fixed } 0 < t < 1}} x_2(t, \varepsilon) = -\frac{1}{1+t}. \tag{10.4.59}$$

It follows that the problem (10.4.55) has two solutions of boundary-layer type as considered here, one with a boundary layer at $t = 0$, and one with the layer at $t = 1$, as indicated in Figure 10.6.

Note that the condition (10.4.16) or (10.4.17) of Coddington and Levinson cannot be further weakened significantly for solutions of the type considered here, because the condition is required for the existence of the proposed approximate solution of boundary-layer type.

We can consider boundary conditions other than the Dirichlet conditions of (10.4.4). For example, the differential equation (10.4.1) (and also coupled systems of such equations) appears as a mathematical model for certain chemical and biochemical flow reactors [see the discussion following (10.1.26) in Section 10.1], subject to various boundary conditions of the type

$$L \begin{bmatrix} x(0, \varepsilon) \\ \dfrac{dx(0, \varepsilon)}{dt} \end{bmatrix} + R \begin{bmatrix} x(1, \varepsilon) \\ \dfrac{dx(1, \varepsilon)}{dt} \end{bmatrix} = \begin{bmatrix} \gamma \\ \delta \end{bmatrix} \tag{10.4.60}$$

for suitable given boundary matrices L and R, and for given constants γ and δ. As usual, there is no difficulty in permitting L, R, γ, and δ to depend on ε with suitable asymptotic expansions, but we shall not do so here. In fact, we shall only discuss briefly the case in which L and R are given as

$$L := \begin{bmatrix} -a & 1 \\ 0 & 0 \end{bmatrix}, \qquad R := \begin{bmatrix} 0 & 0 \\ b & 1 \end{bmatrix}, \qquad (10.4.61)$$

for given constants a and b, so that the boundary conditions become

$$\frac{dx(0, \varepsilon)}{dt} - ax(0, \varepsilon) = \gamma,$$
$$\frac{dx(1, \varepsilon)}{dt} + bx(1, \varepsilon) = \delta. \qquad (10.4.62)$$

The problem (10.1.20)–(10.1.21) of Example 10.1.2 is of this latter type, with

$$a = 1, \qquad b = \tfrac{1}{3}, \qquad \gamma = 1, \qquad \delta = 1, \qquad (10.4.63)$$

and the functions h and k are given there as

$$h(t, x, \varepsilon) = (x - 2)^2, \qquad (10.4.64)$$
$$k(t, x, \varepsilon) = -1 \qquad \text{for all } t, x, \varepsilon.$$

Following O'Malley (1974a), we take as given a fixed, smooth solution X_0 of the reduced differential equation (10.4.6) that satisfies precisely one of the two conditions

$$X_0'(0) - aX_0(0) = \gamma, \quad \text{with } k_0[t, X_0(t)] > 0 \quad \text{for } 0 \le t \le 1,$$
$$\text{and } \mathscr{L}(\alpha_0) = 0 \quad \text{for } \alpha_0 := X_0(0), \qquad (10.4.65)$$

or

$$X_0'(1) + bX_0(1) = \delta, \quad \text{with } k_0[t, X_0(t)] < 0 \quad \text{for } 0 \le t \le 1,$$
$$\text{and } \mathscr{R}(\beta_0) = 0 \quad \text{for } \beta_0 := X_0(1), \qquad (10.4.66)$$

where the quantities $\mathscr{L}$ and $\mathscr{R}$ are defined as

$$\mathscr{L}(\alpha) := k_0(0, \alpha)(\gamma + a\alpha) + h_0(0, \alpha),$$
$$\mathscr{R}(\beta) := k_0(1, \beta)(\delta - b\beta) + h_0(1, \beta), \qquad (10.4.67)$$

for any suitable numbers α and β. If X_0 satisfies (10.4.65), then a boundary layer occurs at the right endpoint $t = 1$, but if X_0 satisfies (10.4.66), the boundary layer occurs at $t = 0$.

For definiteness we consider the case in which (10.4.66) holds, and then the simplest situation occurs if $\beta_0 := X_0(1)$ is a *simple* root of

$\mathscr{R}(\beta)$, with

$$\mathscr{R}(\beta_0) = 0 \quad \text{and} \quad \mathscr{R}'(\beta_0) \neq 0, \quad \text{for } \beta_0 = X_0(1). \quad (10.4.68)$$

In this case there is a corresponding exact solution for the given problem (10.4.1) and (10.4.62) that can be represented asymptotically as

$$x(t, \varepsilon) \sim X(t, \varepsilon) + \varepsilon^* X(\tau, \varepsilon), \quad \text{with } \tau := t/\varepsilon, \quad (10.4.69)$$

where the outer solution $X(t, \varepsilon)$ and the boundary-layer term $^*X(\tau, \varepsilon)$ possess expansions as in (10.4.23). Note that the explicit factor of ε that multiplies $^*X$ in (10.4.69) reflects the relatively weak boundary-layer effect due to the type of boundary condition here, similar to related situations in Exercise 9.3.11 and in Section 10.2.

A straightforward analysis of the sort used earlier for the Dirichlet problem (10.4.1) and (10.4.4) permits a recursive determination of the required coefficients in the expansions (10.4.23) [see O'Malley (1972b, 1974a)], and then the original problem (10.4.1) and (10.4.62) can be linearized about the resulting truncated expansion

$$x^N(t, \varepsilon) := \sum_{j=0}^{N} \left[X_j(t) + {}^*X_{j-1}\left(\frac{t}{\varepsilon}\right) \right] \varepsilon^j, \quad {}^*X_{-1} \equiv 0.$$

$$(10.4.70)$$

If x denotes a solution of the original problem, then the function $\hat{x}$ defined by (10.4.39), along with the auxiliary function $\hat{y} := \varepsilon\, d\hat{x}/dt$, is seen to satisfy the problem (10.4.45)–(10.4.47), with suitable residuals that satisfy estimates related to (10.4.42), and with boundary matrices L and R given in this case as

$$L = L(\varepsilon) := \begin{bmatrix} -\varepsilon a & 1 \\ 0 & 0 \end{bmatrix} \quad \text{and} \quad R = R(\varepsilon) := \begin{bmatrix} 0 & 0 \\ \varepsilon b & 1 \end{bmatrix}.$$

$$(10.4.71)$$

The details are indicated in Exercise 10.4.3 in the case $N = 1$.

The appropriate Green function is easily constructed (see Exercises 10.4.4 and 10.4.5), and the problem can be written then as an integral equation of the type (10.4.49) that permits a direct proof of existence of an exact solution x that is well approximated by the given approximate solution x^N. See O'Malley (1972b, 1974a) for a complete statement and proof of the results. [O'Malley gives a different proof that does not involve the Green function.]

Example 10.4.4: In the case of the example (10.4.63)–(10.4.64), we see that the function $\mathscr{R}$ is given as

$$\mathscr{R}(\beta) = \beta^2 - \tfrac{11}{3}\beta + 3, \quad (10.4.72)$$

with two simple roots β_1 and β_2 given as [see (10.1.26)]

$$\beta_1 = \frac{11 - \sqrt{13}}{6} \quad \text{and} \quad \beta_2 = \frac{11 + \sqrt{13}}{6}. \qquad (10.4.73)$$

In this case there are two different solutions X_0 of the reduced equation (10.4.6) satisfying (10.4.66) and (10.4.68), one for each of these two simple roots β_1 and β_2. Hence, we find two distinct solutions of boundary-layer type for the given boundary-value problem, as indicated further in Exercise 10.4.6.

Suppose now that the function $\mathcal{R} = \mathcal{R}(\beta)$ has a (nonsimple) root β_0 of multiplicity greater than 1, such that the reduced differential equation (10.4.6) has a smooth solution X_0 satisfying $X_0(1) = \beta_0$. In this case it is shown in O'Malley (1972b) that the appropriate asymptotic expansion for a proposed solution takes a somewhat different form than (10.4.69); this point is related to the point discussed earlier in the remarks following (9.3.56). Details are omitted here.

Note that a given boundary-value problem of the type considered here for (10.4.62) may have many solutions of boundary-layer type, possibly including one or more solutions with boundary layers at $t = 1$ for which (10.4.65) hold, and also possibly including one or more solutions with boundary layers at $t = 0$ for which (10.4.66) hold.

Exercises

Exercise 10.4.1: The second-order equation (10.4.32)$_0$ can be integrated once with the appropriate matching condition of (10.4.24) to yield

$$\frac{d}{d\tau} {}^*X(\tau) = \int_0^{{}^*X(\tau)} k_0 \left[0, X_0(0) + z \right] dz. \qquad (10.4.74)$$

Show that this first-order equation (10.4.74) subject to the initial condition (10.4.33)$_0$ has a unique solution satisfying the inequalities of (10.4.34) if the condition (10.4.17) holds. *Hint*: The assumed integral inequality from (10.4.17) leads, with (10.4.74), to a differential inequality that can be integrated by a Gronwall-type argument. [You may wish to consider separately the cases $X_0(0) = \alpha_0$, $X_0(0) > \alpha_0$, and $X_0(0) < \alpha_0$.] Once the first inequality of (10.4.34) is known to be valid, then the second inequality can be shown to follow with (10.4.74).

Exercise 10.4.2: In the case $N = 0$, show that the function $\hat{x}$ of (10.4.39) satisfies the boundary conditions (10.4.40) and the differential equation

(10.4.41), with residuals that satisfy the inequalities (10.4.42). *Hint*: From
(10.4.40), derive the results $\phi_1(\varepsilon) = \alpha(\varepsilon) - \alpha_0$ and $\phi_2(\varepsilon) = \beta(\varepsilon) - \beta_0$,
with which the inequalities for ϕ_1 and ϕ_2 follow directly. Now let ρ be
defined as

$$\rho(t, \varepsilon) := -\varepsilon \frac{d^2 x^N(t, \varepsilon)}{dt^2} + h[t, x^N(t, \varepsilon), \varepsilon] + k[t, x^N(t, \varepsilon), \varepsilon]\frac{dx^N}{dt},$$

$$(10.4.75)$$

and use (10.4.6) and $(10.4.32)_0$ in the case $N = 0$ to obtain

$$\rho(t, \varepsilon) = O(\varepsilon) + O\left(\frac{t}{\varepsilon} e^{-\nu t/\varepsilon}\right) \quad \text{as } \varepsilon \to 0+, \qquad (10.4.76)$$

uniformly for $0 \le t \le 1$, from which the inequality for ρ of (10.4.42)
follows.

Exercise 10.4.3: Consider the boundary-value problem (10.4.1) and
(10.4.62), and let X_0 be a given, fixed smooth solution of the reduced
differential equation (10.4.6) satisfying (10.4.66) and (10.4.68). The next
outer coefficient X_1 is taken to satisfy the terminal-value problem

$$k_0[t, X_0(t)] X_1' + \{h_{0, x}[t, X_0(t)] + k_{0, x}[t, X_0(t)] X_0'(t)\} X_1$$
$$= X_0''(t) - [h_1(t, X_0) + k_1(t, X_0) X_0'] \quad \text{for } 0 \le t \le 1,$$
$$\text{and } X_1'(1) + bX_1(1) = 0, \quad (10.4.77)$$

and the leading coefficient $*X_0(\tau)$ in the boundary-layer term is taken to
satisfy the problem

$$\frac{d^2}{d\tau^2} *X_0 = k_0[0, X_0(0)] \frac{d}{d\tau} *X_0 \quad \text{for } \tau > 0,$$

$$(10.4.78)$$

$$*X_0'(0) = -X_0'(0) + aX_0(0) + \gamma, \quad \text{and} \quad *X_0(\infty) = 0.$$

(a) Show that the value $\beta_1 := X_1(1)$ satisfies

$$\mathcal{R}'[X_0(1)] \beta_1 = X_0''(1) - \{h_1[1, X_0(1)] + k_1[1, X_0(1)] X_0'(1)\},$$

$$(10.4.79)$$

so that (10.4.68) implies that β_1 is well determined. It follows then that
X_1 is also uniquely determined (in terms of the given X_0).

(b) Show that $*X_0$ is uniquely determined by (10.4.78). Give an
explicit expression for $*X_0(\tau)$, and verify that

$$|*X_0(\tau)| \le \text{const.} \, e^{-\nu \tau} \quad \text{for } \tau \ge 0, \quad \text{with } \nu := -k_0[0, X_0(0)] > 0.$$

$$(10.4.80)$$

(c) Let $x^1(t) = x^1(t, \varepsilon)$ be defined by (10.4.70) in the case $N = 1$, and let $\hat{x} := x - x^1$, with $\hat{y} := \varepsilon \, d\hat{x}/dt$. Show that $\hat{x}, \hat{y}$ satisfy (10.4.45)–(10.4.47), with boundary matrices L and R given by (10.4.71), and with residuals satisfying

$$\phi_1(\varepsilon) := \varepsilon\left[\gamma + ax^1(0, \varepsilon) - \frac{dx^1(0, \varepsilon)}{dt}\right] = O(\varepsilon^2),$$

$$\phi_2(\varepsilon) := \varepsilon\left[\delta - bx^1(1, \varepsilon) - \frac{dx^1(1, \varepsilon)}{dt}\right] = O(\varepsilon^3), \quad (10.4.81)$$

$$\rho(t, \varepsilon) := -\varepsilon\frac{d^2x^1}{dt^2} + h(t, x^1, \varepsilon) + k(t, x^1, \varepsilon)\frac{dx^1}{dt}$$

$$= O(\varepsilon^2) + O(te^{-\nu t/\varepsilon}),$$

as $\varepsilon \to 0+$, where the estimate for $\rho(t, \varepsilon)$ holds uniformly for $0 \le t \le 1$.

Exercise 10.4.4: Let X_0 be as in Exercise 10.4.3 for the problem (10.4.1) and (10.4.62), subject to the conditions (10.4.66) and (10.4.68). Let x^N be the proposed approximate solution (10.4.70), and consider the corresponding problem (10.4.45)–(10.4.47) for $\hat{x} = x - x^N$. Show that (for small enough $\varepsilon > 0$) the linearized homogeneous operator of (10.4.45) (with $\rho \equiv 0$ and $E \equiv 0$) has a fundamental solution $Z = Z(t, \varepsilon)$ of the form

$$Z(t, \varepsilon) = \left[Z_{ij}(t, \varepsilon)\right]|_{i, j=1,2}$$

$$= \begin{bmatrix} \exp\left(-\int_0^t T\right) & -S(t, \varepsilon)\exp\left[\dfrac{1}{\varepsilon}\int_0^t(B + \varepsilon T)\right] \\ -\varepsilon T(t, \varepsilon)\exp\left(-\int_0^t T\right) & [1 + \varepsilon T(t, \varepsilon)S(t, \varepsilon)]\exp\left[\dfrac{1}{\varepsilon}\int_0^t(B + \varepsilon T)\right] \end{bmatrix}$$

$$(10.4.82)$$

for suitable smooth functions T and S satisfying

$$T(t, \varepsilon), S(t, \varepsilon) = O(1) \quad \text{as } \varepsilon \to 0+, \quad \text{uniformly for } 0 \le t \le 1,$$

$$(10.4.83)$$

and satisfying also

$$T(0, \varepsilon) = 0, \quad S(1, \varepsilon) = 0, \tag{10.4.84}$$

$$T(1, \varepsilon) = b + \frac{\mathcal{R}'[X_0(1)]}{k_0[1, X_0(1)]} + O(\varepsilon) \quad \text{as } \varepsilon \to 0+,$$

where $\mathcal{R}$ is given by (10.4.67). Moreover, show that the function B

appearing in (10.4.82), given by (10.4.47), is uniformly negative-valued, with

$$B(t, \varepsilon) \le -\nu_0 < 0 \quad \text{as } \varepsilon \to 0+, \quad \text{uniformly for } 0 \le t \le 1,$$

$$(10.4.85)$$

for a fixed positive constant ν_0.

Hint: Show that the Riccati transformation

$$\begin{bmatrix} x \\ y \end{bmatrix} = \begin{bmatrix} 1 & -S \\ -\varepsilon T & 1 + \varepsilon TS \end{bmatrix} \begin{bmatrix} u \\ v \end{bmatrix}$$

transforms the homogeneous system

$$\frac{d}{dt} \begin{bmatrix} x \\ y \end{bmatrix} = \frac{1}{\varepsilon} \begin{bmatrix} 0 & 1 \\ A(t, \varepsilon) & B(t, \varepsilon) \end{bmatrix} \begin{bmatrix} x \\ y \end{bmatrix}$$

into the diagonal form

$$\frac{d}{dt} \begin{bmatrix} u \\ v \end{bmatrix} = \begin{bmatrix} -T(t, \varepsilon) & 0 \\ 0 & \frac{1}{\varepsilon}[B(t, \varepsilon) + \varepsilon T(t, \varepsilon)] \end{bmatrix} \begin{bmatrix} u \\ v \end{bmatrix} \quad (10.4.86)$$

if T and S satisfy

$$\varepsilon \frac{dT}{dt} = B(t, \varepsilon)T - \varepsilon T^2 - \frac{1}{\varepsilon} A(t, \varepsilon),$$

$$(10.4.87)$$

$$\varepsilon \frac{dS}{dt} = -[B(t, \varepsilon) + \varepsilon T(t, \varepsilon)] S - 1.$$

Use (10.4.47) and (10.4.70) to show in this case that A and B satisfy

$$A(t, \varepsilon) = O(\varepsilon) \quad \text{and} \quad B(t, \varepsilon) = k_0[t, X_0(t)] + O(\varepsilon) \quad \text{as } \varepsilon \to 0+,$$

$$(10.4.88)$$

uniformly for $0 \le t \le 1$. This last result, along with (10.4.66), yields (10.4.85). The Riccati equation for T has a bounded solution satisfying the homogeneous initial condition listed in (10.4.84), obtained as the solution of the integral equation

$$T(t, \varepsilon) = T_0(t, \varepsilon) + \int_0^t \left[\exp\left(\frac{1}{\varepsilon} \int_s^t B \right) \right] T(s, \varepsilon)^2 \, ds,$$

$$(10.4.89)$$

$$T_0(t, \varepsilon) := -\frac{1}{\varepsilon^2} \int_0^t \left[\exp\left(\frac{1}{\varepsilon} \int_s^t B \right) \right] A(s, \varepsilon) \, ds.$$

A routine application of the Banach/Picard fixed-point theorem using (10.4.85) and (10.4.88) shows that (10.4.89) has a bounded solution T, and then this solution can be inserted into the equation for S in (10.4.87).

That equation has a bounded solution satisfying the homogeneous terminal condition of (10.4.84), given as

$$S(t, \varepsilon) = \frac{1}{\varepsilon} \int_t^1 \exp\left[\frac{1}{\varepsilon} \int_t^s (B + \varepsilon T)\right] ds. \qquad (10.4.90)$$

This completes the proof of (10.4.83)–(10.4.85), with the exception of the stated result for $T(1, \varepsilon)$ given in (10.4.84). From (10.4.89) it follows that $T(1, \varepsilon) = T_0(1, \varepsilon) + O(\varepsilon)$, so that it suffices to prove the stated result with T replaced by T_0.

The expression for T_0 given by (10.4.89) can be integrated by parts to give $T_0(1, \varepsilon) = [A(1, \varepsilon)/\varepsilon B(1, \varepsilon)] + O(\varepsilon)$, and this result, along with (10.4.47), yields

$$T_0(1, \varepsilon) = \frac{h_{0,x}[1, X_0(1)] + k_{0,x}[1, X_0(1)] X_0'(1)}{k_0[1, X_0(1)]} + O(\varepsilon)$$

$$\text{as } \varepsilon \to 0+. \qquad (10.4.91)$$

Using (10.4.66) and (10.4.67), along with (10.4.91), we find the desired result

$$T_0(1, \varepsilon) = b + \frac{\mathcal{R}'[X_0(1)]}{k_0[1, X_0(1)]} + O(\varepsilon).$$

The fundamental solution (10.4.82) follows directly from an appropriate fundamental solution for (10.4.86) and the given Riccati transformation.

Exercise 10.4.5: Under the same conditions as in Exercise 10.4.4, show that (for small enough $\varepsilon > 0$) the linearized homogeneous operator of (10.4.45) (with $\rho \equiv 0$ and $E \equiv 0$) has a Green function $G = G(t, s, \varepsilon)$ corresponding to the boundary conditions of (10.4.46) with the boundary matrices of (10.4.71). *Hint:* The Green function exists and is given by (0.1.14) and (0.1.16), provided that the matrix $M = M(\varepsilon) = L(\varepsilon)Z(0, \varepsilon) + R(\varepsilon)Z(1, \varepsilon)$ is invertible. Give a direct calculation using (10.4.71) and the fundamental solution Z of Exercise 10.4.4, and obtain the result

$$\det M(\varepsilon) = \varepsilon \frac{\mathcal{R}'[X_0(1)]}{k_0[1, X_0(1)]} \exp\left(-\int_0^1 T\right) + O(\varepsilon^2) \quad \text{as } \varepsilon \to 0+.$$

$$(10.4.92)$$

Given that (10.4.68) holds, conclude that M is invertible.

Exercise 10.4.6: For fixed, sufficiently small $\varepsilon > 0$, sketch the graphs of two distinct solutions for the differential equation

$$\varepsilon \frac{d^2 x}{dt^2} + \frac{dx}{dt} = (x - 2)^2 \quad \text{for } 0 \le t \le 1, \qquad (10.4.93)$$

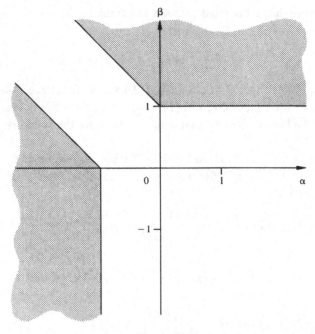

Figure 10.7

subject to the boundary conditions

$$\frac{dx}{dt} - x = 1 \quad \text{at } t = 0,$$

$$\frac{dx}{dt} + \frac{1}{3}x = 1 \quad \text{at } t = 1.$$

(10.4.94)

Exercise 10.4.7: **(a)** Discuss the boundary-value problem

$$\varepsilon \frac{d^2x}{dt^2} + x\frac{dx}{dt} - x = 0 \quad \text{for } 0 \le t \le 1,$$

$$x(0, \varepsilon) = \alpha_0 \quad \text{and} \quad x(1, \varepsilon) = \beta_0,$$

(10.4.95)

for fixed boundary values (α_0, β_0) lying in either of the shaded regions in Figure 10.7, for small values of $\varepsilon > 0$.

(b) Sketch the graphs of solutions for (10.4.95) for small $\varepsilon > 0$, in the two cases $(\alpha_0, \beta_0) = (-1, 3)$ and $(\alpha_0, \beta_0) = (-3, 1)$. [The problem (10.4.95) has often appeared as an example in the literature; see J. Cole (1968, pp. 29–38), Dorr, Parter, and Shampine (1973), and Howes (1978) for a complete discussion for arbitrary boundary values.]

10.5 A problem in the physical theory of semiconductors

The system of differential equations

$$\frac{dn}{dt} = -ne + I_n(t),$$

$$\frac{dp}{dt} = pe - I_p(t), \qquad (10.5.1)$$

$$\varepsilon^2 \frac{de}{dt} = p - n + D(t),$$

appears in the steady-state one-dimensional theory of semiconducting diodes [see Shockley (1949) and van Roosbroeck (1950)], where t denotes a spatial coordinate, $e = e(t)$ represents the electrostatic field, $n = n(t)$ and $p = p(t)$ represent the respective electron and hole densities (of negative and positive charges, respectively), $I_n = I_n(t)$ and $I_p = I_p(t)$ represent the electron and hole currents, $D = D(t)$ is the doping density, which gives the difference between the densities of donor electrons and holes, and the positive parameter ε represents the Debye length, which is a measure of the thickness of the physical diode junction as compared with the length of the diode. In practice, ε is small, typically taking values between 10^{-6} and 10^{-2}. For simplicity we assume here that the currents $I_p = I_p(t)$ and $I_n = I_n(t)$ are given, known functions, which amounts physically to the assumption that the effects of recombination are known a priori.

A commonly used doping model for an abrupt symmetric $p - n$ junction is given as

$$D(t) := \begin{cases} -D_0 & \text{for } -1 \le t < 0, \\ +D_0 & \text{for } \ \ 0 < t \le 1, \end{cases} \qquad (10.5.2)$$

where the junction is placed at the origin $t = 0$, and where D_0 is a given positive constant. We consider here a more general symmetric abrupt doping profile $D = D(t)$ subject to the conditions

$$D(t) = -D(-t) > 0 \quad \text{for } 0 < t \le 1,$$
$$D(0+) > 0, \qquad (10.5.3)$$

where the given function D is assumed to be smooth on each of the subintervals $[-1, 0)$ and $(0, 1]$, and piecewise continuous on $[-1, 1]$.

The condition (10.5.3) implies that the region $0 < t \le 1$ is the n side of the diode, so that the physical currents are expected to flow from left to right, with electron flow dominating hole flow in the n side. Hence, we have the following natural current conditions that are assumed here:

$$I_n(t) \geq I_p(t) \geq 0 \quad \text{for } 0 \leq t \leq 1. \tag{10.5.4}$$

We also have from the physical theory of semiconductors the condition that the total current $I_p(t) + I_n(t)$ is constant, say

$$I_p(t) + I_n(t) \equiv I_0 \quad \text{for all } t, \tag{10.5.5}$$

for some positive constant I_0. For a symmetrically operating symmetric diode as considered here, we also have that the current functions satisfy

$$I_p(t) = I_n(-t) \quad \text{for all } t \neq 0. \tag{10.5.6}$$

Subject to these conditions, the diode is expected to have a symmetric state, corresponding to a solution of (10.5.1) with

$$p(t) = n(-t) \quad \text{and} \quad e(t) = e(-t) \quad \text{for all } t \neq 0. \tag{10.5.7}$$

In the study of such a symmetric state it suffices to consider only one side of the diode, say the n side, and hence we shall consider the system (10.5.1) for the unknown functions n, p, and e on $0 \leq t \leq 1$, subject to the boundary conditions

$$\begin{aligned} n(0, \varepsilon) &= p(0, \varepsilon), \\ n(1, \varepsilon) &= D(1) + \gamma_n, \quad \text{and} \quad p(1, \varepsilon) = \gamma_p, \end{aligned} \tag{10.5.8}$$

for given, fixed constants γ_p and γ_n satisfying

$$D(1) + \gamma_n > 0 \quad \text{and} \quad \gamma_p \geq 0. \tag{10.5.9}$$

These inequalities of (10.5.9) are natural conditions for the given problem, because the *densities* $n(t)$ and $p(t)$ should be nonnegative. The parameters γ_n and γ_p represent source strengths of electrons and holes, respectively, at $t = 1$. The case $\gamma_n = 0$ in (10.5.8) corresponds to the situation in which the electron density n agrees with the intrinsic density of donor electrons at $t = 1$.

If the data satisfy $\gamma_p = 0$ and $I_p \equiv 0$, then we see that the hole charge density vanishes, with $p \equiv 0$. We wish to exclude this degenerate case in the following discussion, and so we make the additional assumption [see (10.5.4) and (10.5.9)]

$$\gamma_p + \int_0^1 I_p > 0. \tag{10.5.10}$$

The boundary-value problem (10.5.1) and (10.5.8) is studied in Vasil'eva, Kardo-Sysoev, and Stel'makh (1976) and in Vasil'eva and Stel'makh (1977) [see also Vasil'eva and Butuzov (1980)] in the special case

$$\gamma_p = \gamma_n = 0, \qquad I_p \equiv I_n \equiv \tfrac{1}{2}I_0, \quad \text{and} \quad D \equiv D_0, \qquad (10.5.11)$$

as in (10.5.2). The more general problem, as described earlier, is studied in Smith (1980), where a number of results of physical interest are obtained within the framework of the given mathematical model. For example, we verify the physical effect of Shockley, Pearson, and Haynes (1949) that "in a semiconductor containing substantially only one type of current carrier, it is impossible to increase the total carrier concentration by injecting carriers of the same type; however, such increases can be produced by injecting the opposite type." We also verify the existence of a suitable "floating potential drop" in the presence of zero current, as described in Bardeen (1950).

The system (10.5.1) is a special case of the previous (vector) system (10.2.3), repeated here as

$$\frac{dx}{dt} = F(t, x, \varepsilon) + G(t, x, \varepsilon)y,$$

$$\varepsilon^2 \frac{dy}{dt} = H(t, x, \varepsilon) + \varepsilon^2 K(t, x, \varepsilon)y \quad \text{for } 0 \le t \le 1, \qquad (10.5.12)$$

for solution functions $x = x(t, \varepsilon)$ and $y = y(t, \varepsilon)$ that are respectively m_1-dimensional and m_2-dimensional real vector-valued functions, and where the given data functions F, G, H, and K are real matrix-valued or vector-valued functions with appropriate compatible orders, subject to suitable smoothness conditions. We can consider various linear or nonlinear boundary conditions for (10.5.12), of which we mention here only the linear boundary condition

$$L\begin{bmatrix} x(0, \varepsilon) \\ y(0, \varepsilon) \end{bmatrix} + R\begin{bmatrix} x(1, \varepsilon) \\ y(1, \varepsilon) \end{bmatrix} = \text{given } (m_1 + m_2)\text{-vector,}$$

$$(10.5.13)$$

for given boundary matrices L and R.

The foregoing semiconductor problem (10.5.1) and (10.5.8) is of the type of (10.5.12)–(10.5.13), with

$$m_1 = 2, \qquad m_2 = 1, \qquad x = \begin{bmatrix} n \\ p \end{bmatrix}, \qquad y = e, \qquad (10.5.14)$$

along with

$$F = \begin{bmatrix} I_n(t) \\ -I_p(t) \end{bmatrix}, \qquad G = \begin{bmatrix} -n \\ p \end{bmatrix}, \qquad H = p - n + D(t), \qquad K \equiv 0,$$

$$(10.5.15)$$

and with boundary matrices

$$L = \begin{bmatrix} 1 & -1 & 0 \\ 0 & 0 & 0 \\ 0 & 0 & 0 \end{bmatrix} \qquad \text{and} \qquad R = \begin{bmatrix} 0 & 0 & 0 \\ 1 & 0 & 0 \\ 0 & 1 & 0 \end{bmatrix}. \qquad (10.5.16)$$

We shall be content here to discuss only this special case described by (10.5.14)–(10.5.16), although in some cases that follow it is convenient to use the general notation in terms of F, G, H, and K as in (10.5.12).

The reduced system obtained by putting $\varepsilon = 0$ in (10.5.12) is given as

$$\frac{dX_0}{dt} = F_0(t, X_0) + G_0(t, X_0)Y_0,$$

$$0 = H_0(t, X_0) \quad \text{for } 0 < t < 1, \qquad (10.5.17)$$

where X_0 and Y_0 denote the leading terms of suitable outer solution expansions, with $F_0(t, x) = F(t, x, 0)$, and so forth. Using an approach similar to that of Section 7.3, we differentiate the second equation of (10.5.17) to give

$$0 = H_{0,t}[t, X_0(t)] + H_{0,x}[t, X_0(t)]\frac{dX_0}{dt}, \qquad (10.5.18)$$

and this result yields, with the first equation of (10.5.17),

$$-H_{0,x}(t, X_0)G_0(t, X_0)Y_0 = H_{0,t}(t, X_0) + H_{0,x}(t, X_0)F_0(t, X_0).$$

$$(10.5.19)$$

In the special case (10.5.14)–(10.5.16) considered here, we find

$$-H_{0,x}G_0 = -[N_0(t) + P_0(t)], \qquad (10.5.20)$$

where N_0 and P_0 are the leading outer terms for the densities n and p. Moreover, in this case we find the result [see Smith (1980)]

$$N_0 + P_0 > 0, \qquad (10.5.21)$$

so that the "matrix" $-H_{0,x}G_0$ is nonsingular with negative eigenvalue. A related linear problem is studied in O'Malley (1979) subject to an assumption that a certain corresponding matrix must be stable (with all eigenvalues having negative real parts).

Suitable leading outer functions X_0 and Y_0 are assumed to be given, satisfying (10.5.17) and (10.5.18), and such that the matrix

$H_{0,x}(t, X_0)G_0(t, X_0)$ is nonsingular. In particular, we are assuming here that $m_1 \geq m_2$, as in (10.5.14). Then (10.5.19) can be solved for Y_0, and the result can be inserted back into (10.5.17) so as to eliminate Y_0 there. In this way we find for X_0 the equation

$$\frac{dX_0}{dt} = F_0(t, X_0) - G_0(t, X_0)[H_{0,x}(t, X_0)G_0(t, X_0)]^{-1}$$
$$\times [H_{0,t}(t, X_0) + H_{0,x}(t, X_0)F_0(t, X_0)]. \tag{10.5.22}$$

Along with this latter system of differential equations, we also have the algebraic system $0 = H_0(t, X_0)$ of (10.5.17). Subject to suitable conditions, we can solve this latter algebraic system for certain components of the m_1-vector X_0 in terms of the other components, and then use these results back in (10.5.22) so as to eliminate certain components of X_0. In general, we see then that X_0 is characterized by an associated system of differential equations with lower effective order not exceeding $m_1 - m_2$. In general, the leading outer solution functions X_0, Y_0 need not satisfy *any* of the original boundary conditions. Suitable boundary conditions for X_0 are obtained only after we have certain information on the boundary-layer correction terms, as illustrated later.

Turning now exclusively to the special case (10.5.14)–(10.5.16), we seek a solution of the original problem for small $\varepsilon \to 0+$ in the asymptotic form

$$n(t, \varepsilon) \sim N(t, \varepsilon) + {}^*N(\tau, \varepsilon) + N^*(\sigma, \varepsilon),$$
$$p(t, \varepsilon) \sim P(t, \varepsilon) + {}^*P(\tau, \varepsilon) + P^*(\sigma, \varepsilon), \tag{10.5.23}$$
$$e(t, \varepsilon) \sim E(t, \varepsilon) + \frac{1}{\varepsilon} {}^*E(\tau, \varepsilon) + \frac{1}{\varepsilon} E^*(\sigma, \varepsilon),$$

with

$$\tau := \frac{t}{\varepsilon} \quad \text{and} \quad \sigma := \frac{1-t}{\varepsilon}, \tag{10.5.24}$$

and where the functions on the right side of (10.5.23) are to be determined in the form of asymptotic expansions as

$$\begin{bmatrix} N(t, \varepsilon) \\ P(t, \varepsilon) \\ E(t, \varepsilon) \end{bmatrix} \sim \sum_{k=0}^{\infty} \begin{bmatrix} N_k(t) \\ P_k(t) \\ E_k(t) \end{bmatrix} \varepsilon^k, \qquad \begin{bmatrix} {}^*N(\tau, \varepsilon) \\ {}^*P(\tau, \varepsilon) \\ {}^*E(\tau, \varepsilon) \end{bmatrix} \sim \sum_{k=0}^{\infty} \begin{bmatrix} {}^*N_k(\tau) \\ {}^*P_k(\tau) \\ {}^*E_k(\tau) \end{bmatrix} \varepsilon^k,$$

$$\begin{bmatrix} N^*(\sigma, \varepsilon) \\ P^*(\sigma, \varepsilon) \\ E^*(\sigma, \varepsilon) \end{bmatrix} \sim \sum_{k=0}^{\infty} \begin{bmatrix} N_k^*(\sigma) \\ P_k^*(\sigma) \\ E_k^*(\sigma) \end{bmatrix} \varepsilon^k, \tag{10.5.25}$$

where the leading outer terms N_0, P_0, and E_0 are as discussed earlier, and where the respective left and right boundary-layer terms are required to satisfy the usual matching conditions,

$$\lim_{\tau \to \infty} \begin{bmatrix} {}^*N_k(\tau) \\ {}^*P_k(\tau) \\ {}^*E_k(\tau) \end{bmatrix} = \begin{bmatrix} 0 \\ 0 \\ 0 \end{bmatrix} \quad \text{and} \quad \lim_{\sigma \to \infty} \begin{bmatrix} N_k^*(\sigma) \\ P_k^*(\sigma) \\ E_k^*(\sigma) \end{bmatrix} = \begin{bmatrix} 0 \\ 0 \\ 0 \end{bmatrix},$$

(10.5.26)

for $k = 0, 1, \ldots$. [We are led to conjecture the asymptotic forms of (10.5.23) by an examination of related linear constant-coefficient problems.]

The boundary conditions (10.5.8), along with (10.5.23)–(10.5.26), lead to the related boundary conditions

$$N_k(0) + {}^*N_k(0) = P_k(0) + {}^*P_k(0) \quad \text{for } k = 0, 1, \ldots, \quad (10.5.27)$$

$$N_k(1) + N_k^*(0) = \begin{cases} D(1) + \gamma_n & \text{for } k = 0, \\ 0 & \text{for } k = 1, 2, \ldots, \end{cases} \quad (10.5.28)$$

and

$$P_k(1) + P_k^*(0) = \begin{cases} \gamma_p & \text{for } k = 0, \\ 0 & \text{for } k = 1, 2, \ldots. \end{cases} \quad (10.5.29)$$

The outer solution functions N, P, and E are required to satisfy the original system (10.5.1) for $0 < t < 1$, and if we insert the outer expansions from (10.5.25) into (10.5.1), we find that the outer coefficients satisfy the equations

$$\frac{dN_0}{dt} = -E_0 N_0 + I_n(t),$$

$$\frac{dP_0}{dt} = E_0 P_0 - I_p(t),$$

(10.5.30)

$$0 = P_0 - N_0 + D(t),$$

for $0 < t < 1$, and

$$\frac{dN_k}{dt} = -E_0 N_k - \sum_{j=0}^{k-1} N_j E_{k-j}, \qquad \frac{dP_k}{dt} = E_0 P_k + \sum_{j=0}^{k-1} P_j E_{k-j},$$

$$0 = P_k - N_k - E'_{k-2}(t) \quad \text{for } k = 1, 2, \ldots, \quad (10.5.31)$$

and for $0 < t < 1$, where $E_{-1} := 0$. As usual, the equations (10.5.31) for the higher-order coefficients are linear equations.

Turning now to the boundary-layer correction terms, we find likewise, with the usual procedure, that these boundary-layer coefficients satisfy certain differential equations, of which we list here only those for the leading terms, as follows:

$$\frac{d}{d\tau}\begin{bmatrix} *N_0(\tau) \\ *P_0(\tau) \\ *E_0(\tau) \end{bmatrix} = \begin{bmatrix} -\left[N_0(0) + *N_0(\tau)\right] *E_0(\tau) \\ \left[P_0(0) + *P_0(\tau)\right] *E_0(\tau) \\ *P_0(\tau) - *N_0(\tau) \end{bmatrix} \quad \text{for } \tau \geq 0,$$

$$(10.5.32)$$

and

$$-\frac{d}{d\sigma}\begin{bmatrix} N_0^*(\sigma) \\ P_0^*(\sigma) \\ E_0^*(\sigma) \end{bmatrix} = \begin{bmatrix} -\left[N_0(1) + N_0^*(\sigma)\right] E_0^*(\sigma) \\ \left[P_0(1) + P_0^*(\sigma)\right] E_0^*(\sigma) \\ P_0^*(\sigma) - N_0^*(\sigma) \end{bmatrix} \quad \text{for } \sigma \geq 0.$$

$$(10.5.33)$$

Again, higher-order coefficients satisfy certain associated linear equations, but we shall not list these here.

We require a certain first integral of the boundary-layer system (10.5.33) in order to obtain the correct boundary conditions at $t = 1$ for the leading outer terms in (10.5.30) and for the leading boundary-layer terms in (10.5.33). We find easily that (10.5.33) has the first integral

$$\left[N_0(1) + N_0^*(\sigma)\right]\left[P_0(1) + P_0^*(\sigma)\right] = N_0(1)P_0(1), \quad (10.5.34)$$

where the appropriate matching condition from (10.5.26) has been used to fix a constant of integration. This result (10.5.34) evaluated at $\sigma = 0$, along with the boundary conditions of (10.5.28) and (10.5.29) (for $k = 0$), yields now the boundary relation

$$N_0(1)P_0(1) = \left[D(1) + \gamma_n\right]\gamma_p. \quad (10.5.35)$$

Finally, this last equation and the final equation of (10.5.30) (evaluated at $t = 1$) can be solved together to yield the boundary values

$$N_0(1) = \frac{D(1) + \left\{D(1)^2 + 4\left[D(1) + \gamma_n\right]\gamma_p\right\}^{1/2}}{2},$$

$$(10.5.36)$$

$$P_0(1) = \frac{-D(1) + \left\{D(1)^2 + 4\left[D(1) + \gamma_n\right]\gamma_p\right\}^{1/2}}{2},$$

for the outer functions N_0 and P_0. Then these values can be used back in

(10.5.28) and (10.5.29) to yield the values

$$N_0^*(0) = \frac{D(1) + 2\gamma_n - \left\{ D(1)^2 + 4[D(1) + \gamma_n]\gamma_p \right\}^{1/2}}{2},$$

(10.5.37)

$$P_0^*(0) = \frac{D(1) + 2\gamma_p - \left\{ D(1)^2 + 4[D(1) + \gamma_n]\gamma_p \right\}^{1/2}}{2}.$$

Similarly, we require a suitable first integral of (10.5.32) in order to fix the correct boundary conditions at the left endpoint. We find that (10.5.32) has the first integral

$$[N_0(0) + {}^*N_0(\tau)][P_0(0) + {}^*P_0(\tau)] = N_0(0)P_0(0), \quad (10.5.38)$$

where again the matching conditions have been used. This last result at $\tau = 0$, along with the boundary condition (10.5.27) (for $k = 0$), yields now the conditions

$$^*N_0(0) = -N_0(0) + \sqrt{N_0(0)P_0(0)},$$
$$^*P_0(0) = -P_0(0) + \sqrt{N_0(0)P_0(0)},$$

(10.5.39)

where positive square roots are taken here and in (10.5.36) in order to provide nonnegative values for the charge densities n and p. The quantity $N_0(0)P_0(0)$ appearing under the root sign in (10.5.39) is positive, as indicated later.

We shall discuss only the leading terms in the expansions of (10.5.25). First, the functions N_0 and P_0 are given as

$$N_0(t) = \frac{u(t) + D(t)}{2} \quad \text{and} \quad P_0(t) = \frac{u(t) - D(t)}{2}, \quad (10.5.40)$$

where D is the given doping density and $u = u(t) := N_0(t) + P_0(t)$ is the solution of the first-order terminal-value problem

$$u\frac{du}{dt} = [I_n(t) - I_p(t)]u - D(t)[I_0 - D'(t)] \quad \text{for } 0 \le t \le 1,$$

(10.5.41)

$$u(1) = \left\{ D(1)^2 + 4[D(1) + \gamma_n]\gamma_p \right\}^{1/2}.$$

This latter problem for u follows directly from (10.5.30) and (10.5.36). It is shown in Smith (1980) that (10.5.41) has a unique solution for $0 \le t \le 1$, and the resulting solution function u satisfies $u(t) > D(t)$ for all such t, so that the outer density N_0 of (10.5.40) exceeds D, and P_0 is positive. The leading outer function E_0 is given by (10.5.19) in terms of

N_0 and P_0 as

$$E_0(t) = \frac{I_0 - D'(t)}{N_0(t) + P_0(t)}, \tag{10.5.42}$$

where (10.5.5) has been used. This completes the determination of the leading outer coefficients. Note that, in general, all of the boundary conditions of (10.5.8) are cancelled for these outer functions N_0, P_0, and E_0.

The boundary-layer terms N_0^*, P_0^*, and E_0^* are determined now as the unique solutions of (10.5.33) subject to the usual matching conditions at ∞ and subject to the initial conditions of (10.5.37). Using the approach of Vasil'eva and Stel'makh (1977), we prove that these boundary-layer correction terms are well defined for $\sigma \geq 0$ and decay exponentially as $\sigma \to \infty$, with estimates of the type

$$|N_0^*(\sigma)|, |P_0^*(\sigma)|, |E_0^*(\sigma)|^2 \leq \text{const.} \ e^{-\nu_1 \sigma} \quad \text{for } \sigma \geq 0, \tag{10.5.43}$$

for a fixed positive constant ν_1 that is itself a constant multiple of $\sqrt{D(1)}$. For example, we can take

$$\nu_1 = 8^{-1/2}\sqrt{D(1)}, \tag{10.5.44}$$

although the constant $8^{-1/2}$ is not always the best possible here.

Similarly, the boundary-layer terms $^*N_0$, $^*P_0$, and $^*E_0$ are determined as the solutions of (10.5.32) subject to the usual matching conditions at ∞ and subject to the initial conditions of (10.5.39). We find that these boundary-layer correction terms are well determined for $\tau \geq 0$ and decay exponentially as

$$|^*N_0(\tau)|, |^*P_0(\tau)|, |^*E_0(\tau)|^2 \leq \text{const.} \ e^{-\nu_0 \tau} \quad \text{for } \tau \geq 0, \tag{10.5.45}$$

for a fixed positive constant ν_0 that is a constant multiple of $\sqrt{P_0(0)}$. For example, we can take

$$\nu_0 = \sqrt{2} \sqrt{P_0(0)}. \tag{10.5.46}$$

The quantity $P_0(0)$ can be estimated in terms of the data, and we see that $P_0(0)$ is positive if (10.5.10) holds. In general, $P_0(0)$ depends on the boundary data γ_n and γ_p at $t = 1$, along with the doping profile $D = D(t)$ and the currents $I_n(t)$ and $I_p(t)$ for $0 \leq t \leq 1$. Further details are given in Smith (1980).

The procedure can be continued recursively to obtain higher-order terms $N_k, P_k, \ldots, P_k^*, E_k^*$ in the expansions of (10.5.23)–(10.5.25) for

$k = 1, 2, \ldots,$ provided, as usual, that the data functions are sufficiently smooth. We can then truncate the resulting expansions so as to provide a proposed approximate solution for the original problem, say n^M, p^M, and e^M for a given nonnegative integer M, with

$$\begin{bmatrix} n^M(t, \varepsilon) \\ p^M(t, \varepsilon) \\ e^M(t, \varepsilon) \end{bmatrix} := \sum_{k=0}^{M} \left\{ \begin{bmatrix} N_k(t) \\ P_k(t) \\ E_k(t) \end{bmatrix} + \begin{bmatrix} *N_k(\tau) \\ *P_k(\tau) \\ \frac{1}{\varepsilon} *E_k(\tau) \end{bmatrix} + \begin{bmatrix} N_k^*(\sigma) \\ P_k^*(\sigma) \\ \frac{1}{\varepsilon} E_k^*(\sigma) \end{bmatrix} \right\} \varepsilon^k,$$

$$(10.5.47)$$

with τ and σ defined by (10.5.24), for $0 \leq t \leq 1$ and for $0 < \varepsilon \leq \varepsilon_0$, for a suitable fixed, sufficiently small $\varepsilon_0 > 0$. The original problem can then be recast as a linearization about (10.5.47), and a Green function approach can be used to prove that the problem has an exact solution close to the given approximate solution. For example, to lowest order, we find for the exact solution functions the results

$$e(t, \varepsilon) = e^0(t, \varepsilon) + O(\varepsilon) + O(e^{-\hat{\nu}_0 t / \varepsilon}) + O(e^{-\hat{\nu}_1 (1 - t) / \varepsilon}),$$

$$n(t, \varepsilon) = n^0(t, \varepsilon) + O(\varepsilon), \qquad (10.5.48)$$

$$p(t, \varepsilon) = p^0(t, \varepsilon) + O(\varepsilon) \quad \text{as } \varepsilon \to 0+$$

uniformly for $0 \leq t \leq 1$, where e^0, n^0, and p^0 are given by (10.5.47) as

$$e^0(t, \varepsilon) := E_0(t) + \frac{1}{\varepsilon} *E_0\left(\frac{t}{\varepsilon}\right) + \frac{1}{\varepsilon} E_0^*\left(\frac{1 - t}{\varepsilon}\right),$$

$$n^0(t, \varepsilon) := N_0(t) + *N_0\left(\frac{t}{\varepsilon}\right) + N_0^*\left(\frac{1 - t}{\varepsilon}\right), \qquad (10.5.49)$$

$$p^0(t, \varepsilon) := P_0(t) + *P_0\left(\frac{t}{\varepsilon}\right) + P_0^*\left(\frac{1 - t}{\varepsilon}\right),$$

for $0 \leq t \leq 1$ and for all sufficiently small $\varepsilon > 0$, and where $\hat{\nu}_0$ and $\hat{\nu}_1$ in (10.5.48) can be taken as any fixed positive numbers less than the respective values ν_0 and ν_1 of (10.5.43) and (10.5.45).

All three solution components n, p, and e have boundary layers at the diode junction $t = 0$, whereas the boundary-layer structure at the endpoint $t = 1$ depends on the actual values of the specified source strengths γ_n and γ_p. For example, the solution components have no boundary layers at $t = 1$ if $\gamma_n = \gamma_p$. On the other hand, if $\gamma_n \neq \gamma_p = 0$, then p has no boundary layer at $t = 1$, but n and e have boundary layers there. Finally, if $\gamma_n \neq \gamma_p > 0$, then all components n, p, and e have boundary layers at $t = 1$.

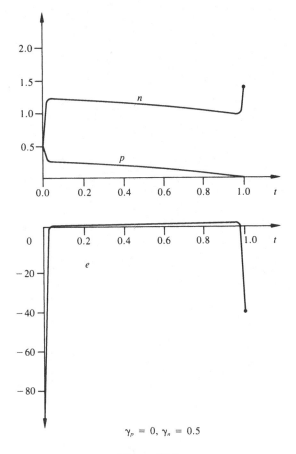

Figure 10.8

Typical results are graphed in Figures 10.8 through 10.11 for various choices of the boundary source strengths γ_n and γ_p. In each case the doping profile, current functions, and Debye length are taken as follows:

$$
\begin{aligned}
D(t) &:= 1.0, \\
I_n(t) &:= 1.5, \\
I_p(t) &:= 0.5 \quad \text{for } 0 \le t \le 1, \\
\varepsilon &:= 10^{-2}.
\end{aligned}
\tag{10.5.50}
$$

The solution values used in the graphs here are taken from Maier and Smith (1981), where a numerical study of the problem is given, based in part on the foregoing asymptotic analysis.

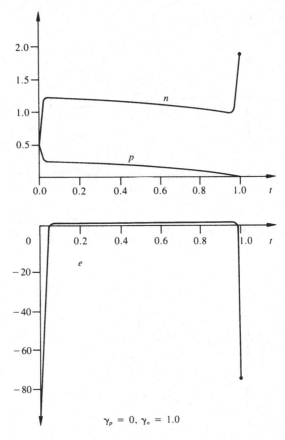

Figure 10.9

Several interesting physical results are confirmed by (10.5.48)–(10.5.49). For example, if we take $\gamma_p = 0$ in (10.5.8) so that no holes are injected into the n-type semiconductor at $t = 1$, then the terminal-value problem (10.5.41) for u is independent of the source strength γ_n of electrons, and the resulting solution $u = u(t)$ is independent of γ_n. Hence, in this case the outer carrier densities $N_0(t)$ and $P_0(t)$ of (10.5.40) cannot be affected by injecting or extracting electrons only, at $t = 1$. The same is then true also, to first order in the small parameter ε, for the total carrier concentrations $n(t, \varepsilon)$ and $p(t, \varepsilon)$ on compact subsets of the semiconductor, as illustrated in Figures 10.8 and 10.9. On the other hand, if we have $\gamma_p > 0$, then we find that $u(t)$ is a strictly increasing function of the boundary source strengths γ_n and γ_p, and then (10.5.40) shows that the

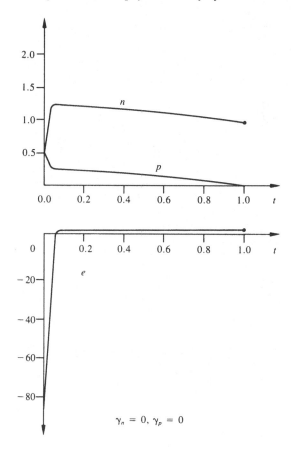

Figure 10.10

outer carrier densities (and hence also the total carrier densities n and p, on compact subsets) are both increasing functions of the boundary source strengths γ_n and γ_p. For example, if $\gamma_p \geq 0$ and $\gamma_n = 0$, then both the hole and electron densities can be increased (on compact subsets) by increasing the source strength of holes at $t = 1$, as illustrated in Figures 10.10 and 10.11. Hence, we verify the physical effect of Shockley, Pearson, and Haynes (1949) that in a semiconductor containing predominantly one type of current carrier, it is not possible to increase the total carrier concentration (on compact proper subsets of the semiconductor) by injecting or extracting only carriers of that same type, whereas such increases can be produced by injecting the opposite type.

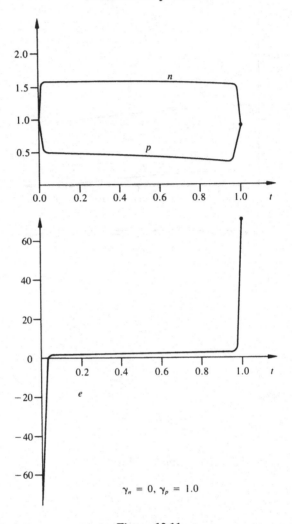

Figure 10.11

Several additional physical results are similarly verified with the foregoing asymptotic results, of which we mention here only a result on the
total potential drop. The *total potential drop* V over the diode for
$0 \le t \le 1$ is given as

$$V = V(\varepsilon): = -\int_0^1 e(t, \varepsilon)\, dt, \qquad (10.5.51)$$

and then it follows from the preceding asymptotic results that $V(\varepsilon)$

satisfies

$$V(\varepsilon) = V_0 + O(\varepsilon) \quad \text{as } \varepsilon \to 0+, \tag{10.5.52}$$

with

$$V_0 = -\int_0^1 E_0(t)\, dt - \int_0^\infty {}^* E_0(\tau)\, d\tau - \int_0^\infty E_0^*(\sigma)\, d\sigma. \tag{10.5.53}$$

The last two terms on the right side here give the boundary-layer contributions to V_0 and are found by direct calculation from (10.5.26), (10.5.32), (10.5.33), (10.5.37), and (10.5.39), to be given as [see Smith (1980)]

$$-\int_0^\infty {}^* E_0(\tau)\, d\tau = \log\left(\frac{u(0) + D(0)}{u(0) - D(0)}\right)^{1/2},$$

$$-\int_0^\infty E_0^*(\sigma)\, d\sigma = \log\left(\frac{2[D(1) + \gamma_n]}{D(1) + \left\{D(1)^2 + 4[D(1) + \gamma_n]\gamma_p\right\}^{1/2}}\right), \tag{10.5.54}$$

where $u(0)$ is obtained from (10.5.41). It follows that even though the boundary-layer correction terms $(1/\varepsilon)\,{}^* E_0(t/\varepsilon)$ and $(1/\varepsilon)\, E_0^*[(1-t)/\varepsilon]$ are themselves negligible on any fixed compact subset of $0 < t < 1$ (for small $\varepsilon \to 0+$), these boundary-layer corrections are nevertheless important for accurate assessment of the total potential drop.

In the case of equal flows of electrons and holes, with

$$I_n(t) = I_p(t) = \tfrac{1}{2}I_0 \quad \text{for } 0 \le t \le 1, \tag{10.5.55}$$

we easily solve the terminal-value problem (10.5.41) explicitly in closed form to find

$$u(t) = \left\{D(t)^2 + 2I_0\int_t^1 D(s)\, ds + 4[D(1) + \gamma_n]\gamma_p\right\}^{1/2}. \tag{10.5.56}$$

In the further special case of zero currents, with

$$I_n = I_p = \tfrac{1}{2}I_0 = 0 \quad \text{for } 0 \le t \le 1, \tag{10.5.57}$$

and with $\gamma_p > 0$ [see (10.5.10)], we find, with (10.5.42) and (10.5.56), the result

$$E_0(t) = \frac{D'(t)}{\left\{D(t)^2 + 4[D(1) + \gamma_n]\gamma_p\right\}^{1/2}}, \tag{10.5.58}$$

which yields, on integration,

$$-\int_0^1 E_0(t)\,dt = \log\left(\frac{D(1) + \left\{D(1)^2 + 4[D(1) + \gamma_n]\gamma_p\right\}^{1/2}}{D(0) + \left\{D(0)^2 + 4[D(1) + \gamma_n]\gamma_p\right\}^{1/2}}\right).$$

(10.5.59)

This last result, along with the earlier results of (10.5.52)–(10.5.54), yields, for the total potential drop,

$$V(\varepsilon) = \tfrac{1}{2}\log\left(\frac{D(1) + \gamma_p}{\gamma_p}\right) + O(\varepsilon) \quad \text{as } \varepsilon \to 0+, \quad (10.5.60)$$

provided that (10.5.57) holds. Such a potential drop in the presence of zero current is called a *floating potential drop* in Bardeen (1950). In the special case of constant doping, as in (10.5.2), we see that the potential drop (10.5.59) due to the leading outer term vanishes, so that the potential drop across the semiconductor is due entirely to boundary-layer effects (to first order).

Exercises

Exercise 10.5.1: Give the details of a calculation leading from (10.5.54) and (10.5.59) to (10.5.60) when (10.5.57) holds.

Exercise 10.5.2: Verify that the system (10.5.33) has the first integral

$$\tfrac{1}{2}E_0^*(\sigma)^2 = N_0^*(\sigma) + P_0^*(\sigma) + D(1)\log\left(\frac{N_0(1)}{N_0(1) + N_0^*(\sigma)}\right),$$

(10.5.61)

where $N_0(1)$ is given by (10.5.36), and where the matching conditions of (10.5.26) have been used. A similar first integral can be given for the system (10.5.32).

Exercise 10.5.3: For small $\varepsilon \to 0+$, show that the field strength $e(t, \varepsilon)$ is uniformly small on any fixed compact subset of $0 < t < 1$ if the doping profile D and the total current I_0 satisfy the relation

$$D'(t) = I_0 \quad \text{for } 0 \le t \le 1. \quad (10.5.62)$$

Also, in this case show that the potential drop across the semiconductor is, to first order in ε, due entirely to boundary-layer effects. *Hint:* The leading term E_0 in the outer field can be shown to vanish if (10.5.62) holds.

10.6 Examples on interior layers

We saw in Section 8.5 several examples of linear boundary-value problems on an interval $[t_1, t_2]$, with solutions exhibiting rapid variation across a transition point that is located in the interior of $[t_1, t_2]$. In this section we shall discuss briefly certain examples involving *nonlinear* problems with solutions exhibiting such interior layers. The literature contains a wide variety of such problems studied by many authors, including Haber and Levinson (1955), Vasil'eva (1963), O'Malley (1970c, 1976), Boglaev (1970), Fife (1974, 1976), Howes (1975, 1978), Lorenz (1982), Lutz and Sari (1982), Chang and Howes (1984), Kelley (1984), O'Donnell (1984), and van Harten and Vader-Burger (1984). We are content here to consider only Examples 10.6.1 and 10.6.2.

Example 10.6.1: The problem (Haber and Levinson 1955)

$$\varepsilon x'' = 1 - (x')^2 \quad \text{for } 0 < t < 1,$$
$$x(0) = \alpha, \qquad x(1) = \beta \qquad (|\alpha - \beta| < 1), \tag{10.6.1}$$

can be solved by quadrature [see Wasow (1970) or Howes (1978)], and we find the result

$$\lim_{\varepsilon \to 0+} x(t, \varepsilon) = \begin{cases} \alpha - t & \text{for } 0 \le t \le T, \\ t + \beta - 1 & \text{for } T \le t \le 1, \end{cases} \tag{10.6.2}$$

with

$$T := \frac{\alpha - \beta + 1}{2}. \tag{10.6.3}$$

Note that the assumption $|\alpha - \beta| < 1$ implies that T is an interior point, $0 < T < 1$. The limiting solution (10.6.2) is continuous, but its derivative has a jump discontinuity at the point T, as indicated in Figure 10.12. The problem (10.6.1) is a special case of a class of problems having *angular* limiting solutions, as studied in Haber and Levinson (1955), Vasil'eva (1963), O'Malley (1970c), and Howes (1978).

Example 10.6.2: Consider the Dirichlet problem

$$\varepsilon \frac{d^2 x}{dt^2} = x \frac{dx}{dt} - x^3 \quad \text{for } 0 < t < 1,$$
$$x(0, \varepsilon) = \tfrac{2}{3} \quad \text{and} \quad x(1, \varepsilon) = -\tfrac{1}{2}. \tag{10.6.4}$$

This problem is of the type previously considered in Section 10.4, where solutions of boundary-layer type were studied, and we see directly from Section 10.4 that (10.6.4) has two distinct solutions x_1 and x_2 of

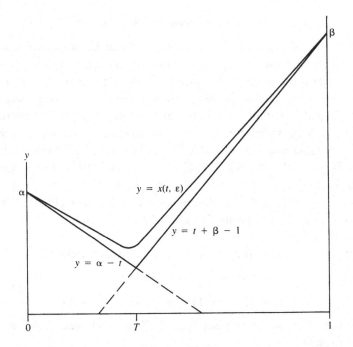

Figure 10.12

boundary-layer type, as indicated in Figure 10.13 (see Exercise 10.6.4).
Moreover, it follows from Theorem 5.5 of Howes (1978) that the problem
(10.6.4) has yet an additional solution x_3, of interior-layer type, as
indicated in Figure 10.14, where in this case the interior layer is centered
at the point

$$T = \tfrac{1}{4}. \qquad (10.6.5)$$

The rest of this section is devoted solely to a discussion of this latter
solution of interior-layer type for (10.6.4). We seek a representation for
this solution in the form

$$x(t, \varepsilon) \sim X(t, \varepsilon) + \hat{X}(\tau, \varepsilon), \qquad (10.6.6)$$

where the layer variable is given as

$$\tau := \frac{t - T}{\varepsilon} \quad \text{with } t = T + \varepsilon\tau, \qquad (10.6.7)$$

and where the layer is centered at $t = T$ for some suitable number
$T \in (0, 1)$ that is to be determined later. [We do not assume, but rather
we shall derive, the result (10.6.5).]

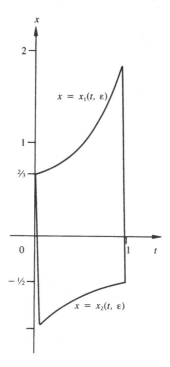

Figure 10.13

The outer solution $X = X(t, \varepsilon)$ in (10.6.6) is expected to have a jump discontinuity at $t = T$, so that each of the outer coefficients $X_k(t)$ in the expansion [see (10.4.23)]

$$X(t, \varepsilon) \sim \sum_{k=0}^{\infty} X_k(t)\varepsilon^k \qquad (10.6.8)$$

is expected to be characterized differently on the two subintervals $[0, T)$ and $(T, 1]$. Similarly, the layer correction term $\hat{X} = \hat{X}(\tau, \varepsilon)$ in (10.6.6) is expected to be characterized differently for $\tau < 0$ and $\tau > 0$, so that the coefficients $\hat{X}_k(\tau)$ in the expansion

$$\hat{X}(\tau, \varepsilon) \sim \sum_{k=0}^{\infty} \hat{X}_k(\tau)\varepsilon^k \qquad (10.6.9)$$

must be characterized differently for $\tau < 0$ and $\tau > 0$. However, the resulting solution $x(t, \varepsilon)$ must be continuous at $t = T$, and this continuity condition will be used later in the determination of T.

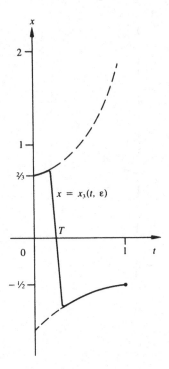

Figure 10.14

From (10.6.6)–(10.6.9) we have the expansions

$$x(t,\varepsilon) \sim \sum_{k=0}^{\infty} \left[X_k(t) + \hat{X}_k(\tau) \right] \varepsilon^k \qquad (10.6.10)$$

and

$$\frac{dx(t,\varepsilon)}{dt} \sim \frac{1}{\varepsilon} \hat{X}_0'(\tau) + \sum_{k=0}^{\infty} \left[X_k'(t) + \hat{X}_{k+1}'(\tau) \right] \varepsilon^k, \quad (10.6.11)$$

where the layer variable τ is given by (10.6.7), and where $X_k'(t) = dX_k(t)/dt$ and $\hat{X}_k'(\tau) = d\hat{X}_k(\tau)/d\tau$.

As mentioned earlier, we require x to be continuous at $t = T$, and in fact we require the coefficient of ε^k on the right side of (10.6.10) to be continuous at $t = T$ (and $\tau = 0$) for each k, which provides the conditions

$$\hat{X}_k(0+) - \hat{X}_k(0-) = -\left[X_k(T+) - X_k(T-) \right] \quad (10.6.12)_k$$

for $k = 0, 1, \ldots$. Similarly, we require x' to be continuous at $t = T$, which leads, with (10.6.11), to the conditions

$$\hat{X}_k'(0+) - \hat{X}_k'(0-) = -\left[X_{k-1}'(T+) - X_{k-1}'(T-)\right] \quad (10.6.13)_k$$

for $k = 0, 1, \ldots$, where $X_{-1} \equiv 0$.

The layer correction terms $\hat{X}_k(\tau)$ are required to vanish in this case as τ tends both to $+\infty$ and to $-\infty$, so that we have the matching conditions

$$\lim_{\tau \to +\infty} \hat{X}_k(\tau) = 0 \quad \text{and} \quad \lim_{\tau \to -\infty} \hat{X}_k(\tau) = 0 \quad (10.6.14)_k$$

for $k = 0, 1, \ldots$. The outer coefficient functions X_k on the right side of (10.6.10) are then required to carry the prescribed boundary conditions from (10.6.4), with

$$X_k(0) = \begin{cases} \frac{2}{3} & \text{for } k = 0, \\ 0 & \text{for } k = 1, 2, \ldots, \end{cases} \quad (10.6.15)_k$$

and

$$X_k(1) = \begin{cases} -\frac{1}{2} & \text{for } k = 0, \\ 0 & \text{for } k = 1, 2, \ldots. \end{cases} \quad (10.6.16)_k$$

The outer solution $X(t, \varepsilon)$ is required to satisfy the full differential equation as in (10.4.25), and then, in particular, the leading term X_0 in the expansion (10.6.8) is taken to satisfy the following reduced problem on the appropriate subinterval:

$$X_0 \frac{dX_0}{dt} = X_0^3 \quad \text{for } 0 < t < T, \quad \text{with } X_0(0) = \tfrac{2}{3} \quad \text{at } t = 0,$$

$$(10.6.17)$$

and

$$X_0 \frac{dX_0}{dt} = X_0^3 \quad \text{for } T < t < 1, \quad \text{with } X_0(1) = -\tfrac{1}{2} \quad \text{at } t = 1.$$

$$(10.6.18)$$

These problems (10.6.17) and (10.6.18) have unique solutions leading to the result

$$X_0(t) = \begin{cases} \dfrac{2}{3 - 2t} & \text{for } 0 \le t < T, \\ -\dfrac{1}{1 + t} & \text{for } T < t \le 1. \end{cases} \quad (10.6.19)$$

The higher-order outer coefficients X_k satisfy related linear problems that need not be given here [see (10.4.27)].

Turning now to the layer correction $\hat{X}(\tau, \varepsilon)$, we require that $x = X + \hat{X}$ satisfy the full differential equation of (10.6.4), from which we find the differential equation [see (10.4.30)]

$$\frac{d^2\hat{X}(\tau,\varepsilon)}{d\tau^2} = [X(T+\varepsilon\tau,\varepsilon) + \hat{X}(\tau,\varepsilon)]\frac{d\hat{X}(\tau,\varepsilon)}{d\tau}$$

$$+\varepsilon\{X(T+\varepsilon\tau,\varepsilon)^3 - [X(T+\varepsilon\tau,\varepsilon) + \hat{X}(\tau,\varepsilon)]^3\}$$

$$+\varepsilon\hat{X}(\tau,\varepsilon)X'(T+\varepsilon\tau,\varepsilon), \tag{10.6.20}$$

both for $\tau > 0$ and for $\tau < 0$. We insert the appropriate asymptotic expansions into (10.6.20) and find, in particular, for the leading layer term $\hat{X}_0(\tau)$ of (10.6.9), the equations

$$\frac{d^2\hat{X}_0}{d\tau^2} = \begin{cases} [X_0(T-) + \hat{X}_0(\tau)]\dfrac{d\hat{X}_0}{d\tau} & \text{for } \tau < 0, \\[2mm] [X_0(T+) + \hat{X}_0(\tau)]\dfrac{d\hat{X}_0}{d\tau} & \text{for } \tau > 0. \end{cases} \tag{10.6.21}$$

Along with (10.6.21), we have the appropriate matching conditions of $(10.6.14)_0$ and the continuity conditions of $(10.6.12)_0$ and $(10.6.13)_0$. The higher-order layer coefficients $\hat{X}_k$ satisfy related linear problems that we shall omit here [see (10.4.32)].

The differential equations of (10.6.21) can be integrated once, along with the appropriate matching conditions $(10.6.14)_0$, to give [see (10.6.37)–(10.6.38)]

$$\frac{d\hat{X}_0(\tau)}{d\tau} = \begin{cases} X_0(T-)\hat{X}_0(\tau) + \frac{1}{2}\hat{X}_0(\tau)^2 & \text{for } \tau < 0, \\[2mm] X_0(T+)\hat{X}_0(\tau) + \frac{1}{2}\hat{X}_0(\tau)^2 & \text{for } \tau > 0. \end{cases} \tag{10.6.22}$$

We can pass to the limits $\tau \to 0-$ and $\tau \to 0+$ in (10.6.22) and use the continuity condition $(10.6.13)_0$ to find

$$[X_0(T-) + \tfrac{1}{2}\hat{X}_0(0-)]\hat{X}_0(0-) = [X_0(T+) + \tfrac{1}{2}\hat{X}_0(0+)]\hat{X}_0(0+), \tag{10.6.23}$$

which, with the continuity condition $(10.6.12)_0$, implies the result [see (10.6.36)]

$$X_0(T+)^2 = X_0(T-)^2. \tag{10.6.24}$$

This last result, along with (10.6.19), implies $T = \frac{1}{4}$, as listed earlier in (10.6.5).

This result $T = \frac{1}{4}$ can be used with (10.6.19) to yield now

$$X_0(T-) = X_0(\tfrac{1}{4}-) = \tfrac{4}{5} \quad \text{and} \quad X_0(T+) = X_0(\tfrac{1}{4}+) = -\tfrac{4}{5}, \tag{10.6.25}$$

and then the differential equations of (10.6.22) can be integrated with

(10.6.25) to give

$$
\hat{X}_0(\tau) = \begin{cases} \dfrac{8\hat{X}_0(0-)e^{4\tau/5}}{8 + 5\hat{X}_0(0-)[1 - e^{4\tau/5}]} & \text{for } \tau < 0, \\[3ex] \dfrac{8\hat{X}_0(0+)e^{-4\tau/5}}{8 + 5\hat{X}_0(0+)[e^{-4\tau/5} - 1]} & \text{for } \tau > 0. \end{cases} \tag{10.6.26}
$$

The unknown limiting values $\hat{X}_0(0+)$ and $\hat{X}_0(0-)$ satisfy the jump condition

$$
\hat{X}_0(0+) - \hat{X}_0(0-) = \tfrac{8}{5}, \tag{10.6.27}
$$

obtained directly from $(10.6.12)_0$ and (10.6.25).

In order to complete the determination of $\hat{X}_0(0\pm)$, we impose the requirement that the residual

$$
\rho(t) \equiv \rho(t, \varepsilon) := -\varepsilon \frac{d^2 x^0}{dt^2} + x^0 \frac{dx^0}{dt} - (x^0)^3
$$

$$
\left(\text{with } x^0(t) \equiv x^0(t, \varepsilon) := X_0(t) + \hat{X}_0[(t - 0.25)/\varepsilon] \right)
$$

should be suitably small, with $\int_0^1 |\rho(t, \varepsilon)| dt$ of order(ε). In particular, the requirement that $\rho(t, \varepsilon)$ be of order(ε) at the layer point $T = 0.25$,

$$
\rho(T-, \varepsilon), \ \rho(T+, \varepsilon) = O(\varepsilon), \tag{10.6.28}
$$

leads in the present case directly to the results [see (10.6.41)]

$$
\hat{X}_0(0-) = -\tfrac{4}{5}, \qquad \hat{X}_0(0+) = +\tfrac{4}{5}. \tag{10.6.29}
$$

These last values, with (10.6.26), now determine the leading layer coefficient as

$$
\hat{X}_0(\tau) = \begin{cases} -\dfrac{8}{5} \dfrac{e^{4\tau/5}}{1 + e^{4\tau/5}} & \text{for } \tau < 0, \\[3ex] \dfrac{8}{5} \dfrac{e^{-4\tau/5}}{1 + e^{-4\tau/5}} & \text{for } \tau > 0. \end{cases} \tag{10.6.30}
$$

The procedure can be continued recursively to provide higher-order coefficients X_k and $\hat{X}_k$ in the representation (10.6.10), and we obtain the proposed approximate solution $x^N(t, \varepsilon)$ of interior-layer type defined as

$$
x^N(t, \varepsilon) := \sum_{k=0}^{N} \left[X_k(t) + \hat{X}_k\left(\frac{t - 0.25}{\varepsilon} \right) \right] \varepsilon^k, \tag{10.6.31}
$$

where the layer is centered at $t = T = 0.25$, and where the leading terms X_0 and $\hat{X}_0$ are given by (10.6.19) and (10.6.30). The original problem (10.6.4) can now be recast as a linearization about x^N, and we can use a Green function approach to prove that x^N provides a useful approximation to the exact solution of Howes (1978).

The entire procedure can be used effectively to study a wide class of problems with interior layers. For example, this procedure is used in Wan (1980) and Parker and Wan (1984) to study the dimpling of dome-shaped thin elastic shells under various pressure distributions and in van Harten and Vader-Burger (1984) to study the distribution of two competing ecological species. Van Harten and Vader-Burger use an *approximate Green function* approach for the proofs.

We refer the reader to Howes (1978) and Chang and Howes (1984) for general discussions of nonlinear singularly perturbed boundary-value problems with interior layers, along with many interesting examples. Howes uses differential inequalities for the proofs, as first outlined by Nagumo (1937). See also Lutz and Sari (1982), where *nonstandard analysis* is applied to such problems.

Exercises

Exercise 10.6.1: For the problem (10.6.4), show that the reduced solutions $X_0(t) = 2/(3 - 2t)$ and $X_0(t) = -1/(1 + t)$ satisfy the respective conditions (10.4.16) and (10.4.17), so that the results of Section 10.4 imply the existence of two distinct exact solutions x_1 and x_2 of boundary-layer type, as indicated in Figure 10.13.

Exercise 10.6.2: The Dirichlet problem

$$\varepsilon \frac{d^2x}{dt^2} = x\frac{dx}{dt} - x^3 \quad \text{for } 0 < t < 1,$$

$$x(0, \varepsilon) = \tfrac{3}{5} \quad \text{and} \quad x(1, \varepsilon) = -\tfrac{2}{3}, \tag{10.6.32}$$

has a solution of interior-layer type for $\varepsilon \to 0+$. Find the point $t = T$ at which the interior layer is centered. *Hint:* Use the condition (10.6.24) of Example 10.6.2 with a suitable leading outer function corresponding to (but different from) (10.6.19).

Exercise 10.6.3: Consider interior-layer solutions [of *shock* type; cf. Howes (1978)] for the Dirichlet problem

$$\varepsilon \frac{d^2x}{dt^2} = h(t, x) + k(t, x)\frac{dx}{dt} \quad \text{for } 0 < t < 1,$$

$$x = \alpha \quad \text{at } t = 0, \qquad x = \beta \quad \text{at } t = 1. \tag{10.6.33}$$

Assume that the data functions h and k are sufficiently smooth, and assume that the reduced problems

$$h(t, L) + k(t, L)\frac{dL}{dt} = 0 \quad \text{for } 0 \le t \le T,$$
$$L(0) = \alpha \quad \text{at } t = 0,$$
$$(10.6.34)$$

and

$$h(t, R) + k(t, R)\frac{dR}{dt} = 0 \quad \text{for } T \le t \le 1,$$
$$R(1) = \beta \quad \text{at } t = 1,$$
$$(10.6.35)$$

have smooth solution functions $L = L(t)$ and $R = R(t)$ satisfying the condition

$$\int_{L(T)}^{R(T)} k(T, s)\, ds = 0 \quad \text{for some fixed } T \in (0,1). \quad (10.6.36)$$

Moreover, assume that the terminal-value problem

$$\frac{d\hat{L}(\tau)}{d\tau} = \int_0^{\hat{L}(\tau)} k(T, L(T) + s)\, ds \quad \text{for } \tau < 0,$$
$$\hat{L} = \kappa_- \quad \text{at } \tau = 0,$$
$$(10.6.37)$$

and the initial-value problem

$$\frac{d\hat{R}(\tau)}{d\tau} = \int_0^{\hat{R}(\tau)} k(T, R(T) + s)\, ds \quad \text{for } \tau > 0,$$
$$\hat{R} = \kappa_+ \quad \text{at } \tau = 0,$$
$$(10.6.38)$$

have smooth solutions for suitable constant κ_- and κ_+ satisfying the conditions

$$L(T) + \kappa_- = R(T) + \kappa_+, \quad (10.6.39)$$
$$\int_0^{\kappa_-} k(T, L(T) + s)\, ds \ne 0, \quad \int_0^{\kappa_+} k(T, R(T) + s)\, ds \ne 0, \quad (10.6.40)$$

and

$$h(T, L(T) + \kappa_-) = k(T, L(T) + \kappa_-)k(T, L(T))^{-1}h(T, L(T)),$$
$$h(T, R(T) + \kappa_+) = k(T, R(T) + \kappa_+)k(T, R(T))^{-1}h(T, R(T)).$$
$$(10.6.41)$$

(a) Show that the solution $\hat{L}$ of (10.6.37) exists globally for all $\tau < 0$ and decays exponentially as

$$|\hat{L}(\tau)| \le |\kappa_-|e^{\nu\tau} \quad \text{for } \tau \le 0 \quad (10.6.42)$$

if there holds [compare with (10.4.16)]

$$x \cdot \left[\int_0^1 k(T, L(T) + sx) \, ds \right] x \geq \nu |x|^2 \qquad (10.6.43)$$

for all x between 0 and κ_-, for a fixed constant $\nu > 0$. We see similarly that the solution $\hat{R}$ of (10.6.38) exists globally and satisfies

$$|\hat{R}(\tau)| \leq |\kappa_+| e^{-\nu \tau} \quad \text{for } \tau \geq 0 \qquad (10.6.44)$$

if there holds [compare with (10.4.17) and (10.4.34)]

$$x \cdot \left[\int_0^1 k(T, R(T) + sx) \, ds \right] x \leq -\nu |x|^2 \qquad (10.6.45)$$

for all x between 0 and κ_+. Note that (10.6.40) rules out the particular constant solutions $\hat{L} \equiv \kappa_-$ or $\hat{R} \equiv \kappa_+$.

(b) Define the leading terms $X_0(t)$ and $\hat{X}_0(\tau)$ in the expansion (10.6.10) as

$$X_0(t) := \begin{cases} L(t) & \text{for } 0 \leq t < T, \\ R(T) & \text{for } T < t \leq 1, \end{cases} \qquad (10.6.46)$$

and

$$\hat{X}_0(\tau) := \begin{cases} \hat{L}(\tau) & \text{for } \tau < 0, \\ \hat{R}(\tau) & \text{for } \tau > 0, \end{cases} \qquad (10.6.47)$$

and let x^0 be the lowest-order proposed approximate solution defined as $x^0(t, \varepsilon) := X_0(t) + \hat{X}_0((t - T)/\varepsilon)$, so that x^0 is a smooth function of t on each of the subintervals $[0, T)$ and $(T, 1]$. Show with (10.6.39) that x^0 can be defined at $t = T$ so as to be of class C^0 uniformly for $0 \leq t \leq 1$, even though $X_0(t)$ and $\hat{X}_0((t - T)/\varepsilon)$ suffer discontinuities at $t = T$.

(c) Let ρ be the (lowest-order) residual defined as

$$\rho(t) \equiv \rho(t, \varepsilon) := -\varepsilon \frac{d^2 x^0}{dt^2} + h(t, x^0(t, \varepsilon)) + k(t, x^0(t, \varepsilon)) \frac{dx^0}{dt} \qquad (10.6.48)$$

for $t \in [0, T) \cup (T, 1]$ and for all small enough $\varepsilon > 0$. Show that ρ satisfies

$$\int_0^1 |\rho(t, \varepsilon)| \, dt \leq \text{const. } \varepsilon \qquad (10.6.49)$$

for a fixed constant, as $\varepsilon \to 0+$. *Hint:* Use the constructions of X_0 and $\hat{X}_0$ to find

$$\rho(t, \varepsilon) = \rho_1(t, \varepsilon) + \rho_2(t, \varepsilon) \qquad (10.6.50)$$

with

$$\rho_1(t, \varepsilon) = \begin{cases} \dfrac{1}{\varepsilon}[k(t, L(t) + \hat{L}(\tau)) - k(T, L(T) + \hat{L}(\tau))]\dfrac{d\hat{L}(\tau)}{d\tau} \\ \qquad\qquad\qquad \text{for } t \in [0, T), \\[2ex] \dfrac{1}{\varepsilon}[k(t, R(t) + \hat{R}(\tau)) - k(T, R(T) + \hat{R}(\tau))]\dfrac{d\hat{R}(\tau)}{d\tau} \\ \qquad\qquad\qquad \text{for } t \in (T, 1], \end{cases}$$

$$(10.6.51)$$

with $\tau = (t - T)/\varepsilon$, and

$$\rho_2(t, \varepsilon) = -\varepsilon L''(t) + [h(t, L(t) + \hat{L}(\tau)) - h(t, L(t))]$$
$$+ [k(t, L(t) + \hat{L}(\tau)) - k(t, L(t))]L'(t)$$
$$\text{for } t \in [0, T), \quad (10.6.52)$$

and a similar result for $t \in (T, 1]$ with L and $\hat{L}$ replaced by R and $\hat{R}$ on the right side of (10.6.52). Now use (10.6.51) along with such results as

$$k(T + \varepsilon\tau, L(T + \varepsilon\tau) + \hat{L}) - k(T, L(T) + \hat{L})$$
$$= \int_0^1 \frac{d}{ds}k(T + s\varepsilon\tau, L(T + s\varepsilon\tau) + \hat{L})\, ds$$
$$= \varepsilon O(\tau) \qquad\qquad (10.6.53)$$

to show that (10.6.49) holds with ρ replaced there by ρ_1. A similar argument with (10.6.52) shows also that (10.6.49) holds with ρ replaced by ρ_2, from which the stated result follows. Note also that (10.6.41) implies that $\rho(t, \varepsilon)$ is order(ε) at $t = T$,

$$\rho(T-, \varepsilon), \rho(T+, \varepsilon) = O(\varepsilon) \quad \text{as } \varepsilon \to 0+. \qquad (10.6.54)$$

The analysis can be pursued further so as to demonstrate the existence of an exact solution for the problem (10.6.33) that is well approximated by x^0, but we need not do so here. As usual the data h, k, α, and β can be permitted to depend regularly on ε, as in (10.4.5).

10.7 A numerical algorithm of Maier

It was noted in Section 6.4 that many classical numerical initial-value algorithms fail for singularly perturbed initial-value problems because of the multiscale character of solutions, whereas the method of Miranker, which is based on ideas from singular perturbation theory, is well suited for such problems. Similarly, most classical boundary-value solvers fail for singularly perturbed boundary-value problems as the small parameter tends toward zero.

Such singularly perturbed boundary-value problems arise in numerous important areas of applications, and there has thus been a strong impetus toward the development of effective algorithms for such problems. Indeed, there is a large literature on the numerical study of such stiff boundary-value problems, of which we mention here only the works of Pearson (1968a, 1968b), Abrahamsson, Keller, and Kreiss (1974), and Kreiss (1978) based on finite-difference methods, the works of Flaherty and Mathon (1980), Ascher (1980), Weiss (1984), and Maier (1984b) based on collocation methods, the work of Osher (1981) based on the convergence to a steady state of solutions of auxiliary systems, the work of Flaherty and O'Malley (1977, 1980) and O'Malley (1984) based on singular-perturbation theory, and the work of Maier (1982, 1984a) and Mattheij and O'Malley (1984) employing multiple shooting; see also Hemker and Miller (1979) and Miller (1980, 1984). In fact, we are content here to describe briefly only the method of Maier (1982, 1984a), which is based on multiple shooting and which provides an accurate and efficient method for numerical study of solutions of boundary-layer type (with possible boundary layers at the endpoints). Interior layers are not considered in Maier (1982, 1984a), although boundary-layer turning points are permitted within a boundary layer.

Consider a boundary-value problem of the form

$$\frac{dy}{dt} = f(t, y) \quad \text{for } t \in [a, b], \quad \text{with } r[y(a), y(b)] = 0, \quad (10.7.1)$$

with $a, b \in \mathbb{R}$, $y = (y_1, y_2, \ldots, y_n)^{\mathrm{T}}$, $y_i \colon [a, b] \to \mathbb{R}$, $f = (f_1, \ldots, f_n)^{\mathrm{T}}$, $f_i \colon [a, b] \times \mathbb{R}^n \to \mathbb{R}$, $r = (r_1, \ldots, r_n)^{T}$, and $r_i \colon \mathbb{R}^n \times \mathbb{R}^n \to \mathbb{R}$, and where the various data functions are assumed to be sufficiently smooth. The algorithm of Maier does not require any explicit a priori identification of any small parameter or parameters such as those that typically appear in singularly perturbed problems, although the algorithm is explicitly constructed so as to handle possible solutions of boundary-layer type. The algorithm automatically decomposes the interval $[a, b]$ into an outer interval in which the solution behaves regularly, and one or two boundary-layer intervals, each adjoining an endpoint a or b, where the solution may undergo rapid variation.

The algorithm employs multiple shooting [see Stoer and Bulirsch (1980)], which replaces the given boundary-value problem with a certain collection of initial-value problems. For the problems considered here, precisely one of the resulting initial-value problems lives within the outer interval, and the remaining initial-value problems live within boundary layers. The multiscale character of solutions of such problems is a reflection of the fact that the Jacobian functional matrix $D_y f$ of the right side of the differential equation has some eigenvalues with absolute real parts that are large compared with other eigenvalues. The automatic decomposition of the given interval $[a, b]$ into the outer and boundary-layer subintervals is based on a certain numerically computed measure of the sizes of the real parts of such eigenvalues of large absolute real parts.

The eigenvalues with large absolute real parts tend to produce very small step sizes during the numerical solution of the *initial-value problems* in multiple shooting, and this is ordinarily the case also for the initial-value problem living within the outer interval, where the exact solution is regular and where we would wish to be permitted to use large step sizes. To overcome this difficulty, Maier uses for the initial-value problem living in the outer interval a special Runge/Kutta method with a suitable stability property that permits large step sizes. The remaining initial-value problems that live within boundary layers are solved with a different, high-order Runge/Kutta method. Multiple shooting connects the various initial-value problems in a suitable manner, yielding a certain nonlinear algebraic system that is solved using a modified Newton method in conjunction with Householder transformations. The entire algorithm is iterative, and nodes (subintervals) are removed and/or inserted automatically as needed, prior to each new (global) iterative cycle.

Using numerically computed measures of the eigenvalues of the functional matrix $D_y f$, the algorithm first automatically chooses initial thicknesses for the boundary-layer subintervals, along with an initial number m of nodes and an initial distribution of nodes t_j ($j = 1, 2, \ldots, m$) so as to subdivide the interval $[a, b]$ into $m - 1$ subintervals, with many nodes placed densely in the boundary layers and with no nodes placed in the outer subinterval, as illustrated in Figure 10.15. Each boundary layer is initially required to have at least five nodes, and the initial distribution of nodes in the boundary layers is taken to be uniform, with equidistant nodes there. After each iterative cycle, the boundary-layer thicknesses and node distribution are automatically tested numerically and modified when needed before the next iteration. Hence, the decomposition of $[a, b]$ into outer and boundary-layer intervals is generally modified somewhat during the course of the calculation until a convergence criterion has been achieved. The final distribution of nodes is generally not uniform, but rather is governed by the local values of the eigenvalues of $D_y f(t, y)$.

At the nodes t_j, n-dimensional starting values s_j are prescribed, for $j = 1, \ldots, m$. If $y(t, t_j, s_j)$ denotes the solution of the initial-value problem

$$\frac{dy}{dt} = f(t, y) \quad \text{for } t \in \left[t_j, t_{j+1} \right),$$
$$y = s_j \quad \text{at } t = t_j,$$

$$(10.7.2)_j$$

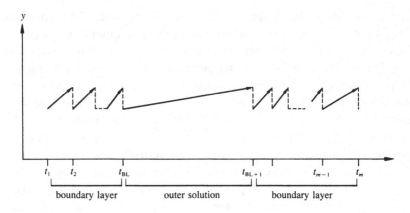

Figure 10.15

for $j = 1, 2, \ldots, m - 1$, then s_j must be determined so that the piece-wise-composed function

$$y(t) := \begin{cases} y(t, t_j, s_j) & \text{for } t \in [t_j, t_{j+1}), \; j = 1, \ldots, m - 1, \\ s_m & \text{at } t = t_m, \end{cases}$$

(10.7.3)

is continuous and satisfies the boundary conditions. This yields the generally nonlinear system

$$F(s) \equiv \begin{bmatrix} F_1(s_1, s_2) \\ F_2(s_2, s_3) \\ \vdots \\ F_{m-2}(s_{m-2}, s_{m-1}) \\ F_{m-1}(s_1, s_{m-1}) \end{bmatrix} := \begin{bmatrix} y(t_2, t_1, s_1) - s_2 \\ y(t_3, t_2, s_2) - s_3 \\ \vdots \\ y(t_{m-1}, t_{m-2}, s_{m-2}) - s_{m-1} \\ r[s_1, y(t_m, t_{m-1}, s_{m-1})] \end{bmatrix} = 0,$$

(10.7.4)

where $s = (s_1, \ldots, s_{m-1})^T$.

As mentioned earlier, the initial-value problem $(10.7.2)_j$ is solved numerically with an integration method that depends on whether or not the subinterval $[t_j, t_{j+1})$ lies in a boundary layer. A specially chosen, stable integrator is used if this subinterval coincides with the outer interval, whereas a high-order method is used if this subinterval is in a boundary layer. The nonlinear system (10.7.4) is solved iteratively with a modified Newton method that also employs Householder transformations. Details are omitted; see Maier (1984a).

Example 10.7.1: Consider again the problem (10.2.61), rewritten here for convenience as

$$\varepsilon^2 \frac{d^2x}{dt^2} = (x^2 - 1)(x^2 - 4) \quad \text{for } 0 < t < 1,$$

$$\frac{dx}{dt} = 0 \quad \text{at } t = 0, \quad \text{and} \quad x = \beta \quad \text{at } t = 1. \tag{10.7.5}$$

As in Section 10.2, let β_1 and β_2 be the real roots of the respective cubics $q_1(\beta)$ and $q_2(\beta)$ of (10.2.62), with

$$\beta_1 \doteq -2.40569 \quad \text{and} \quad \beta_2 \doteq 0.275279. \tag{10.7.6}$$

For any fixed $\beta > \beta_1$, it follows from Theorem 10.2.1 that the problem (10.7.5) has a solution x_1 of boundary-layer type satisfying

$$\lim_{\substack{\varepsilon \to 0+ \\ \text{fixed } 0 < t < 1}} x_1(t, \varepsilon) = -1. \tag{10.7.7}$$

Similarly, for fixed $\beta > \beta_2$, it follows from Theorem 10.2.1 that (10.7.5) has a solution x_2 of boundary-layer type satisfying

$$\lim_{\substack{\varepsilon \to 0+ \\ \text{fixed } 0 < t < 1}} x_2(t, \varepsilon) = +2. \tag{10.7.8}$$

The problem (10.7.5) can be written in the form (10.7.1), with $n = 2$, and

$$y = \begin{bmatrix} y_1 \\ y_2 \end{bmatrix} := \begin{bmatrix} x \\ \varepsilon x' \end{bmatrix}, \quad f = \begin{bmatrix} f_1 \\ f_2 \end{bmatrix} := \frac{1}{\varepsilon} \begin{bmatrix} y_2 \\ (y_1^2 - 1)(y_1^2 - 4) \end{bmatrix},$$

$$r = \begin{bmatrix} r_1 \\ r_2 \end{bmatrix} := Ly(0) + Ry(1) - \begin{bmatrix} 0 \\ \beta \end{bmatrix}, \tag{10.7.9}$$

$$L := \begin{bmatrix} 0 & 1 \\ 0 & 0 \end{bmatrix}, \quad R := \begin{bmatrix} 0 & 0 \\ 1 & 0 \end{bmatrix}.$$

The algorithm was applied to this problem for the solutions x_1 and x_2 of (10.7.7) and (10.7.8), for different values of the boundary value β. The numerical calculation was started in each case with the two nodes $t_1 = 0$ and $t_2 = 1$ ($m = 2$), with corresponding s_j taken from the respective exact outer solutions for x_1 and x_2, as

$$s_1 = s_2 = \begin{bmatrix} -1 \\ 0 \end{bmatrix} \quad \text{for the solution } x_1 \tag{10.7.10}$$

and

$$s_1 = s_2 = \begin{bmatrix} +2 \\ 0 \end{bmatrix} \quad \text{for the solution } x_2. \tag{10.7.11}$$

Table 10.1

	β						
	-3.0	-2.41	-2.4	-1.0	$+0.27$	$+0.28$	$+0.5$
x_1	$-$	$-$	$+$	$+$	$+$	$+$	$+$
x_2	$-$	$-$	$-$	$-$	$-$	$+$	$+$

Note: Plus sign indicates that accurate values are given by the numerical algorithm; minus sign means that the numerical algorithm indicates that a solution does not exist.

For example, in the case (10.7.10) for x_1 with $\beta: = -2.4$ and with $\varepsilon = 10^{-5}$ [$\varepsilon^2 = 10^{-10}$ in (10.7.5)], the algorithm produces accurate solution values in less than two seconds of computer time using six nodes in a boundary layer adjoining the right endpoint $t = 1$, with about 70,000 function calls of the right side of the differential equation. The calculations were performed in FORTRAN IV with single precision (48-bit mantissa) on the CDC CYBER 175 of the Leibniz Rechenzentrum der Bayerischen Akademie der Wissenschaften (Maier, private communication).

The results regarding "numerical existence" for x_1 and x_2 are summarized in Table 10.1 for the values $\beta = -3.0$, -2.41, -2.4, -1.0, $+0.27$, $+0.28$, and $+0.5$. In each case the calculation was performed for the different values $\varepsilon^2 = 10^{-3}, 10^{-4}, \ldots, 10^{-11}, 10^{-12}$. We find that the algorithm produces accurate solution values for x_1 in the cases $\beta = -2.4$, -1.0, $+0.27$, $+0.28$, and $+0.5$, but the algorithm fails to converge for x_1 in the cases $\beta = -3.0$ and -2.41. For x_2, the algorithm produces accurate solution values in the cases $\beta = +0.28$ and $\beta = +0.5$, but the algorithm fails to converge for the other values of β listed in Table 10.1.

When exact solutions x_1 and/or x_2 exist, the algorithm produces accurate approximate-solution values efficiently, and the accuracy is good throughout the interval, including within the boundary layer. On the other hand, in each case in which exact solutions x_1 and/or x_2 fail to exist, then the algorithm automatically gives an indication that a corresponding solution fails to exist. In every case listed in Table 10.1 the algorithm is in agreement with the theoretical results of Section 10.2.

Note that, as expected from general theory, suitable starting values are required in the use of the numerical algorithm. As in Example 10.7.1, it is often possible to compute an appropriate outer solution initially, and that outer solution can then be used to provide good starting values for the s_j. Additional techniques are available for choosing or improving the starting values [see Maier (1982, 1984a), where nontrivial applications of the algorithm to several important physical problems are also given].

References

Abrahamsson, L., Keller, H. B., and Kreiss, H.-O. (1974). "Difference approximations for singular perturbations of systems of ordinary differential equations," *Numer. Math.* **22**, 367–91.

Ackerberg, R. C., and O'Malley, R. E., Jr. (1970). "Boundary layer problems exhibiting resonance," *Studies in Appl. Math.* **49**, 277–95.

Aiken, R. C., and Lapidus, L. (1974). "An effective numerical integration method for typical stiff systems," *Amer. Inst. Chem. Eng. Journal* **20**, 368–75.

Aris, R. (1969). "On stability criteria of chemical reaction engineering," *Chem. Engineering Science* **24**, 149–69.

——— (1975). *The Mathematical Theory of Diffusion and Reaction in Permeable Catalysts. Vol. 1: The Theory of the Steady State* (Oxford: Clarendon Press).

Ascher, U. (1980). "Solving boundary value problems with a spline-collocation code," *J. Comp. Phys.* **34**, 401–13.

Atkinson, F. V. (1954). "The asymptotic solution of second order differential equations," *Ann. Mat. Pura Appl.* **37**, 347–78.

Bardeen, J. (1950). "Theory of relation between hole concentration and characteristics of germanium points contacts," *Bell System Tech. J.* **29**, 469–95.

Baum, H. R. (1972). "On the weakly damped harmonic oscillator," *Quart. Applied Mathematics* **29**, 573–6.

Bellman, R. (1953). *Stability Theory of Differential Equations* (New York: McGraw-Hill).

Bensoussan, A., Lions, J. L., and Papanicolaou, G. (1978). *Asymptotic Analysis for Periodic Structures* (Amsterdam: North-Holland).

Berger, M. S., and Fraenkel, L. E. (1969–70). "On the asymptotic solution of a nonlinear Dirichlet problem," *J. Math. Mech.* (*Indiana Univ. Math. J.*) **19**, 553–85.

Birkhoff, G. D. (1908). "On the asymptotic character of the solutions of certain linear differential equations containing a parameter," *Trans. Amer. Math. Soc.* **9**, 219–31.

Birkhoff, G., and Rota, G.-C. (1960). *Ordinary Differential Equations* (Boston: Ginn and Company).

——— (1978). *Ordinary Differential Equations*, 3rd edition (New York: Wiley).

Bobisud, L. (1966). "Degeneration of the solutions of certain well-posed systems of partial differential equations depending on a small parameter," *J. Math. Anal. Appl.* **16**, 419–54.

Boglaev, Y. P. (1970). "The two-point problem for a class of ordinary differential equations with a small parameter coefficient of the derivative," *USSR Comp. Math. Phys.* **10**, 191–204.

Bogoliubov, N. N., and Mitropolosky, Y. A. (1961). *Asymptotic Methods in the Theory of Nonlinear Oscillations* (Delhi: Hindustan Publishing).

Boyce, W. E., and Handelman, G. H. (1961). "Vibrations of rotating beams with tip mass," *Z. Angew. Math. Phys.* **12**, 369–92.

Brillouin, L. (1926). "Remarques sur la mecanique ondultoire," *J. Phys. Radium* **7**, 353–68.

Brish, N. I. (1954). "On boundary value problems for the equation $\varepsilon \cdot d^2y/dx^2 = f(x, y, dy/dx)$ for small ε," *Dokl. Akad. Nauk. SSSR* **95**, 429–32.

Bromberg, E. (1956). "Nonlinear bending of a circular plate under normal pressure," *Comm. Pure Appl. Math.* **9**, 633–59.

Bush, W. B. (1971). "On the Lagerstrom mathematical model for viscous flow at low Reynolds number," *SIAM J. Appl. Math.* **20**, 279–87.

Butuzov, V. F. (1965). "Asymptotic behavior of solutions of differential equations with a small parameter in the derivative on a semi-infinite interval," *Vestnik Moskov. Univ. Ser. I Mat. Meh.*, no. 1, 16–25.

Butuzov, V. F., and Vasil'eva, A. B. (1970). "Differential and difference equation systems with a small parameter for the case in which the unperturbed (singular) system is in the spectrum," *Diff. Equations* **6**, 499–510.

Campbell, S. L. (1980). *Singular Systems of Differential Equations* (San Francisco: Pitman). (1982). *Singular Systems of Differential Equations II* (San Francisco: Pitman).

Campbell, S. L., and Rose, N. J. (1979). "Singular perturbation of autonomous linear systems," *SIAM J. Math. Anal.* **10**, 542–51.

Carlini, F. (1817). "Richerche sulla convergenza della serie che serve alla soluzione del problema di Keplero," *Memoria di Francesco Carlini* (Milan); *Jacobi's Gesammelte Werke* **7**, 189–245; *Astronom. Nach.* **30**(1850), 197–254.

Carrier, G. F. (1970). "Singular perturbation theory and geophysics," *SIAM Review* **12**, 175–93.

Carrier, G. F., and Pearson, C. (1968). *Ordinary Differential Equations* (Waltham, Mass.: Blaisdell).

Chang, K. W. (1969*a*). "Two problems in singular perturbations of differential equations," *J. Australian Math. Soc.* **10**, 33–50.

(1969*b*). "Remarks on a certain hypothesis in singular perturbations," *Proc. Amer. Math. Soc.* **23**, 41–45.

(1972). "Singular perturbations of a general boundary value problem," *SIAM J. Math. Anal.* **3**, 520–6.

(1974*a*). "Approximate solutions of nonlinear boundary value problems involving a small parameter," *SIAM J. Appl. Math.* **26**, 554–67.

(1974*b*). "Diagonalization method for a vector boundary problem of singular perturbation type," *J. Math. Anal. Appl.* **48**, 652–65.

(1975). "Diagonalization methods in singular perturbations," *International Conference on Differential Equations*, ed. H. A. Antosiewicz (New York: Academic Press), pp. 164–84.

(1976). "Singular perturbations of a boundary value problem for a vector second order differential equation," *SIAM J. Appl. Math.* **30**, 42–54.

Chang, K. W., and Coppel, W. S. (1969). "Singular perturbations of initial value problems over a finite interval," *Arch. Rational Mech. Anal.* **32**, 268–80.

Chang, K. W., and Howes, F. A. (1984). *Nonlinear Singular Perturbation Phenomena: Theory and Applications* (Berlin: Springer).

Chen, J. (1972). "Asymptotic solutions of certain singularly perturbed boundary value problems arising in chemical reactor theory" (New York University, doctoral dissertation).

Chen, J., and O'Malley, R. E., Jr. (1974). "On the asymptotic solution of a two-parameter boundary value problem arising in chemical reactor theory," *SIAM J. Appl. Math.* **26**, 717–29.

Chow, P. L. (1972). "Asymptotic solutions of inhomogeneous initial boundary value problems for weakly nonlinear partial differential equations," *SIAM J. Appl. Math.* **22**, 629–47.

Cochran, J. A. (1962). "Problems in singular perturbation theory" (Stanford University, doctoral dissertation).

——— (1968). "On the uniqueness of solutions of linear differential equations," *J. Math. Anal. Appl.* **22**, 418–26.

Coddington, E. A., and Levinson, N. (1952). "A boundary value problem for a nonlinear differential equation with a small parameter," *Proc. Amer. Math. Soc.* **3**, 73–81.

Cole, J. D. (1968). *Perturbation Methods in Applied Mathematics* (Waltham, Mass.: Blaisdell).

Cole, J. D., and Kevorkian, J. (1963). "Uniformly valid asymptotic approximations for certain non-linear differential equations," *Proceedings of an International Symposium on Non-linear Differential Equations and Non-linear Mechanics* (New York: Academic Press), pp. 113–20.

Cole, R. H. (1968). *Theory of Ordinary Differential Equations* (New York: Appleton-Century-Crofts).

Cook, L. P., and Eckhaus, W. (1973). "Resonance in a boundary value problem of singular perturbation type," *Studies in Appl. Math.* **52**, 129–39.

Coppel, W. A. (1965). *Stability and Asymptotic Behavior of Differential Equations* (Boston: D. C. Heath).

——— (1967). "Dichotomies and reducibility," *J. Diff. Equations* **3**, 500–21.

——— (1978). *Dichotomies in Stability Theory. Lecture Notes in Mathematics* **629** (Berlin: Springer-Verlag).

Cotton, E. (1910). "Sur lés solutions asymptotiques des équations différentielles," *Paris Compt. Rend. Acad. Sci.*, **150**, 511–13.

Cronin, J. (1964). *Fixed Points and Topological Degree in Nonlinear Analysis* (Providence, R.I.: American Mathematical Society).

Danby, J. M. A. (1962). *Fundamentals of Celestial Mechanics* (New York: Macmillan).

Davis, H. T. (1962). *Introduction to Nonlinear Differential and Integral Equations* (New York: Dover).

De Villiers, J. M. (1973). "A uniform asymptotic expansion of the positive solution of a non-linear Dirichlet problem," *Proc. London Math. Soc.* **27**, 701–22.

Dorr, F. W. (1970). "Some examples of singular perturbation problems with turning points," *SIAM J. Math. Anal.* **1**, 141–6.

Dorr, F. W., Parter, S. V., and Shampine, L. F. (1973). "Applications of the maximum principle to singular perturbation problems," *SIAM Review* **15**, 43–88.

Eckhaus, W. (1973). *Matched Asymptotic Expansions and Singular Perturbations* (Amsterdam: North-Holland).

——— (1979). *Asymptotic Analysis of Singular Perturbations* (Amsterdam: North-Holland).

Eckhaus, W., and de Jager, E. M. (editors) (1982). *Theory and Applications of Singular Perturbations* (proceedings of a conference held in Oberwolfach, August 16–22, 1981), *Lecture Notes in Mathematics* **942** (Berlin: Springer-Verlag).

Erdélyi, A. (1956). *Asymptotic Expansions* (New York: Dover).

——— (1962). "On a nonlinear boundary value problem involving a small parameter," *J. Australian Math. Soc.* **2**, 425–39.

——— (1968a). "Approximate solutions of a nonlinear boundary value problem," *Arch. Rational Mech. Anal.* **29**, 1–17.

(1968b). "Two-variable expansions for singular perturbations," *J. Inst. Math. Appl.*
4, 113–19.

(1975). "A case history in singular perturbations," *International Conference on Differential Equations*, ed. H. A. Antosiewicz (New York: Academic Press), pp. 266–86.

Fenichel, N. (1979). "Geometric singular perturbation theory for ordinary differential equations," *J. Diff. Equations* **31**, 53–98.

Ferguson, W. E., Jr. (1975). "A Singularly perturbed linear two-point boundary-value problem" (California Institute of Technology, doctoral dissertation).

Fife, P. C. (1972). "Singular perturbation problems whose degenerate forms have many solutions," *Applicable Analysis* **1**, 331–58.

(1973a). "Semilinear elliptic boundary value problem with small parameters," *Arch. Rational Mech. Anal.* **52**, 205–32.

(1973b). "Singular perturbation by a quasilinear operator," *Lecture Notes in Mathematics* **322** (Berlin: Springer-Verlag), pp. 87–100.

(1974). "Transition layers in singular perturbation problems," *J. Diff. Equations* **15**, 77–105.

(1976). "Boundary and interior transition layer phenomena for pairs of second-order differential equations," *J. Math. Anal. Appl.* **54**, 497–521.

Flaherty, J. E., and Mathon, W. (1980). "Collocation with polynomial and tension splines for singularly-perturbed boundary value problems," *SIAM J. Sci. Stat. Comput.* **1**, 260–89.

Flaherty, J. E., and O'Malley, R. E., Jr. (1977). "The numerical solution of boundary value problems for stiff differential equations," *Math. Comput.* **31**, 66–93.

(1980). "On the numerical integration of two-point boundary value problems for stiff systems of ordinary differential equations," *Boundary and Interior Layers: Computational and Asymptotic Methods*, ed. J. J. H. Miller (Dublin: Boole Press), pp. 93–102.

(1982). "Singularly perturbed boundary value problems for nonlinear systems, including a challenging problem for a nonlinear beam," *Lecture Notes in Mathematics* **942** (Berlin: Springer-Verlag), pp. 170–91.

Flatto, L., and Levinson, N. (1955). "Periodic solutions of singularly perturbed systems," *J. Math. Mech.* (*J. Rational Mech. Anal.*) **4**, 943–50.

Gans, R. (1915). "Propagation of light through an inhomogeneous medium," *Ann. Physik* **47**, 709–36.

Garabedian, P. (1964). *Partial Differential Equations* (New York: Wiley Interscience).

Geel, R. (1978). "Singular perturbations of hyperbolic type" (University of Amsterdam, doctoral dissertation).

(1981a). "Linear initial value problems with a singular perturbation of hyperbolic type," *Proc. Roy. Soc. Edinburgh* **87A**, 167–87.

(1981b). "Nonlinear initial value problems with a singular perturbation of hyperbolic type," *Proc. Roy. Soc. Edinburgh* **89A**, 333–45.

Geel, R., and de Jager, E. M. (1979). "Initial value problems for singularly perturbed nonlinear ordinary differential equations," *Math. Chronicle* **8**, 25–38.

Gordon, N. (1975). "Matched asymptotic expansion solutions of nonlinear partial differential equations with a small parameter," *SIAM J. Math. Anal.* **6**, 1007–16.

Green, G. (1837). "On the motion of waves in a variable canal of small depth," *Trans. Cambridge Philos. Soc.* **6**, 457–62.

Greenlee, W. M., and Snow, R. E. (1975). "Two-timing on the half line for damped oscillation equations," *J. Math. Anal. Appl.* **51**, 394–428.

de Groen, P. P. N. (1980). "The singularly perturbed turning-point problem: a spectral approach," *Singular Perturbations and Asymptotics* (proceedings of an advanced seminar conducted by the Mathematics Research Center, University of Wisconsin-Madison, May 28–30, 1980), ed. R. E. Meyer and S. V. Parter (New York: Academic Press), pp. 149–72.

Gronwall, T. H. (1919). "Note on the derivative with respect to a parameter of the solutions of a system of differential equations," *Annals of Math.* **20**, 292–6.

Haber, S., and Levinson, N. (1955). "A boundary value problem for a singularly perturbed differential equation," *Proc. Amer. Math. Soc.* **6**, 866–72.

Habets, P. (1974). "Singular perturbations of a vector boundary value problem," *Lecture Notes in Mathematics* **415** (Berlin: Springer-Verlag), pp. 149–54.

(1983). "Singular perturbations in nonlinear systems and optimal control," *International Centre for Mechanical Sciences, Courses and Lectures, No. 280: Singular Perturbations in Systems and Control*, ed. M. D. Ardema (Berlin: Springer-Verlag), pp. 103–42.

Handelman, G. H., and Keller, J. B. (1962). "Small vibrations of a slightly stiff pendulum," *Proceedings of the 4th U.S. National Congress on Applied Mechanics* **1** (New York: American Society of Mechanical Engineers), pp. 195–202.

Handelman, G. H., Keller, J. B., and O'Malley, R. E., Jr. (1968). "Loss of boundary conditions in the asymptotic solution of linear ordinary differential equations, I. Eigenvalue problems," *Comm. Pure Appl. Math.* **21**, 243–61.

Harris, W. A., Jr. (1960). "Singular perturbations of two point boundary problems for linear systems of ordinary differential equations," *Arch. Rational Mech. Anal.* **5**, 212–25.

(1961). "Singular perturbations of eigenvalue problems," *Arch. Rational Mech. Anal.* **7**, 224–41.

(1962*a*). "Singular perturbations of two point boundary problems," *J. Math. Mech.* **11**, 371–82.

(1962*b*). "Singular perturbations of a boundary value problem for a nonlinear system of differential equations," *Duke Math. J.* **29**, 429–45.

(1973). "Singularly perturbed boundary value problems revisited," *Lecture Notes in Math.* **312** (Berlin: Springer-Verlag), pp. 54–64.

Harris, W. A., Jr., and Lutz, D. A. (1974). "On the asymptotic integration of linear differential systems," *J. Math. Anal. Appl.* **48**, 1–16.

(1977). "A unified theory of asymptotic integration," *J. Math. Anal. Appl.* **57**, 571–86.

van Harten, A. (1978*a*). "Nonlinear singular perturbation problems: proofs of correctness of a formal approximation based on a contraction principle in a Banach space," *J. Math. Anal. Appl.* **65**, 126–68.

(1978*b*). "Singular perturbations for nonlinear second-order ODE with nonlinear b.c. of Neumann or mixed type," *J. Math. Anal. Appl.* **65**, 169–83.

van Harten, A., and Vader-Burger, E. (1984). "Approximate Green functions as a tool to prove correctness of a formal approximation in a model of competing and diffusing species," preprint No. 343, Department of Mathematics, University of Utrecht.

Hartman, P., and Wintner, A. (1955). "Asymptotic integrations of linear differential equations," *Amer. J. Math.* **77**, 45–86, 932.

Heineken, F. G., Tsuchiya, H. M., and Aris, R. (1967). "On the mathematical status of the pseudo-steady state hypothesis of biochemical kinetics," *Math. Biosciences* **1**, 95–113.

Hemker, P. W., and Miller, J. J. H. (editors) (1979). *Numerical Analysis of Singular Perturbation Problems* (proceedings of a conference held at the University of Nijmegen, The Netherlands, May 30–June 2, 1978) (New York: Academic Press).

Hirsch, M. W., and Smale, S. (1974). *Differential Equations, Dynamical Systems, and Linear Algebra* (New York: Academic Press).

Hoppensteadt, F. (1966). "Singular perturbations on the infinite interval," *Trans. Amer. Math. Soc.* **123**, 521–35.

(1971*a*). "On quasi-linear parabolic equations with a small parameter," *Comm. Pure Appl. Math.* **24**, 17–38.

(1971*b*). "Properties of solutions of ordinary differential equations with small parameters," *Comm. Pure Appl. Math.* **24**, 807–40.

(editor) (1979). *Nonlinear Oscillations in Biology* (Providence, R.I.: American Mathematical Society).

Hoppensteadt, F. C., and Miranker, W. L. (1976). "Differential equations having rapidly changing solutions: analytic methods for weakly nonlinear systems," *J. Diff. Equations* **22**, 237–49.

Howes, F. A. (1975). "Singularly perturbed nonlinear boundary value problems with turning points," *SIAM J. Math. Anal.* **6**, 644–60.

(1976*a*). "Effective characterization of the asymptotic behavior of solutions of singularly perturbed boundary value problems," *SIAM J. Appl. Math.* **30**, 296–306.

(1976*b*). "Differential inequalities and applications to nonlinear singular perturbation problems," *J. Diff. Equations* **20**, 133–49.

(1978). "Boundary-interior layer interactions in nonlinear singular perturbation theory," *Memoirs Amer. Math. Soc.* **203**, 1–108.

Howes, F. A., and O'Malley, R. E., Jr. (1980). "Singular perturbations of semilinear second order systems," *Lecture Notes in Mathematics* **827** (Berlin: Springer-Verlag), pp. 130–50.

Hsiao, G. C., and Weinacht, R. J. (1983). "Singular perturbations for a semi-linear hyperbolic equation," *SIAM J. Math. Anal.* **14**, 1168–79.

Ince, E. L. (1927). *Ordinary Differential Equations* (London: Longmans Green); Dover edition (1956) (New York: Dover).

de Jager, E. M. (1975). "Singular perturbations of hyperbolic type," *Nieuw Arch. Wisk.* **23**, 145–72.

Jeffreys, H. (1924*a*). "On certain approximate solutions of linear differential equations of the second order," *Proc. London Math. Soc.* **23**, 428–36.

(1924*b*). "On certain solutions of Mathieu's equation," *Proc. London Math. Soc.* **23**, 437–76.

Kalman, R. E. (1960). "Contributions to the theory of optimal control," *Boletin Sociedad Matematico Mexicana* **5**, 102–19.

(1963). "The theory of optimal control and the calculus of variations," *Mathematical Optimization Techniques* (proceedings of a 1960 symposium on mathematical optimization), ed. R. Bellman (Santa Monica, Calif.; Rand Corp.), pp. 309–31.

von Kármán, T., and Biot, M. A. (1940). *Mathematical Methods in Engineering* (New York: McGraw-Hill).

Keller, H. B. (1972). "Existence theory for multiple solutions of a singular perturbation problem," *SIAM J. Math. Anal.* **3**, 86–92.

Keller, J. B., and Kogelman, S. (1970). "Asymptotic solutions of initial value problems for nonlinear partial differential equations," *SIAM J. Appl. Math.* **18**, 748–58.

Kelley, W. G. (1979). "A nonlinear singular perturbation problem for second order systems," *SIAM J. Math. Anal.* **10**, 32–7.

(1984). "Boundary and interior layer phenomena for singularly perturbed systems," *SIAM J. Math. Anal.* **15**, 635–41.

Kevorkian, J. (1961). "The uniformly valid asymptotic representation of the solutions of certain nonlinear ordinary differential equations" (California Institute of Technology, doctoral dissertation).

(1966). "The two variable expansion procedure for the approximate solution of certain nonlinear differential equations," *Lectures in Applied Mathematics* **7** (Providence, R.I.: American Mathematical Society), pp. 206–75.

Kevorkian, J., and Cole, J. D. (1981). *Perturbation Methods in Applied Mathematics*, *Applied Mathematical Sciences Series* **34** (Berlin: Springer-Verlag).

Kirchhoff, G. (1877). "Zur Theorie des Condensators," *Berlin Akad. Monatsber.*, 144–62.

Kline, M. (1972). *Mathematical Thought from Ancient to Modern Times* (New York: Oxford University Press).

References		485

Kokotović, P. V. (1975). "A Riccati equation for block-diagonalization of ill-conditioned systems," *IEEE Trans. Automatic Control* **20**, 812–14.
(1984). "Applications of singular perturbation techniques to control problems," *SIAM Review* **26**, 501–50.
Kokotović, P. V., and Haddad, A. H. (1975). "Controllability and time-optimal control of systems with slow and fast modes," *IEEE Trans. Automatic Control* **20**, 111–13.
Kokotović, P. V., and Sannuti, P. (1968). "Singular perturbation method for reducing the model order in optimal control design," *IEEE Trans. Automatic Control* **13**, 377–84.
Kollett, F. W. (1974). "Two-timing methods valid on expanding intervals," *SIAM J. Math. Anal.* **5**, 613–25.
Kopell, N. (1980). "The singularly perturbed turning-point problem: a geometric approach," *Singular Perturbations and Asymptotics* (proceedings of an advanced seminar conducted by the Mathematics Research Center, University of Wisconsin-Madison, May 28–30, 1980), ed. R. E. Meyer and S. V. Parter (New York: Academic Press), pp. 173–90.
Kramers, H. A. (1926). "Wellenmechanik und halbzahlige Quantisierung," *Z. Physik* **39**, 829–40.
Krein, S. G. (1971). *Linear Differential Equations in Banach Space* (translation of 1963 Russian edition) (Providence, R.I.: American Mathematical Society).
Kreiss, H.-O. (1978). "Difference methods for stiff ordinary differential equations," *SIAM J. Numer. Anal.* **15**, 21–58.
Kreiss, H.-O., and Nichols, N. (1975). "Numerical methods for singular perturbation problems," report No. 57, Department of Computer Science, Uppsala University.
Kreiss, H.-O., and Parter, S. V. (1974). "Remarks on singular perturbations with turning points," *SIAM J. Math. Anal.* **5**, 230–51.
Krylov, N. M., and Bogoliubov, N. N. (1934). "The application of the methods of nonlinear mechanics to the theory of stationary oscillations," *Ukran. Akad. Sci. (Kiev)*, **8**.
(1947). *Introduction to Nonlinear Mechanics, Annals of Mathematics Studies* **11** (Princeton University Press).
Kurzweil, J. (1963). "The averaging principle in certain partial differential boundary problems," *Casopis Pest. Mat.* **88**, 444–56.
(1966). "Exponentially stable integral manifolds, averaging principle and continuous dependence on a parameter," *Czech. Math. J.* **16**, 463–91.
(1967). "van der Pol perturbation of the equation for a vibrating string," *Czech. Math. J.* **17**, 558–608.
Kuzmak, G. E. (1959). "Asymptotic solutions of nonlinear second order differential equations with variable coefficients," *J. Appl. Math. (PPM)* **23**, 730–44.
Lagerstrom, P. A. (1961). "Méthodes asymptotiques pour l'étude des équations de Navier-Stokes," *Lecture Notes*, Institut Henri Poincaré, Paris.
Lagerstrom, P. A., and Casten, R. G. (1972). "Basic concepts underlying singular perturbation techniques," *SIAM Review* **14**, 63–120.
Lakin, W. D. (1972). "Boundary value problems with a turning point," *Studies in Appl. Math.* **51**, 261–75.
Langer, R. E. (1949). "The asymptotic solutions of ordinary linear differential equations of the second order, with special reference to a turning point," *Trans. Amer. Math. Soc.* **67**, 461–90.
Lapidus, L., and Amundson, N. R. (editors) (1977). *Chemical Reactor Theory, A Review* (Englewood Cliffs, N.J.: Prentice-Hall).
Laplace, P. S. (1805). *Traite de Mechanique Celeste, Vol. 4, Supplement to Book V* (Paris: Gauthier-Villars), pp. 349–495.
Latta, G. E. (1951). "Singular perturbation problems" (California Institute of Technology, doctoral dissertation).

Levey, H. C., and Mahony, J. J. (1968). "Resonance in almost linear systems," *J. Inst. Math. Appl.* **4**, 282–94.

Levin, J. J. (1956). "Singular perturbations of nonlinear systems of differential equations related to conditional stability," *Duke Math. J.* **23**, 609–20.

——— (1957). "The asymptotic behavior of the stable initial manifolds of a system of nonlinear differential equations," *Trans. Amer. Math. Soc.* **85**, 357–68.

Levin, J. J., and Levinson, N. (1954). "Singular perturbations of non-linear systems of differential equations and an associated boundary layer equation," *J. Math. Mech.* (*J. Rational Mech. Anal.*) **3**, 247–70.

Levinson, N. (1948). "The asymptotic nature of solutions of linear systems of differential equations," *Duke Math. J.* **15**, 111–26.

——— (1950). "The first boundary value problem for $\varepsilon \cdot \Delta u + A(x, y) \cdot u_x + B(x, y) \cdot u_y + C(x, y) \cdot u = D(x, y)$ for small ε," *Ann. Math.* **51**, 428–45.

Lighthill, M. J. (1949). "A technique for rendering approximate solutions to physical problems uniformly valid," *Philos. Mag.* **40**, 1179–201.

Lindstedt, A. (1882). "Ueber die Integration einer für die Störungstheorie wichtigen Differentialgleichung," *Astron. Nachr.* **103**, 211–20.

——— (1883). "Beitrag zur Integration der Differentialgleichungen der Störungstheorie," *Mem. Acad. Sci. St. Petersburg* **31**, ser. 7, no. 4.

Liouville, J. (1837). "Second memoire sur le developpement des fonctions en series dont divers termes sont assujettis a satisfaire a une meme equation differentielle du second ordre contenant un parametre variable," *J. Math. Pure Appl.* **2**, 16–35.

Lorenz, J. (1982). "Nonlinear boundary value problems with turning points and properties of difference schemes," *Lecture Notes in Mathematics* **942** (Berlin: Springer-Verlag), pp. 150–69.

Lutz, R., and Sari, T. (1982). "Applications of nonstandard analysis to boundary value problems in singular perturbation theory," *Lecture Notes in Mathematics* **942** (Berlin: Springer-Verlag), pp. 113–35.

Macki, J. W. (1967). "Singular perturbation of a boundary value problem for a system of nonlinear ordinary differential equations," *Arch. Rational Mech. Anal.* **24**, 219–32.

Mahony, J. J. (1961–2). "An expansion method for singular perturbation problems," *J. Australian Math. Soc.* **2**, 440–63.

——— (1972). "Validity of averaging methods for certain systems with periodic solutions," *Proc. Roy. Soc. London* **A/330**, 349–71.

Maier, M. R. (1982). "Numerische Lösung singulär gestörter Randwertprobleme mit Anwendung auf Halbleitermodelle" (Mathematisches Institut der Technischen Universität München, doctoral dissertation).

——— (1984*a*). "An adaptive shooting method for singularly perturbed boundary value problems," accepted for publication in *SIAM J. Sci. Stat. Comput.*

——— (1984*b*). "Numerical solution of singular perturbed boundary value problems using a collocation method with tension splines," report presented at Workshop on Numerical Boundary Value ODEs, University of British Columbia, Vancouver, July 10–13, 1984.

Maier, M. R., and Smith, D. R. (1981). "Numerical solution of a symmetric one-dimensional diode model," *J. Comp. Physics* **42**, 308–26.

Matkowsky, B. J. (1975). "On boundary layer problems exhibiting resonance," *SIAM Review* **17**, 82–100.

——— (1976). "Errata: On boundary layer problems exhibiting resonance," *SIAM Review* **18**, 112.

Mattheij, R. M. M. (1982). "Riccati-type transformations and decoupling of singularly perturbed ODE," *Proceedings of an International Conference on Stiff Computation*, Park City.

Mattheij, R. M. M., and O'Malley, R. E., Jr. (1984). "On solving boundary value problems for multi-scale systems using asymptotic approximations and multiple shooting," *BIT* (in press).

Maxwell, J. C. (1866). "On the viscosity or internal friction of air and other gases," *Philos. Trans. Roy. Soc. London* **A/156**, 249–68.

Mika, J. (1982). "Singular-singularly perturbed linear equations in Banach spaces," *Lecture Notes in Mathematics* **942** (Berlin: Springer-Verlag), pp. 72–83.

Miller, J. J. H. (editor) (1980). *Boundary and Interior Layers, Computational and Asymptotic Methods* (proceedings of a conference held at Trinity College, Ireland, June 3–6, 1980) (Dublin: Boole Press).

— (editor) (1984). *Boundary and Interior Layers III* (proceedings of a conference held at Trinity College, Ireland, June 20–22, 1984) (Dublin: Boole Press).

Minorsky, N. (1962). *Nonlinear Oscillations* (Princeton, N.J.: D. Van Nostrand).

Miranker, W. L. (1963). "Singular perturbation of eigenvalues by a method of undetermined coefficients," *J. Math. Phys.* **42**, 47–58.

— (1973). "Numerical methods of boundary layer type for stiff systems of differential equations," *Computing* **11**, 221–34.

— (1981). *Numerical Methods for Stiff Equations and Singular Perturbation Problems* (London: D. Reidel).

von Mises, R. (1950). "Die Grenzschichte in der Theorie der gewöhnlichen Differentialgleichungen," *Acta Univ. Szeged.* **12**, 29–34.

Morrison, J. A. (1966). "Comparison of the modified method of averaging and the two variable expansion procedure," *SIAM Review* **8**, 66–85.

Moser, J. (1955). "Singular perturbation of eigenvalue problems for linear differential equations of even order," *Comm. Pure Appl. Math.* **8**, 251–78.

Munk, W. H. (1950). "On the wind-driven ocean circulation," *J. Meteorology* **7**, 79-94; reprinted (1963) in *Wind-Driven Ocean Circulation*, ed. A. R. Robinson (New York: Blaisdell), pp. 23–56.

Munk, W. H., and Carrier, G. F. (1950). "The wind-driven circulation in ocean basins of various shapes," *Tellus* **2**, 158–67; reprinted (1963) in *Wind-Driven Ocean Circulation*, ed. A. R. Robinson (New York: Blaisdell), pp. 57–68.

Nagumo, M. (1937). "Ueber die Differentialgleichung $y'' = f(x, y, y')$," *Proc. Phys. Math. Soc. Japan* **19**, 861–6.

Nayfeh, A. H. (1973). *Perturbation Methods* (New York: Wiley).

— (1981). *Introduction to Perturbation Techniques* (New York: Wiley).

Nayfeh, A. H., and Mook, D. T. (1979). *Nonlinear Oscillations* (New York: Wiley).

Nipp, K. (1978). "A formal method for matched asymptotic expansions applied to the Field-Noyes model of the Belousov-Zhabotinskii reaction," *Mech. Res. Commun.* **5** (4).

— (1980). "An algorithmic approach to singular perturbation problems in ordinary differential equations with an application to the Belousov-Zhabotinskii reaction" (Swiss Federal Institute of Technology, doctoral dissertation ETH 6643).

Noaillon, P. (1912). "Développements asymptotiques dans les équations différentielles linéaires à paramètre variable," *Mem. Soc. Roy. Sci. Liège* **9** (ser. 3), 197.

O'Donnell, M. A. (1984). "Boundary and corner layer behavior in singularly perturbed semilinear systems of boundary value problems," *SIAM J. Math. Anal.* **15**, 317–32.

Olver, F. W. J. (1961). "Error bounds for the Liouville-Green (or WKB) approximation," *Proc. Cambridge Philos. Soc.* **57**, 790–810.

— (1974). *Asymptotics and Special Functions* (New York: Academic Press).

— (1978). "Sufficient conditions for Ackerberg-O'Malley resonance," *SIAM J. Math. Anal.* **9**, 328–55.

(1980). "General connection formulae for Liouville-Green approximations in the complex plane," *Philos. Trans. Roy. Soc. London* A /**289**, 501–48.

O'Malley, R. E., Jr. (1968a). "A boundary value problem for certain nonlinear second order differential equations with a small parameter," *Arch. Rational Mech. Anal.* **29**, 66–74.

(1968b). "Topics in singular perturbations," *Adv. Math.* **2**, 365–470; reprinted (1970) in *Lectures in Ordinary Differential Equations*, ed. R. W. McKelvey (New York: Academic Press), pp. 155–260.

(1969a). "On a boundary value problem for a nonlinear differential equation with a small parameter," *SIAM J. Appl. Math.* **17**, 569–81.

(1969b). "On the asymptotic solution of boundary value problems for nonhomogeneous ordinary differential equations containing a parameter," *J. Math. Anal. Appl.* **28**, 450–60.

(1969c). "Boundary value problems for linear systems of ordinary differential equations involving many small parameters," *J. Math. Mech.* **18**, 835–55.

(1970a). "Singular perturbations of a boundary value problem for a system of nonlinear differential equations," *J. Diff. Equations* **8**, 431–47.

(1970b). "On boundary value problems for a singularly perturbed differential equation with a turning point," *SIAM J. Math. Anal.* **1**, 479–90.

(1970c). "On nonlinear singular perturbation problems with interior nonuniformities," *J. Math. Mechanics (Indiana Univ. Math. J.)* **19**, 1103–12.

(1971a). "Boundary layer methods for nonlinear initial value problems," *SIAM Review* **13**, 425–34.

(1971b). "On initial value problems for nonlinear systems of differential equations with two small parameters," *Arch. Rational Mech. Anal.* **40**, 209–22.

(1972a). "The singularly perturbed linear state regulator problem," *SIAM J. Control* **10**, 399–413.

(1972b). "On multiple solutions of a singular perturbation problem," *Arch. Rational Mech. Anal.* **49**, 89–98.

(1974a). *Introduction to Singular Perturbations* (New York: Academic Press).

(1974b). "Boundary layer methods for ordinary differential equations with small coefficients multiplying the highest derivatives," *Lecture Notes in Mathematics* **430** (proceedings of a symposium on constructive and computational methods for differential and integral equations) (Berlin: Springer-Verlag), pp. 363–89.

(1976). "Phase-plane solutions to some singular perturbation problems," *J. Math. Anal. Appl.* **54**, 449–66.

(1978a). "Singular perturbations and optimal control," *Lecture Notes in Mathematics* **680** (Berlin: Springer-Verlag), pp. 170–218.

(1978b). "On singular singularly-perturbed initial value problems," *Applicable Analysis* **8**, 71–81.

(1979). "A singular singularly-perturbed linear boundary value problem," *SIAM J. Math. Anal.* **10**, 695–708.

(1980). "On multiple solutions of singularly perturbed systems in the conditionally stable case," *Singular Perturbations and Asymptotics* (proceedings of a 1980 seminar held in Madison, Wisconsin), ed. R. E. Meyer and S. Parter (New York: Academic Press), pp. 87–108.

(1983a). "On nonlinear optimal control problems," *International Centre for Mechanical Sciences, Courses and Lectures, No. 280: Singular Perturbations in Systems and Control*, ed. M. D. Ardema (Berlin: Springer-Verlag), pp. 93–101.

(1983b). "Slow/fast decoupling–analytical and numerical aspects," *International Centre for Mechanical Sciences, Courses and Lectures, No. 280: Singular Perturbations in Systems and Control*, ed. M. D. Ardema (Berlin: Springer-Verlag), pp. 143–60.

(1984). "On the simultaneous use of asymptotic and numerical methods to solve nonlinear two point problems with boundary and interior layers," presented at Workshop on Numerical Boundary Value ODEs, University of British Columbia, Vancouver, July 10-13, 1984.

O'Malley, R. E., Jr., and Anderson, L. R. (1982). "Time-scale decoupling and order reduction for linear time-varying systems," *Optimal Control Appl. Methods* **3**, 133-53.

O'Malley, R. E., Jr., and Flaherty, J. E. (1977). "Singular singular-perturbation problems," *Lecture Notes in Mathematics* **594** (Berlin: Springer-Verlag), pp. 422-436.

(1980). "Analytical and numerical methods for nonlinear singular singularly-perturbed initial value problems," *SIAM J. Appl. Math.* **38**, 225-48.

O'Malley, R. E., Jr., and Keller, J. B. (1968). "Loss of boundary conditions in the asymptotic solution of linear ordinary differential equations, II. Boundary value problems," *Comm. Pure Appl. Math.* **21**, 263-70.

Osher, S. (1981). "Nonlinear singular perturbation problems and one sided difference schemes," *SIAM J. Numer. Anal.* **18**, 129-44.

Pannekoek, A. (1961). *A History of Astronomy* (London: Allen & Unwin).

Parker, D. F., and Wan, F. Y. M. (1984). "Finite polar dimpling of shallow caps under sub-buckling axisymmetric pressure distributions," *SIAM J. Appl. Math.* **44**, 301-26.

Parter, S. (1972). "Remarks on the existence theory for multiple solutions of a singular perturbation problem," *SIAM J. Math. Anal.* **3**, 496-505.

(1982). "On the swirling flow between rotating coaxial disks: a survey," *Lecture Notes in Mathematics* **942** (Berlin: Springer-Verlag), pp. 258-280.

Pearson, C. E. (1968a). "On a differential equation of boundary layer type," *J. Math. Phys.* (*Studies in Applied Math.*) **47**, 134-54.

(1968b). "On nonlinear differential equations of boundary layer type," *J. Math. Phys.* (*Studies in Applied Math.*) **47**, 351-8.

Perko, L. M. (1968). "Higher order averaging and related methods for perturbed periodic and quasi-periodic systems," *SIAM J. Appl. Math.* **17**, 698-724.

Perron, O. (1929). "Ueber Stabilität und asymptotisches Verhalten der Integrale von Differentialgleichungssystemen," *Math. Zeitschrift* **29**, 129-60.

(1930). Ueber ein vermeintliches Stabilitätskriterium," *Nachr. Ges. Wiss.* (Göttingen: Math.-physik. Kl. Fachgruppe I).

Poincaré, H. (1892-3). *Les Methodes Nouvelles de la Mecanique Celeste 1, 2* (Paris: Gauthier-Villars) (Dover reprint, 1957).

van der Pol, B. (1926). "On 'relaxation-oscillations'," *Phil. Mag.* **2** (ser. 7), 978-92.

(1927). "Forced oscillations in a circuit with non-linear resistance," *Phil. Mag.* **3** (ser. 7), 65-80.

(1935). *Non-linear Theory of Electrical Oscillations* (Moscow: Communications Publishing House).

Prandtl, L. (1905). "Ueber Flussigkeiten bei sehr kleiner Reibung," *Verh. 3rd Intern. Kongr., Heidelberg* (Leipzig: Teubner), pp. 484-91.

Protter, M. H., and Weinberger, H. F. (1967). *Maximum Principles in Differential Equations* (Englewood Cliffs, N.J.: Prentice-Hall).

Rayleigh (J. W. Strutt) (1877-8). *The Theory of Sound 1, 2* (London) (American edition, 1945, New York: Dover).

(1912). "On the propagation of waves through a stratified medium, with special reference to the question of reflection," *Proc. Roy. Soc. London* A/**86**, 208-26.

Reid, W. T. (1972). *Riccati Differential Equations* (New York: Academic Press).

Reiss, E. L. (1971). "On multivariable asymptotic expansions," *SIAM Review* **13**, 189-96.

Riccati, J. F. (1724). "Animadversiones in aequationes differentiales secundi gradus," *Actorum Eruditorum quae Lipsiae publicantur. Supplementa 8*, 66-73.

van Roosbroeck, W. (1950). "Theory of the flow of electrons and holes in germanium and other semiconductors," *Bell System Tech. J.* **29**, 560–607.

Rubenfeld, L. A. (1978). "On a derivative-expansion technique and some comments on multiple scaling in the asymptotic approximation of solutions of certain differential equations," *SIAM Review* **20**, 79–105.

Sacker, R. J., and Sell, G. R. (1980). "Singular perturbations and conditional stability," *J. Math. Anal. Appl.* **76**, 406–31.

Saksena, V. R., O'Reilly, J., and Kokotović, P. V. (1984). "Singular perturbations and time-scale methods in control theory: survey 1976–1983," *Automatica* **20**, 273–93.

Sannuti, P., and Kokotović, P. (1969). "Singular perturbation method for near optimum design of high-order non-linear systems," *Automatica* **5**, 773–9.

Schrödinger, E. (1926). "Quantisierung als Eigenwertproblem," *Ann. Physik* **80**, 437–90.

Searl, J. W. (1971). "Expansions for singular perturbations," *J. Inst. Math. Appl.* **8**, 131–8.

Shockley, W. (1949). "The theory of *p-n* junctions in semiconductors and *p-n* junction transistors," *Bell System Tech. J.* **28**, 435–89.

Shockley, W., Pearson, G. L., and Haynes, J. R. (1949). "Hole injection in germanium–quantitative studies and filamentary transistors," *Bell System Tech. J.* **28**, 344–66.

Sibuya, Y. (1958). "Sur réduction analytique d'un système d'équations différentielles ordinaires linéaires contenant un paramètre," *J. Fac. Sci. Univ. Tokyo (Section 1)* **7**, 527–40.

(1963a). "Simplification of a linear ordinary differential equation of the nth order at a turning point," *Arch. Rational Mech. Anal.* **13**, 206–21.

(1963b). "Asymptotic solutions of initial value problems of ordinary differential equations with a small parameter in the derivative, I," *Arch. Rational Mech Anal.* **14**, 304–11.

(1964a). "On the problem of turning points for a system of linear ordinary differential equations of higher orders," *Proceedings, Symposium Mathematical Research Center, University of Wisconsin (Madison)* (New York: Wiley), pp. 145–62.

(1964b). "Asymptotic solutions of initial value problems of ordinary differential equations with a small parameter in the derivative, II," *Arch. Rational Mech. Anal.* **15**, 247–62.

(1966). "A block-diagonalization theorem for systems of linear ordinary differential equations and its applications," *SIAM J. Appl. Math.* **14**, 468–75.

Smith, D. R. (1971). "An asymptotic analysis of a certain hyperbolic Cauchy problem," *SIAM J. Math. Anal.* **2**, 375–92.

(1974). *Variational Methods in Optimization* (Englewood Cliffs, N.J.: Prentice-Hall).

(1975a). "The multivariable method in singular perturbation analysis," *SIAM Review* **17**, 221–73.

(1975b). "A nonlinear boundary value problem on an unbounded interval," *SIAM J. Math. Anal.* **6**, 601–15.

(1980). "On a singularly perturbed boundary value problem arising in the physical theory of semiconductors," Technical Report TUM-M8021, Institut für Mathematik der Technischen Universität München.

(1981). "On boundary layer approximations in the numerical solution of certain stiff boundary value problems," *SIAM J. Numer. Anal.* **18**, 377–80.

(1984a). "Decoupling and order reduction via the Riccati transformation," submitted for publication in *SIAM Review*.

(1984b). "Liouville/Green approximations via the Riccati transformation," accepted for publication in *J. Math. Anal. Appl.*

(1984c). "A Green function for a singularly perturbed Dirichlet problem," technical report, Department of Mathematics, University of California, San Diego.

Smith, D. R., and Palmer, J. T. (1970). "On the behavior of the solution of the telegraphist's equation for large absorption," *Arch. Rational Mech. Anal.* **39**, 146–57.

Smith, D. R., and Weinstein, M. B. (1976). "Sturm transformations and linear hyperbolic differential equations in two variables," *J. Math. Anal. Appl.* **56**, 548–66.

Snow, R. E. (1976). "Two-timing for forced oscillation equations," *J. Math. Anal. Appl.* **54**, 5–25.

Stewartson, K. (1953). "On the flow between two rotating coaxial disks," *Proc. Cambridge Phil. Soc.* **49**, 333–41.

Stoer, J., and Bulirsch, R. (1980). *Introduction to Numerical Analysis* (Berlin: Springer-Verlag).

Stoker, J. J. (1950). *Nonlinear Vibrations* (New York: Wiley Interscience).

Struble, R. A. (1961). "An application of the method of averaging in the theory of satellite motion," *J. Math. Mech.* **10**, 691–704.

Strutt, J. W.: see Rayleigh.

Sturm, C. (1836). "Sur les équations différentielles linéaires du second ordre," *J. Mathematiques Pures et Appl.* **I**, 106–86.

Sturrock, P. A. (1957). "Nonlinear effects in electron plasmas," *Proc. Roy. Soc. London* **A/242**, 277–99.

Tang, M. M. (1972). "Singular perturbation of some quasilinear elliptic boundary value problems given in divergence form," *J. Math. Anal. Appl.* **39**, 208–26.

Tang, M. M., and Fife, P. C. (1975). "Asymptotic expansions of solutions of quasi-linear elliptic equations with small parameter," *SIAM J. Math. Anal.* **6**, 523–32.

Taylor, J. G. (1978). "Error bounds for the Liouville-Green approximation to initial-value problems," *Z. Angew. Math. Mech.* **58**, 529–37.

   (1982). "Improved error bounds for the Liouville-Green (or WKB) approximation," *J. Math. Anal. Appl.* **85**, 79–89.

Tikhonov, A. N. (1950). "On a system of differential equations containing parameters" (in Russian), *Mat. Sb.* **27**, 147–56.

   (1952). "Systems of differential equations containing a small parameter multiplying the highest derivatives" (in Russian), *Mat. Sb.* **31**, 575–86.

Trenogin, V. A. (1963). "Asymptotic behavior and existence of a solution of the Cauchy problem for a first order differential equation with a small parameter in a Banach space," *Soviet Math. Dokl.* **4**, 1261–4.

   (1970). "The development and application of the asymptotic method of Lyusternik and Vishik," *Russian Math. Surveys* **25**, 119–56.

Tschen, Y.-W. (1935). "Ueber das Verhalten der Lösungen einer Folge von Differentialgleichungen, welche im Limes ausarten," *Compositio Math.* **2**, 378–401.

Tupčiev, V. A. (1962). "Asymptotic behavior of the solution of a boundary problem for systems of differential equations of first order with a small parameter in the derivative," *Soviet Math. Dokl.* **3**, 612–16.

Turrittin, H. L. (1936). "Asymptotic solutions of certain ordinary differential equations associated with multiple roots of the characteristic equation," *Amer. J. Math.* **58**, 364–78.

   (1973). "My mathematical expectations," *Lecture Notes in Mathematics* **312** (Berlin: Springer-Verlag), pp. 1–22.

Van Dyke, M. (1964). *Perturbation Methods in Fluid Dynamics* (New York: Academic Press).

   (1975). *Perturbation Methods in Fluid Dynamics*, annotated edition (Palo Alto, Calif: Parabolic Press).

Varma, A., and Aris, R. (1977). "Stirred pots and empty tubes," *Chemical Reactor Theory, A Review*, ed. L. Lapidus and N. R. Amundson (Englewood Cliffs, N.J.: Prentice-Hall).

Vasil'eva, A. B. (1959). "Uniform approximation to the solution of a system of differential
   equations with a small parameter multiplying some of the derivatives and an
   application to boundary-value problems," *Dokl. Akad. Nauk SSSR* **124**, 509–12.
   (1963). "Asymptotic behavior of solutions to certain problems involving nonlinear
   differential equations containing a small parameter multiplying the highest
   derivatives," *Russian Math. Surveys* **18**, 13–84.
   (1972). "The influence of local perturbations on the solution of a boundary value
   problem," *Differential Equations* **8**, 437–43.
   (1975a). "Conditionally stable singularly perturbed systems with singularities in the
   boundary conditions," *Differential Equations* **11**, 171–80.
   (1975b). "A similarity between conditionally stable singularly perturbed systems and
   singularly perturbed systems with zero eigenvalues," *Differential Equations* **11**,
   1307–15.
   (1975c). "Singularly perturbed systems containing indeterminancy in the case of
   degeneracy," *Soviet Math. Dokl.* **16**, 1121–4.
   (1976). "Singularly perturbed systems with an indeterminancy in their degenerate
   equations," *Differential Equations* **12**, 1227–35.
Vasil'eva, A. B., and Butuzov, V. F. (1973). *Asymptotic Expansions of Solutions of Singularly
   Perturbed Equations* (Moscow: Nauka) (in Russian).
   (1980). *Singularly Perturbed Equations in the Critical Case*, MRC Technical Summary
   Report 2039 (Mathematics Research Center, University of Wisconsin-Madison)
   (originally published in 1978 by Moscow State University; translated from Russian
   by F. A. Howes).
Vasil'eva, A. B., Kardo-Sysoev, A. F., and Stel'makh, V. G. (1976). "Boundary layer in
   pn-junction theory," *Soviet Phys. Semicond.* **10**, 784–6.
Vasil'eva, A. B., and Stel'makh, V. G. (1977). "Singularly disturbed systems of the theory
   of semiconductor devices," *USSR Comp. Math. Phys.* **17**, 48–58.
Vasil'eva, A. B., and Tupčiev, V. A. (1960). "Asymptotic formulae for the solution of a
   boundary value problem in the case of a second order equation containing a small
   parameter in the term containing the highest derivative," *Soviet Math. Dokl.* **1**,
   1333–5.
Vishik, M. I., and Lyusternik, L. A. (1957). "Regular degeneration and boundary layer
   for linear differential equations with small parameter" (in Russian), *Uspehi. Mat.
   Nauk* **12**, 3–122; *Amer. Math. Soc. Transl.* **20**, 239–364, 1962.
Wan, F. Y. M. (1980). "The dimpling of spherical caps," *Mech. Today* **5**, 495–508.
Wasow, W. (1941). "On boundary layer problems in the theory of ordinary differential
   equations" (New York University, doctoral dissertation).
   (1944). "On the asymptotic solution of boundary value problems of ordinary differential
   equations containing a parameter," *J. Math. Phys.* **23**, 173–83.
   (1956). "Singular perturbations of boundary value problems for nonlinear differential
   equations of the second order," *Comm. Pure Appl. Math.* **9**, 93–113.
   (1965). *Asymptotic Expansions for Ordinary Differential Equations* (New York: Wiley).
   (1966). "On turning point problems for systems with almost diagonal coefficient matrix,"
   Funkcialaj Ekvacioj **8**, 143–70.
   (1970). "The capriciousness of singular perturbations," *Nieuw Arch. Wisk.* **18**, 190–210.
   (1971). "The central connection problem at turning points of linear differential
   equations," *Lecture Notes in Mathematics* **183** (Berlin: Springer-Verlag), pp. 158–64.
   (1976). *Asymptotic Expansions for Ordinary Differential Equations*, reprint edition (New
   York: Robert E. Krieger Publishing).
   (1985). *Linear Turning Point Theory* (Berlin: Springer).
Watts, A. M. (1971). "A singular perturbation problem with a turning point," *Bull.
   Australian Math. Soc.* **5**, 61–73.

Weinstein, M. B., and Smith, D. R. (1975). "Comparison techniques for overdamped systems," *SIAM Review* **17**, 520–40.

(1976). "Comparison techniques for certain overdamped hyperbolic partial differential equations," *Rocky Mountain J. Math.* **6**, 731–42.

Weiss, R. (1984). "An analysis of the box and trapezoidal schemes for linear singularly perturbed boundary value problems," *Math. Computation* **42**, 41–67.

Wentzel, G. (1926). "Eine Verallgemeinerung der Quantenbedingung für die Zwecke der Wellenmechanik," *Z. Physik* **38**, 518–29.

Whittaker, E. T. (1914). "On the general solution of Mathieu's equation," *Proc. Edinburgh Math. Soc.* **32**, 75–80.

Wilde, R. R., and Kokotović, P. V. (1972). "A dichotomy in linear control theory," *IEEE Trans. Automatic Control* **17**, 382–3.

(1973). "Optimal open- and closed-loop control of singularly perturbed linear systems," *IEEE Trans. Automatic Control* **18**, 616–26.

Willett, D. (1966). "On a nonlinear boundary value problem with a small parameter multiplying the highest derivative," *Arch. Rational Mech. Anal.* **23**, 276–87.

Yarmish, J. (1975). "Newton's method techniques for singular perturbations," *SIAM J. Math. Anal.* **6**, 661–80.

Zauderer, E. (1972). "Boundary value problems for a second order differential equation with a turning point," *Studies in Appl. Math.* **51**, 411–13.

Name index

Abrahamsson, Leif, 474
Ackerberg, Robert C., 327–8
Aiken, Richard C., 210
Amundson, N. R., 387, 403
Anderson, Leonard R., 340
Aris, Rutherford, 165, 194, 196, 387, 403
Ascher, Uri M., 471
Atkinson, F. V., 228

Banach, Stephen, 8
Bardeen, John, 449, 462
Baum, Howard R., 82, 119
Bellman, Richard, 94, 228, 232, 258–9, 353
Bensoussan, A., 45, 62
Berger, M. S., 400
Bernoulli, Jakob (James), xv, 303
Bernoulli, Johann, xv, 2
Bessel, Friedrich Wilhelm, 5, 10, 155
Biot, M. A., 304
Birkhoff, Garrett, 116, 265
Birkhoff, George David, 61
Bobisud, Larry E., 162
Boglaev, Y. P., 385, 463
Bogoliubov, N. N., 40, 42–4, 61
Boyce, William E., 271
Brillouin, Léon, 61, 283
Brish, N. I., 285, 385, 386
Bromberg, Eleazer, 61
Bulirsch, Roland Z., xvi, 374, 474
Bush, William B., 427
Butuzov, V. F., 205, 227, 337, 385, 386, 448

Campbell, Stephen L., 173, 227
Carlini, Francesco, 61
Carrier, George F., 276, 284, 286, 305, 311, 402
Casten, Richard G., 428

Cauchy, Augustin-Louis, xv, 98, 147
Chang, K. W., 205, 227, 300, 302, 330, 332, 336, 338, 340, 342–3, 345, 348, 351, 355, 359, 385, 386, 388, 391, 429, 463, 470
Chen, Jye, 387
Chow, Pao-Liu, 76, 82
Cochran, James Alan, 9, 62, 267, 275, 294, 300, 328, 386
Coddington, Earl A., 288, 300, 386, 388, 428, 438
Cole, Julian D., 61, 62, 76, 82, 134, 288, 304, 446
Cole, Randal H., 4
Cook, L. Pamela, 328
Coppel, William A., 205, 227–8, 232, 258, 332, 346, 351, 353
Coriolis, Gaspard Gustave, 305
Cotton, Émile, 228, 232
Coulomb, Charles, 54
Cronin, Jane, 20

Danby, J. M. A., 30
Davis, Harold T., 32, 340
De Villiers, J. M., 400
Dirichlet, Peter Gustav Lejeune, 9, 11, 226
Dorr, F. W., 18, 285, 300, 319, 446

Eckhaus, Wiktor, 141, 271, 300, 328, 386, 400, 437
Einstein, Albert, 29–34
Erdélyi, Arthur, 14, 62, 279, 300, 386
Euler, Leonhard, xv, 2, 196, 303, 372

Faraday, Michael, 54
Fenichel, Neil, 228
Ferguson, Warren E., Jr., 171, 332, 336
Fife, Paul C., 149–50, 227, 245, 271, 285, 385, 390, 395, 397, 400, 403, 404, 407,

Fife, Paul C., (*cont.*)
    409, 413, 420, 429, 463
Flaherty, Joseph E., 156, 227, 245, 364,
    388, 474
Flatto, L., 170–1, 175–6, 264, 332, 350
Fourier, Joseph, 72
Fraenkel, L. E., 400
Fredholm, Erik Ivar, 4

Gans, R., 61
Garabedian, Paul, 159
Geel, R., 161–2
Gordon, Noam, 227
Green, George, 4, 17–18, 61, 219, 232, 259,
    353
Greenlee, Wilfred M., 34, 51, 53, 62, 75–6,
    78, 82–3, 101–5, 117–25, 130
de Groen, Pieter P. N., 328
Gronwall, T. H., 7

Haber, S., 463
Habets, Patrick A., 340, 386
Haddad, A. H., 340
Halley, Edmund, 31
Hamilton, William R., 373, 379
Handelman, George H., 271
Harris, William A., Jr., 173, 228, 264, 271,
    330, 332, 336, 338, 340–3, 345, 348,
    353, 355, 359
van Harten, Aart, 271, 285, 300, 385, 386,
    390, 399, 400, 403, 412, 428, 435, 437,
    463, 470
Hartman, Philip, 228, 353
Haynes, J. R., 449, 459
Heineken, F. G., 165, 194, 196
Hemker, Pieter W., 474
Henry, Joseph, 54
Hirsch, Morris W., 129, 175
Hölder, Otto, 175
Hoppensteadt, Frank C., 45, 82, 145, 151,
    164, 177, 205, 227, 300
Householder, Alston S., 475, 476
Howes, Frederick A., 285, 300, 385, 386,
    388, 391, 402, 428, 429, 446, 463, 464,
    470
Hsiao, George C., 162

Ince, Edward Lindsay, 24

Jacobi, Carl Gustav Jacob, 32, 379
de Jager, E. M., 161–2, 271
Jeffreys, Harold, 61, 283

Kalman, Rudolf Emil, 370, 373, 379
Kardo-Sysoev, A. F., 448
von Kármán, Theodor, 304
Keller, Herbert B., 387, 474

Keller, Joseph B., 75, 271, 315
Kelley, Walter G., 226, 285, 302, 385, 463
Kevorkian, Jerry K., 61, 62, 76, 84
Kirchhoff, Gustav R., 54, 61
Kline, Morris, 8
Kogelman, Stanley, 75
Kokotović, Petar V., 340, 374, 378
Kollett, Francis W., 62, 74, 75–6, 92–3,
    101–2, 104–5
Kopell, Nancy J., 328
Kramers, Hendrik A., 61, 283
Krein, S. G., 151
Kreiss, Heinz-Otto, 177, 328, 474
Krylov, N. M., 40, 42–4, 61
Kurzweil, Jaroslav, 45, 76
Kutta, M. W., 196
Kuzmak, G. E., 61

Lagerstrom, Paco A., 427, 428
Lagrange, Joseph-Louis, xv, 2, 372
Lakin, William D., 319, 328
Langer, Rudolph E., 14
Lapidus, Leon, 210, 387, 403
Laplace, Pierre-Simon, 61
Latta, Gordon Eric, 61
Leibniz, Gottfried, xv, 2
Levey, H. C., 61
Levin, Jacob J., 175, 181, 205, 226–8,
    231–2, 336
Levinson, Norman, 61, 170–1, 175–6, 181,
    205, 226–8, 232, 258–9, 271, 288, 300,
    332, 350, 386, 388, 428, 437, 438, 463
Lighthill, M. J., 61
Lindstedt, A., 60–2, 65–6
Lions, Jacques L., 45, 62
Liouville, Joseph, 8, 61, 219, 259
Lorenz, Jens, 463
Lutz, Donald A., 173, 228, 353
Lutz, Robert, 463, 470
Lyusternik, L. A., 61, 145, 151, 264, 271

Macki, Jack W., 300
Mahony, John J., 61, 93, 101–2, 105
Maier, Maximilian R., 382, 457, 473–8
Mathon, William, 474
Matkowsky, Bernard J., 328
Mattheij, Robert M. M., 340, 364
Maxwell, James Clerk, 61
Mika, J., 227
Milgram, R. James, xv
Miller, John J. H., 474
Minorsky, N., 35, 44
Miranker, Willard L., 163, 209, 211–6,
    473
von Mises, Richard, 300, 386
Mitropolsky, Y. A., 44, 61
Mook, Dean T., 45, 61, 76
Morrison, John A., 61, 78, 82, 84

Moser, Jurgen K., 271
Munk, Walter H., 304–5, 311

Nagumo, Mitio, 470
Nayfeh, Ali Hasan, 45, 61, 76
Neumann, Carl Gottfried, 8, 235, 269
Newton, Isaac, xv, 2, 29–31, 211
Nichols, Nancy, 177
Nipp, Kaspar, 137
Noaillon, P., 61

O'Donnell, Mark A., 463
Olver, Frank W. J., 14, 22, 259, 265, 283, 328, 353
O'Malley, Robert E., Jr., xv, 62, 141, 145, 156, 164–5, 177, 180, 185, 189–90, 205, 227–8, 245, 264, 271, 279, 300, 303, 308, 315, 317, 319, 322, 327–8, 336–7, 340, 359, 364, 374–5, 377–8, 385–6, 402, 426, 428, 430, 433, 439–41, 450, 463, 474
O'Reilly, J., 340, 378
Osher, Stanley J., 474

Palmer, James T., 151, 161
Pannekoek, A., 30, 31
Papanicolaou, George C., 45, 62
Parker, David F., 470
Parter, Seymour V., 18, 285, 300, 325, 328, 387, 446
Pearson, Carl E., 276, 284, 286, 402, 474
Pearson, G. L., 449, 459
Perko, Lawrence M., 45, 61, 102
Perron, Oskar, 94, 228, 232, 258–9
Picard, (Charles) Émile, 8
Poincaré, Henri, xv, 35, 61–2, 65–6
Poisson, Siméon-Denis, 34
van der Pol, Balth., 41–3
Prandtl, Ludwig, 61, 137
Protter, Murray H., 18

Rayleigh, Lord (John William Strutt), 41, 61
Reid, William Thomas, 340
Reiss, Edward L., 62–3, 75, 76, 117
Riccati, Jacopo Francesco, 173, 228, 259, 340, 348
van Roosbroeck, W., 447
Rose, Nicholas J., 227
Rota, Gian-Carlo, 116, 265
Rubenfeld, L. A., 84, 130
Runge, C., 196

Sacker, Robert J., 227–8
Saksena, V. R., 340, 378
Sannuti, Peddapullaiah, 374
Sari, Tewfik, 463, 470

Schrödinger, Erwin, 61
Searl, J. W., 279, 300
Sell, George R., 227–8
Shampine, Lawrence F., 18, 285, 300, 446
Shockley, William Bradford, 447, 449, 459
Sibuya, Yasutaka, 141, 145, 340, 348
Smale, Stephen, 129, 175
Smith, Donald R., 14, 16, 62, 74, 76, 102, 105, 108, 151, 156, 161–2, 205, 227, 245, 279, 283, 294, 340, 353, 372, 412, 427, 435–6, 449, 454–5, 457, 461
Snow, R. E., 34, 51, 53, 62, 75–6, 78, 82–3, 101–5, 117–25, 130
Stel'makh, V. G., 448, 455
Stewartson, K., 325
Stoer, Josef A., 374, 474
Stoker, James J., 41
Struble, Raimond A., 45
Strutt, John William, see Rayleigh
Sturm, Charles, 14
Sturrock, P. A., 61

Tang, Min Ming, 271, 400
Taylor, Brook, 98, 147
Taylor, James G., 283, 353
Tikhonov, A. N., 181, 205, 226–7
Trenogin, V. A., 151
Tschen, Yü-Why, 151, 264, 300, 303
Tsuchiya, H. M., 165, 194, 196
Tupčiev, V. A., 285, 337
Turrittin, Hugh L., 24

Vader-Burger, Els, 399, 403, 463, 470
Van Dyke, Milton, 139–40
Varma, A., 387
Vasil'eva, A. B., 141, 145, 227, 276, 285, 300, 337, 385, 386, 413, 448, 455, 463
Vishik, M. I., 61, 145, 151, 264, 271
Volterra, Vito, 8

Wan, Frederic Y. M., 470
Wasow, Wolfgang, 141, 264, 300, 302, 308, 315, 319, 328, 386, 388, 463
Watts, A. M., 319, 328
Weinacht, Richard J., 162
Weinberger, Hans F., 18
Weinstein, Mills B., 14, 16, 161, 294
Weiss, Richard, 340, 474
Wentzel, Gregor, 61, 283
Whittaker, E. T., 61
Wilde, Robert R., 340
Willett, Douglas, 300, 386
Wintner, Aurel, 228, 353

Yarmish, Joshua, 279, 300, 385, 386, 412

Zauderer, Erich, 328

Subject index

Abel's equation, 93
Ackerberg/O'Malley resonance, 327–8
angular velocity of viscous fluid, 319, 325
asymptotic matching principle of
    Van Dyke, 139–41
averaging technique, 40–59
    equivalence with two-timing, 78, 82

Banach/Picard fixed-point theorem, 8,
    19–21, 37, 99, 201, 203–4, 207, 233,
    340, 343, 348, 352, 399, 409, 421, 423,
    425, 436, 444
Banach space, 19–21, 99, 151, 201, 235,
    240, 348
beats, 58, 75, 82–3
Bessel function, 5, 10, 155
biochemical kinetics, 163, 387, 403, 438
biological oscillation, 45, 82
boundary condition
    mixed (spatially coupled), 331, 361–4,
        368
    separated, 263–4, 269, 308
boundary-layer
    characteristic equation, 313–4
    correction equation, 307, 309–12
    correction function, 133–5, 145–8,
        177–80, 246–53, 284, 291, 300, 306,
        310, 362, 365, 392–6, 432–3, 453–5
    impulse, 276, 288, 412–3
    jump, 227, 386, 391
    region, 133, 289
    turning point, 413, 430, 474
    variable, 277, 294, 306, 308, 377
boundary-value problem, 3–4, 9, 263–476

$C^k[t_1, t_2]$, 16
calculus of variations, 370
    Hamiltonian formulation, 373, 379

cancellation rule, 313–7, 336, 357–8, 361–4,
    377
Cauchy problem, 8
chemical kinetics, 387, 403
classical perturbation approach, 34–9
Cochran construction, 267, 275, 294
comparison
    function, 16
    system, 420
comparison techniques, 18, 116
composite expansion, 61, 140, 143–4
compound pendulum, 59, 84–9, 112–15
conditional stability, 181, 219–60, 285,
    332, 335, 341
contraction mapping, 20–21
control function, 371
Coriolis friction wave-number, 305
Coulomb model on circuit capacitance,
    54

Debye length, 447, 457
differential inequality technique, 226, 470
Dirichlet problem, 9, 11, 226, 269, 288,
    356–9, 388, 426
divergent series, 22–5, 64
domain of dependence, 154
doping density, 447
Duffing oscillator, see oscillator

eigenvalue problem, 271
Einstein equation for Mercury, 29–41, 43,
    47, 69, 95–7, 108–10
elasticity, 303, 388, 470
electric circuit, 54–8, 90, 116, 119, 125
electron density, 447
electrostatic field, 447
elliptic function (Jacobi), 32, 36, 49
energy method, 31, 33–4, 93, 120
enzyme reactions, kinetic theory, 165–6,

182–3, 194–6
error estimate
 boundary-value problem, 282–3, 294–5,
  358, 368–9, 400, 437
 oscillatory initial-value problem, 49,
  57, 64, 91–130
 overdamped initial-value problem, 141–5,
  185, 187–9
Euler/Bernoulli beam theory, 303
Euler/Lagrange equation, 372
Euler's method with extrapolation, 196,
 208, 211, 215
expanding interval, 38, 61–2, 91–2, 99,
 188–9
exponential dichotomy, 332, 336–7, 340–1,
 343, 345, 349, 354, 364, 381

Faraday/Henry model on circuit
 inductance, 54
fast
 dynamics, 374
 time, 60, 62–3, 134, 137, 157–8
 variable, 167, 207, 215–6, 331, 363
feedback rule, 378
Fourier expansion, 72, 79
Fredholm
 alternative, 4, 266, 335
 integral equation, 268
fundamental solution, 168–9, 172, 201,
 207, 228, 257–60, 331–2, 334–5, 338,
 345, 350, 354–5, 360, 366, 416, 418,
 422, 443

Green function, 4, 17–8, 169, 232, 267,
 269, 271–4, 294, 336, 353–69, 398–9,
 411–25, 435–6, 440, 445
Gronwall's inequality, 7–10, 19, 115, 167
Gronwall-type argument, 33, 37, 69, 105–7,
 112, 116, 161, 171, 174, 188, 194, 207,
 441
Gulf Stream, 305

Hamilton/Jacobi equation, 379
higher-order
 (Krylov/Bogoliubov/Mitropolsky)
  averaging, 44–5, 102
Hölder continuity, 175–6
hole density, 447
Householder transformation, 475, 476
hyperbolic splitting, 336, 364

infinity (maximum, supremum) norm, 16–7,
 21, 167
inequality
 differential, 7–8, 12–19, 34, 115, 121,
  182, 184, 226

integral, 106, 109, 112, 114–5
initial-layer correction, 246–53
inner approximation, 138–40
interior layer, 263, 319–24, 463–73
 angular type, 388, 463
 shock type, 470

Jacobi elliptic function, 32, 36, 49
JWKB approximation, 283

Keplerian orbit, 29–30, 49
Kirchhoff's model on circuit voltage, 54
Krylov/Bogoliubov averaging, 40–60
 equivalence with two-timing, 78, 82
Kuroshio Current, 305

Landau order symbol, 24
Laplace transform, 10, 23, 155
limit cycle, 52, 80
Lindstedt/Poincaré condition, 65–7, 71,
 73, 77–8, 80, 88
linear state regulator, 370–81
Liouville/Green approximation, 219, 259,
 283–5

Maier algorithm (multiple shooting), 473–8
matching, 136–41
 conditions, 146, 178, 184–5, 246, 300,
  306, 309, 392, 431, 452
 principle of Van Dyke, 139–40
maximum (infinity, supremum) norm, 16,
 17, 21
maximum/minimum principle, 16–8, 136,
 265, 267, 283, 285
mean value, 302
Mercury (planetary orbit), 29–39, 48–9,
 69, 109
midpoint rule, 211
Miranker approximation, 212–3
Moore/Penrose pseudoinverse, 260
multiple shooting, 474–6
multiscale (multivariable, two-timing)
 technique
 boundary-value problems, 277, 289, 304,
  306, 358, 362, 364–5, 399
 oscillatory initial-value problems, 60–90
 equivalence with averaging, 78, 82
 error estimate, 103–30
 overdamped initial-value problems,
  145–62
 error estimate, 149–51

Neumann
 problem, 269, 274, 286–7, 295, 360–1,
  367, 412
 series, 235, 365
Newtonian physics, 29–30, 49, 85

Newton's method, 211, 475, 476
nonstandard analysis, 470
numerical methods
  boundary value problem, 318, 473–8
  overdamped initial-value problem,
    207–18

ocean circulation, 305
O'Malley construction, 437
O'Malley/Hoppensteadt
  approximate solution, 197–8, 204
  construction (method), 141, 145–8, 163,
    177–89, 196–8, 228
operator
  differential, 1, 12, 18, 19, 92, 98, 141
  integral, 99
optimal control, 165, 340, 370–81
order symbol (Landau), 24
oscillator (equation)
  cubically damped, 42, 50–2, 74, 80–2,
    110–12, 119
  Duffing, 52, 74, 82
  van der Pol, 41–2, 46, 52, 74, 78–80
  Rayleigh, 41–3, 74
outer (solution) approximation, 133–5,
  138–40, 145–7, 177–81, 246–53, 370,
  392–3, 431, 452–5

partial differential equation, 45, 75–6, 151,
  153–62, 271, 305, 400, 437
Picard iteration (successive approximation),
  6, 8, 10
population (ecological) kinetics, 403
potential drop, 460–2
  floating, 449, 462
precession of perihelion, 49
projection matrix (operator), 222, 230, 332,
  334–5, 355, 377
Prüfer substitution, 93

quasi-linear system, 191, 216, 426

Rayleigh oscillator, *see* oscillator
relaxation oscillations, 41
residual, 95–7, 104, 142, 198, 255, 281–2,
  293–4, 358, 397, 407–8, 434, 442, 443,
  469, 472–3
resolvent kernel (Volterra), 8, 10, 13, 268
resolvent matrix (Green function for
  initial-value problem), 168–70, 172,
  232, 371
Riccati
  equation, 339–41, 343, 378, 424, 444
  transformation, 173, 228, 259, 330, 332,
    334, 336, 338–54, 356, 360, 367, 373,
    378–9, 381, 417, 444–5
Riemann function, 155

Robin problem, 286, 298, 386
Runge/Kutta method, 196, 208, 211, 215,
  475

saddle, 243
scaled variable, 60–2, 283, 299
secular effect, 31
secular term, spurious, 38–9, 61, 63–5,
  72–3, 127, 279, 290
semiconductor, 263, 337, 447–62
singular singularly perturbed problem,
  156, 219, 227
slow
  time, 60, 62–3, 75, 134
  variable, 167, 216, 331, 363
speed of light, 31
stable initial manifold, 181, 222, 224–5,
  229–45
stability condition, 391, 402, 429
state equation, 370
stiff problem, 208, 474
stretched variable, 137, 146
Sturm
  comparison theorem, 265
  transformation, 12–4, 136, 264, 288,
    319
subcharacteristic, 154–5, 162
successive approximation, 8, 10, 20
supremum (maximum, infinity) norm, 16,
  17, 21

Taylor/Cauchy formula, 98, 147, 149, 199,
  397
Taylor expansion, 22–3, 38, 76, 178–9
telegraph equation, 153–62, 219, 227
thermodynamic-chemical process, 276
transference of vibrations, 84–9
transition (turning) point, 263, 265, 318–29,
  332
  angular, 388
  inside boundary layer, 413, 430, 474
two-body equation, 30
two-timing technique, *see* multiscale
  technique

unimodular matrix, 339, 347

van der Pol oscillator, *see* oscillator
variation of parameters, 1–3, 42–3, 46,
  56, 113, 168–9, 176, 371, 407, 421
Volterra integral equation, 6, 7–10, 268

wave equation (perturbed)
  oscillatory, 45, 75–6
  overdamped, 153–62
wind stress, 305
wronskian, 93, 407